D0023039

WELDING TECHNOLOGY TODAY

Principles and Practices

Craig Stinchcomb

Terra Technical College

PRENTICE HALL, Englewood Cliffs, New Jersey 07632

Library of Congress Cataloging-in-Publication Data

Stinchcomb, Craig.
Welding technology today : principles and practices / Craig Stinchcomb.
p. cm.
Includes index.
ISBN 0-13-924416-6
1. Welding. I. Title.
TS227.S78 1989
671.5'2--dc19 88-29263
CIP

Editorial/production supervision: **Kathryn Pavelec**
Interior design: **Maureen Eide**
Cover design: **Maureen Eide**
Manufacturing buyer: **Bob Anderson**
Page layout: **Maureen Eide**

Cover photo: General Electric Research and Development Center

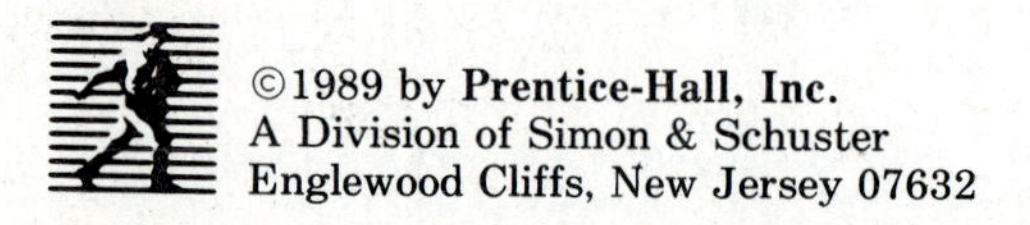

A Division of Simon & Schuster
Englewood Cliffs, New Jersey 07632

Printed in the United States of America

10 9 8 7 6 5 4 3 2 1

ISBN 0-13-924416-6

Prentice-Hall International (UK) Limited, *London*
Prentice-Hall of Australia Pty. Limited, *Sydney*
Prentice-Hall Canada Inc., *Toronto*
Prentice-Hall Hispanoamericana, S.A., *Mexico*
Prentice-Hall of India Private Limited, *New Delhi*
Prentice-Hall of Japan, Inc., *Tokyo*
Simon & Schuster Asia Pte. Ltd., *Singapore*
Editora Prentice-Hall do Brasil, Ltda., *Rio de Janeiro*

CONTENTS

5

OXYFUEL WELDING (OFW) 50

6

SHIELDED METAL ARC WELDING (SMAW) 76

SMAW ELECTRODES 91

8

GAS TUNGSTEN ARC WELDING (GTAW) 123

PULSED GTAW 173

PLASMA ARC WELDING (PAW) 178

11

GAS METAL ARC WELDING (GMAW) 188

PULSED GMAW 207

FLUX-CORED ARC WELDING (FCAW) 220

SUBMERGED ARC WELDING (SAW) 224

15

SELECTING FILLER WIRE 232

16

PIPE WELDING 240

17

OXYFUEL CUTTING 272

18

PLASMA ARC CUTTING 282

19

METALLURGY 289

WELDING CAST IRON 321

DRAWINGS AND BLUEPRINTS 328

WELDING SYMBOLS 347

23

WELDING CODES AND STANDARDS 371

24

TESTING AND CERTIFICATION 377

25

DESTRUCTIVE TESTING 388

26

NONDESTRUCTIVE TESTING (NDT) 396

27

WELDING INSPECTION AND QUALITY CONTROL 408

PREFACE

Welding Technology Today is dedicated to the welding student, welder, welding instructor, welding inspector, or engineer who wants to learn, reference, or upgrade his or her knowledge in the welding field in a clear, straight-to-the-point fashion.

The entire welding field and all related subjects are covered. We explain not only how it works, but why it works. Technical terms are well defined and principles and theories are explained in simple non-technical terms.

You will note as you work your way through that we do not talk a subject to pieces; instead, we go straight to the point of what is critical regarding the subject matter. If we can show it better with a drawing or photo, that is what we use.

We have included the most comprehensive, state-of-the-art information on today's welding field, yet we have attempted to keep it simple enough for the beginning student who has no prior welding background.

Whether you are reading the book from cover to cover, or just referencing an interesting topic, we hope you will find it interesting, informative, and enjoyable.

I would like to thank the following people for their help in advancing the science and art of welding, and freely sharing their knowledge.

- William McCleese
- Jerry Kissell
- Sharron Middleton
- Eric Magnuson
- Stan Roberts
- Chip Rathwell
- Kathy Hess
- Craig Wood
- Arand Bolen
- Jim Harris
- Sally Stinchcomb

I would like to give special thanks to the American Welding Society for their many welding charts and tables, and their continuing effort to advance the science and art of welding. Special thanks also to the following companies for their contributions to *Welding Technology Today:*

- Miller Electric Mfg. Co.
- Lincoln Electric Co.
- Airco Welding Products
- Rockwell International
- ESAB Welding Products, Inc.
- McDonnell Douglas Corp.
- Thermal Dynamics Corp.
- Standard Oil
- The Aluminum Association
- Jetline Engineering
- Toledo Edison Co.
- Smith Welding Equipment
- Sellstrom Manufacturing Co.
- Webster Manufacturing Co.
- Continental-Fremont Co.
- Pettibone-Hansen
- Stinchcomb Associates
- The American Iron & Steel Institute
- U.S. Steel Corp.
- Tempil Division
- Tech/Ops, Inc.
- Advanced Robotics Corp.
- Staveley NDT Technologies, Inc.
- Controlled Systems Division of A.M.H. of Canada, Ltd.
- The American Petroleum Institute
- Ferranti Sciaky Inc.
- Franklin Manufacturing Inc.
- KN Aronson, Inc.
- SATEC Systems, Inc.
- C & J Welding Technical Services
- Lumonics, Inc.
- BP America Inc.
- Amersham Corporation

CRAIG STINCHCOMB

1

INTRODUCING TODAY'S HIGH-TECH WELDING TECHNOLOGY

If you have decided to pursue a career in the field of welding, you have just chosen to learn the technology that provides the fastest, strongest, and most economical method of joining metals, and this technology is here to stay! The "high-tech" welding field has moved from coal-fired furnaces and hammers used to forge iron to modern methods such as the concentrated, accelerated free electrons of the electron beam process, and the use of robots and lasers. The newer high-tech welding devices are used on aircraft, space vehicles, and nuclear reactors. Without modern welding technology, the development of these high-tech devices would still be many years off. With all of the new metals recently introduced into the industrial world, we need the newer welding processes to join these metals.

In addition to arc and oxyacetylene fuel welding, today's welding technology includes pulsed GTAW, plasma welding and cutting, submerged arc, pulsed GMAW, and electron beam and laser welding. This is not to say that basic arc and oxyacetylene processes are not as important in industry as ever. However, the new processes have in many applications improved the properties of the finished product, increased productivity, decreased weld defects and rejects, and made the application of welding easier.

It would be difficult, if not impossible, to go through a day without touching a dozen or so things around you that have been welded. Cars, furniture (the chair you may be sitting in), coffee percolators,

(*continued on page 7*)

PHOTOS 1-1 TO 1-4
Today's aircraft industry uses many welding processes, including automatic GTAW and GMAW and electron beam.

PHOTO 1-1
Automatic welding in the aircraft industry. (Courtesy of McDonnell Douglas Corp.)

PHOTO 1-2
Welding makes the McDonnell Douglas MD-80 the most efficient aircraft in its class. (Courtesy of McDonnell Douglas Corp.)

PHOTO 1-3
Welding plays a major role in the construction of the F-15 Eagle. It has a speed of Mach 2.5 and is designed to outperform *any* enemy aircraft. (Courtesy of McDonnell Douglas Corp.)

PHOTO 1-4
With a speed of Mach 1.8, the F/A-18 Hornet relies heavily on welding for its construction. (Courtesy of McDonnell Douglas Corp.)

PHOTOS 1-5 TO 1-7
Welding in the aerospace industry.

PHOTO 1-5
The space shuttle orbiter, external tanks, and solid fuel rocket motors use miles of welds for their construction. (Courtesy of Rockwell International.)

PHOTO 1-6
Welding plays a major role in the construction of the space shuttle main engines. These engines produce 513,000 pounds of thrust, and are the most advanced rocket engines in the world. (Courtesy of Rockwell International.)

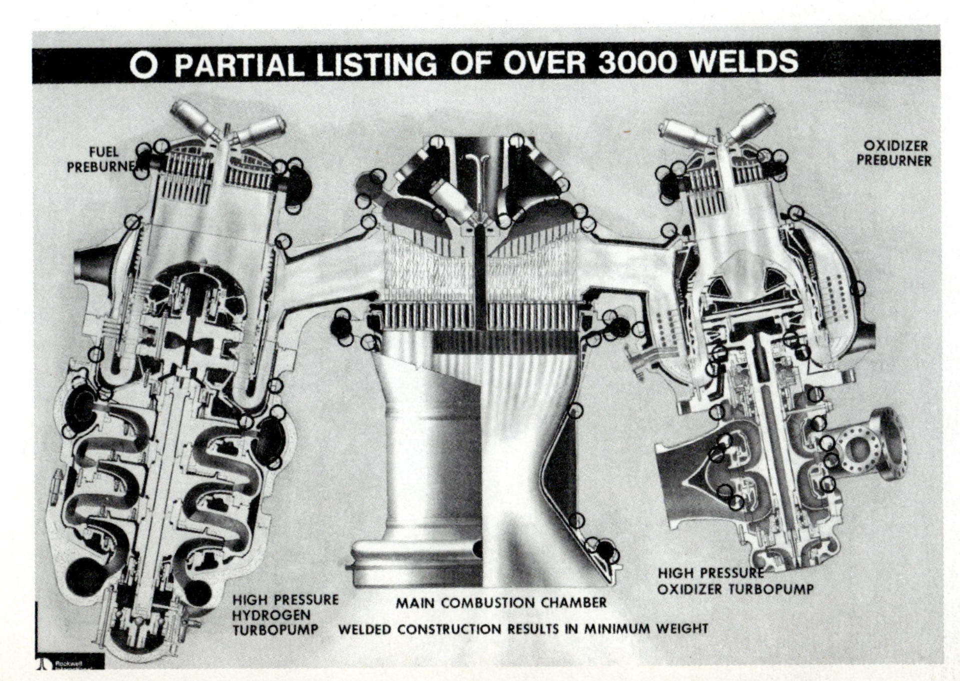

PHOTO 1-7
The circles on the cross-sectional view of the space shuttle main engine indicate some points that are welded. There are actually 3801 GTAW, 18 GMAW, 249 electron beam, 600 inertia, and 6000 spot welds. This adds up to a total of 40,307 inches of weld. (Courtesy of Rockwell International.)

PHOTOS 1-8 TO 1-11
Welding in the production of energy.

PHOTO 1-8
The nuclear power industry would be difficult, if not impossible, without welding. (Courtesy of Toledo Edison.)

PHOTO 1-9
Nuclear power plants contain miles of welded pipe. (Courtesy of Toledo Edison.)

PHOTO 1-10
The oil industry relies on welding for drilling, piping, and storage. Here welders complete a pipe section on the Alaskan pipe line. (Courtesy of B P America.)

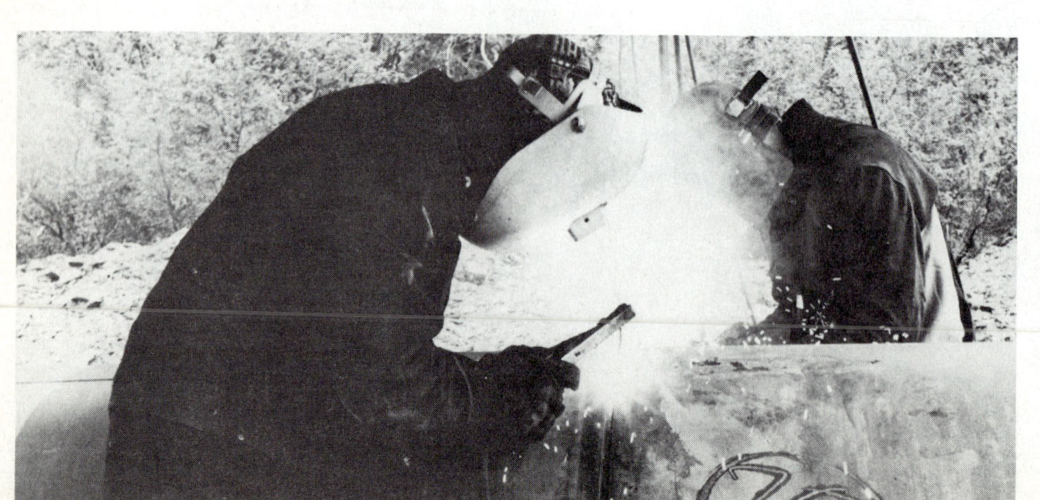

PHOTO 1-11
Welders begin the "hot pass" on 48-in.-diameter pipe for the trans-Alaska pipeline. Six to seven passes are required for a completed weld on the ½-in.-thick pipe. (Courtesy of Alyeska Pipeline Service Company.)

PHOTO 1–12
Heavy construction machinery uses welding for strength and economy. This construction equipment would be considerably heavier, use much more fuel, and appear much different if it were not for today's modern welding systems. (Courtesy of Pettibone Hansen.)

PHOTOS 1-13 TO 1-16
Production machinery using welded construction for high-speed, high-quality products.

PHOTO 1-13
Conveyors being FCAW and GMAW welded. (Courtesy of Webster Manufacturing.)

PHOTO 1-14
The final product always looks smoother and more streamlined when welded.

PHOTO 1-15
All-welded tube conveyors. (Courtesy of Continental Fremont Co.)

PHOTO 1-16
Completed tube conveyor. (Courtesy of Continental Fremont Co.)

PHOTO 1–17
The sky is the limit to welding opportunities. Here a welder completes the top flange of a spandrel beam on the 52nd floor of the Georgia Pacific Building in Atlanta, Georgia, using the Innershield process. (Courtesy of The Lincoln Electric Company.)

buildings and bridges, and yes, even your kitchen sink may have stainless steel welds in it. Aircraft, cars, and space vehicles would be just about impossible without welding, as would many other necessities and luxuries that we take for granted.

HOW STRONG ARE WELDS?

You may have heard a welder, engineer, or inspector say that "the weld is stronger than the base metal." This saying is, in fact, quite true. Let us look first at the process used to deposit a weld. If it is an electric arc process (such as SMAW, GMAW, or GTAW), which the majority of the welding done today is, we can expect a high-quality weld. This is due to the low heat input but high penetration into the weld and surrounding base metal. The arc processes also tend to produce a smaller heat-affected zone (see Chapter 8).

Chemically and metallurgically a weld is also superior to the base metal. The filler used in the weld is a high-quality metal with added deoxidizers. This gives a very clean and dense weld deposit.

Most mild and low-carbon steels used in fabrication today have a tensile strength of about 58,000 psi. Most mild and low-carbon steel electrodes used to weld these steels have been tested to a tensile strength of 62,000 to 72,000 psi (depending on the electrode classification). So we see an advantage of 4000 to 14,000 psi in favor of the weld metal.

Fluxes added to the welding process help improve the mechanical and chemical properties of a weld. They also help deoxidize and act as scavengers in the weld. Scavengers and deoxidizers remove oxygen (in the form of porosity) and other contaminants from the weld. The fluxes turn into slag and blanket the weld, allowing it to cool slowly and prevent the buildup of stress and possible cracking. This all adds up to the highest-quality, yet most economical methods of joining metals.

WELDING VERSUS OTHER FORMS OF CONSTRUCTION

The idea for bringing welding into construction and general fabrication was obtained from the extensive welding done on ships during World War II. Welding was observed to be much faster than the old riveting methods, as well as being incredibly strong. Ships that were welded could take a considerable number of blows to the hull and stay afloat. Engineers started considering welding tanks, buildings, bridges, aircraft, and other items that needed to be strong, yet as light as possible. This was the beginning of the influence of welding on our society.

Figure 1-1 shows a comparison of welding versus riveted or bolted construction. The advantages of welding are as follows:

1. Considerably smaller structural members can be used to provide the additional joint strength.
2. Buildings, bridges, and other structures are lighter and can be built higher due to the reduced weight.
3. Buildings, bridges, and other structures can be built more cheaply due to the reduction in weight and material cost.
4. Architects are free to provide much more streamlined and graceful-looking designs, due to the compact-looking joints (riveting requires additional fastener plates and rivets for each joint).
5. Welds are much more corrosion resistant than rivets and bolts.

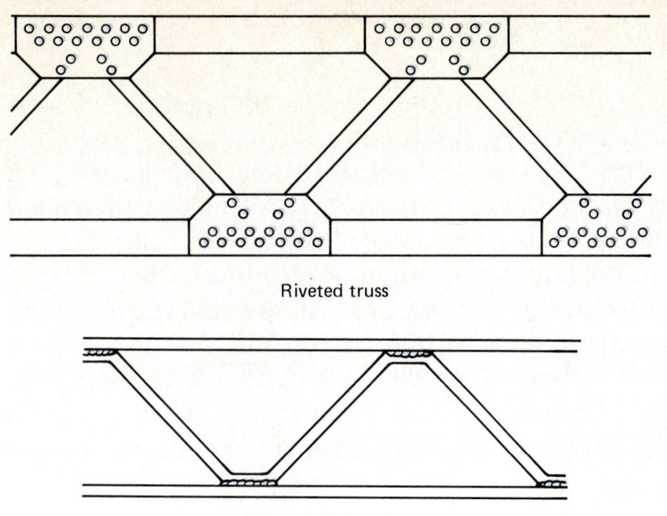

FIGURE 1–1
Riveted and welded trusses.

6. Welded construction allows for five different types of joints, whereas riveting and bolting utilize only one, the lap joint.
7. Welds make a fluid-tight joint on storage tanks and piping systems. Rivets can be a problem in this area.
8. Welded structures can be altered easily and economically.

OPPORTUNITIES IN THE WELDING FIELD

The possibilities for employment in the welding field are almost limitless (see Figure 1–2). How far you go and how much you earn is up to you. You will be limited only by your level of education and your willingness to learn.

FIGURE 1–2
Occupations in the welding field.

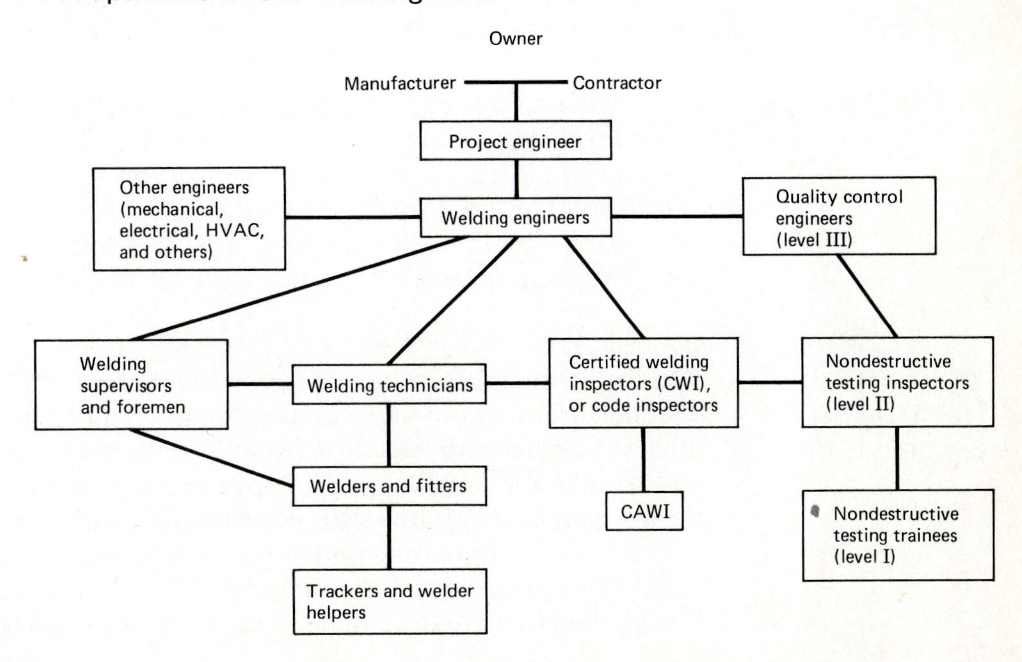

Today's welders must (depending on how far they want to go in the welding field) learn not only the basic oxyfuel and arc welding processes, but all the modern methods as well. These processes include gas tungsten arc welding (GTAW) and gas metal arc welding (GMAW), and at least a theoretical background in plasma welding and welding robotics, and manual and automatic flame cutting. If they wish to advance in the welding field, they must learn to read blueprints and welding symbols, and may also want to study some of the welding codes.

Remember, the more you advance your knowledge in the welding industry, the more you will be sought after by employers. Plainly stated, the more you know, the more you are worth.

Combination Welder

Combination welders are proficient in two or more welding processes. This skill level sometimes takes years to develop effectively, but it can be a rewarding journey. Having the ability in many processes keeps you continuously moving into new jobs. This also keeps the learning process going and the job interesting. Combination welders are well paid for their versatility.

Specialist Welder

Specialist welders are highly skilled in one or two welding processes. They thoroughly understand their chosen process and can get the most out of it. Pay scales may range widely, depending on the skill necessary for the given processes.

Welder-Fitter

Welder-fitters are usually welders who have an excellent mathematical aptitude. They must know how to read and interpret blueprints and welding symbols. In addition to their welding duties, they cut, fit, and tack parts together precisely as the prints specify. Using their math skills, they are required to hold very close tolerances on the parts they fabricate. Welder-fitter pay scales are high due to the additional responsibilities involved.

Welder-Tacker

The welder-tacker classification provides an excellent place for the potential welder to start. Tackers assist welder-fitters or welders in joint preparation, fitting, and blueprint interpretation. They are specifically responsible for tack welding together prepared parts. Persons working as tackers should utilize this important training time to observe carefully the tradespersons they are working under.

Welding Foreman/ Supervisor

Welding foremen or supervisors must know and understand all the welding processes in their departments. Especially important is their ability to communicate in a friendly and effective manner. They are specifically responsible for planning and assigning duties for the welding personnel. They must know the skill levels and certification levels of each welder and are responsible for scheduling and keeping welders' certifications current. They must be able to communicate with welding technicians and welding inspectors to monitor production welds.

Welding Technician

The welding technician is probably the most overlooked position in the welding industry. Their use could save contractors and manufacturers thousands of dollars and much wasted time.

Welding technicians can weld, troubleshoot welding-related problems, write welding procedures, interpret most welding codes, and are blueprinting and welding symbol experts. They act as middlemen between the engineer and the welder: perfect utility men who can weld, if needed, or do basic engineering functions such as selecting a filler metal for a given condition.

Welding technicians can solve welding-related problems before they become the welding inspector's problem. They check the setup of the welding machines, the joint fit-up, and make sure that the correct electrode or filler metal is being used on the job as specified. Since welding technicians have practical welding experience, they help engineers work and design prototypes before a project is put into production. They may also be in production, where they may work on drawings or interpret drawings and welding symbols on the blueprints. When utilized, they can be *key* figures in a company's productivity and future growth.

Welding Inspector

Welding inspectors are key persons in any quality program. They are knowledgeable not only in welding processes themselves, but also the possible defects associated with each process. They must know destructive and nondestructive testing processes and their limitations. They must be able to write clear, accurate reports and keep good records. They must know how to use and interpret codes and standards, and most important, must be able to communicate effectively with welders, engineers, and technicians. Knowing the language and terminology of welding as well as welding and nondestructive testing symbols is absolutely essential.

Some welding inspectors are specifically nondestructive testing inspectors who actually perform tests, such as x-ray, ultrasonics, dye penetrant, and magnetic particle inspection.

In other classifications, such as CWI (certified welding inspector), welding inspectors perform visual inspection before, during, and after welding to see if it meets the code or job standards for a particular project. Many welding inspector classifications require documented experience, testing, and inspection certification.

People pursuing a career in welding inspection should not be afraid of doing paperwork and a lot of reading. They must keep up to date with welding codes and the changes made from year to year. They must also be capable of assuming responsibility, as well as having good judgment. Although welding inspectors do work from written standards, they may be called upon to make decisions on code interpretations or borderline cases.

Welding Engineer

Welding engineers are the designers or originators of project welding specifications. They specify filler metals, weld sizes, and joint designs based on calculations and strength requirements. Welding engineers have extensive training in mathematics, drafting and design, metallurgy, and chemical and mechanical engineering. Four years of college in an accredited engineering institution may also be required, or a two-year associate's degree from a technical college in welding engineering technology, along with welding design experience.

Welding Business Owner

The welding field offers a variety of opportunities for starting your own business. The possibilities are almost limitless!

- Job shop
- Contractor
- Manufacturer
- Pipeline contractor
- Engineering firm (engineering contractor)
- Welding inspection contractor (NDT contractor)
- Weld testing lab (testing contractor)

You can start as a small job-shop owner and grow into a major manufacturer or contractor. You can be a contractor in a number of areas, such as manufacture, inspection, or testing lab.

Education and ambition are the only limitations. So remember that the more you are capable of doing, the greater your opportunities will be.

PRODUCTION FLOW OF A TYPICAL CONSTRUCTION PROJECT

1. Ideas are formulated by the architect and owners, who consider the project requirements and end use, size, cost, location, building codes, and so on.
2. Perspective designs are produced by the architect and considered by the owners.
3. Designs are furnished and agreed upon by the owners.
4. Working drawings are developed by the appropriate engineer (electrical, HVAC, welding) and coordinated by the project engineer.
5. Welding procedures and welders are qualified.
6. Final drawings are approved and construction begins.
7. As construction progresses, inspectors are continuously with the project, monitoring the workmanship. They must also assure that the project is constructed as specified on the original drawings, and that applicable codes are being followed.
8. During construction, welders and fitters study the blueprints, with the assistance of welding technicians.
9. As various phases of construction are completed, examinations

such as radiograph and ultrasonic tests are made to assure the quality of critical welds and connections. These test reports are forwarded to the certified welding inspector for approval.

10. Any required repairs or alteration procedures are established by engineers, with the assistance of technicians and inspectors.
11. Final construction and repairs are completed.
12. Final inspections are then made for approval so that the project can now be used.
13. The structure is signed off and turned over to the owners.

REVIEW QUESTIONS

1. Which is usually stronger: the base metal, or the weld metal?
2. What is the purpose of fluxes in the welding process?
3. List 5 advantages of welded construction over other forms of construction.
4. List 5 applications of welding in modern industry.
5. What are your goals, and what would you like to be doing in the welding industry in a few years? (Please explain.)

WELDING AND CUTTING SAFETY

"An ounce of prevention is worth a pound of cure." Taking just a few extra minutes to follow safeguards can possibly mean the difference between life and death in the welding industry. In this field you will be working with explosive gases, burn hazards, electrical hazards, and possibly high altitudes. We are not trying to scare you into another career, but lack of knowledge and being in too big a hurry could be fatal to you and possibly those around you. As long as you have the knowledge and take the time to follow safeguards, you will never get into a dangerous situation. You would not think of climbing a high-tension pole and working on an electrical transformer without the proper training, or climbing into a F-14 jet fighter without taking time to strap yourself in, just as in welding you would not weld on a fuel tank without first using one of the safeguarding techniques. Please take the time to read and understand the following chapter; you will be respected for your knowledge in these areas and actually more productive. A safe person is also more productive, because jobs do not have to be reworked or repaired due to accidents.

WELDING RADIATION

Do not confuse welding radiation with hard ionizing radiation such as that produced from an x-ray tube; it is actually very different. Welding radiation has a much longer wavelength and therefore does not have the penetrating ability of dangerous x-rays or gamma rays. Welding does produce some radiation or rays, however, they are very similar to that which the sun produces. We break these rays down into three general categories: (1) visible light rays, (2) infrared rays, and (3) ultraviolet rays. Visible light rays are those that you can see with your eyes. To unprotected eyes these rays can cause eye strain and general discomfort. They also reflect off surrounding objects and can cause spontaneous, temporary blindness from very intense flashes of light such as a welding arc.

Ultraviolet rays are invisible and difficult for welders to detect. On unprotected skin these rays cause burns similar to a sunburn. They tend to reflect like visible rays. Ever notice how fast you tan or burn at the beach compared to other places? This is due to ultraviolet rays reflecting off the water. Most welding helmets go through an ultraviolet test before being approved for packaging. The key with ultraviolet rays is to keep all areas of normally exposed skin covered up. Watch especially such areas as cuffs, neck and upper chest with front-buttoning jackets, and belt areas. Helmets can now be purchased with leather neck drops to protect the neck and upper chest. Aprons and jackets should be long enough to cover the belt area. Sleeves and cuffs can be taped or wrapped with a rubber band to keep them from hanging open.

Infrared rays have a longer wavelength than do ultraviolet or visible light rays. They actually produce heat when they strike and are absorbed into a surface. Prolonged exposure can cause skin burns, and using an improper shade in the welding hood can cause eye damage. The key here is the proper selection of a helmet shade. Figure 2–1 shows the suggested minimum shade number, although each person has a different tolerance to these rays. Cutting and OFW use shade 5 or 6, depending on personal tolerance. It is suggested that welders use the darkest shade possible, provided that they can still see the puddle.

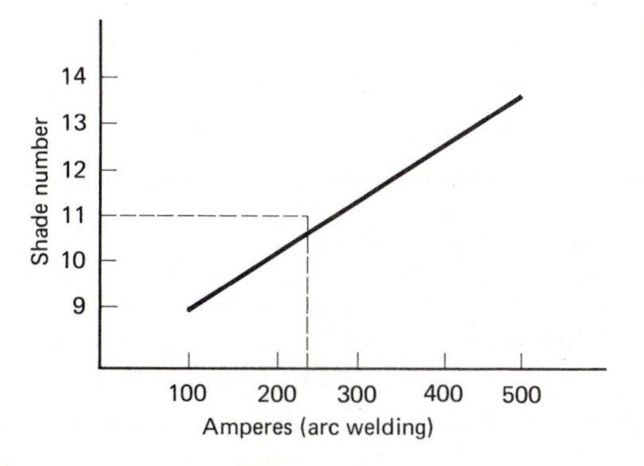

FIGURE 2–1
Helmet shade selection.

After all, it is not necessary to see much more than the puddle to produce a good weld.

BURNS

Daily handling of hot metal requires special care to prevent burns. The following safety rules will help to prevent burns to yourself and others.

1. After welding or cutting pieces of metal, mark all pieces "hot" with soapstone or other suitable metal-marking device. If you leave the area, other workers will not accidentally grab the hot metal.
2. Always use leather or other suitable gloves when arc or OFW welding or hand cutting.
3. Do not handle hot metal with gloves only; use vise grips, channel locks, or pliers. Handling hot metal with your gloves will bake out the natural oils in the leather and make the gloves stiff and difficult to use.
4. As mentioned earlier, leave no skin exposed during welding or cutting.
5. Always wear high-top boots when arc welding, to prevent hot sparks from becoming trapped in your shoes and against your skin.

Treatment of Burns

Burns are classified into three categories: first degree, second degree, and third degree. A first-degree burn is quite painful, although it is not usually serious if treated promptly. This type of burn is usually a surface burn and does not penetrate into other layers of skin. Cold water is the best treatment for first-degree burns. A first-degree burn can be identified by redness and mild swelling.

Second-degree burns penetrate into additional layers of skin. This type of burn requires medical attention for cleansing. Immediate first aid would require a dry sterile dressing carefully applied. A second-degree burn can be identified by considerable swelling and usually blisters.

Third-degree burns penetrate deep into skin tissue. If clothing is stuck to burned area, do not attempt to remove it. Cover the burn with a thick sterile dressing; if possible, elevate the affected area above the victim's heart. Get medical attention promptly.

EXPLOSION AND FIRE

Enclosed Objects (Gas Tanks, Drums, Vessels)

Fires and explosions within enclosed objects account for more deaths among welders than does any other hazard. This is all avoidable if just a few safeguards are taken. If you plan to cut or weld on any type of enclosed object, such as a 55-gallon drum, oil drum, empty propane tank, or gas tank and are not sure what it contained previously, assume

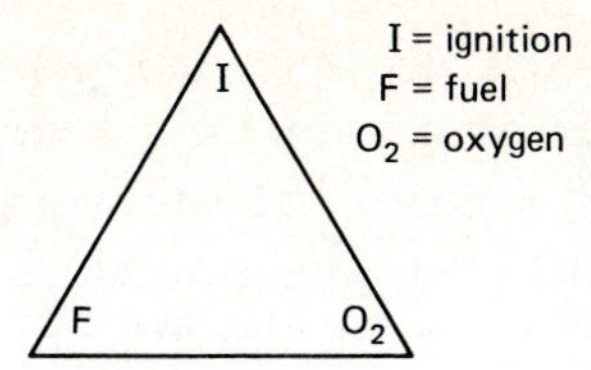

FIGURE 2-2
Fire triangle.

that the substance was volatile. These vapors will remain in seams and pores long after the container is empty. When these vapors are ignited by a welder's torch or electric arc, they expand many times their normal volume. Small holes or vents will not vent this pressure fast enough. The result can be a violent explosion.

There are three prerequisites for any fire or explosion to occur. They are fuel, oxygen, and some type of ignition. These three elements make up what is known as the fire triangle (Figure 2-2). If we can eliminate any one of these elements, we are safe from fire or explosion. One of the surest ways to eliminate one of these elements is filling the tank with water (Figure 2-3). The tank should be positioned so that the water can be filled right up to the area to be welded or cut. The water should remain in the tank while welding or cutting is being done. Filling and immediately emptying the water will not wash out the flammables. As soon as the water is emptied, the fumes will come right back out of the metal pores and seams of the tank.

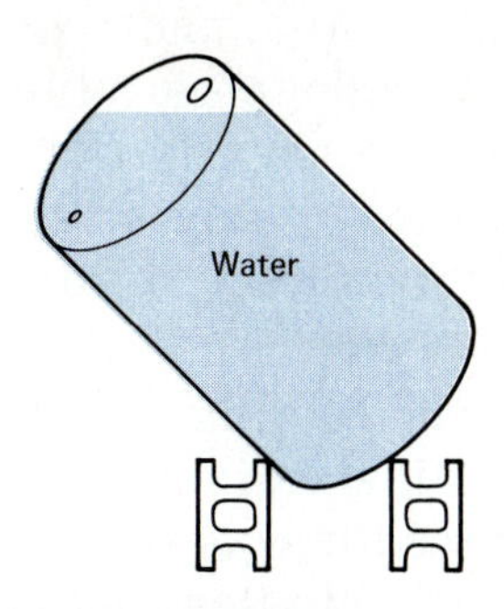

FIGURE 2-3
Safeguarding with water while welding an enclosed vessel.

Purging tanks with inert gas will displace the oxygen and safeguard the tank for welding or cutting (Figure 2-4). Argon is the safest gas to use because it is heavier than air. A volatile-gas-sensing device should be used to make sure that all flammable vapors have been displaced.

Steam cleaning the inside of tanks is one method of permanently removing the flammable vapor from pores and seams. It must be done thoroughly and completely to be effective. Again, volatile gas sensors should be used before welding or cutting to verify that they are safe.

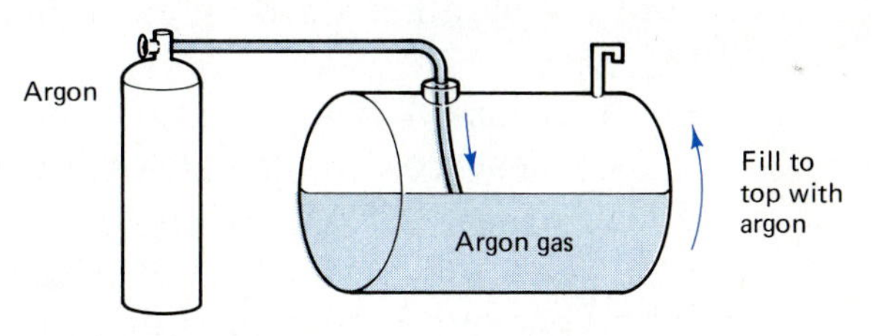

FIGURE 2-4
Safeguarding with inert gas while welding an enclosed vessel.

Tire Bombs

Occasionally, tire rims or tractor hubs need to be welded or heated to remove lug nuts. If the tires are still on, they already contain anywhere from 30 to 100 lb. of pressure. If you input heat into the rim, the heat will transfer to the tire. The hotter you get the rim, the higher will be the pressure in the tire. If you have ever seen a tire explode, you will never forget it! It can flip cars, tractors, and trucks completely over. Removing the tire from the rim is the best choice. Some people deflate the tire, but if the tire seal is not broken, the tire will quickly reinflate when heated.

Fires

Welders also get cited for starting more industrial fires than any other personnel. This is avoidable if the correct safety precautions are taken. Most shops and plants where welding and cutting are performed, even if just occasional maintenance or repair welding, use the "hot work permit" system. This permit is issued when welding or cutting must be performed in an area that is not usually used for welding. If the item cannot be moved into an approved, safe welding area, the hot work permit system must be employed.

1. Have the official safety officer, yourself, and persons who work in, or know the area well, survey the area for potential fire hazards.
2. Take all necessary precautions to safeguard the area, and post fire watch persons if necessary. (You will not be able to see a fire under your hood while welding.) In some cases, wetting the floor down may stop sparks from equipment such as cutting or air carbon arc cutting from traveling too far. Use common sense if using water around electric arc welding processes.
3. Issue the permit and proceed with welding or cutting operations. The permit should be good only for the time estimated to complete the job, and only in the area for which it was issued. Permits should never be issued in areas that contain flammable vapors, liquids, or heavy dust.

Even if all precautions are taken, fires may still occur, so let's examine what to do if a fire breaks out. It is important that we know what fire extinguishers are designed for which types of fires. Using the wrong extinguisher may make the situation worse by spreading the fire, or causing electrical shock. Let's examine the four classes of fires and the symbols used on extinguishers.

1. (A) Class A fires are the wood and paper variety. They are usually easily extinguished with water.
2. [B] Class B fires are the petroleum, grease, or oil type. The fire is burning the vapor off the top of the fluids. Water will be more harmful than good, because it tends to spread the fluid around and will rise above the water and continue burning. Foam works well because it is light and will smother the oxygen that is supporting the combustion. Powder extinguishers will also extinguish class B fires.
3. (C) Class C fires are electrical fires, burning electrical equipment or insulation. Obviously, using water or any electrical-conducting material would create an electrical shock hazard. Dry chemical or powder extinguishers are commonly used on class C fires.
4. (D) Class D fires are flammable metal fires. Although unusual, some metals, such as magnesium, can be ignited, especially small dust particles or shavings of these metals. Many flammable metals generate their own oxygen as they burn, so smothering may not work. Class D extinguishers simply lower the metal's temperature below the flash point to extinguish the fire.

Most fire extinguishers today are of the multiple fire classification type. For example, an ABC-labeled extinguisher will work for all

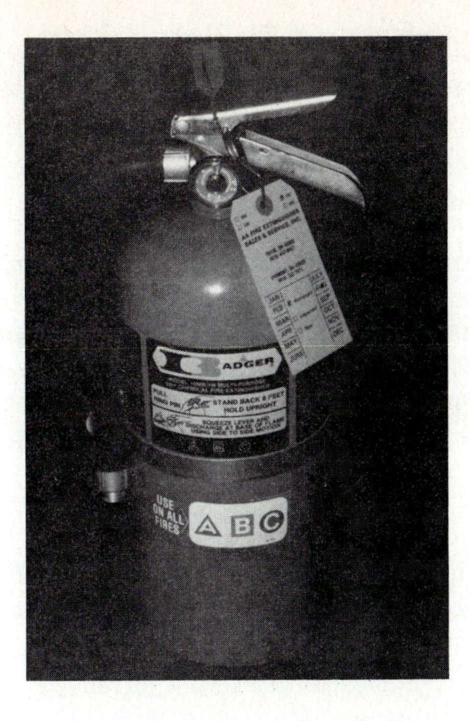

PHOTO 2-1
ABC fire extinguisher.

three classifications of fires. As stated earlier, do not use an extinguisher on a fire for which it is not classified.

Fire Safety Checklist

1. Never weld or cut in an unapproved area without a hot work permit.
2. Always check behind or inside a bulkhead or wall on which you are cutting or welding.
3. Ask yourself: "Where will the sparks be traveling?" Sparks have been known to travel down chutes and elevator shafts, long distances from their point of origin.
4. Use a "fire watch person" if a potential hazard may exist. Do *not* take chances.
5. Never weld on enclosed vessels or tanks without taking the precautions described previously in this chapter.
6. Check the area for smoldering material before you leave. Items can smolder for hours before actual ignition.

FUMES

Some metals may produce dangerous fumes and vapors when being welded or cut. The best policy for any welding or cutting operation is to maintain adequate ventilation at all times. It should be understood that research is continuing on the effects of many different types of elements being produced by welding and cutting. We will limit this discussion to the known hazardous fumes, but we suggest that you keep abreast of the latest research findings. The American Welding Society has published books on the effects of welding on health. Their findings are both interesting and encouraging when comparing the welding trade with other areas of industry.

Lead Fumes

Lead fumes can come from welding on surfaces that were painted with lead-based paints. Although these paints are rare today, you may come across them doing maintenance welding on old parts or machinery. Lead has been a known health hazard for years, and may accumulate in the body. The safest procedure is to clear off painted surfaces with a grinder and wire brush, using a mask for filtering particulate matter, or a respirator.

Cadmium Fumes

Cadmium fumes usually come from cadmium-plated metals. Enough cadmium inhalation can be deadly. Adequate ventilation is the best protection.

Zinc Fumes

Zinc fumes are produced from welding galvanized metals. These fumes can produce symptoms known as metal fume fever. It is a temporary condition that makes victims feel like they have stomach flu. Again, prevention lies in adequate ventilation.

Fume Safety Checklist

1. If ventilation is poor where an item must be welded, move the item outside, if practicable.
2. Use a respirator if you know you are working in toxic fumes.
3. Grind off any metal coatings using a face breathing mask designed for particulate matter. These are readily available where most paint supplies are sold.
4. In all cases keep adequate ventilation in your work area.

Safety Clothing and Equipment

Welders must protect themselves from sparks, hot metal, ultraviolet, infrared, and visible light rays, welding fumes, and other hazards found in the industrial workplace. Wearing the proper clothing will allow you to tackle jobs in difficult positions, such as welding overhead. Clothing for arc welding and many oxyfuel jobs includes the following:

1. A welding jacket or sleeves, usually made of leather or denim; overalls of the same material are okay
2. Leather leggings or denim pants without cuffs
3. Leather welding gloves that fit tightly up to the jacket sleeves
4. High-top boots with steel-toe shoes

Clothes should fit tight enough so that no bare skin is exposed to sparks or ultraviolet rays.

Do not even think of entering a shop or construction site with-

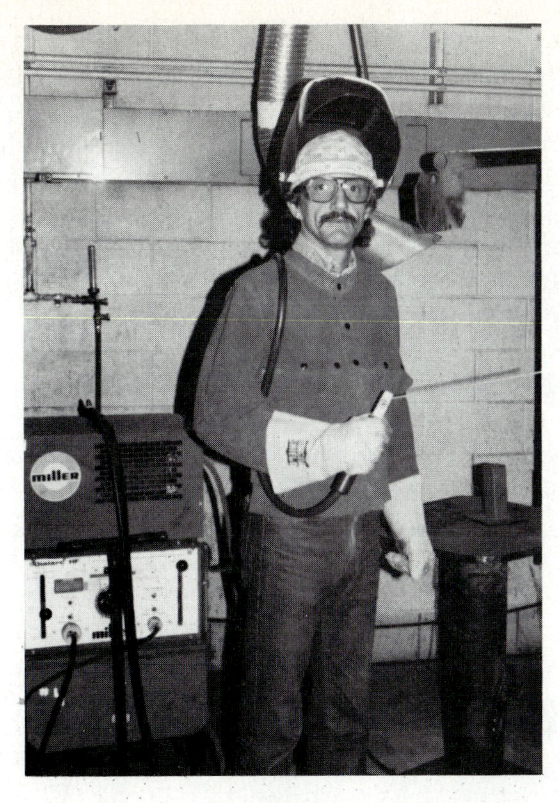

PHOTO 2-2
Well-dressed arc welder.

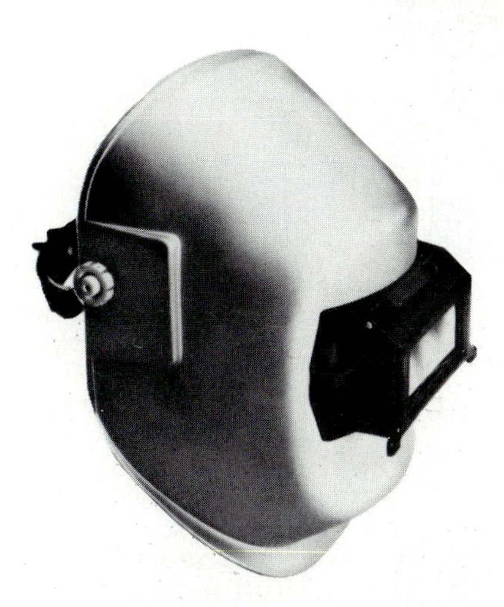

PHOTO 2-3
Welding helmet. (Courtesy of Sellstrom Safeguards.)

PHOTO 2-4
Safety glasses. (Courtesy of Sellstrom Safeguards.)

PHOTO 2-5
Hard hat. (Courtesy of Sellstrom Safeguards.)

out safety glasses. If you wear glasses with safety lenses, use side-shields.

Head gear should include a hard hat if shop or construction conditions warrant. Most hard-hat manufacturers have welding hood conversion clips that allow you to attach your welding hood to your hard hat. A full welding hood that covers your full face and neck area should be used. Many welding hoods are now equipped with a leather flap that drops down over the neck area. The welding hood should have a filter lens of the shade suggested in the welding rays section of this chapter.

Where hard hats are not required, some welders use small skull caps to protect themselves from hot sparks flying over their hoods.

Breathing equipment such as respirators should be used where welders will be working around or on known hazardous metals or fumes. Many respirators are especially made to fit under welding helmets.

For oxyfuel welding and clothing, tight-fitting goggles of shade 5 or 6 should be used. A good tight fit will protect your eyes from sparks from popping or backfiring torches, as well as light rays from the flame.

Additional personal gear may include:

1. Earplugs to keep sparks out of the inner ear as well as for sound prevention.
2. Heat-reflective pads for arms or hands that must rest on hot surfaces.

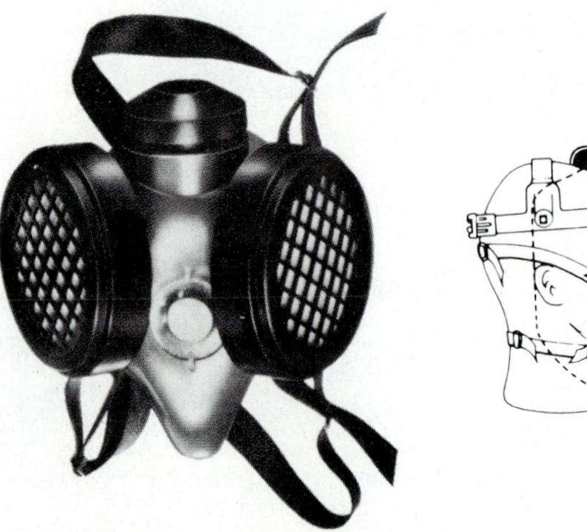

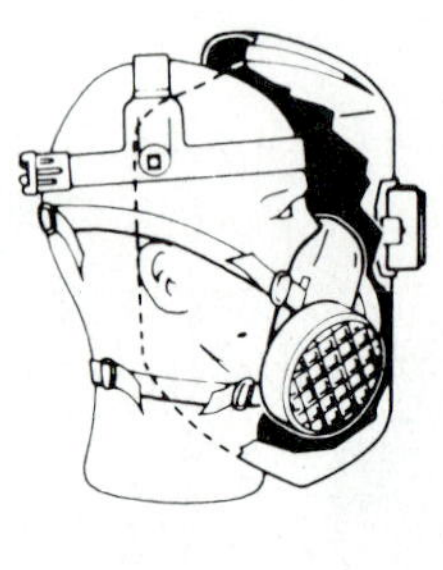

PHOTO 2-6 Respirators. (Courtesy of Sellstrom Safeguards.)

PHOTO 2-7 Well-dressed oxyfuel welder.

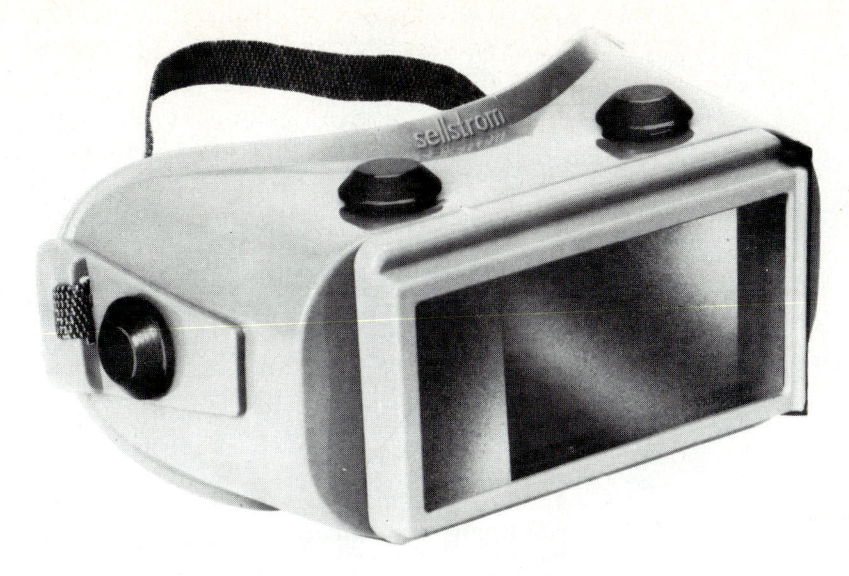

PHOTO 2-8
Oxyfuel goggles. (Courtesy of Sellstrom Safeguards.)

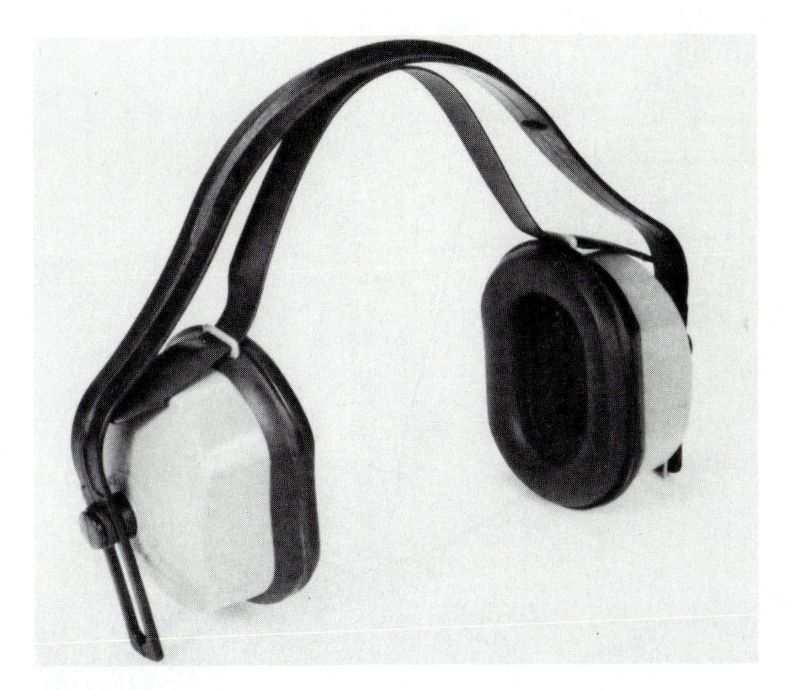

PHOTO 2-9
Ear protection. (Courtesy of Sellstrom Safeguards.)

PHOTO 2-10
Earplugs. (Courtesy of Sellstrom Safeguards.)

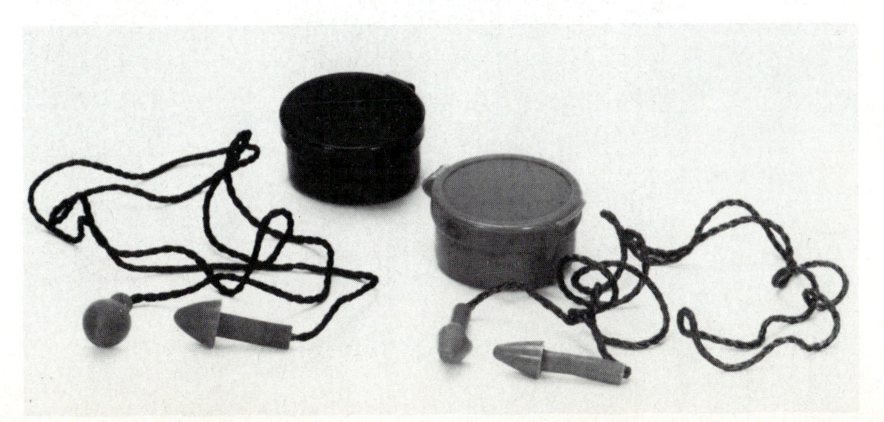

WELDING REPAIR SHOP
DO NOT WELD ON CONCRETE
OIL
OXYGEN
OXYGEN ON LEFT
FUEL ON RIGHT
HARD HAT AREA
CADMIUM PRODUCTS INC.
GALVANIZING MFG CO.
Smith's
INDUSTRIAL PARK FOUL UPS
ROBERT MARTIN

Answers:

1 Keep equipment in good working condition. (1.3.)

2 Light torch with friction lighter or stationary pilot flame. (3.1.3.3.)

3 Point away from persons and combustibles when lighting torch. (3.1.3.3.)

4 Keep oxygen away from combustibles; don't use oxygen for compressed air; never allow oxygen jet to strike oily or greasy surfaces; don't use oxygen to blow away dust. (3.1.6.1.)

5 Store and use acetylene and LP gas cylinders with valve end up. (3.2.1.1.)

6 Store oxygen cylinders apart from fuel gas cylinders. (3.2.3.3.)

7 Store cylinders in protected place. (3.2.3.5.1.)

8 Handle cylinders with care, to avoid damage to valves, safety devices, or the cylinder itself. (3.2.4.1.1.)

9 Use valve protection caps on cylinders that will accept them. (3.2.4.3.)

10 Secure or wedge cylinders into position when transporting by motor vehicle. (3.2.4.6.)

11 Only the cylinder owner, or persons authorized by him, shall refill a cylinder. (3.2.5.1.1.)

12 Use a pressure regulator! (3.2.5.5.)

13 Secure cylinders, to keep them from being knocked over while in use. (3.2.5.8.1.)

14 When a cylinder is empty, close its valve and mark it "EMPTY". (3.2.5.9.4.)

15 Replace or repair damaged hose. (3.5.6.4.)

16 Use the proper regulator for each gas and pressure range. (3.5.7.1.)

17 Never force connections that do not fit. Watch out for connectors with faulty seats; get rid of them. (3.5.7.2.)

18 Weld or cut only in a nonflammable atmosphere. (6.1.1.)

19 Take steps to prevent sparks from falling through floor cracks, etc. Remove or protect combustible materials in "falling spark" zones. (9.4.2.3.)

20 Keep suitable fire extinguishing equipment close at hand. (6.2.1.)

21 Before cutting or welding on an "empty" container, be sure it does not contain flammable vapors or any residues that might burn or give off flammable or toxic vapors. (6.4. Also see AWS A6.0-65.)

22 Keep flame and hot slag off of concrete; intense heat may cause flying fragments.

23 Wear gloves (7.3.2.) and goggles or other suitable eye protection (7.2.1.2.)

24 Keep sleeves and collars buttoned. *Pants uncuffed.* (7.3.3.)

25 Choose protective clothing to suit the work to be done. (7.3.3.)

26 Provide adequate ventilation whenever welding or cutting around cadmium, zinc, lead, fluorine compounds, or other toxic materials. (8.0. Also see ANSI Z49.1.)

27 Protect stored gasoline, oil, grease, etc. from sparks and slag. (9.4.2.2.1.)

28 Store reserve oxygen and fuel gas cylinders away from sparks and slag. (9.4.2.2.2.)

29 Wear a safety hat in designated areas.

30 Check connections for gas-tightness. Use soapy water or its equivalent – never a flame. (3.1.3.1.)

31 Post a warning near a leaking fuel-gas cylinder and promptly notify the supplier. (3.2.5.6.4.)

32 Don't tamper with or attempt to repair cylinder valves. Notify supplier. (3.2.5.7.)

33 Never leave a lighted torch unattended.

34 Place cylinders in a cradle or on a suitable platform for lifting. Don't use slings or electromagnets. (3.2.4.5.)

35 Keep hoses, cable, and other equipment clear of passageways, stairs, ladders, etc. (7.1.1.)

36 Use only approved manifolds. (3.3.1.1.)

37 Call fuel gases by their proper names. (3.1.5.2.)

PHOTO 2–11
Thirty-seven violations of welding safety rules are shown in the photo on the previous page. How many can you identify? (Courtesy of Smith Welding Equipment.)

1 "I never release the adjusting screw, takes too much time. Oops!"
Safety Rule: Always release regulator adjusting screw before opening valve.

2 "I told you, some of the welders were using oil on their regulators."
Safety Rule: Never use oil or petroleum base grease on regulators, inlet and outlet connections, cylinder valves, torches, fittings or other equipment in contact with oxygen.

3 "When you light up, you don't miss much, do you Jonesy?"
Safety Rule: Keep heat, flames and sparks away from combustibles.

4 "Blast! I must have forgotten to chain my oxygen cylinder."
Safety Rule: Secure and locate all cylinders to prevent them from falling, dropping or being knocked over.

5 Can you see the 37 (or more) safety violations shown here?
Safety Rule: Know and follow all safety rules and practices during welding, heating and cutting operations.

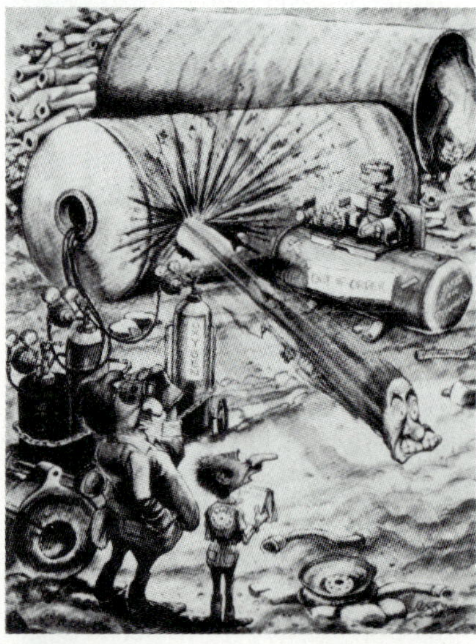

6 "There goes Lumis now . . . and his theory on using oxygen for ventilation."
Safety Rule: Never use oxygen as a substitute for compressed air.

PHOTO 2–12
Safety tips. (Courtesy of Smith Welding Equipment.)

REVIEW QUESTIONS

1. What are the three types of welding radiation?
2. Which of the three types cause burns to unprotected skin?
3. How should you mark metal that has just been welded or cut?
4. What are the three prerequisites for a fire or an explosion?
5. How can you safeguard an enclosed tank, drum, or vessel for welding or cutting?
6. What type of fire extinguisher will extinguish oil, wood, and electrical fires?
7. What should you do before welding in an area that is not specifically designed for welding?
8. What type of coated surfaces may present a health hazard for welders?
9. What action might you take if you know you might be generating toxic fumes?
10. What action might prevent most accidents?

3

WELDING TERMS AND DEFINITIONS

Air Carbon Arc Cutting (AAC): Cutting process using a carbon graphite electrode and compressed air to cut, bevel, and remove defective welds (Figure 3-1).

Arc Blow: Problem encountered mainly when applying shielded metal arc welding on direct current (dc). It is caused by a magnetic field that deflects the arc force and puddle from its normal desired direction. The following steps will help overcome the effects of arc blow:

1. Use alternating current (ac). Arc blow will still exist on ac, but the effects are neutralized due to the continuous reversing flow on ac.

FIGURE 3-1
Air carbon arc.

2. Use opposite electrode angle. When arc blow is occurring, the puddle will tend to flow to one side; angling the electrode to the opposite side will help counter this force.
3. Increase the amperage. More amperage will increase the arc force, which will help offset the magnetic forces of arc blow.
4. Wrap the ground cable around the work or change the location of the ground cable. Moving the ground cable will rearrange the magnetic field, usually removing the arc blow. Always try to weld away from the ground clamp.

Arc Cutting: This process, as it relates to shielded metal arc cutting, uses special electrodes with high amperages. The arc force and high amperage cause a cutting or gouging action. It is usually used where air carbon arc is not available.

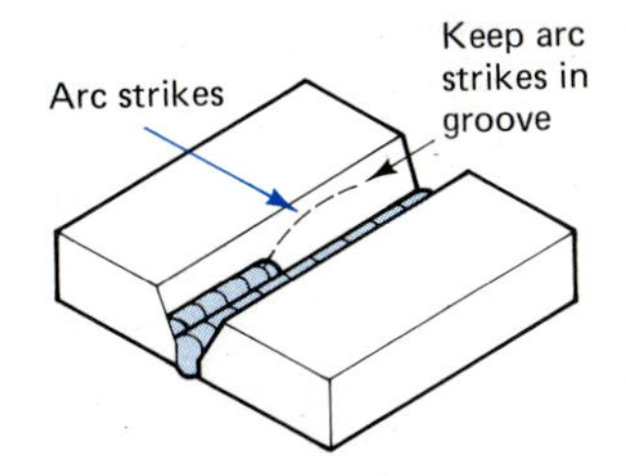

FIGURE 3–2
Arc strikes.

Arc Strikes: Undesirable melted spots left from striking SMAW electrodes on the base metal caused by the arc (Figure 3–2). It is best to strike electrodes in the groove, where they will be remelted when the groove is filled.

Arc Welding: Group of welding processes wherein coalescence is produced by heating with an electric arc, with or without the application of pressure, and with or without the use of filler metal.

As-welded: Condition of weld metal, welded joints, and weldments after welding to any period of time, by either thermal, mechanical, or chemical treatment of the welded area.

Axis of a Weld: Line through the length of a weld, perpendicular to the cross section at its center of gravity.

Backfire: This results from combustion of the fuel mixture *in* the oxy-fuel torch tip. It causes a popping sound and can blow molten metal from the weld puddle.

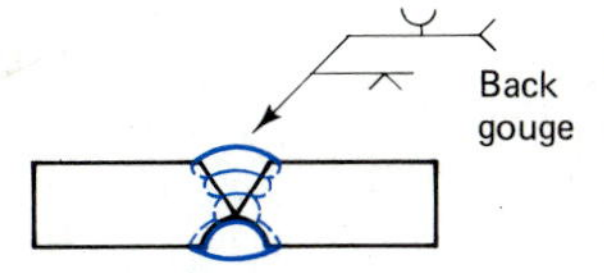

FIGURE 3–3
Back gouge groove weld.

Back Gouging: Process of removing the root pass from the backside of a groove weld, usually with air carbon arc (Figure 3–3). It is done to assure 100% penetration at the root side of groove welds.

Backing: Material applied to the back of groove welds to keep root passes uniform (Figure 3–4).

Backing (consumable): Consumable backing or inserts are root backings that are designed to fuse into the root and remain as a permanent part of the weldment (Figure 3–5).

FIGURE 3–4
Backings.

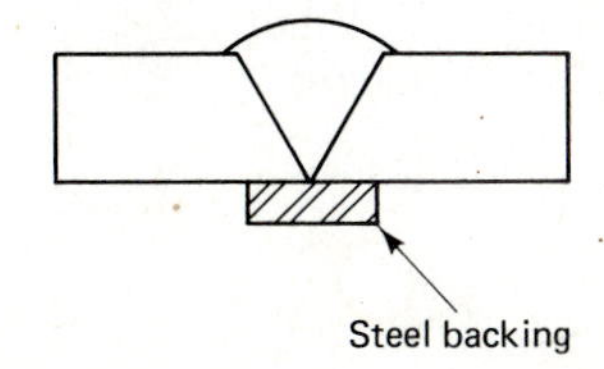

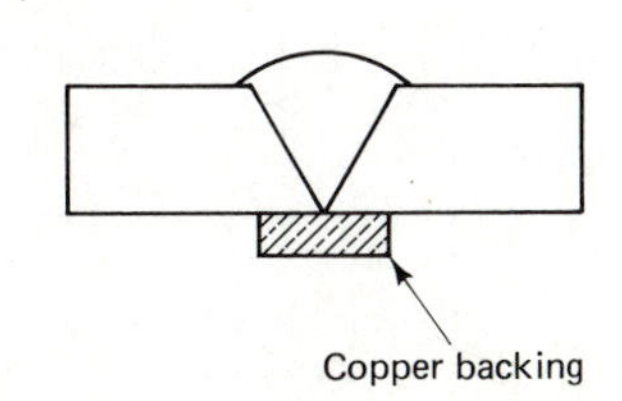

FIGURE 3–5
Consumable insert.

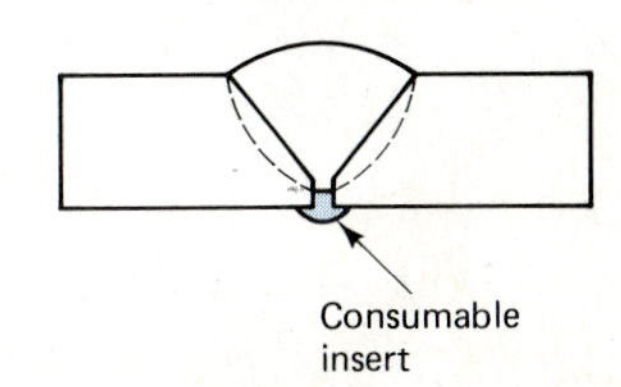

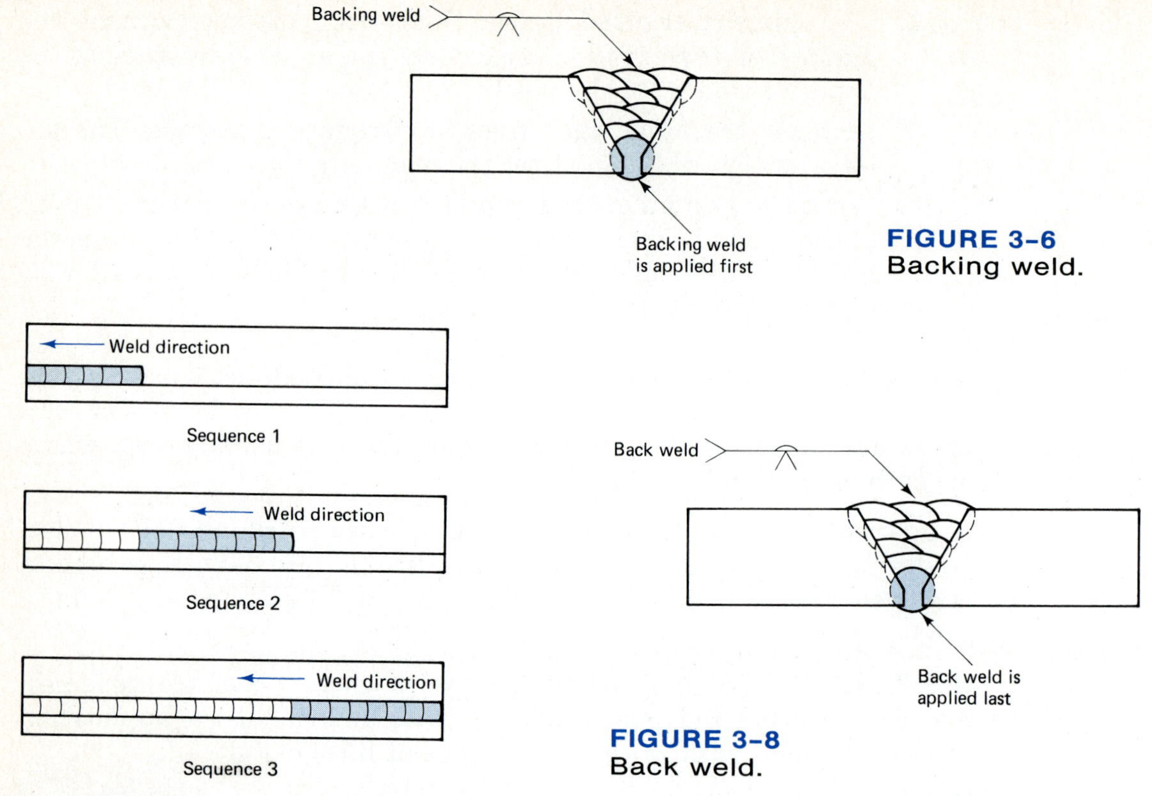

FIGURE 3–6
Backing weld.

FIGURE 3–7
Backstepping.

FIGURE 3–8
Back weld.

Backing Welds: Welds applied to the back of the groove joint before the joint is filled on the face side of the joint, intended to act as backing (Figure 3–6).

Back-stepping: Welding sequence used to reduce distortion and lower heat input into parts to be welded; welds are spaced and run backward to the overall progression of the joint (Figure 3–7).

Back Welds: Weld beads applied to the back side of a groove joint after the groove is filled (Figure 3–8).

Base Metal: Parent metal that is to be welded.

Bead: Refers to a weld or braze.

Bevel Angle: Angle at which the metal is prepared in a groove joint for welding (Figure 3–9).

Bevel Groove: Groove with only one side prepared (Figure 3–10).

Block Sequence Welds: Multiple-pass welds staggered so as to avoid stops and starts at the same point (Figure 3–11).

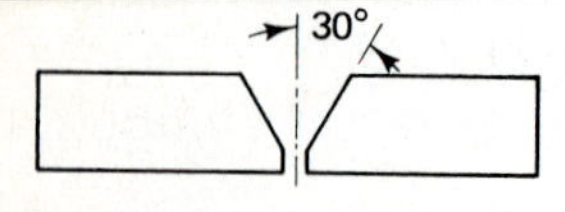

FIGURE 3–9
30° bevel angle.

FIGURE 3–10
Bevel groove.

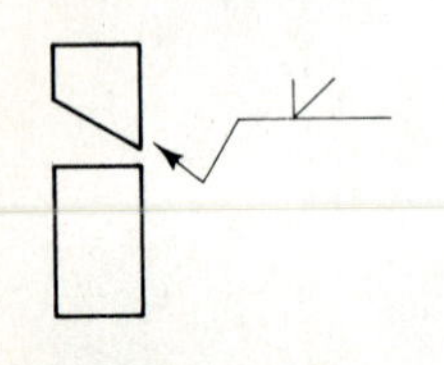

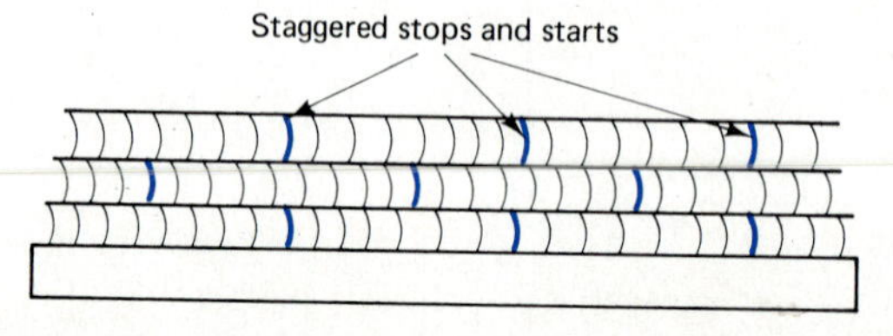

FIGURE 3–11
Block cascade welds.

Brazing: Capillary action process rather than fusion welding. Bonding is obtained by melting brass into the pores of the parts to be welded.

Capillary Action: Principal on which brazing and soldering rely for bonding. Molten metal flows into the open capillaries or pores of the base metal, while the base metal is in the solid state.

Carbon Arc Welding (CAW): Old arc welding process that used a carbon-graphite electrode to maintain an arc and heat the metals to be welded.

Carburizing Flame: Recognized by a feather in the flame. It is the result of too much fuel gas or not enough oxygen.

Cascading Beads: Beads that slope down toward the crater; continuous starting and stopping in the same place on multiple-pass welds that may result in cascading beads (Figure 3–12).

Chain Intermittent Welds: Intermittent welds that are directly opposite to each other; used for fillet welds (Figure 3–13).

Complete Penetration: One hundred percent fusion at the root of the weld. This means no unfused metal or factory edge visible at the root side of groove-type welds.

Concave Welds: Welds that have a sunken surface at the face of the weld (Figure 3–14).

Contact Tip: Device that transmits the current from the cable to the filler wire electrode in GMAW. They are usually made of copper.

Convex Welds: Welds that have a protruding surface or reinforcement at the face of the weld (Figure 3–15).

Craters: Depressions found at the end of the weld bead and are the result of high heat and the force of the arc.

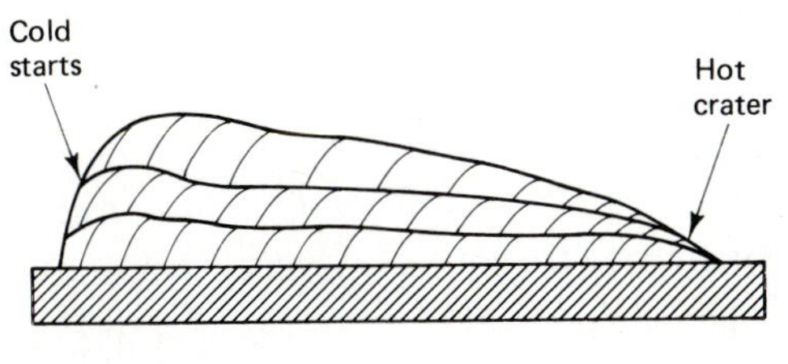

FIGURE 3–12
Cascading beads.

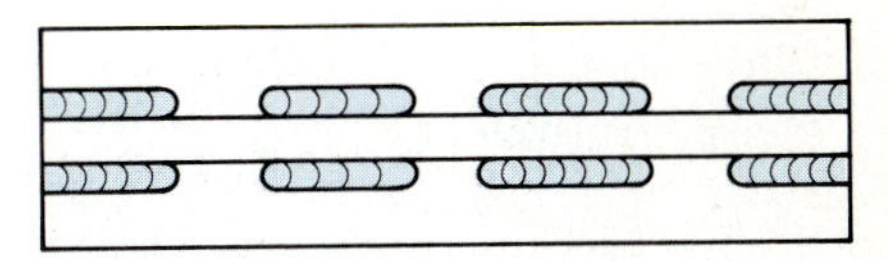

FIGURE 3–13
Chain intermittent welds.

FIGURE 3–14
Concave weld.

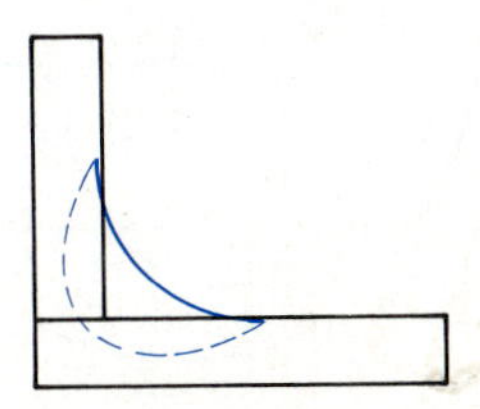

FIGURE 3–15
Convex weld.

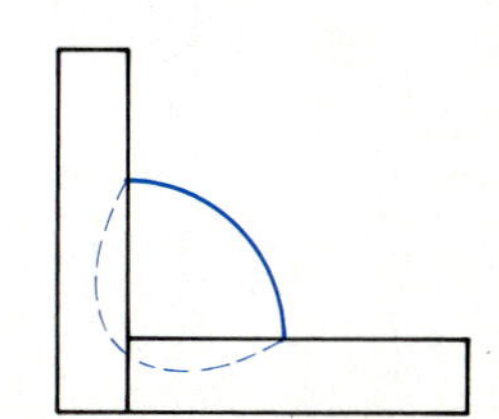

Defects: Flaws that because of their type, size, or location are considered defective. When a flaw is called a defect, it has been rejected.

Deposition Rate: Speed that filler metal is deposited into the joint.

Dilution: Mixing action or diluting of filler metal into the base metal.

Direct-Current Reverse Polarity (DCRP): Also known as direct-current electrode positive (DCEP). On this polarity the work is negative and the electrode is positive. The current flow is from the work to the electrode (negative to positive) (Figure 3–16).

Direct-Current Straight Polarity (DCSP): Also known as direct-current electrode negative (DCEN). On this polarity the electrode is negative and the work is positive. The current flow is from the electrode to the work (negative to positive) (Figure 3–17).

Discontinuity: Flaw or interruption in sound metal. A discontinuity

FIGURE 3–16
DCRP (DCEP) (current flow).

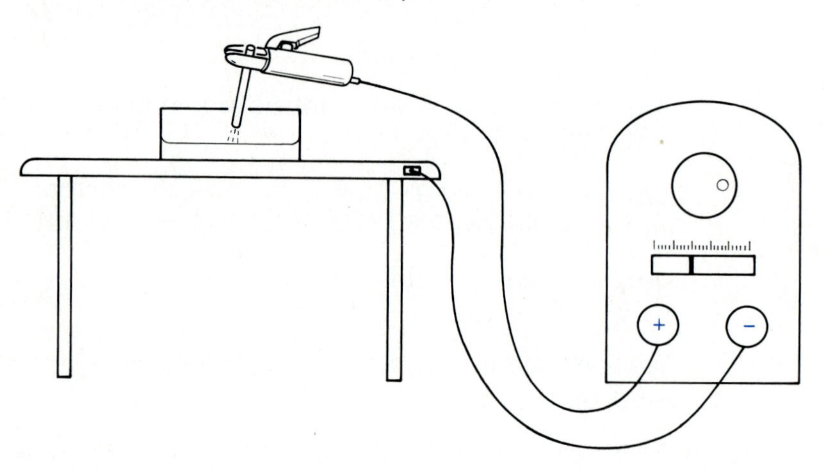

FIGURE 3–17
DCSP (DCEN) (current flow).

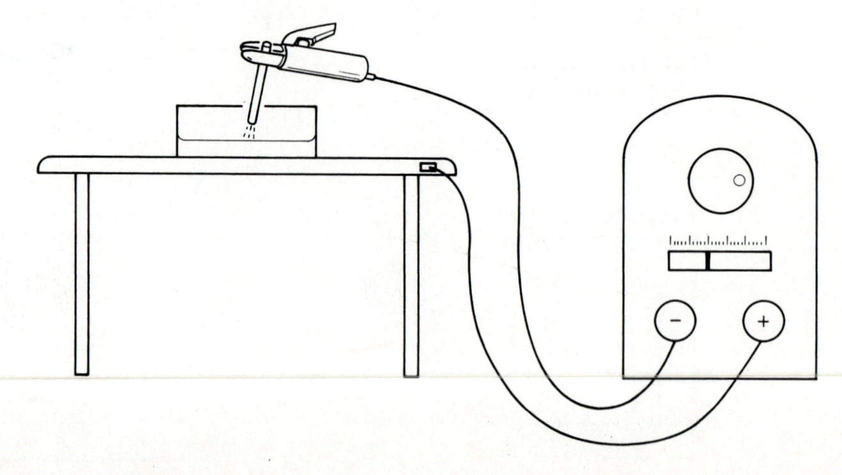

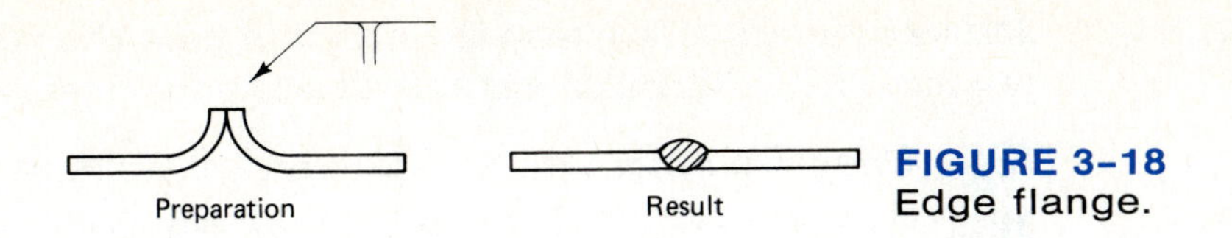

FIGURE 3-18
Edge flange.

may or may not be a defect. Whether it is a defect depends on the size, type, and location of the discontinuity.

Downhand Position: Older term also used to mean flat position.

Duty Cycle: Welding power source capability to operate safely under load. It is based on the percentage of time it can operate under load in a 10-minute period. Example: A 60% duty cycle means that the power source can operate for 6 minutes out of 10.

Edge Flange Joints: Joints that are designed for butting sheet metal without filler metal. The result is a very smooth uniform bead (Figure 3-18).

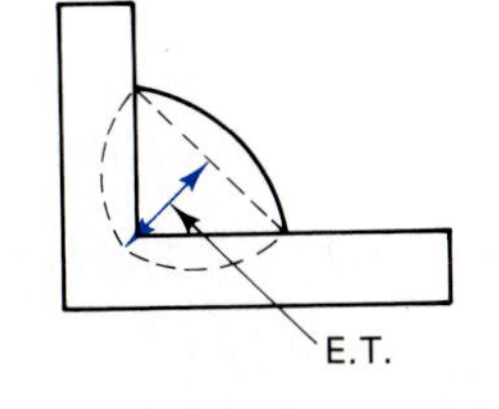

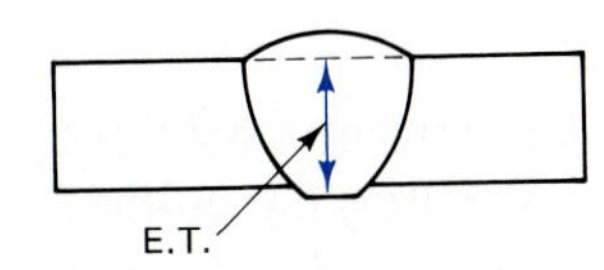

FIGURE 3-19
Effective throat.

Effective Throat: Distance through the center of the weld from a line parallel to the plate surface down to the deepest point of penetration (Figure 3-19). Weld reinforcement is not included as part of the effective throat.

Effective Weld Area: Effective throat times the effective weld length (ET × EWL = EWA; Figure 3-20).

Effective Weld Length: Length of the weld (as long as the full cross section is consistent throughout this length).

Electrode: The electrode in any electric arc welding process is the current-carrying conductor which carries the current to the arc. This electrode conductor may be a flux-covered wire (SMAW), a bare wire (GMAW), or tungsten (GTAW).

Electrode Holders: Insulated "grippers" that both hold the electrode in place and conduct the current into the electrode in shielded metal arc welding (SMAW).

Electron Beam Welding (EBW): Advanced welding system that hurls high-velocity free electrons at the seam to be welded. The electron beam

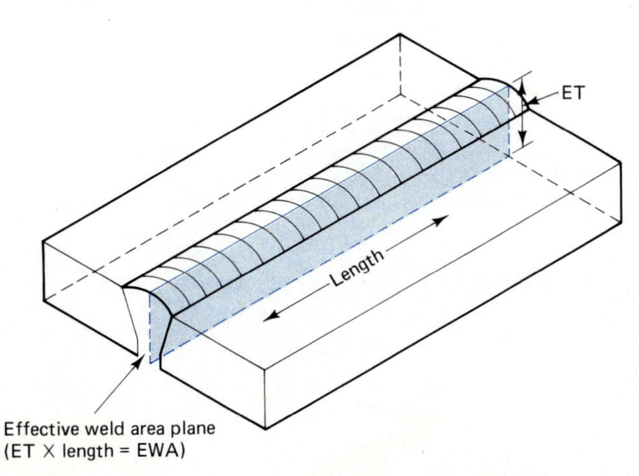

FIGURE 3-20
Affected weld area.

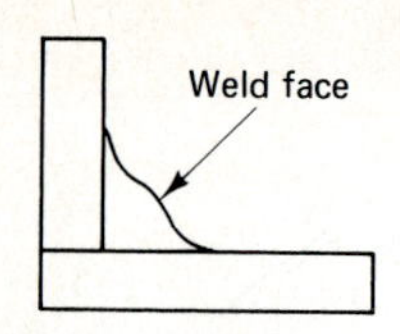

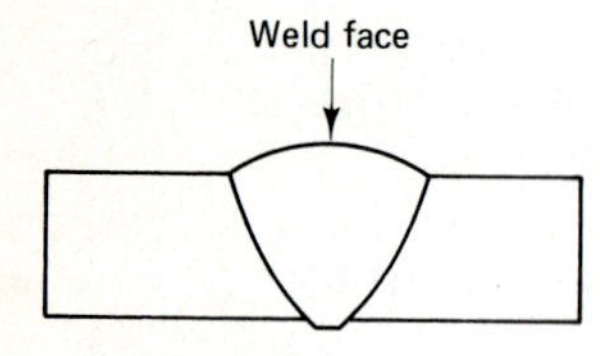

FIGURE 3-21
Weld face.

actually pierces a hole into the part and uses the keyhole technique to produce the weld. Filler metal is usually not required.

Electroslag Welding: Automatic vertical welding process that feeds an electrically charged filler wire into hot molten slag. The weld progresses upward, with the hot molten slag preheating the joint and the wire providing filler. The weld metal and slag are held in the joint by water-cooled retaining shoes.

Face (Weld Face): Front or outer surface of the weld on the side that was welded (Figure 3-21).

Filler Metals: Includes all the additional metal added to the weld deposit. Filler metals may be in the form of flux-covered or -cored electrodes that are transmitted through the arc, or they may be solid wires that are dipped into the molten weld puddle while welding.

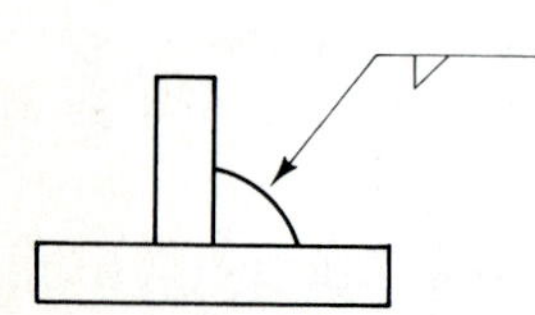

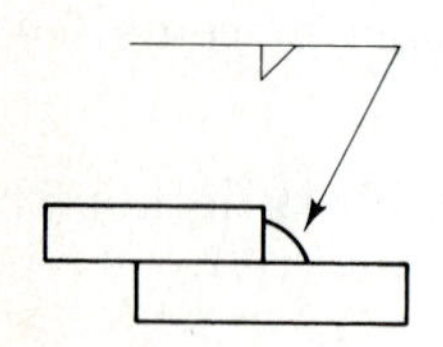

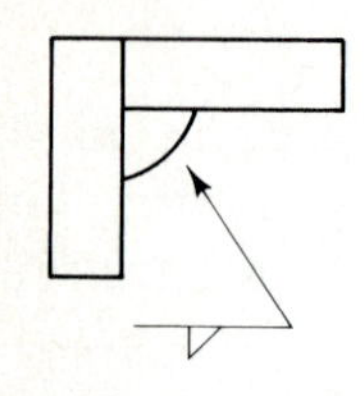

FIGURE 3-22
Fillets.

Fillet Welds: Welds with two legs at approximately 90° to each other (Figure 3-22). Fillets may be tee, lap, or corner joints.

Fissures: Small cracks. Microfissures cannot be seen without magnification. Macro fissures are visible to the eye without magnification.

Flair Bevel: Weld groove joint usually formed when welding pipe to plate (pipe sideways; Figure 3-23).

Flair Vee: Weld groove joint usually formed when welding two pipes side to side (Figure 3-24).

Flash Welds: Classified as resistance welds, but the current induced into the part actually causes an arc between the two parts to be welded. After a short delay for the arc to build up heat, the parts are pushed together and fused (Figure 3-25).

Flat Position: Position where gravity is pulling straight down onto the plate (Figure 3-26). It is one of the easiest positions in which to do welding.

Flaw: Discontinuity or imperfection in the weld or base metal. A flaw may or may not be a defect.

Flux: Substance (usually powder or granular) introduced into a mol-

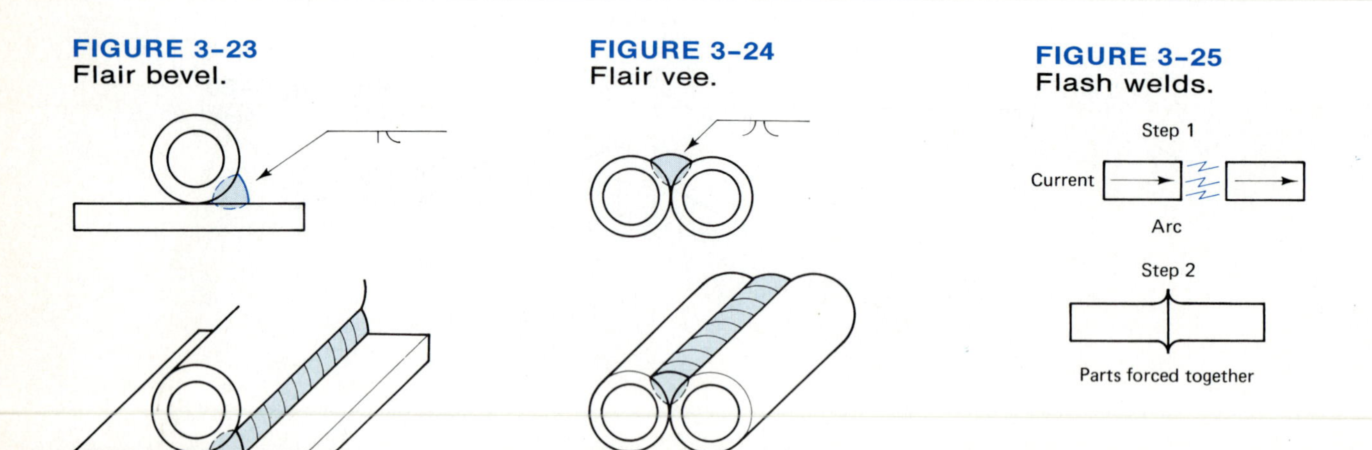

FIGURE 3-23
Flair bevel.

FIGURE 3-24
Flair vee.

FIGURE 3-25
Flash welds.

ten weld deposit to clean, remove oxides, and slow cool the weld as it solidifies and cools.

Flux-Cored Arc Welding (FCAW): Welding process similar to gas metal arc welding (GMAW) but tubular, flux-cored wire is used instead of solid wire. The process uses a constant-voltage power source.

Fusion Zone: Area of the weld where the base metal and filler metal mix and fuse to form the weld deposit.

Gas Metal Arc Welding (GMAW): Arc welding process that feeds a consumable wire electrode filler wire continuously into the molten puddle. A shielding gas is supplied to the weld to protect the arc from atmosphere contaminants. This gas flows from a portable cylinder through the hose and comes out at the cup on the gun.

Gas Tungsten Arc Welding (GTAW): Arc welding process that creates an arc between a tungsten electrode and the work or the part to be welded. Shielding gas comes from the cup and shields the molten puddle. Filler, if needed, is fed into the puddle from the side.

Gas Welding: Also known by the preferred term "oxyfuel welding." It uses a fuel gas and oxygen (to accelerate combustion) as a heating medium to melt parts for welding. Filler may be dipped into the puddle if needed.

Groove Angle: The groove angle is also known as the included angle (given in degrees). It is the total angle of the groove, including both prepared parts (Figure 3-27).

Groove Weld Joints: Joints that require some form of preparation (grooving) of the edges for proper fusion and penetration. They may be butt, tee, or corner joints (Figure 3-28).

Groove Weld Parts: (*See Figure 3-29*).

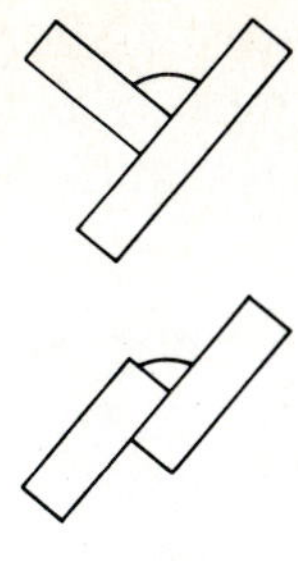

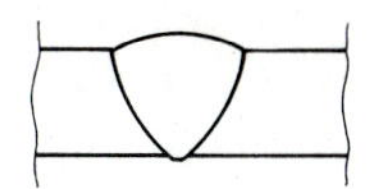

FIGURE 3-26
Flat fillets and flat groove.

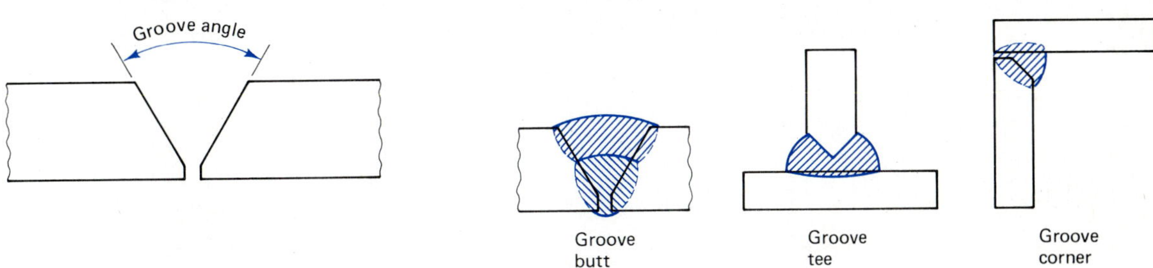

FIGURE 3-27
Groove angle.

FIGURE 3-28
Groove welds.

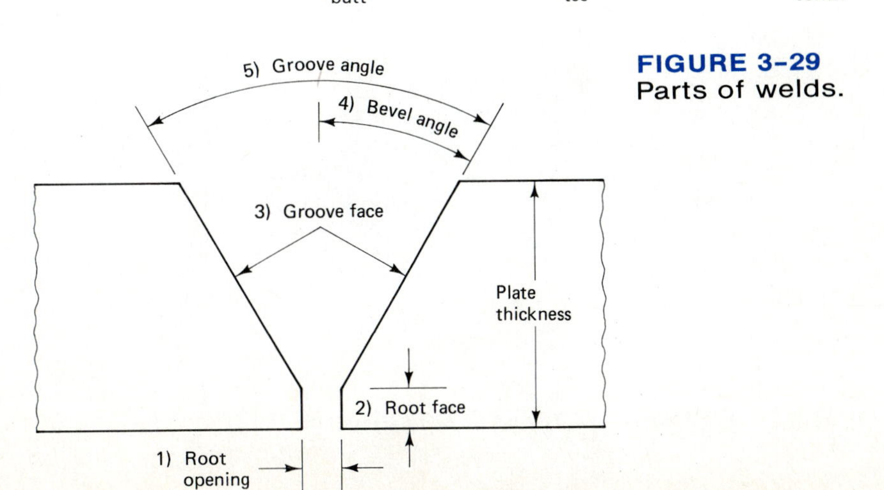

FIGURE 3-29
Parts of welds.

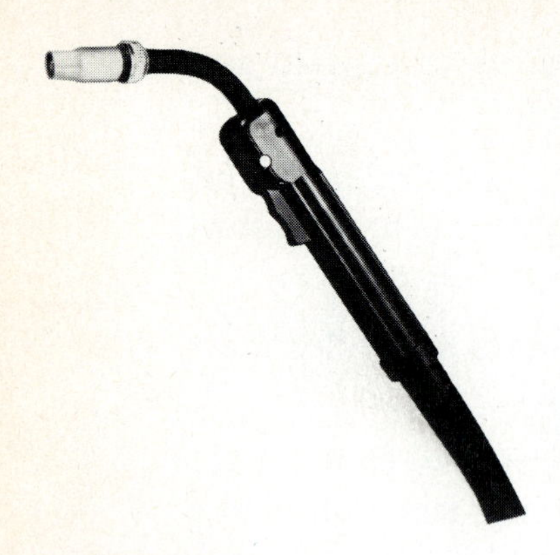

PHOTO 3-1
GMAW welding gun.

Gun (Welding Gun): Welding gun used in GMAW, it contains the contact tip, cup, microswitch for pulling in the contactor, and the shielding gas and electric cable junction block. Other welding processes, such as thermal spray surfacing, also use this type of gun.

Hardfacing (also known as Surfacing): Welding system that deposits a wear-resistant layer of weld metal. Special alloy electrodes or wires can be arc welded or thermal sprayed onto high-wear surfaces.

Heat-Affected Zone: Area that was superheated by the weld metal and arc but did not actually melt (Figure 3-30). The cool metal on the opposite side of the weld causes quick cooling in this zone; the result is a large grain microstructure. When a properly produced weld is pulled to its ultimate tensile strength, it usually fails in the heat-affected zone.

Horizontal Weld Position: When a weld is to be deposited on a plate or pipe seam in the horizontal axis (Figure 3-31).

Inert Gas: Gas that is chemically inactive. Inert gases do not break down when ionized by electric arc. Argon and helium are typical inert gases used in welding as shielding gases.

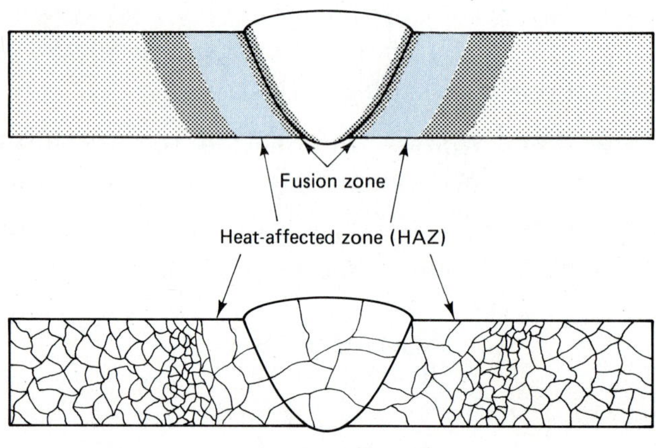

FIGURE 3-30
Weld metal zones.

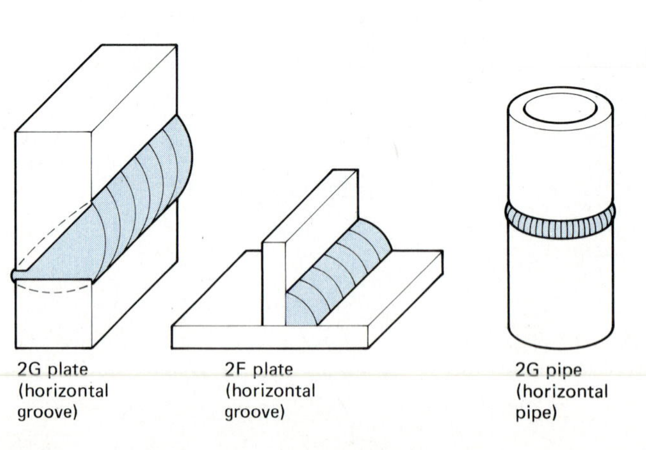

FIGURE 3-31
Horizontal positions.

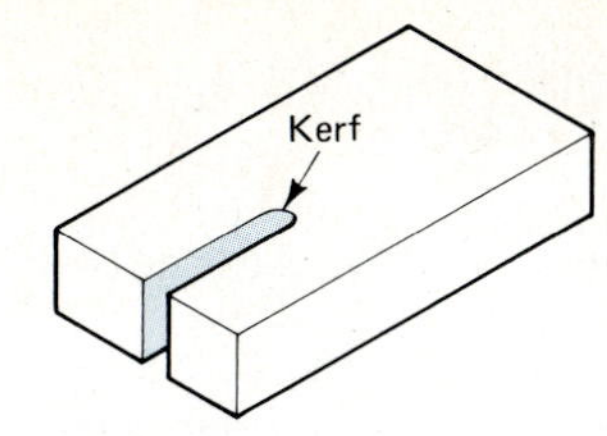

FIGURE 3–32
Kerf.

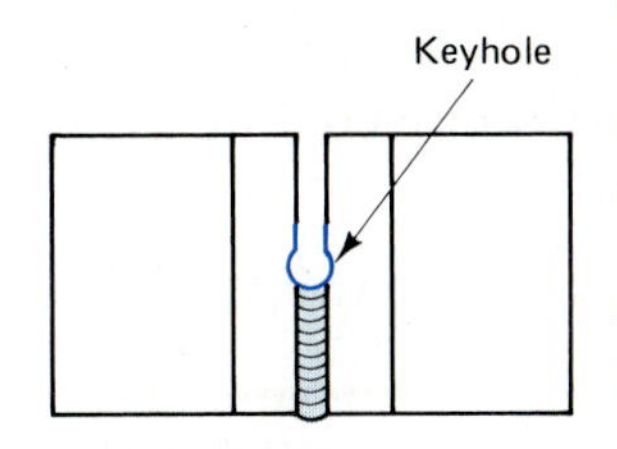

FIGURE 3–33
Keyhole.

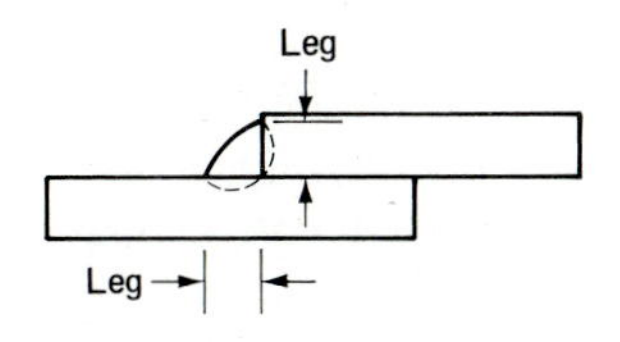

FIGURE 3–34
Weld legs.

Interpass Temperature: Temperature measured between passes, immediately next to (but not on) the weld.

Joint Design: Type of weld joint used. Typical joint designs include V-grooves, U-grooves, bevel grooves, and J-grooves.

Kerf: Metal removed from a cutting process (Figure 3–32).

Keyhole: The keyhole is the hole produced by fused metal at the root of an open butt groove weld (Figure 3–33).

Lack of Fusion: Flaw that is usually found in the side of the joint where the metal has not fully fused into the sides of the joint. Lack of fusion can also result where multipass beads are not well fused. It is usually the result of low amperage or poor electrode angle.

Land: Also known as the root face. It is the flat parallel surface on a groove joint.

Lap Joint: Fairly strong joints that contain two overlapping plates. Fillet welds are usually used on lap joints, but plug or slot welds can also be used.

Laser Welding and Cutting: Uses coherent, concentrated light waves to melt and weld or melt and cut metal and alloy parts. Laser stands for "light-amplified stimulated emission of radiation."

Leg (Fillet Weld Legs): The leg length is also the weld size on fillet welds. The legs are the two perpendicular lines of fusion from the root to the toes (Figure 3–34).

Melting Point: Temperature at which a material's atoms approach the random state (liquid state).

Mixing Chamber: The mixing chamber, or mixer, is where the oxygen and fuel gas are mixed in an oxyfuel torch.

Multipass Weld: Welds made up of a series of "tied-in" passes to make up one large weld (Figure 3–35).

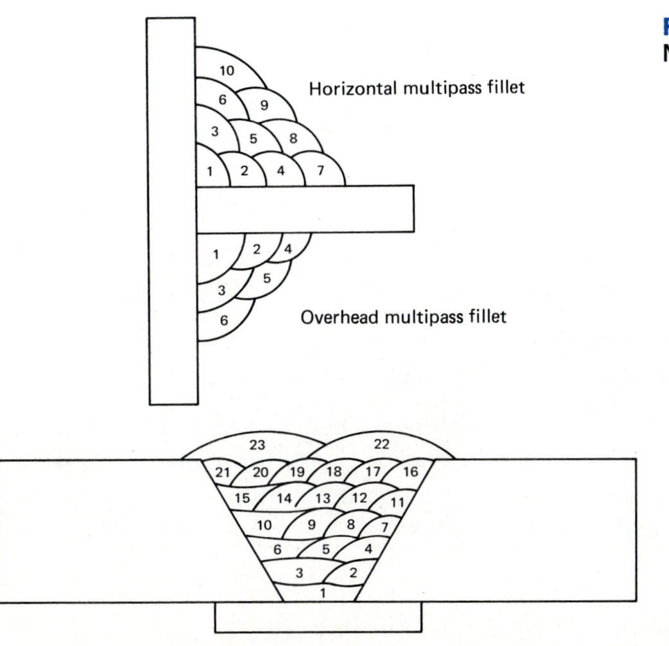

FIGURE 3–35
Multipass welds.

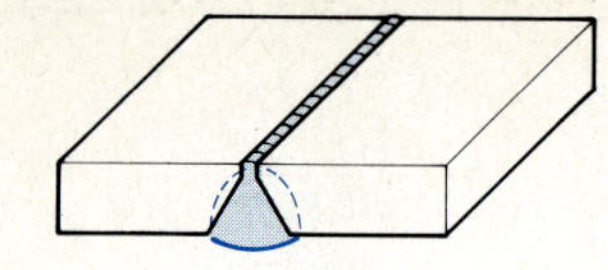

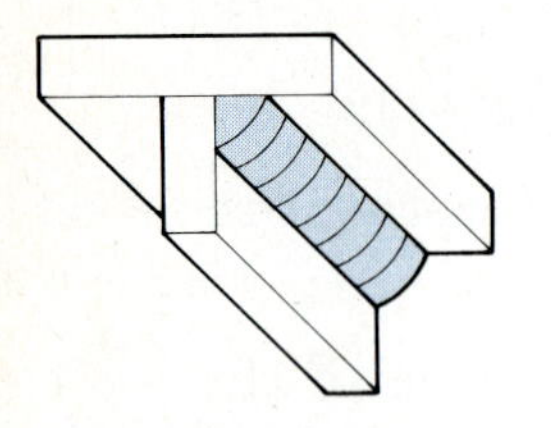

FIGURE 3-36 Overhead position.

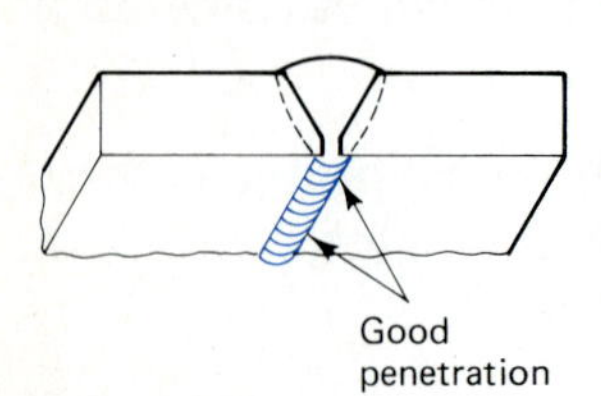

FIGURE 3-37 Penetration.

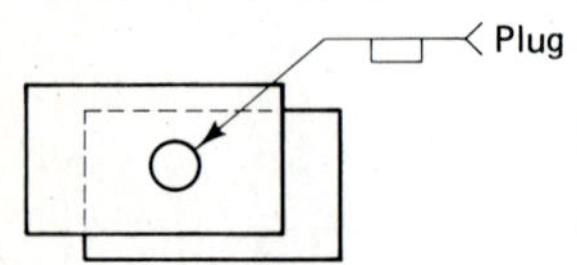

FIGURE 3-38 Plug welds.

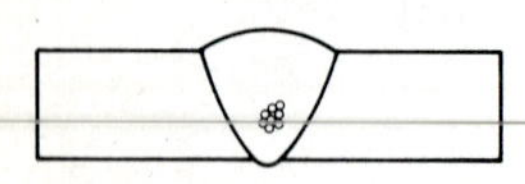

FIGURE 3-39 Porosity.

Neutral Flame: Recognized by a soft inner cone, with no feather showing around the inner cone.

Overhead Weld Positions: Position where the weld is to be deposited on a plate or pipe seam in the overhead axis (Figure 3-36).

Oxyfuel Cutting: Using a fuel gas mixed with pure oxygen to produce an extremely hot flame. The flame heats the steel as a jet of pure oxygen pierces the steel. The oxidation process keeps the cut advancing rapidly.

Peening: Method of reducing residual stress in the weld area. A ball peen hammer is used to give mild blows to just-produced hot weld. These strikes send out shock waves that relieve residual stress.

Penetration: Amount of depth and fusion at the root of a weld (Figure 3-37).

Performance Qualification: Qualification test, or welder certification test, designed to test a welder's ability to make sound welds to the specifications of a welding procedure.

Plasma Arc Cutting (PAC): Uses a high-velocity plasma gas (ionized gas) to melt and pierce metals to be cut. Plasma is very popular for cutting nonferrous metals that cannot easily be cut with the oxyfuel cutting process.

Plasma Arc Welding (PAW): Uses a high-velocity plasma gas to melt and fuse metals to be welded. The plasma gas is forced through a narrow orifice which produces a narrow bead with deep penetration. Filler metal may or may not be used.

Plasma Gas: Gas that is ionized (conducting electricity).

Plug Weld: Welds made on lap joints by drilling, punching, or cutting a hole into one of the plates, and then filling the holes with weld (Figure 3-38).

Porosity: Discontinuity produced by air or some gas being trapped into the weld as it is solidifying (Figure 3-39).

Positions (Weld Positions): Weld joint positions for plate and pipe are identified with the following system (Figure 3-40):

G	Groove	1	Flat
F	Fillet	2	Horizontal
		3	Vertical
		4	Overhead

Postweld Heat Treatment: Heat treatment performed on welds after their completion, usually to relieve stress, anneal, or normalize.

Preheating: Process of heating the metal before welding, usually to improve weldability.

Procedures (Welding Procedures): Document prepared by the manufacturer or contractor describing all the essential details of a welded joint. Most companies have many procedures for various joint designs and welding processes.

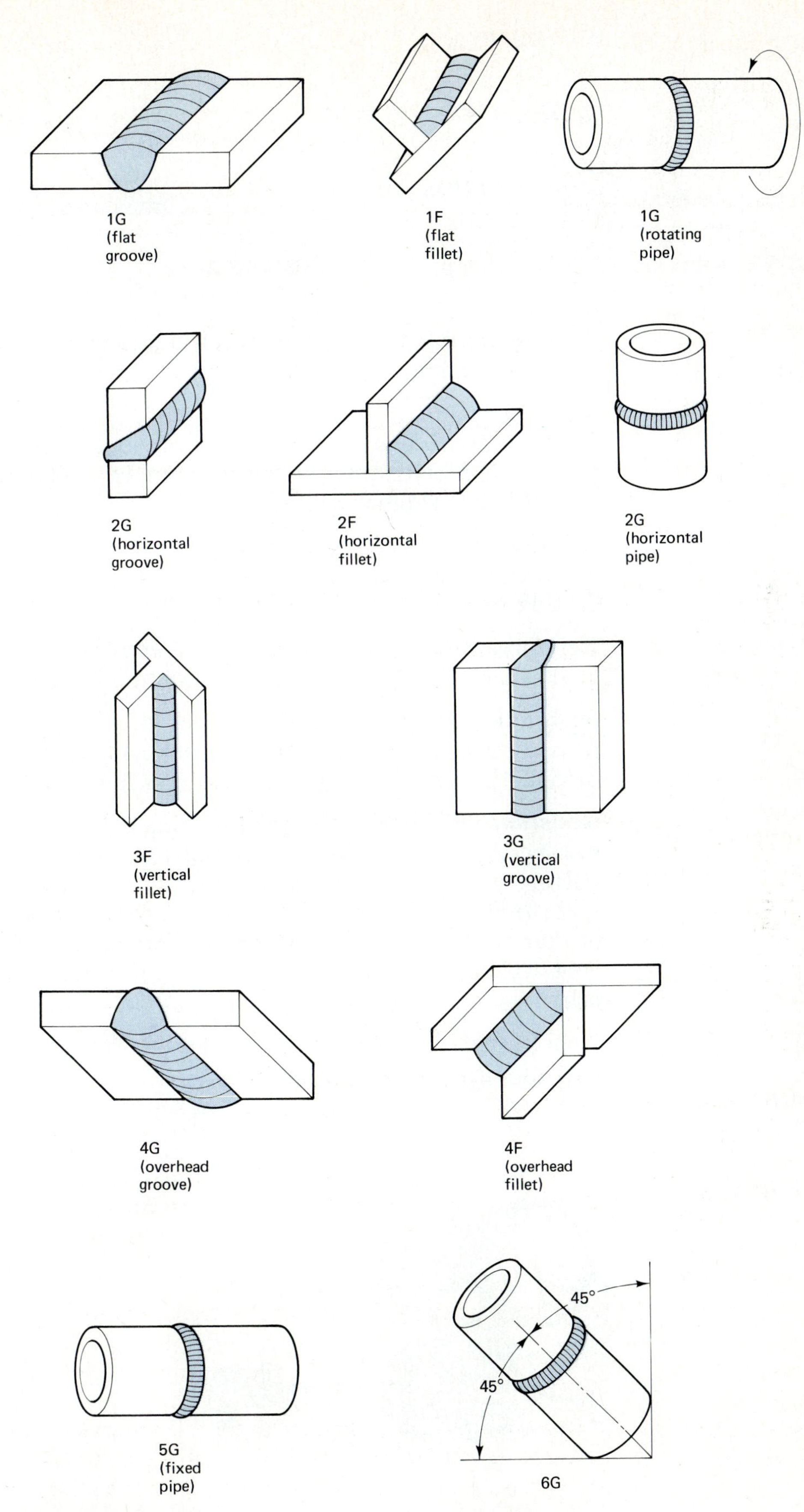

FIGURE 3-40
Weld positions.

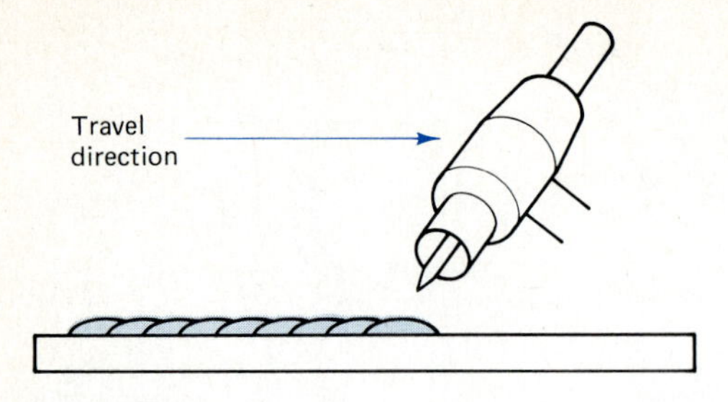

FIGURE 3–41
Torch position for pulling.

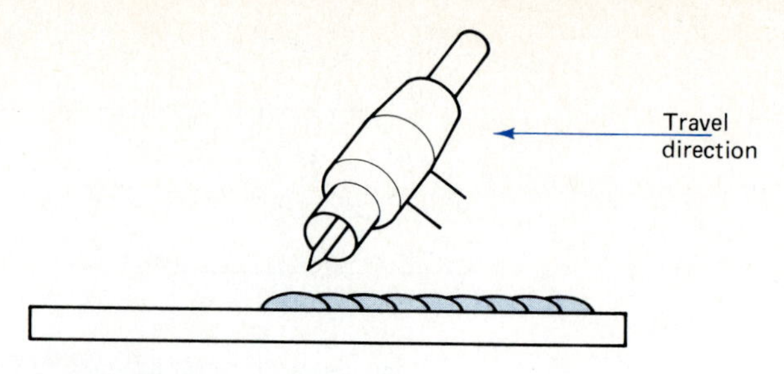

FIGURE 3–42
Torch position for pushing.

Pulling technique: Gun or torch angle and the direction of travel. With the pulling technique, the torch or gun is angled in the same direction as the travel direction (Figure 3–41).

Pushing Techinque: Gun or torch angle and direction of travel. With the pushing technique, the torch or gun is angled away from the direction of travel (Figure 3–42).

Regulator: Gas-controlling device that reduces gas bottle pressures down to the working pressure for the oxyfuel and other welding and cutting processes.

Reinforcement: Extra metal in the convex face of the weld (Figure 3–43). It is usually limited to a maximum dimension of $\frac{1}{8}$ in.

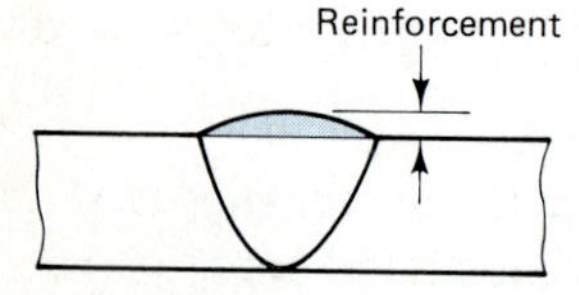

FIGURE 3–43
Reinforcement.

Residual Stress: Internal or "locked-in" stress that is commonly caused by welding due to cooling and contraction of the weld. A common method of relieving residual stress is stress-relief heat-treatment.

Resistance Welding: Group of welding processes that use electrical resistance to heat and weld parts together. Most resistance welding involves heating the metal to the plastic state combined with an upset force (pressure) to produce a weld that can develop strength up to 90% of that of the base metal. Typical resistance welding processes include spot welding, projection welding, seam welding, upset welding, and others.

Root Face: Also known as the land. It is the parallel surface at the root of a groove joint (Figure 3–44).

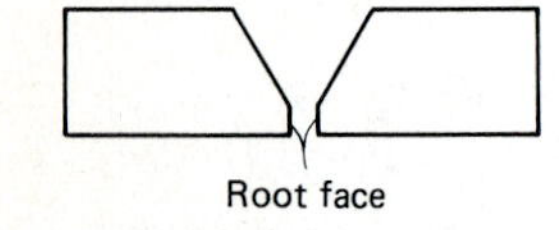

FIGURE 3–44
Root face.

PHOTO 3–2
Regulators. (Courtesy of Smith Welding Equipment.)

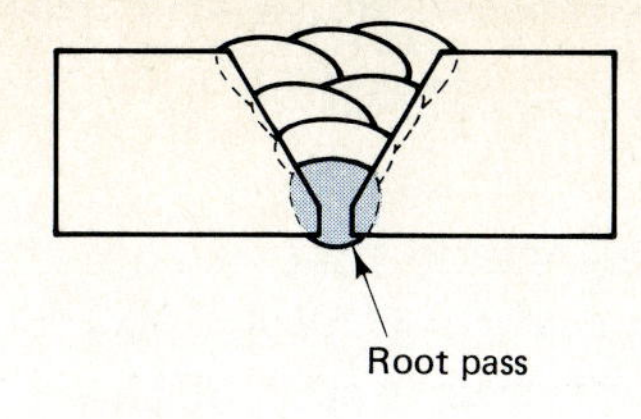

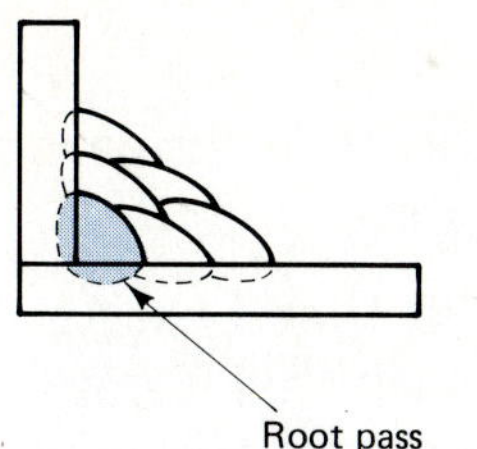

FIGURE 3-45
Root pass.

Root Opening: Also known as the gap. It is the open or gapped area between the two plates in a groove weld.

Root Pass: Usually the first pass on a weld joint, deposited into the root of the joint (Figure 3-45).

Scarf Joint: Butt joint specifically designed for brazing. Its design allows for a larger surface area, for a stronger braze (Figure 3-46).

Shielded Metal Arc Welding (SMAW): Commonly referred to as "arc welding." The process uses a flux covered consumable electrode to heat and melt the base metal. As the electrode is consumed, both filler from the core wire and shielding from the flux covering are provided to the molten puddle.

Shielding Gases: Gases that both provide a path for the electrical current and shield the molten puddle from the atmosphere. Shielding gases can be inert or active. Commonly used shielding gases for welding are argon, helium, and carbon dioxide (CO_2).

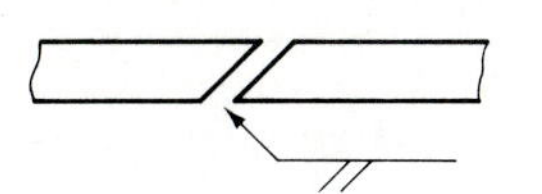

FIGURE 3-46
Scarf joint.

Short-Circuit Transfer: Cold, low-voltage, low-amperage mode of GMAW transfer. The cycle actually causes a short circuit against the work, melts off, and short circuits again.

Slot Weld: Weld produced by cutting a slot into the top plate of a lap joint and filling it with weld (Figure 3-47). Commonly used where welds need to be flush or concealed.

Spatter: Small metal particles that drop off and deposit on either side of the weld bead (Figure 3-48). It is usually the result of too long an arc, too much current or voltage, or the wrong electrode angle.

Spray Transfer: Hot, high-amperage, high-voltage mode of metal transfer. The energy is so high that it actually breaks the filler wire down into small particles as it transfers through the arc. Argon or argon mixtures, such as argon oxygen (98% argon, 2% O_2), is popularly used in GMAW spray transfer.

Stack Cutting: Method that saves time and money by stacking a number of plates on top of each other when thermal cutting (Figure 3-49). With oxyfuel cutting, Mapp gas is recommended for stack cutting.

FIGURE 3-47
Slot welds.

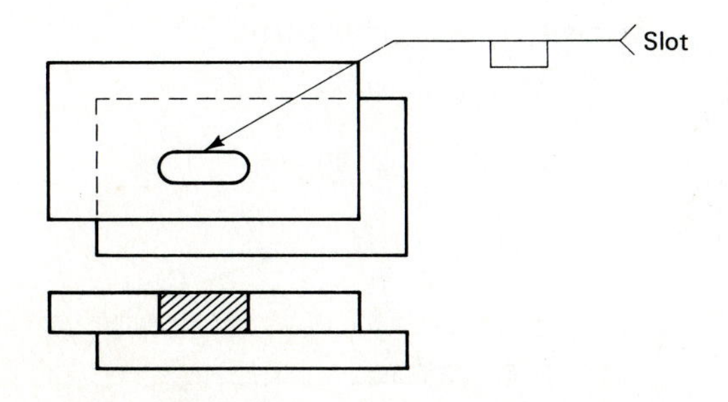

FIGURE 3-48
Weld spatter.

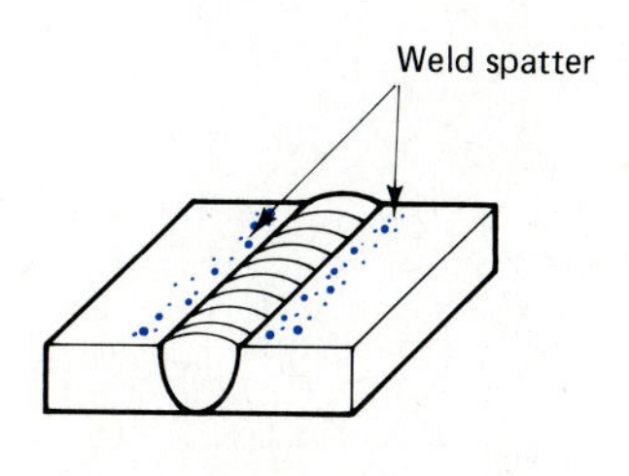

FIGURE 3-49
Stack cutting multilayers of plate.

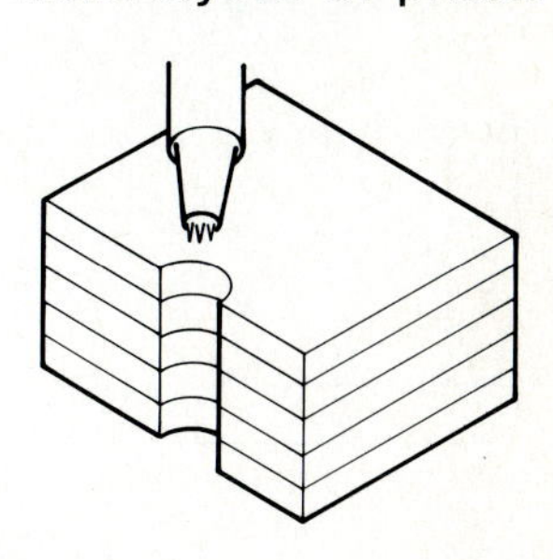

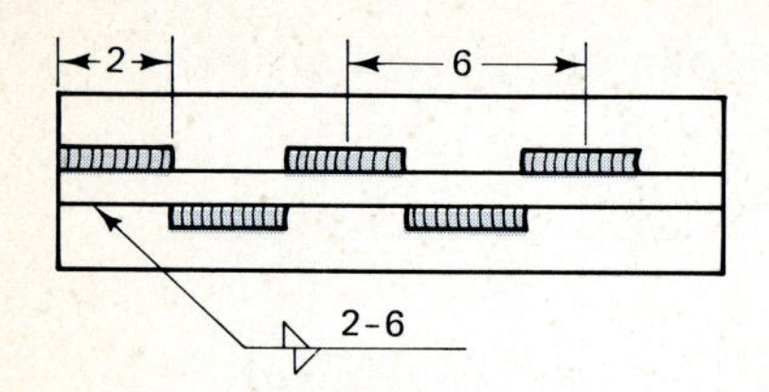

FIGURE 3-50 Staggered intermittent welds.

Staggered Intermittent Welds: Two strings of fillet welds that are opposite and offset from each other (Figure 3-50). A good rule to keep the stagger even is to take one-half the pitch for the starting point of the opposite side of the staggered weld.

Stand Off: Distance between cup or tip and the work. It is similar to the arc length but does not necessarily have to be an arc.

Stress-Relief Heat Treatment: Heat-treating method that heats the metal below the critical temperature, holds or soaks at that temperature, then is allowed to cool. At the elevated temperature the metal tends to soften and relieve any residual stress that may have built up by welding.

Submerged Arc Welding (SAW): Welding process that produces large welds similar to that of GMAW, but does so beneath a blanket of flux. It can be done manually or fully automatically.

Tack Welds: Small welds designed to hold parts together temporarily until the welder can deposit the permanent weld. Tacks can be from $\frac{1}{4}$ to 6 in. long, depending on the size of the plate or pipe being welded.

TIG (Tungsten Inert Gas): Another (unofficial) name given to the gas tungsten arc welding (GTAW) process.

Toe: Edge at the top of a weld, where the filler metal meets the base metal; there are two toes on the fillet and groove welds (Figure 3-51).

Torch: Device used to carry, mix, and contain fuel and oxidizer gases in oxyfuel welding and cutting. But torches are also used in GTAW and PAW to carry current, shield, and plasma gas to the arc.

Travel Angle: Angle of lean either into, or away from, the direction of travel (Figure 3-52). The terms "pushing" and "pulling" denote which direction of travel angle is being used.

Tungstens: Used in GTAW, PAW, PAC (gas tungsten arc welding, plasma arc welding, and plasma arc cutting) as the nonconsumable electrode for carrying the high currents required by these processes. Tungsten has a melting point of almost 6000°F.

Ultrasonic Testing (UT): Method of testing metal and welds using high-frequency sound waves, produced and received by a crystal, then plotted on a CRT (cathode-ray tube) or paper printout. UT is popular because it can detect deep-seated discontinuities fairly easily and accurately.

FIGURE 3-51 Toes.

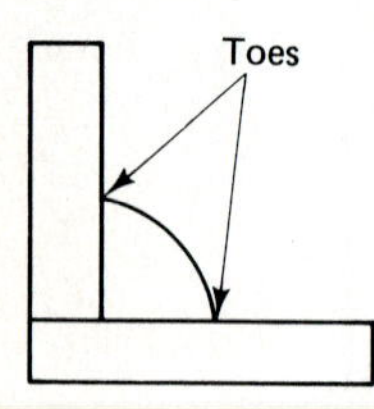

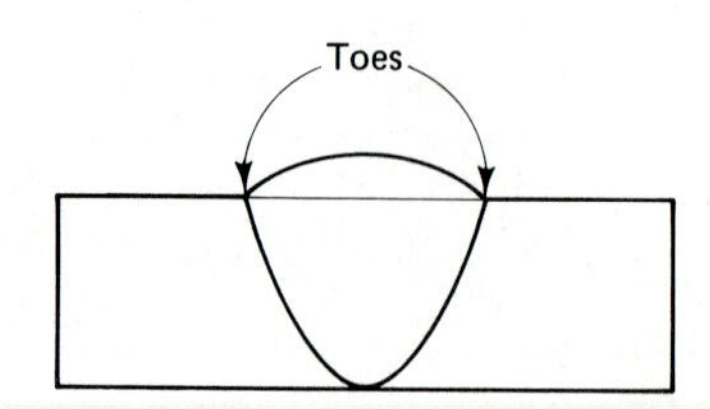

FIGURE 3-52 Travel angle, 15° pull.

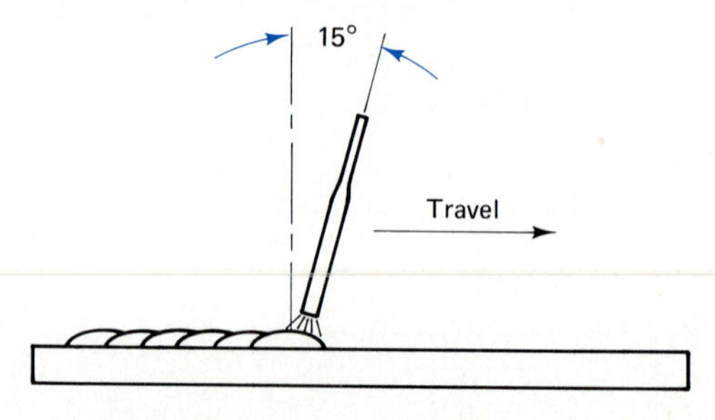

Ultrasonic Welding (USW): Solid-state welding process that uses mechanical energy produced by ultrasonic sound waves to fuse together thin sections of material. The weld is the result of pressure and mechanical energy, not fusion created by high heat.

Underbead Cracks: May also be called hydrogen cracks, because they may result from hydrogen in the weld. Underbead cracks occur beneath the weld in the heat-affected zone.

Upset Welding (UW): Resistance welding process that uses electrical resistance and an upset pressure to produce a weld that will develop up to 90% of the strength of the base metal.

Vacuum Chamber: Pressure-tight container designed to maintain a high, medium, or low vacuum for electron beam welding.

Weldability: Ability of materials to be welded. Weldability depends on metallurgical and chemical properties and joint designs.

Welder: Person capable of producing welds.

Welder Certification: Written document describing all of the essential variables and test results of a welder qualification test.

Weldment: The whole product fabricated by welding.

Weldor: Machine designed for producing welds.

Weld Parts (Parts of the Weld): It is important for welding technicians to know *all* parts of the weld, and their correct terminology (Figure 3–53):

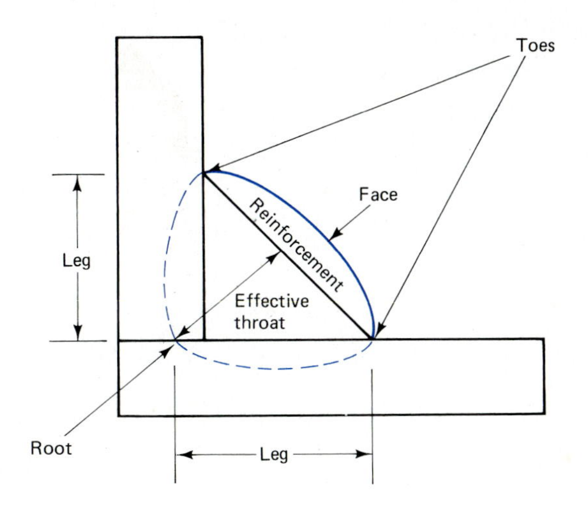

FIGURE 3–53
Parts of the weld.

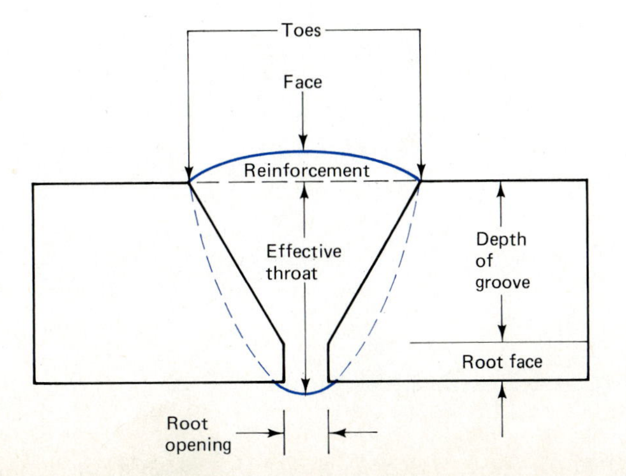

1. reinforcement
2. effective throat
3. root opening
4. root face
5. face
6. toes
7. depth of groove
8. legs
9. root

Wetting Action: Ability of the filler metal to flow easily out into the joint.

REVIEW QUESTIONS

1. What is arc blow, and how might you correct it?
2. How should you avoid arc strikes?
3. What causes backfire with oxyfuel welding?
4. What does the contact tip do in GMAW?
5. What is the fusion zone?
6. What is the heat-affected zone?
7. What is an inert gas?
8. What is the keyhole on a weld?
9. What is porosity in a weld?
10. What is welder certification?

JOINT DESIGN FOR WELDING

Before welding became available, design engineers had only a few choices in designing their steel connections. They were usually lap joints with heavy, bulky fastener plates. Welded and riveted lap joints are shown in Figure 4–1.

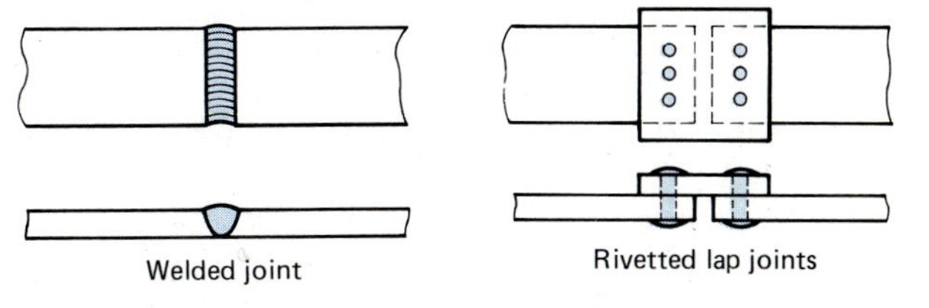

FIGURE 4–1 Welded and riveted joints.

BASIC JOINT DESIGNS

Welding construction gives designers and engineers the added flexibility of five basic joint designs:

1. butt joint
2. tee joint
3. lap joint
4. corner joint
5. edge joint

Butt Joint

The butt joint, when properly welded, is the strongest of all the welded joints. It is used where the weld cross section must carry the same load as the base metal. The butt is the most difficult to weld, as it takes more skill, is more time consuming, and is therefore more expensive than other joints.

In most cases butt welds have to be fully penetrated. Butt welds can use the groove designs shown in Figure 4-2 to obtain full penetration. Vee and bevel grooves can be produced with the cutting torch and grinder. U-grooves and J-grooves are produced with the air carbon arc cutting and gouging system.

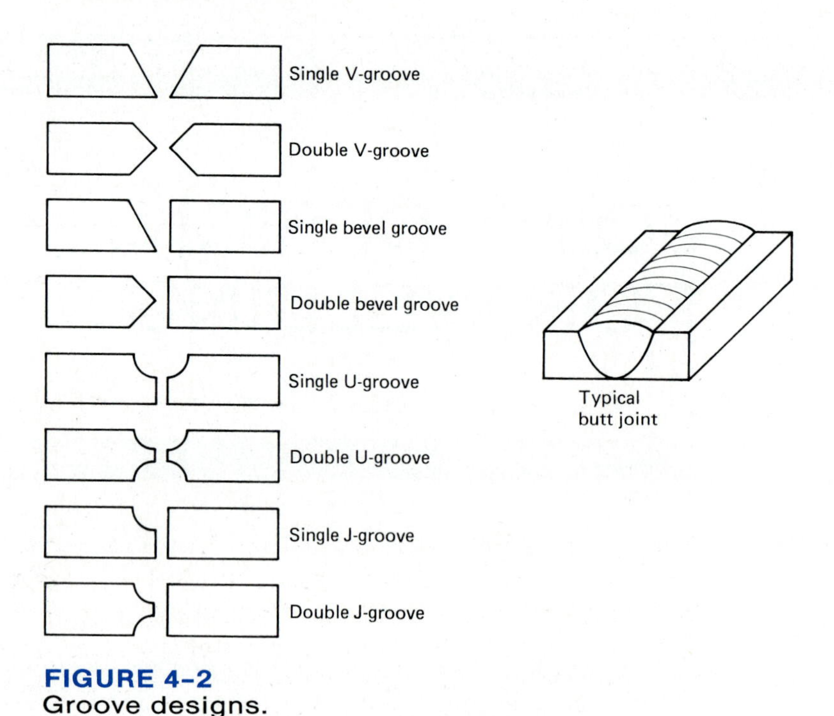

FIGURE 4-2
Groove designs.

Tee Joint

The tee joint is common in structural welding (Figure 4-3). It usually falls into the category of "fillet weld." However, some tee joints can be made up of a combination of fillets and grooves. The leg length of the fillet weld on the tee joint should not exceed the thickness of the base metal.

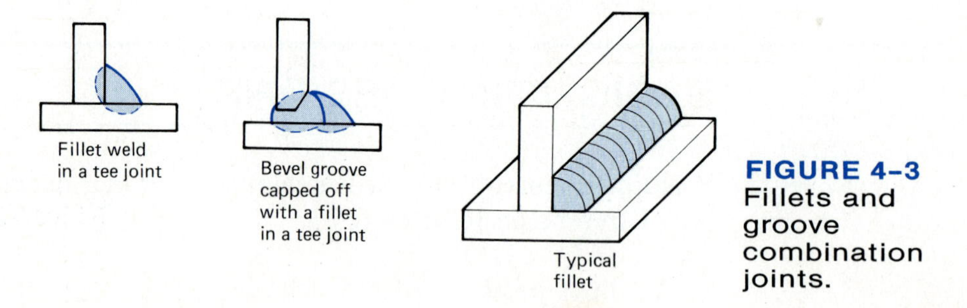

FIGURE 4-3
Fillets and groove combination joints.

Lap Joint

The lap joint (Figure 4-4) is also commonly used in structural welding. It too may use a fillet weld or a fillet-groove weld combination. For maximum strength, fillet welds should be overlapped at least five times

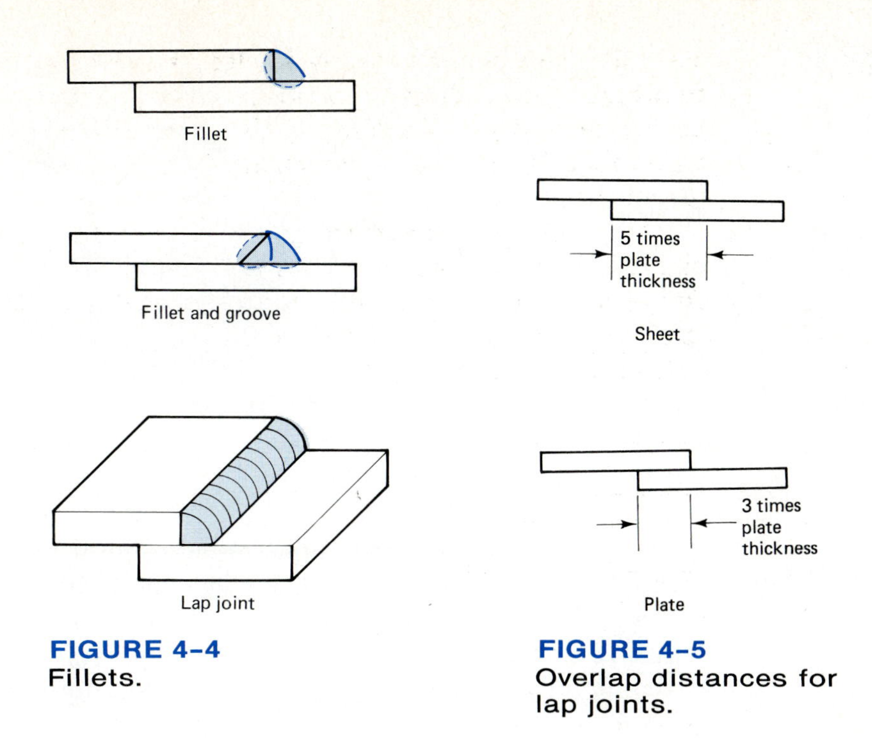

FIGURE 4–4
Fillets.

FIGURE 4–5
Overlap distances for lap joints.

the metal thickness for sheet metal, and three times the metal thickness for plate (Figure 4–5).

Corner Joint

The corner joint can be welded as an inside corner, an outside corner, or both. As an inside corner it is usually a fillet weld but can be a combination of grooves and fillets. The outside corner may be a square or a fillet weld, depending on the type of fit-up. Corners over $\frac{3}{16}$ in. thick should use a root opening for full penetration. The various corner joint possibilities are shown in Figure 4–6.

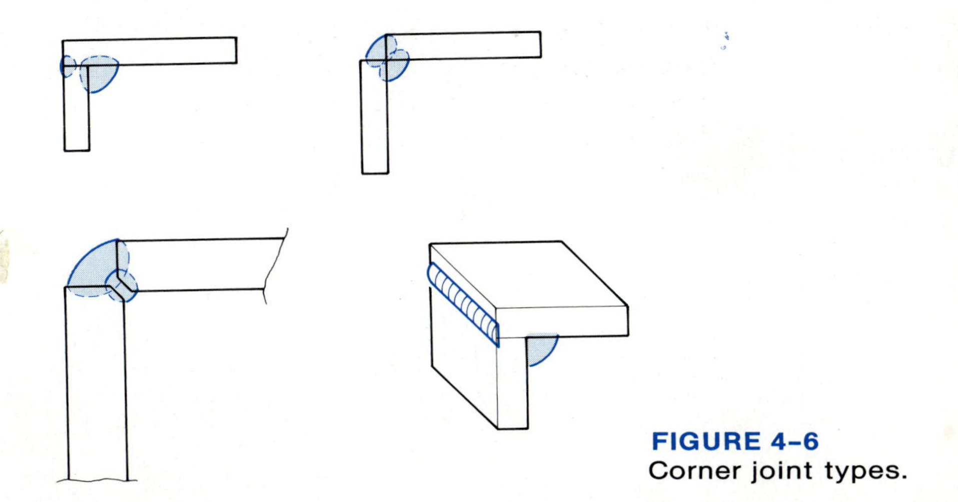

FIGURE 4–6
Corner joint types.

Edge Joint

The edge joint (Figure 4–7a) is specifically for sheet metal joints ($\frac{3}{16}$ in. and under). The edge joint is considered the weakest of all the five basic joints; therefore, its use is limited to sheet metal in low-stress applications. Although these limitations do exist, the edge joint has some

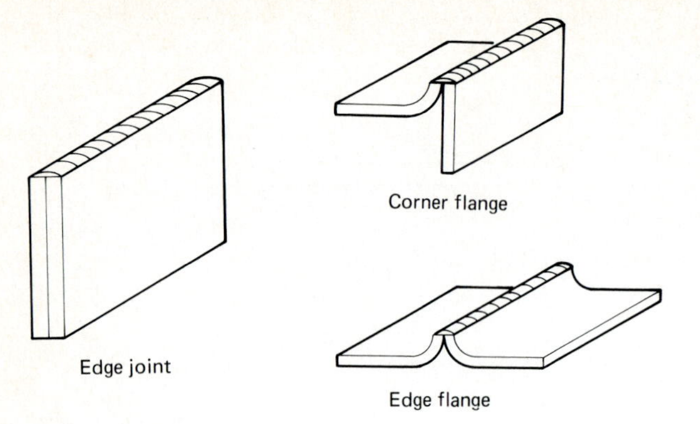

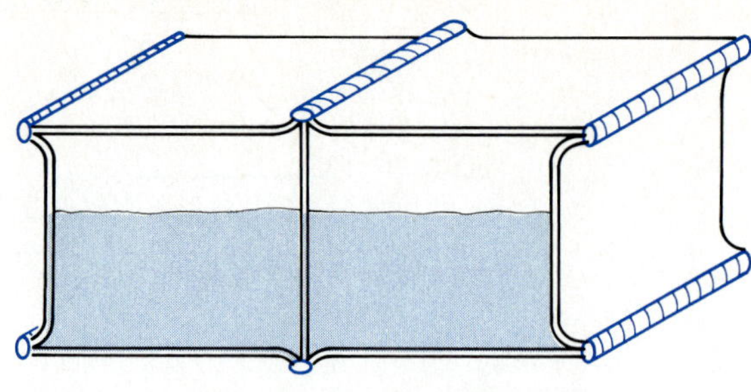

FIGURE 4-7
Edge joints.

unique advantages; for example, an edge joint can be run effectively without the use of filler metal. It also can be used as butt and corner joints in the form of corner and edge flanges (Figure 4-7b and c). The edge and corner flange joints make fairly strong fluid-tight joints that are commonly used on small fuel, water, and liquid storage tanks (Figure 4-7d).

PIPE AND TUBING JOINTS

Pipe and tubing structures are very popular, especially where extreme rigidity and strength are required, as in aircraft, space vehicles, roll cages in race cars, and some modern buildings. These joints require highly skilled welders who have had considerable experience and a lot of practice. Each joint must be strong, yet not overwelded, due to weight concerns. GTAW and OFW are the most popular welding processes used on thin pipe and tubing joints, whereas SMAW, GMAW, and some FCAW are used on heavier pipes. Typical designs include the joints shown in Figure 4-8.

FIGURE 4-8
Pipe joints.

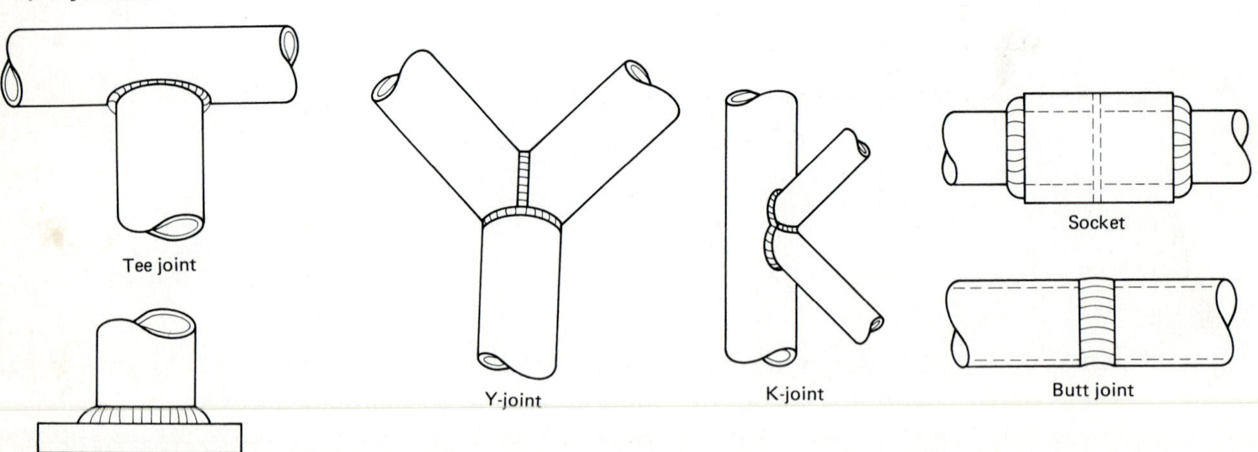

REVIEW QUESTIONS

1. List the 5 basic joints we can use for welding.
2. Which joints fall into the fillet category, and which fall into the groove category?
3. What variations can be made to the corner joint?
4. What is the thickness limitation for edge joints?
5. What are typical pipe and tubing applications?

5

OXYFUEL WELDING

Although oxyfuel welding (OFW) is thought to be an "old," seldom used welding process, looking around the welding industry indicates quite the contrary. Some of today's applications of oxyfuel welding include plant maintenance and repair, small-diameter pipe and aircraft tubing, root pass welds on some moderate-size water, oil, and natural gas transmission-line pipe, and many brazing and soldering operations, such as on bicycles, tubing, radiators, and so on.

The oxyfuel welding process is also one of the best coordination-building processes to lead you smoothly into other welding processes, such as GTAW (which can be very difficult without previous OFW training, due to the required coordination of filler metal, torch, and foot pedal). OFW will help you learn how to "read the puddle" and make corrections and adjustments and to develop good hand–eye coordination as well as a good comfortable dip technique.

OFW THEORY

The OFW process was for many years referred to as oxyacetylene, due to the common use of acetylene as a fuel gas. But today, with the introduction of so many additional gases for both welding and cutting, it is more accurately now called oxyfuel welding (OFW).

This process uses the heat from the burning of a fuel gas, accelerated by pure oxygen to reach temperatures of about 5650°F. This heat is, of course, sufficient to heat and melt most metals (mild steel melts at about 2700°F). This welding process utilizes no external shielding to protect it from the atmosphere. Instead, it has an inherent heat shield, produced by the neutral flame, that surrounds the molten metal. A neutral flame uses all the oxygen for flame combustion, therefore there is no oxygen in the area surrounding the flame for contamination of the weld puddle. Remember, molten steel unshielded from the atmosphere will absorb oxygen, nitrogen, and hydrogen (see Chapter 19).

OFW EQUIPMENT

A typical OFW setup would include the equipment shown in Figure 5-1.

Torches

There are basically two types of OFW torches (or blowpipes, as they are often called): (1) the injector and (2) equal-pressure type. The *injector type* uses the oxygen pressure, traveling around the "injector noz-

FIGURE 5-1
OFW equipment.

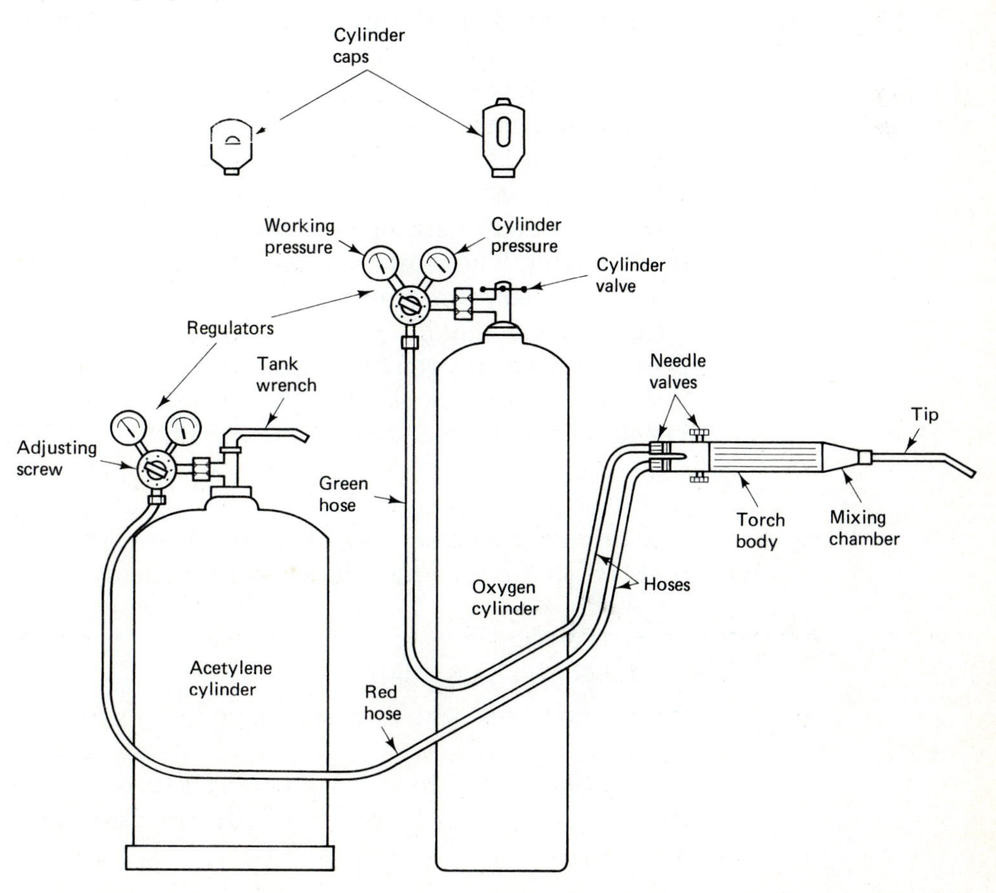

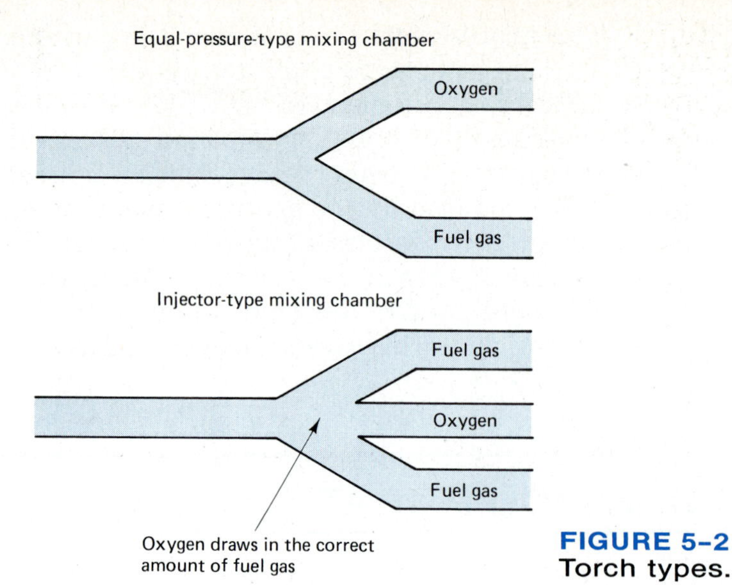

FIGURE 5–2
Torch types.

zle'' to draw the acetylene out into the mixing chamber. This drawing action allows for very accurate, steady mixing of the fuel gas and oxygen, even as tank pressures drop. Usually, with equal-pressure torches when tank pressures start to drop, the torch has to be readjusted continually to compensate for the change in flame due to the loss of tank pressure. However, with the injector torch, this readjusting is reduced due to the fact that acetylene is only drawn out from the velocity of oxygen flow past the injector nozzle. This torch will produce a very steady oxyfuel flame, especially for soft, low-pressure flame settings. The injector torch works well on thin-wall tubing and aircraft tubing and other sensitive applications due to the injector torch's smooth, steady flame. The *equal-pressure torch* is designed primarily for general medium- or high-pressure gas work. Some projects require higher volumes of heat to accomplish the weld, such as heavy sections of metal or cast iron. These metals absorb the heat away from the weld area very fast; therefore, more volume of heat is required. The equal-pressure torch can usually provide this heat (Figure 5–2a). This torch feeds acetylene and oxygen through a mixer nozzle. The acetylene pressure is controlled by the needle valve and not drawn into the mixing chamber as in the injector torch (Figure 5–2b).

Torch Tips

The torch tip is one of the most important equipment decisions you will make with OFW. The following are considerations when selecting an OFW tip.

1. Gases being used
2. Type and size of materials

The type of gas may change the type of tip altogether; for example, some manufacturers use a different design tip for Mapp and natu-

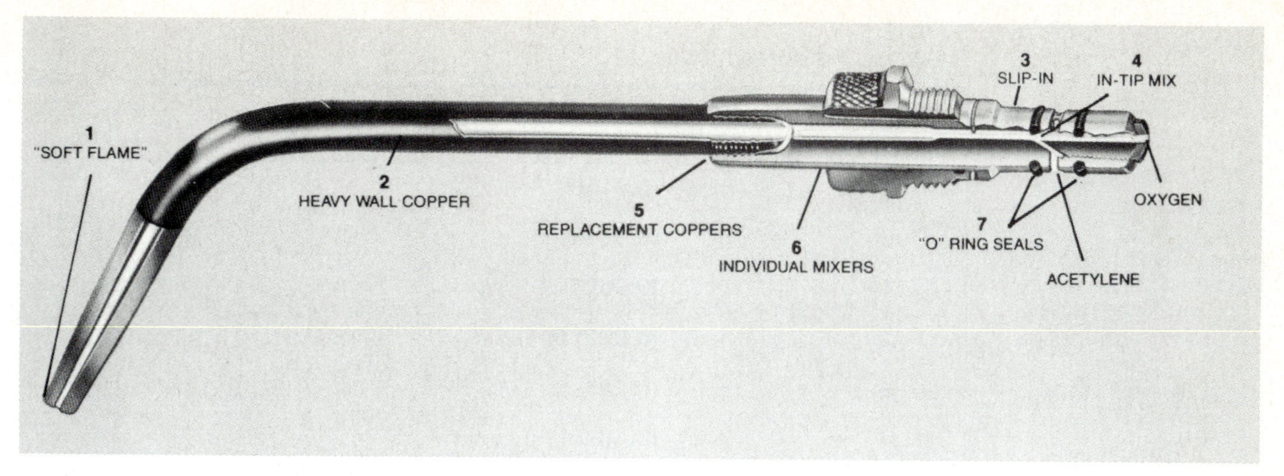

PHOTO 5-1
Injector torch tip. (Courtesy of Smith Welding Equipment.)

ral gas than they do for acetylene. Consult the manufacturers of your torch to make sure that a fuel gas change also requires tip change.

The type and thickness of the material being welded or brazed is probably the most important factor in the tip decision. If the tip orifice is too large for the volume of gas required to make the weld, the torch will backfire. This is also a result of a dirty tip and too low gas pressure for the tip size. What actually happens during backfire is that the tip overheats due to iron particles building up at the orifice, causing combustion inside the orifice instead of the end of the tip. Fuel mixture pressure and the "pop" from backfire dislodge some iron particles from the tip, and proper combustion of the tip is resumed.

If the tip is too small for the material to be welded, the volume of heat required to melt the material will be slowly reached, if ever. When this happens, many welders first try to turn up the oxygen. This results in an oxidizing flame and can boil the material. This results in a very slag and scaly heavy weld. If both acetylene and oxygen are increased with an undersized tip, the puddle will tend to get blown, leaving deep craters in each ripple. The solution is the proper-size tip for the type of gas, material, and thickness of material (Table 5-1).

The mixing chamber is usually a separate chamber where the torch body fits on at one end and the torch tip screws in at the other end. However, with some torch manufacturers the torch tip and mixing chamber are one composite unit. At the torch body end of the mixing chamber, oxygen and fuel gases enter a separate line. In the chamber they are mixed, and this highly flammable mixture enters the tip, where it will travel to the end of the tip (or orifice) for combustion.

Torch tips are usually constructed of copper; this allows for rapid heat dissipation. If a torch tip were constructed of material such as steel, the tip would melt off because of its lower rate of thermal (heat) conductivity. Copper has a very high rate of thermal conductivity, and even though its melting point is lower, it never melts off the tip because it is immediately dissipated back into the copper tip and torch body.

TABLE 5-1
OFW Tip Selection Chart

TIP SIZE[a]	METAL THICKNESS (IN.)	OXYGEN (PSI)	ACETYLENE (PSI)
00	$\frac{1}{64}$ (0.0156)	1	1
0	$\frac{1}{32}$ (0.03125)	1	1
1	$\frac{1}{16}$ (0.0625)	1	2
2	$\frac{3}{32}$ (0.09375)	2	2
3	$\frac{1}{8}$ (0.125)	3	3
4	$\frac{3}{16}$ (0.1875)	4	4
5	$\frac{1}{4}$ (0.250)	5	5
6	$\frac{5}{16}$ (0.3125)	6	6
7	$\frac{3}{8}$ (0.375)	7	7
8	$\frac{1}{2}$ (0.500)	8	8

[a]The lower the tip number, the smaller the tip orifice. Tip sizes may vary between manufacturers. It is suggested that you consult your tip manufacturer's recommendations for exact settings and applications.

Regulators

OFW regulators are used to reduce the high tank pressure to a usable pressure at the torch. Regulators come in two basic types, single and double stage. The single-stage regulator has only a bottle (tank) pressure gage and a single-pressure reducing diaphragm and seat. When tank pressure reduces, the flame will change slightly with single-stage regulation. For many applications this slight change is acceptable and does not justify the cost of the more expensive two-stage regulator. The two-stage regulator has two diaphragms and seats. This double regulation produces a steady flame even as the tank pressure drops. Not until the tank is almost completely empty will the flame with a two-stage regulator show any change. These regulators are superior for most sensitive OFW applications.

Problems with Regulators.

OFW regulators are high-quality, reliable devices which, properly taken care of, will last many years. However, some safety precautions must be observed.

1. Never use grease or oil to lubricate a regulator. The grease or oil in combination with oxygen produces an explosive mixture. Also, the oil can work its way into the bottle or hoses, and even into the torch itself.
2. Once the OFW station is installed and ready to go, open the tanks slowly. This will prevent the high pressure from blowing out the diaphragm in the regulator.
3. If a regulator appears frozen or frosted, remove the regulator and move it to a warm location until it defrosts. Do not heat it with the torch! This is dangerous and fuel gases may be moving

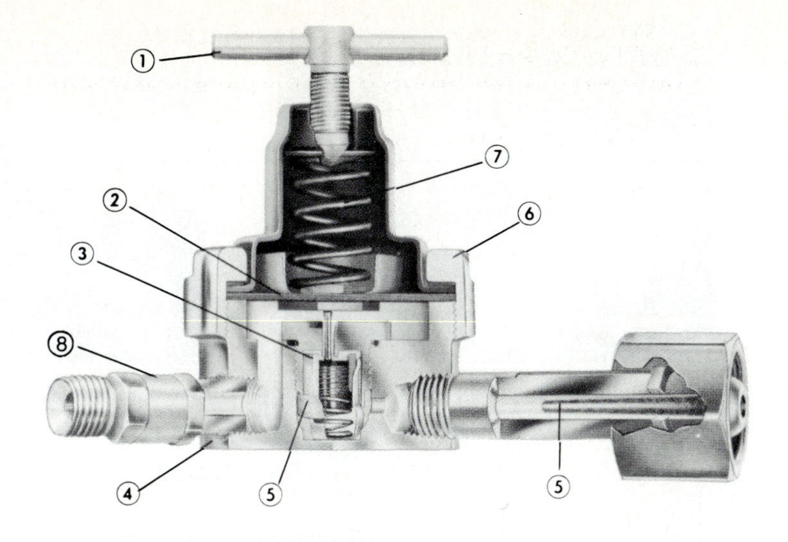

1 Large "T" handle provides fast, accurate pressure settings.
2 Sensitive neoprene rubber diaphragm responds quickly to pressure changes.
3 New elastomer seat sluffs off dirt and dust which are common causes of seat leaks.
4 Solid brass body, forged brass bonnet ring provide the strength needed for normal welding and cutting pressures.
5 Extra protection of TWO FILTERS. Internal filter and the inlet filter guard against dust and other impurities which can cause dangerous seat ignition.
6 Rugged forged brass bonnet ring.
7 Adjusting spring ends are ground for squareness to aid the proper alignment of seat and nozzle.
8 Integral back-flow check valve

PHOTO 5-2
Cutaway of regulator. (Courtesy of Smith Welding Equipment.)

PHOTO 5-3
Regulators. (Courtesy of Airco Welding Products, a division of The BOC Group Inc.)

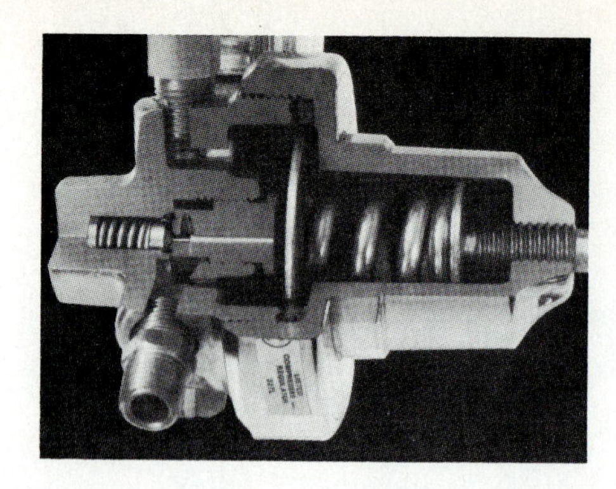

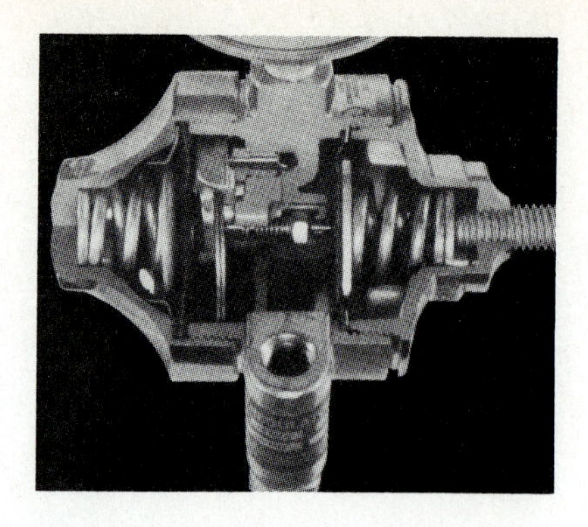

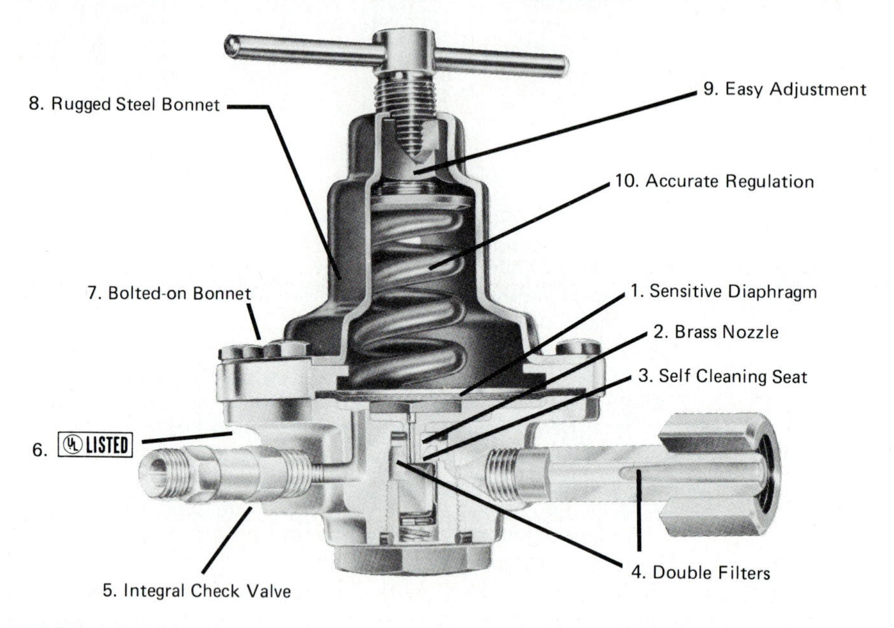

PHOTO 5-4
Cutaways of regulators. (Courtesy of Airco Welding Products, a division of The BOC Group Inc.)

through that regulator, and this will also distort and damage the regulator. A frosted regulator may still be operable for awhile, but might eventually freeze up. When this occurs there will be no regulation because the seat and diaphragm will be frozen. This action of frosting and freezing regulators is due to rapid expansion of the gas as it hits the regulator. For the gas to vaporize, it must draw more heat. This heat usually comes from the regulator, which in turn frosts the regulator.

Hoses, Connections, and Check Valves

The hoses used for OFW are usually made of oil resistant reinforced rubber and will withstand considerable abuse. However, care should be taken not to lay hot metal or allow metal cutoffs to drop onto the hoses. Allowing this situation to occur could cause hose flashback. This is usually the result of the hot metal burning through the hose and igniting the fuel gases. If this does occur, shut off the fuel tank immediately (at the tank, not the regulator). OFW hoses are red (fuel gas) and green (oxygen). Years ago fuel gas hoses were black, but most of these have

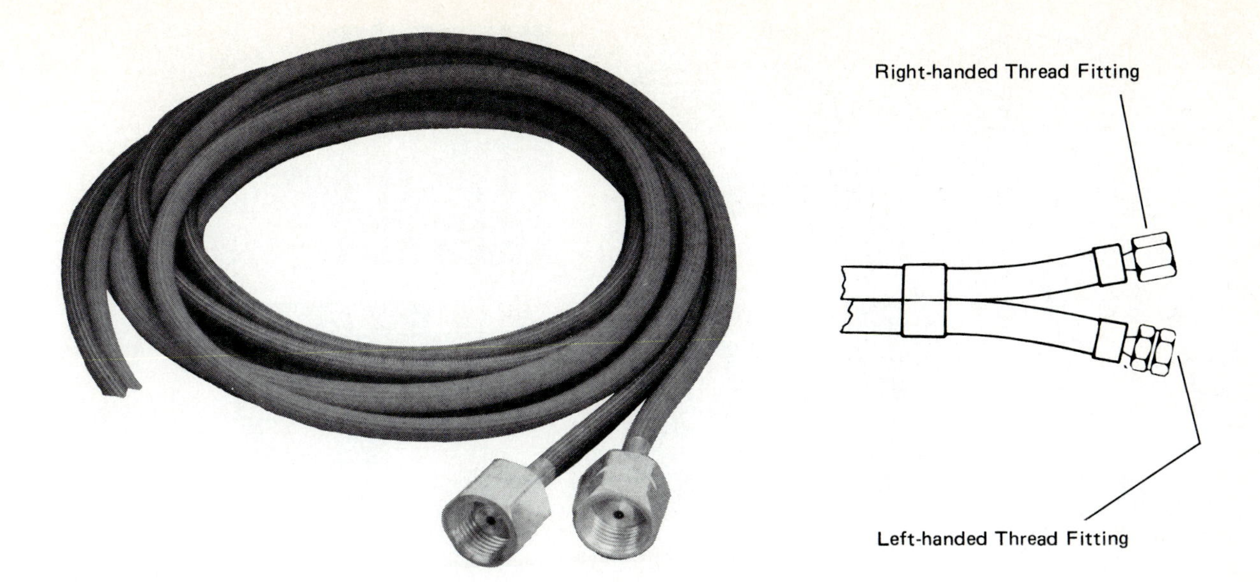

PHOTO 5–5
Hoses.

been removed from service. The fittings on the fuel gas hose (red) will be left-handed; this is to prevent accidental connection to the oxygen line.

The oxygen hose (green) is connected with right-handed fittings, therefore preventing accidental connection to the fuel gas line. The left-handed fuel gas fitting is a casted double-nut type of fitting and is clearly different from the standard right-handed fittings.

Check valves are an important safety addition on any OFW system, but are especially critical for portable OFW stations. These valves allow flow in only one direction, thus preventing any back pressure from flame or gas from moving past them and to the tank. Check valves can be connected at the torch or tank end, or both. Both oxygen and fuel gas should have check valves. If only one set is going to be used, they should be attached to the tank end; this safeguards against hose or regulator flashback, which would be ineffective if check valves were located at the torch connection.

FIGURE 5–3
Check valve operation.

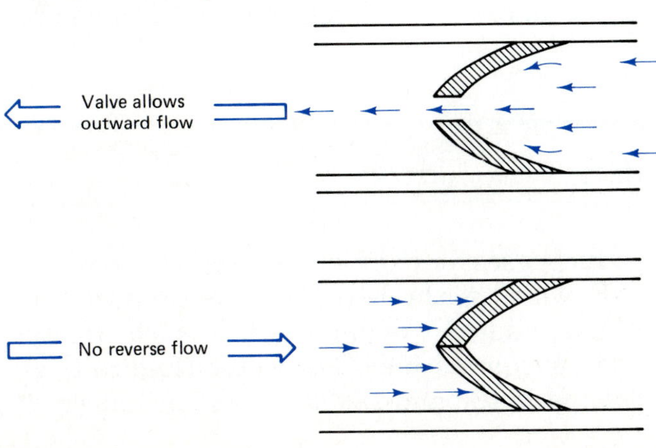

PHOTO 5–6
Check valves: (a) oxygen line; (b) fuel gas line (double-nut fitting); (c) section view. (Courtesy of Smith Welding Equipment.).

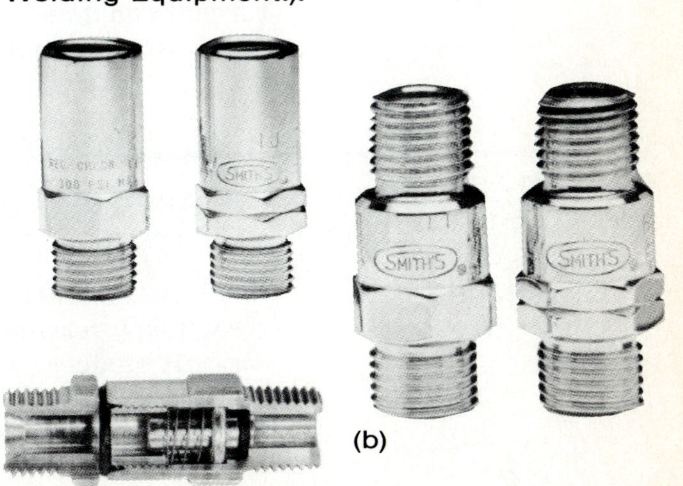

PHOTO 5-7
Bottle carts and straps. (Courtesy of Airco Welding Products, a division of The BOC Group Inc.)

Cylinders

Most high-pressure industrial gas cylinders are constructed of rigid high-grade steel plate and must meet ICC and DOT specifications. Oxygen cylinders contain 99% pure oxygen at about 2200 lb of pressure. There are special precautions that must be taken when using any high-pressure cylinders:

1. Transport only in cylinder carts with cylinders chained.
2. Transport only with cylinder caps on.
3. Never leave cylinders freestanding; chain them to a wall or stationary fixture.
4. Open slowly and always crack the valve first. (This removes any dirt that might blow into and damage the regulator diaphragm before connecting to a regulator.)
5. Never weld, strike, or use the cylinder for anything but that for which it was intended.

Should you ever attempt to move a cylinder without the cap on, and the cylinder tips over and the neck is broken off, the cylinder will turn into an unguided missile and do plenty of damage before it finally comes to rest. So don't take chances; use the cap and chain them up.

OFW FUEL GASES

Acetylene

Acetylene is one of the best fuel gases for oxyfuel welding. The oxyacetylene flame reaches 5900°F, which is the hottest flame produced on earth. It produces a quick fluid puddle for welding carbon steels. Acetylene is produced by mixing calcium carbide and water (Figure 5-4). When acetylene was discovered in the mid-1800s it was not known at

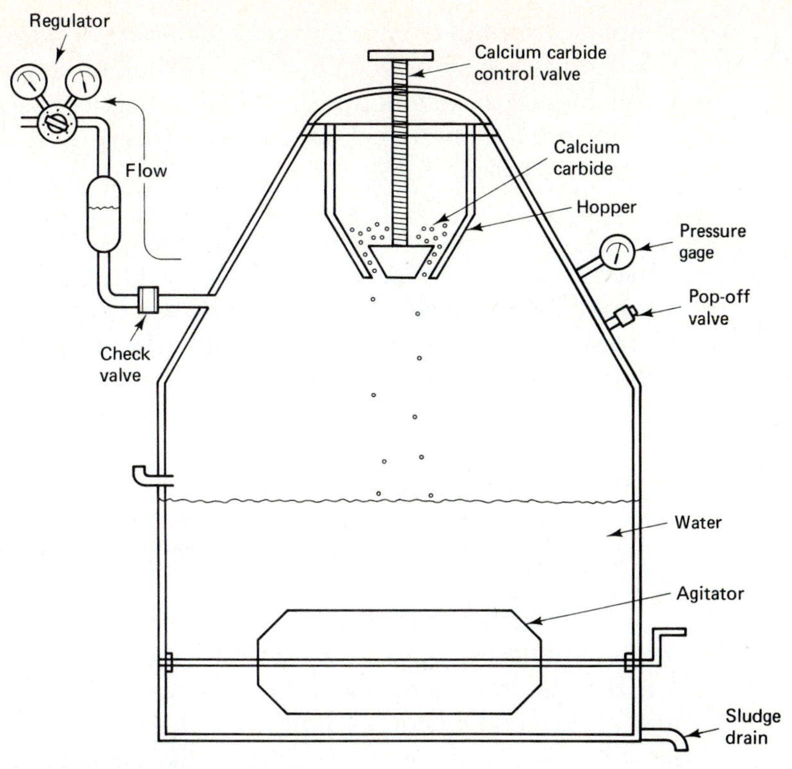

FIGURE 5–4
Acetylene generator.

FIGURE 5–5
Acetylene tank.

the time that it could be safely put into cylinders, so acetylene generators and portable acetylene generators were used to produce on-the-spot acetylene.

In the early 1900s a method for storing acetylene in a stable manner was developed. It was found that acetone would absorb 25 times its own volume of acetylene, and if pressurized, acetone would absorb even more acetylene. Today, cylinders contain about 250 psi, so the acetone has absorbed about 400 times its own volume of acetylene. Acetylene by itself is unstable if compressed over 15 psi. It was observed that a porous substance such as balsa wood or asbestos put into the cylinder would absorb and stabilize the acetone and acetylene combination (Figure 5–5).

Mapp Gas

Mapp gas is composed of methylacetylene and propadiene. The oxy-Mapp flame can reach 5300°F. It is not as high as acetylene, but it does have a higher Btu rating than acetylene. Mapp's real claim to fame is its high efficiency for cutting. Oxy-Mapp brazing works quite well, but fusion welding quality is only fair.

PHOTO 5–8
Mapp gas cylinder and bulk storage tanks. (Courtesy of Airco Welding Products, a division of The BOC Group Inc.)

Propane

Propane is popular for its extra-clean oxypropane cuts. Its flame temperature is considerably lower than that of oxyacetylene Mapp, so it takes longer to form a puddle when fusion welding, and it takes longer to preheat for cutting. Oxypropane does produce good-quality fusion welds and brazed joints.

Natural Gas

Natural gas can be used for oxyfuel welding and brazing, but like propane, it has a lower flame temperature than the oxyacetylene flame. Natural gas can also be cheaper than other fuel gases, especially if large volumes of gas will be used.

Oxygen

Pure oxygen itself does not burn, but when mixed with almost any other flammable substances it will accelerate the rate of combustion. Oxygen is mixed with fuel gases to increase the flame temperature for welding or cutting.

Oxygen is produced by two basic methods, liquefication of air or electrolysis of water. The most popular method is by liquefication of air. In this process air is compressed and cooled until it changes physical states and becomes liquid. It is then allowed to warm back up slowly and change back to gases. The various elements in air all boil off to the gas state, at different temperatures. Each of these gases is captured as it boils off. Note the various temperatures at which gases will boil off:

Oxygen: −297°F
Nitrogen: −320°F
Argon: −302°F
Helium: −452°F

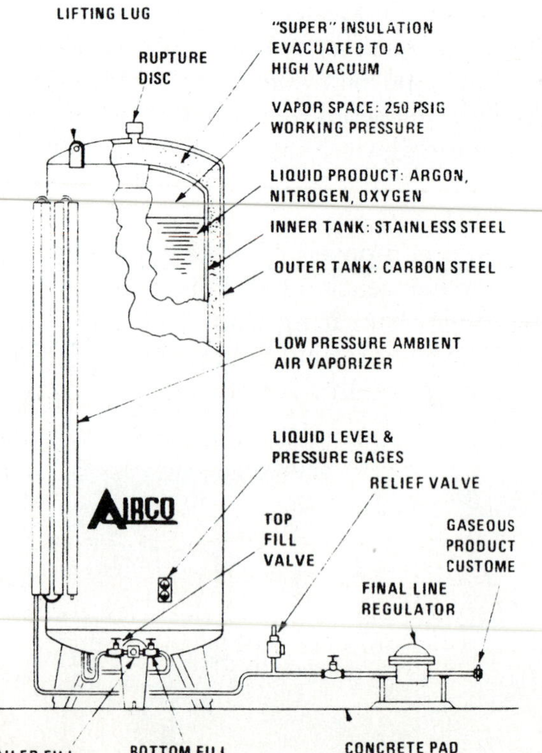

PHOTO 5-9
Bulk oxygen station. (Courtesy of Airco Welding Products, a division of the BOC Group Inc.)

This electrolysis of water system passes dc current through a water and alkaline mixture. When electrodes are charged and inserted into the solution, a chemical reaction occurs and decomposes the water. The liberated hydrogen goes to the negative pole, and oxygen to the positive pole. Remember, water is made up of hydrogen and oxygen. The oxygen and hydrogen gases are then drawn off the top of the solution.

OFW FILLER METALS

Oxyfuel filler metals come in standard lengths of 36 in. and have a thin copper coating to prevent them from rusting during storage. Diameters range from $\frac{1}{16}$ (0.063) to $\frac{1}{4}$ (0.250) in. The classification system is based on the deposited weld metal's ultimate strength. Figure 5-6 shows an example.

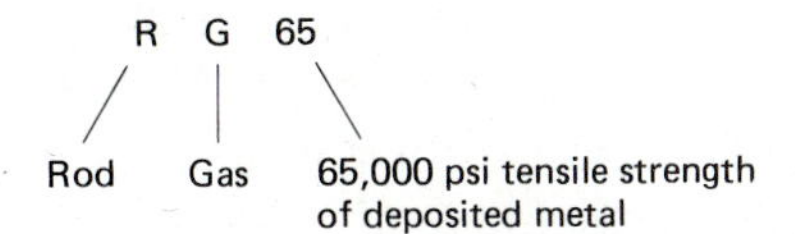

FIGURE 5-6 Classification system for OFW filler metals.

In your selection of the correct filler metal for oxyfuel welding, you must first analyze your strength requirements. In most cases you want the filler rod that has a tensile strength closest to the tensile strength of the base metal. This will give your weldment the most desirable mechanical properties.

Filler rod diameters should be selected based on the base metal thickness properties and welder preference. Thicker filler metals would require larger-diameter filler rods, whereas the thin metals use smaller diameters.

Using small diameters such as the $\frac{1}{16}$ and $\frac{3}{32}$ in., you will notice that you must feed them rapidly, as they consume quickly into the puddle. These smaller diameters may require that the welder stop and regrip the filler rod as it consumes. Some welders prefer using a filler rod one size larger than would normally be used. This reduces the amount of feed as the filler rod is dipped and consumed as the weld progresses.

PREPARING FOR OFW

The first step is to make sure that all fittings and connections are secure. If there is any doubt, soap-check the lines. This should be accomplished with the system pressurized. First open the cylinder valves, starting with the oxygen; open the needle valve at the torch. On the oxygen bottle slowly open the valve until bottle pressure is read on the cylinder pressure gage. Continue to open the valve all the way until slight pressure is felt against the seat. You will recall that oxygen cylinders seat when they are open and seat when they are closed. Turn the regulator adjusting screw inward until pressure is felt against the diaphragm. Continue to turn inward, watching for about 5 to 7 psi or 15 to 40 psi for cutting on the working pressure gage. Close the oxygen needle valve on the torch.

Open the acetylene needle valve on the torch. Go to the acetylene cylinder and slowly crack open the valve until pressure is read on the tank pressure gage. This should only require $\frac{1}{4}$ to $\frac{1}{2}$ turn. Once tank pressure is read, no further opening is required for acetylene. This allows for quick shutoff in an emergency situation, such as flashback.

Now go to the acetylene regulator; turn inward until pressure is felt against the diaphragm, watching the working pressure gage continue turning until 3 to 5 psi for welding or 5 to 7 psi for cutting is established. Turn off the acetylene needle valve on the torch.

Lighting the Torch

Locate a striker, tip cleaner, and any other equipment you plan to use. Get into a comfortable position with torch and striker in hand. You are now ready to light the torch.

Crack open the acetylene needle valve at the torch about $\frac{1}{8}$ to $\frac{1}{4}$ turn for working pressure. Squeeze the spark lighter (striker) in front of the torch until acetylene burning in air flame has been established (see below). If the torch does not light right away, either open the needle valve farther or, if the torch lights and goes out rapidly, indicating that you may have too much acetylene, back the needle valve off slightly.

Once the acetylene is lit, adjust this flame until one-half the length of flame is producing laminar flow and one-half is producing turbulent flow (Figure 5-7). Now open the oxygen needle valve to blend oxygen in with the acetylene flame. As the oxygen is being opened, a long feather flame will start to appear. As more oxygen is added, this feather will get shorter; keep adding oxygen until this feather disappears into the inner cone (Figure 5-8).

If when initially adding oxygen the flame was extinguished, you may have added oxygen too rapidly. This can also be a result of a

PHOTO 5-10
(a) Striker; (b) tip cleaners; (c) tank wrench.

(a)

(b)

(c)

PHOTO 5-11
Oxy torch body. (Courtesy of Airco Welding Products, a division of The BOC Group Inc.)

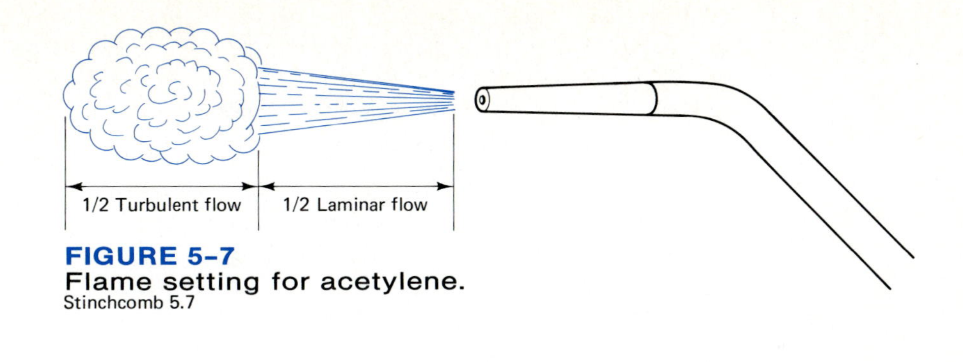

FIGURE 5-7
Flame setting for acetylene.
Stinchcomb 5.7

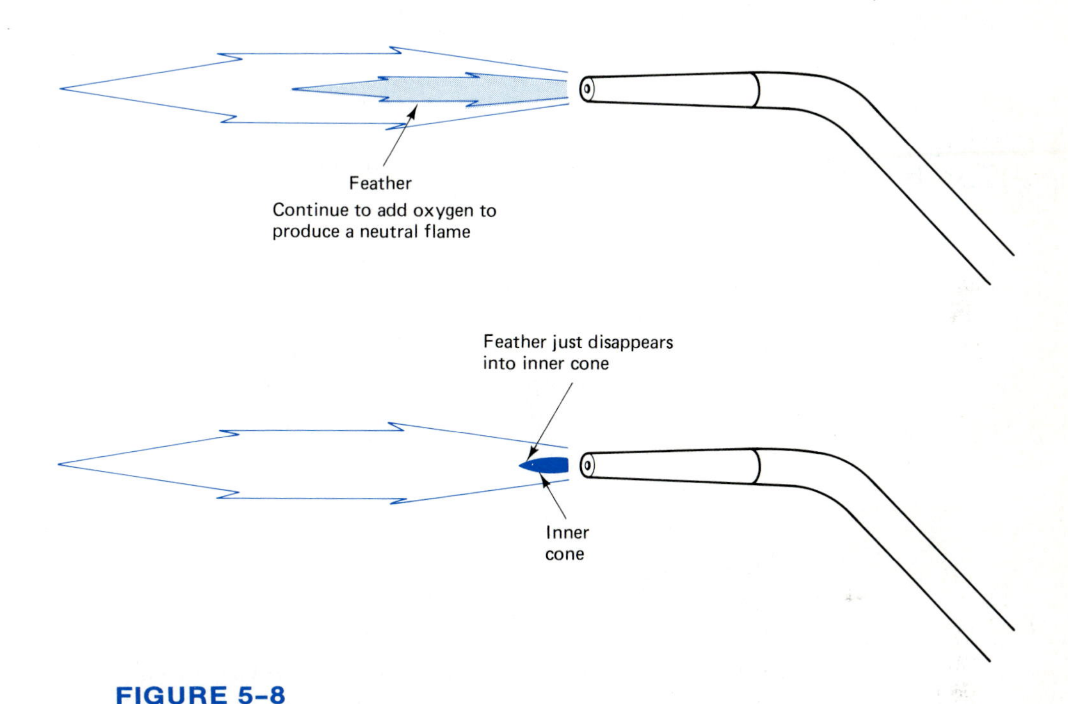

FIGURE 5-8
Setting the flame.

touchy or loose needle valve. See the manufacturer's recommendations for tightening the needle valve (this may require sending the torch out for service). A touchy needle valve can be frustrating, being that any time it is bumped it will disturb gas flow through the torch and change the flame.

Let's experiment with different oxygen settings. The three basic flames used in the welding and cutting industry are (1) neutral flame, (2) oxydizing flame, and (3) carburizing flame (Figure 5-9). With the torch lit and cylinders and regulators adjusted, increase the oxygen flow with the needle valve so that the torch produces a blowing-type sound along with a very sharp cone. This is the oxidizing flame because it is rich in oxygen. This is an extremely hot flame because the oxygen accelerates the combustion. During welding this flame setting can actually boil the puddle, producing a high amount of slag and an undesirable weld. However, in some cases, to produce enough heat for thick sections of material and some joints, such as tee joints, you can use a slight oxydizing flame. Also, many welders prefer a slight oxidizing flame for brazing.

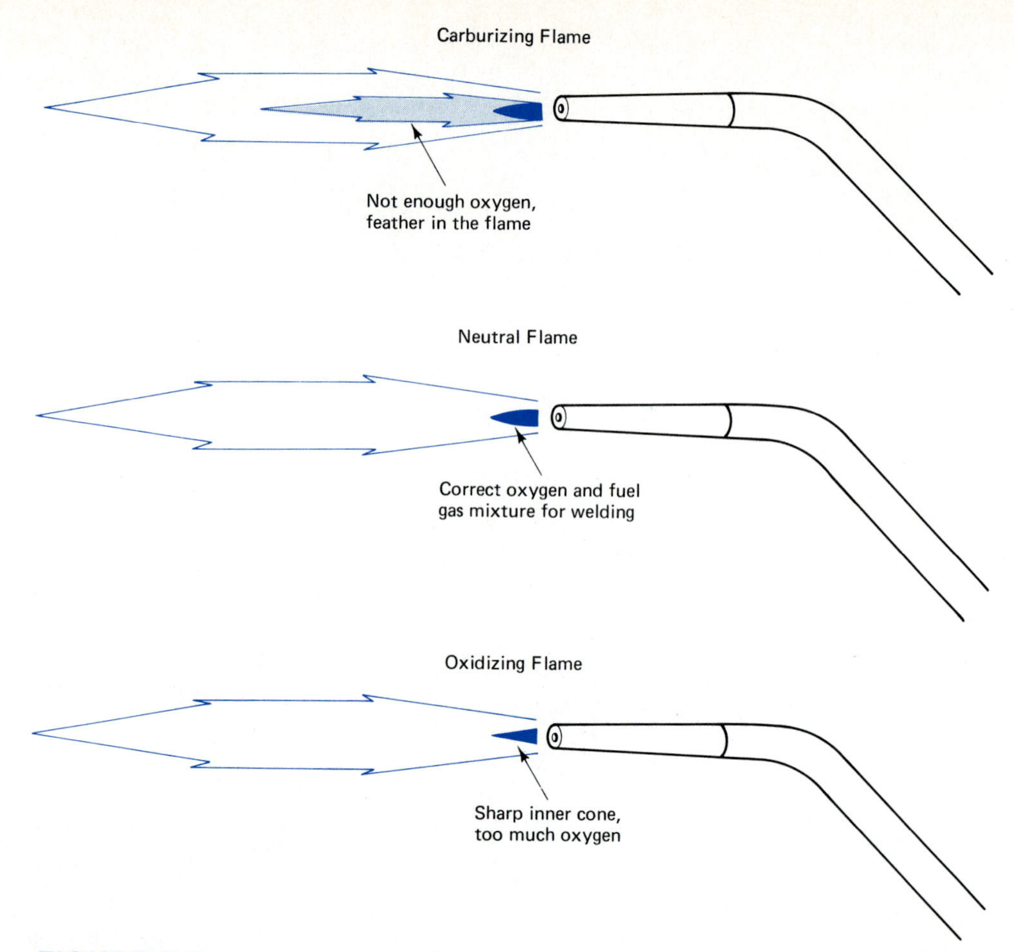

FIGURE 5-9
Three basic flame settings.

Now turn the oxygen needle valve down so that a slight feather appears over the inner cone. This is the *carburizing flame.* This fuel mixture has an overbalance of acetylene and not enough oxygen. Running a carburizing flame over your plate will usually deposit carbon on the plate; this is because the low temperature of this flame does not burn out all the carbon that acetylene produces. Some welders prefer a slight carburizing flame for very thin sections of material. This is a cooler flame than the oxydizing or neutral flames.

The most widely used flame for most welding applications is the **neutral flame.** If we now take our previously adjusted carburizing flame and start to add oxygen until the feather just disappears inside the inner cone, we have a neutral flame. This flame setting is the best setting for most welding applications. When troubleshooting during OFW, be sure to check and experiment with torch settings. Always start with the neutral flame setting; it works best for about 90% of all OFW applications.

Shutting Down the Torch

When shutting off the oxyfuel flame, start with the fuel gas needle valve first. This may make a slight "pop" but will prevent the tip from becoming carboned up, as is the result if the oxygen is extinguished first. Start with the cylinders and turn off the oxygen and acetylene valves, then crack open the acetylene needle valve first and bleed down all acetylene pressure to zero on both cylinder and working pressure gages. Repeat for the oxygen lines. Back the regulator adjusting screws off (counterclockwise) until they are free to spin. This will indicate that there is no pressure on the diaphragm. Be sure to continue practicing all the steps from startup to shutdown, adjusting to a neutral flame, until you feel comfortable with the procedure.

Cleaning the Tip

As you oxyfuel weld, you will notice that the tip will start to collect small particles of iron. This is relatively normal but can be accelerated by too much oxygen in the mixture and boiling the slag in the puddle. As the iron builds up on the torch, it has more trouble dissipating heat because the iron particles hold the heat at the tip. This results in backfire (or popping). The surest way to eliminate this is to clean the tip. Tip cleaners are inexpensive and add many years to the life of a tip if used properly. The procedure for tip cleaning is as follows:

1. Start by shutting down the flame but leaving a small amount of oxygen flowing through the tip. This will prevent iron particles from getting inside the mixing chamber during cleaning.
2. Start with the round serrated cleaner that will just slide into the orifice freely. It should not fit tightly; you do not want to ream out the end. This might over-enlarge the orifice and make the tip backfire. The idea is to work the cleaner straight up and down just enough to knock off loose iron particles.
3. Use the flat file to clean off the very end of the tip. Do not round the end; keep the file flat when filing.
4. Carefully, use a soft-bristle wire brush to clean carbon and iron particles off the outside surface near the orifice. Your tip should now operate smoothly. If not, or the inner cone appears slanted or whistles, repeat step 2.

OXYFUEL BRAZING

Brazing is much different from fusion welding in that capillary action produces the adhesion in this process. In fusion welding the metal must be heated until the atoms reach the random state (see Chapter 19). This is about 2700 to 2800°F in steel. In brazing the temperature reaches only about 1600 to 1800°F. Therefore, in brazing of steel the base metal never reaches its melting temperature; only the brass filler metal melts. In this process the strength is achieved by brass flowing into and gripping the pores of the base metal.

For ultimate weld strength, the surface to be brazed must be clean. Wire brushing will help but may not completely clean out the pores. Flux used in brazing will deep-clean the surface while brazing. Flux-coated filler rods or bare rods dipped into flux are available. The flux actually flows into the pores before the brass does. The sticky flux picks up any dirt or impurities in the pores. When the heavier brass flows into the pores, it flushes the flux out and to the top, allowing the brass a clean grip onto the base metal. This process is excellent for any low-temperature application.

Techniques for oxyfuel brazing are very similar to those for oxyfuel welding. However, there are a few points to remember.

1. Most experienced welders prefer the neutral flame; however, it is sometimes advisable to use a slight oxidizing flame for thick sections or cast irons to build up enough heat into the material.
2. Preheat is required for most materials to be brazed. Brass follows heat. If preheat is not used, the deposited brass will tend to ball up and flow poorly. Too much preheat will make the brass too fluid and no filler metal build will result. Preheating to a dull red is recommended, and preheating helps the brass flow.
3. Brass is made up of copper and zinc and zinc boils at 1600°F. Therefore, always keep the torch moving; failing to do this may result in boiling the zinc out of the brass. A red braze with a crusty porous appearance indicates that zinc boiling has occurred.

Remember that brass will follow the heat. A uniform preheat will usually produce a uniform braze. Another tip is to keep your dip technique uniform. The more uniform the dip and torch motion, the more uniform the appearance of the braze.

Brazing is best suited for thin sections of steel, cast irons, high-carbon steels or any application that would require a low heat input. With practice, brazing can be done in all positions. In the vertical and overhead positions you must make sure that the trailing puddle solidifies before trying to advance the puddle (Figure 5-10).

Bronze welding and silver brazing are done with a similar setup and technique. Bronze is made up of copper and tin. Silver brazing is an ultralow-temperature process. For some applications a slight carbonizing flame would be recommended. This process would be used where a large amount of heat would destroy the properties of the material to be welded, such as carbide tipping saw blades, heat-treated materials, or stainless steels to prevent distortion. Silver is used on the copper lines for the heating and cooling systems and electrical connections.

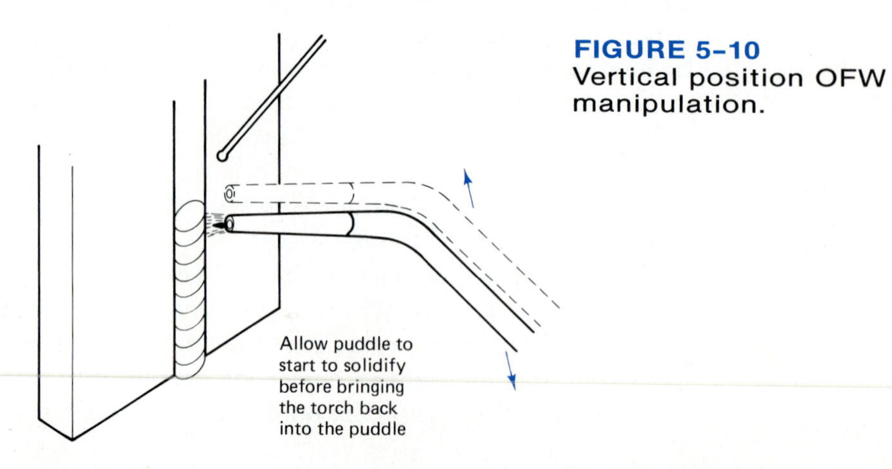

FIGURE 5-10 Vertical position OFW manipulation.

TABLE 5-2
Problems and Solutions for Brazing

PROBLEM	SOLUTIONS
Filler metal balls up and will not flow	Use more preheat; clean base metal; add more flux
Brass flows too much and will not build up into joint	Use less preheat; reduce flame setting; use a faster travel speed
Brass breaks off upon cooling	Clean and prepare the surface by roughing it with grinding or using sandblasting
Brass drips on vertical joints	Use less preheat; keep the torch moving once brass flows, quickly get out of the puddle to reduce heat buildup (whip in and out of puddle)
Brass boils and turns red	Reduce torch setting; keep the torch moving, using a circular, figure-eight, or in-and-out motion

Common problems encountered in brazing and solutions for these problems are listed in Table 5-2.

TECHNIQUES FOR OFW PUDDLE CONTROL FOR FUSION WELDING

The torch adjustment is the single most important factor in producing an acceptable OFW weld (Figure 5-11). If an oxidizing flame is produced, the puddle may boil and become full of slag due to the metal

FIGURE 5-11
Techniques for puddle control.

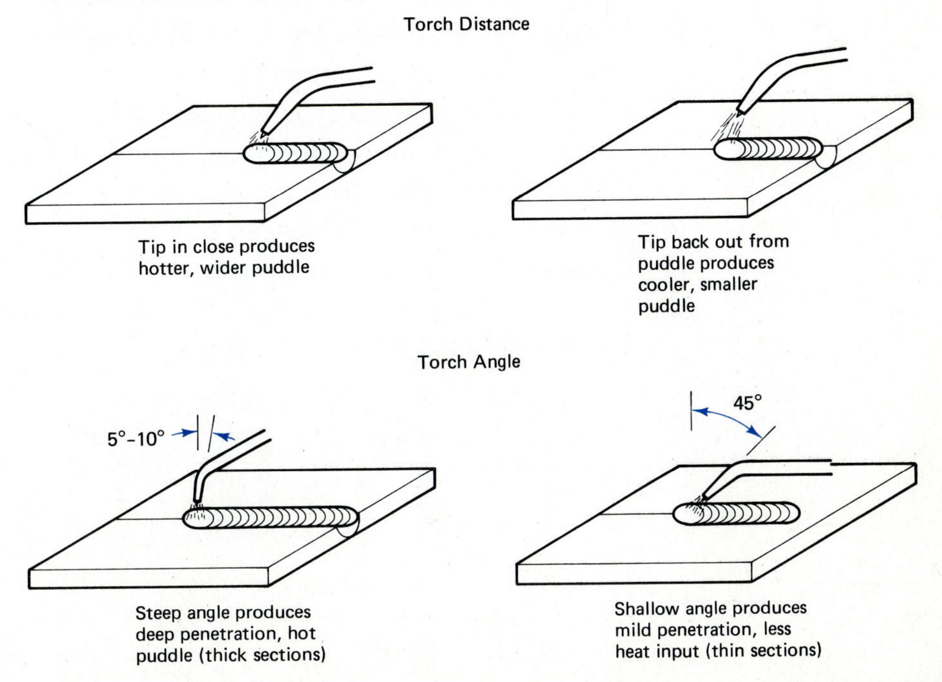

oxidizing. If a carbonizing flame is used, the puddle will develop very slowly and may absorb carbon into the weld, making the weld brittle.

Torch manipulation can also have an effect on the bead. Moving the torch in close to the puddle will heat and widen the puddle. Moving the torch back out from the puddle will cool and shrink the molten puddle. This gives the welder puddle control without having to stop and adjust the gas flow rate. *Caution:* Getting the flame too close to the puddle may oxidize the metal.

Changing the torch angle can also change the action of the puddle without stopping to change the flame adjustment. A steep straight-in torch angle will increase penetration and produce a hotter puddle. A shallow torch angle produces mild penetration and heat input because the angle allows much of the flame heat to reflect off the plate and into the air.

REVIEW QUESTIONS

1. What are typical OFW welding applications?
2. What are the two types of OFW torches, and what are their applications?
3. What must be considered when selecting an OFW torch tip?
4. What is the difference between flashback and backfire?
5. What is the purpose of OFW oxygen and acetylene regulators?
6. What do check valves do?
7. What must be remembered when handling high-pressure cylinders?
8. What does the filler rod classification "RG-45" stand for?
9. What should be considered when selecting a filler wire?
10. What is the maximum working acetylene pressure?
11. How might you check for leaks in an oxyfuel system?
12. What is the most common OFW flame?
13. Excess oxygen or acetylene causes what type of flame?
14. Brazing uses what type of bonding for strength?
15. What purpose does brazing flux serve?
16. If regulators freeze, what can be done?
17. What OFW gas produces the hottest flame?
18. What color and what thread direction are oxygen and acetylene hoses and fittings?

OFW PROJECT 1

Edge Joint

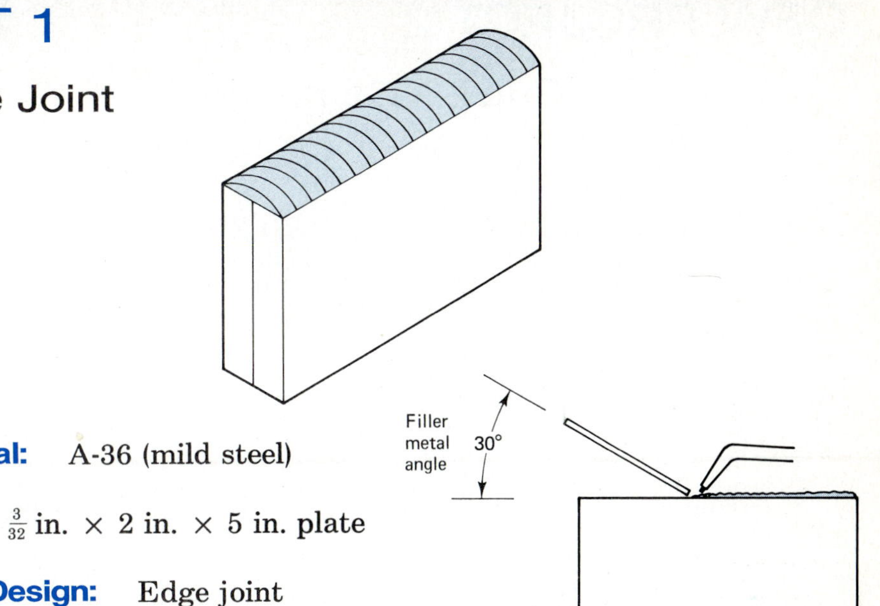

Material: A-36 (mild steel)

Size: $\frac{3}{32}$ in. × 2 in. × 5 in. plate

Joint Design: Edge joint

Position: Flat

Technique: Stringers

Filler Rod: None

Procedure: Cut and tack your two rectangular plates ($\frac{3}{32}$ in. × 2 in. × 5 in.) at the four corners. Position firmly into the flat position for welding. Adjust the torch to a neutral flame. Move the inner cone of the flame about $\frac{1}{4}$ in. from the edges of the joint. Once the puddle is established, move the torch forward and backward to stimulate forward puddle movement, bringing the flame back into the puddle each time. This will keep the beads overlapped. If there is any deviation from a neutral flame, readjust the flame setting. Weld all four sides.

Visual Inspection: Beads should be smooth, uniform, and free of visual discontinuities or unfused areas.

OFW PROJECT 2

Outside Corner Joint

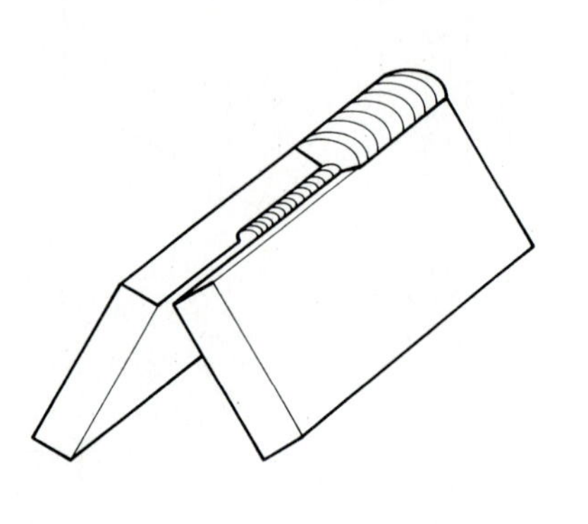

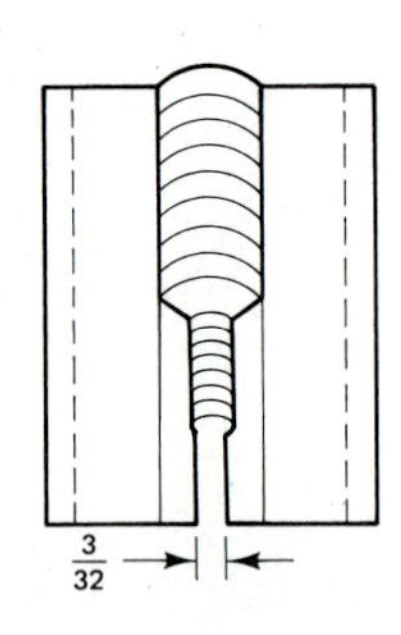

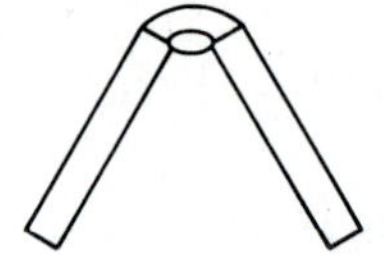

Material: A-36 (mild steel)

Size: $\frac{3}{32}$ in. × 2 in. × 5 in. plate

Joint Design: Corner joint

Position: Flat

Technique: Stringers

Filler Rod: RG 45, $\frac{3}{32}$-in. diameter

Procedure: Cut and tack your two rectangular plates ($\frac{3}{32}$ in. × 2 in. × 5 in.) at the two ends. About a $\frac{1}{16}$-in. root opening should be maintained for full penetration. Adjust the torch flame to a neutral flame. Move the inner cone into a position about $\frac{1}{4}$ in. from the plate. Once the two edges start to melt, slowly dip the filler rod into the center of the puddle. Move the torch forward and backward, then dip the filler rod. Repeat this forward, back, and dip technique until the joint is completed.

Visual Inspection: The face of the weld should be smooth, uniform, and free of visible discontinuities or unfused areas. The root of the joint should show complete penetration and no visible unfused edges.

OFW PROJECT 3

Lap Joint

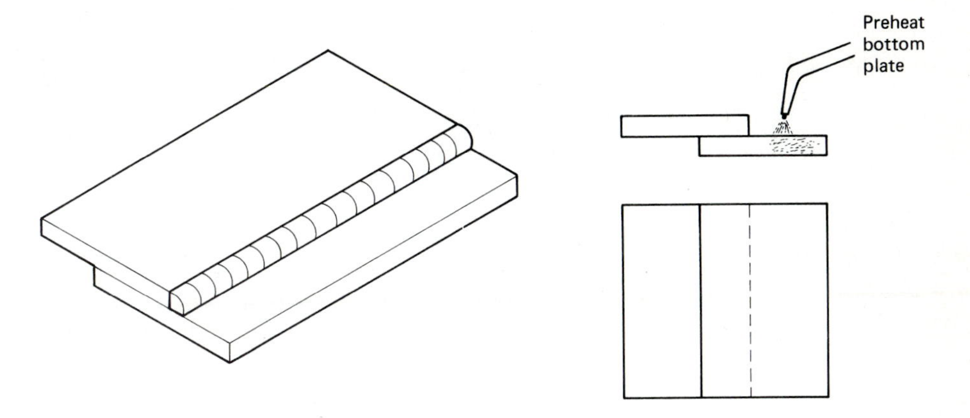

Material: A-36 (mild steel)

Size: $\frac{3}{32}$ in. × 2 in. × 5 in. plate

Joint Design: Lap joint

Position: Horizontal (2F)

Technique: Stringers

Filler Rod: RG 45, $\frac{3}{32}$ -in. diameter

Procedure: Cut and tack your two rectangular plates ($\frac{3}{32}$ in. × 2 in. × 5 in.) at both ends. Adjust to a neutral flame. Preheat the flat side plate slightly, and angle the torch at a 45° angle of attack. Once the two edges start to melt, slowly dip the filler rod into the center of the puddle. Move forward and backward, dipping the filler rod into the center of the puddle. Move the torch forward and backward, dipping the filler rod on the back stroke.

Visual Inspection: The weld legs should be within $\frac{1}{8}$ in. of each other. The leg lengths should be within $\frac{1}{8}$ in. of each other. The faces should be clear of discontinuities and should not have more than 0.010 in. undercut on the top toe.

OFW PROJECT 4

Tee Joint

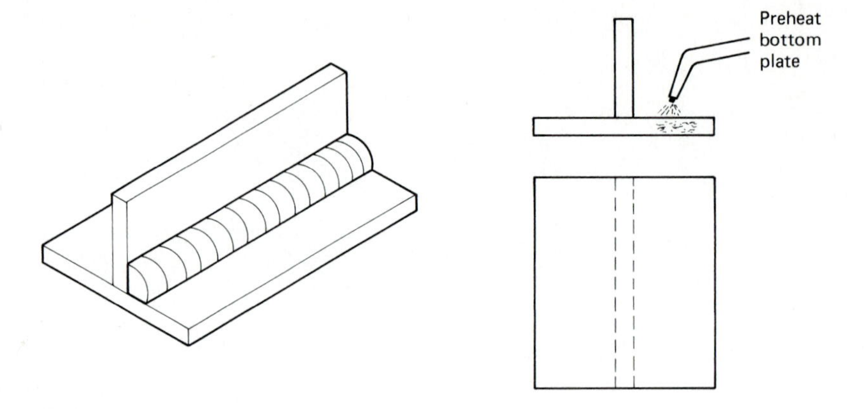

Material: A-36 (mild steel)

Size: $\frac{3}{16}$ in. × 2 in. × 5 in. plate

Joint Design: Tee joint

Position: Horizontal (2F)

Technique: Stringers

Filler Rod: RG 45, $\frac{3}{32}$ -in. diameter

Procedure: Cut and tack your two rectangular plates ($\frac{3}{16}$ in. × 2 in. × 5 in.) into a tee-joint configuration. Adjust to a neutral flame. Preheat the bottom plate slightly. Using a 45° angle of attack, heat the edges until the puddle forms. Slowly move the torch forward and backward in line with the joint, dipping the filler metal into the puddle on the backward stroke.

Visual Inspection: The weld legs should be within $\frac{1}{8}$ in. of each other. The face should be clear of discontinuities and should not have more than 0.010 in. undercut on the top toe.

OFW PROJECT 5

Butt Joint

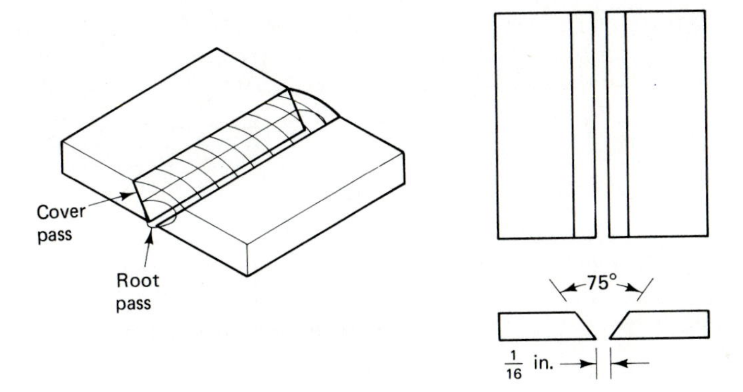

Material: A-36 (mild steel)

Size: $\frac{3}{16}$ in. × 2 in. × 5 in. plate

Joint Design: V-groove, butt joint, 75° included angle

Position: Flat (1G)

Technique: Stringer and weave

Filler Rod: RG 45, $\frac{1}{16}$-in. and $\frac{3}{32}$-in. diameters

Procedure: Bevel the edges of your plates each to about $3\frac{1}{2}$° to produce a 5° included angle. No root face should be used for this joint. Tack plates together, leaving about a $\frac{1}{16}$-in. root opening. Use the keyhole technique for positive penetration. The filler pass should use a normal neutral flame. Use a side-to-side torch motion, dipping filler metal into the puddle as the flame approaches each side.

Visual Inspection: The root side should show complete penetration with no unfused edges. The face of the weld should be free of visible discontinuities.

OFW PROJECT 6

Lap Joint Braze

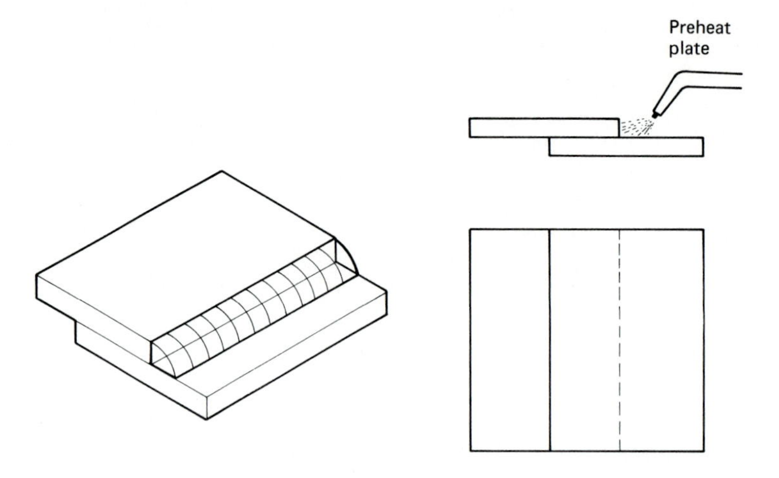

Material: A-36 (mild steel)

Size: $\frac{3}{16}$ in. × 2 in. × 5 in. plate

Joint Design: Horizontal fillet, lap joint

Position: Horizontal (2F)

Technique: Braze

Filler Rod: RCuZn-C

Procedure: Cut and position plates into a lap joint configuration. Preheat your plate to a dull-red color (do not overdo it). Using a flux-covered or flux-dipped brazing rod, lay the rod onto the heated metal, bringing the flame into the joint. As the braze starts to melt into the joint, move the torch back and forth, in and out of the joint to control the brass flow rate. You will find that the more heat that is put into the joint, the more brass will flow. As you approach the end of the joint, you will want to use less flame because of the buildup of heat already in the joint.

Visual Inspection: Brazes must be cleaned before examination, by chipping and wire brushing off any remaining flux. Brazes should appear slightly flatter than fusion welds and exhibit complete adhesion on both legs. The braze should be uniform throughout its length.

OFW PROJECT 7

Tee-Joint Braze

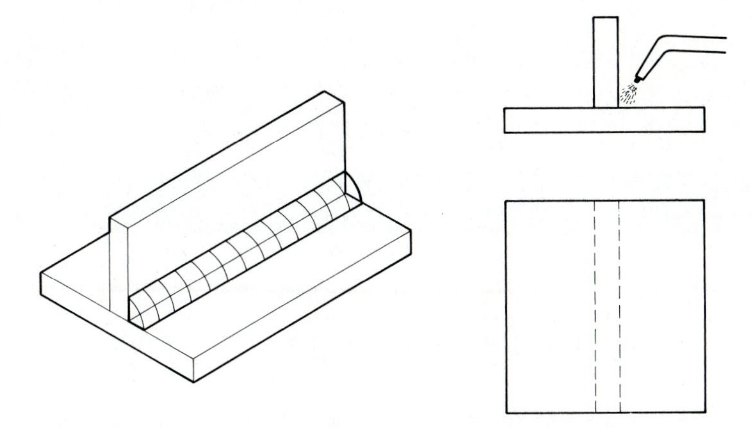

Material: A-36 (mild steel)

Size: $\frac{3}{16}$ in. × 2 in. × 5 in. plate

Joint Design: Horizontal fillet, tee joint

Position: Horizontal (2F)

Technique: Braze

Filler Rod: RCuZn-C

Procedure: Cut and position plates into a tee-joint configuration. Preheat your plate to a dull-red color (do not overdo it). Using a flux-covered or flux-dipped brazing rod, lay the rod onto the heated metal, bringing the flame into the joint. As the braze starts to melt into the joint, move the torch back and forth, in and out of the joint to control the brass flow rate. You will find that the more heat that is put into the joint, the more brass will flow. As you approach the end of the joint, you will want to use less flame because of the buildup of heat already in the joint.

Visual Inspection: Brazes must be cleaned before examination, by chipping and wire brushing off any remaining flux. Brazes should appear slightly flatter than fusion welds and exhibit complete adhesion on both legs. The braze should be uniform throughout its length.

6

SMAW SHIELDED METAL ARC WELDING

Shielded metal arc welding (SMAW), known for many years as "arc welding," has been the industry's mainstay and will remain a factor in many areas of welding for years to come. Today, SMAW is used very heavily in construction because of its quality and portability. In the past it was a major factor in production, using fast-fill electrodes such as E7024, E7027, and E6020. But the GMAW and FCAW processes are rapidly overtaking SMAW in this area of the industry. However, most welding done today is still SMAW because of its versatility. SMAW can weld most metals, such as stainless steels and carbon steels, and with specialty electrodes can weld high-carbon steels, copper, brasses, and even aluminums. SMAW can be used for hardfacing and buildup of high-wear parts and surfaces as well as other applications.

SMAW THEORY

Once the arc is established, the core wire and the flux coating begin to consume. The filler wire forms molten droplets which deposit into the weld (Figure 6-1). The flux forms a gas that shields the molten weld

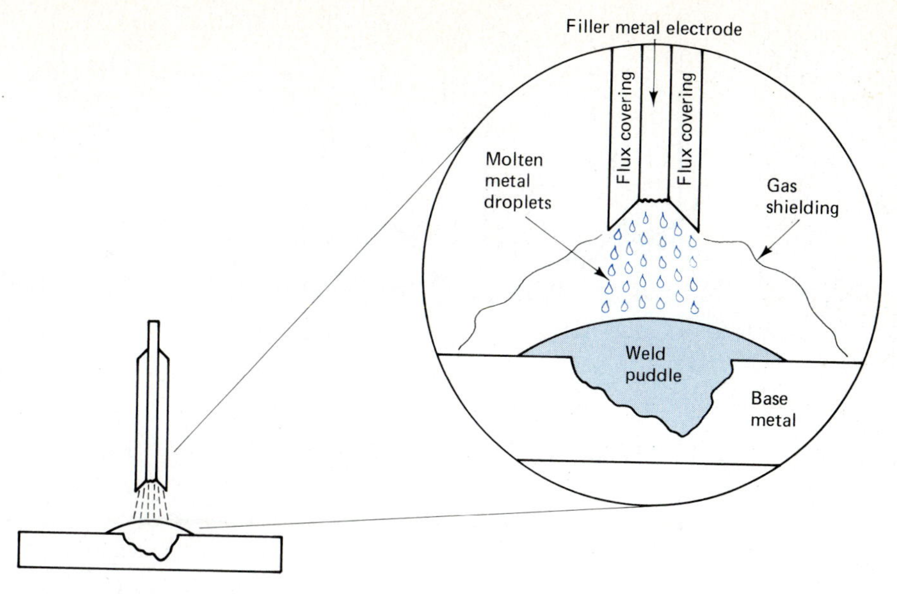

FIGURE 6–1
SMAW process diagram.

puddle. The arc force provides the digging action into the base metal for penetration. This process continues as the weld widens and the electrode continues across the joint.

Electricity and Polarities for SMAW

Shielded metal arc welding (SMAW) requires a completed electrical circuit (Figure 6–2). The type and direction of the current will affect the way the process runs, the fusion, and the appearance of the bead.

There are three possible circuits (Figure 6–3): (1) direct current straight polarity (DCSP), (2) direct current reverse polarity (DCRP), and (3) alternating current (ac). Recently, to simplify the understanding

FIGURE 6–2
Complete SMAW circuit.

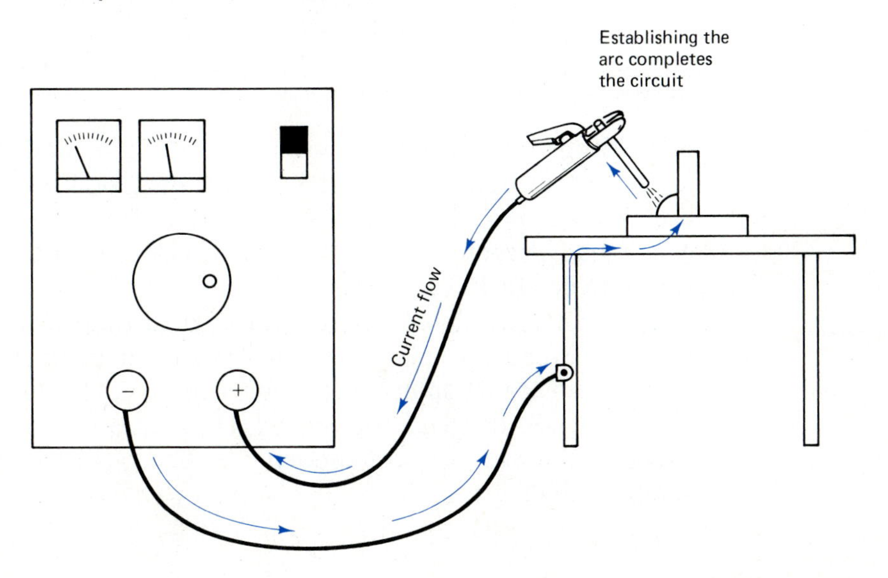

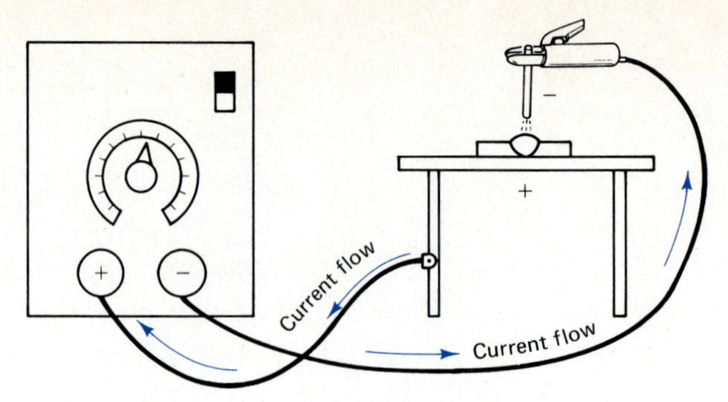

DCSP (DCEN): direct current, electrode negative-work positive

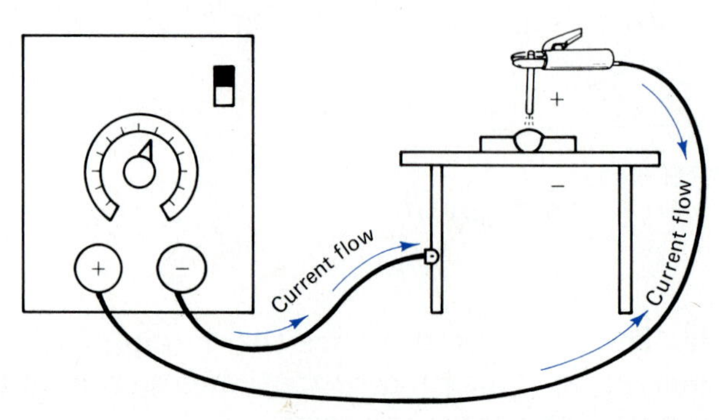

DCRP (DCEP): direct current, electrode positive-work negative

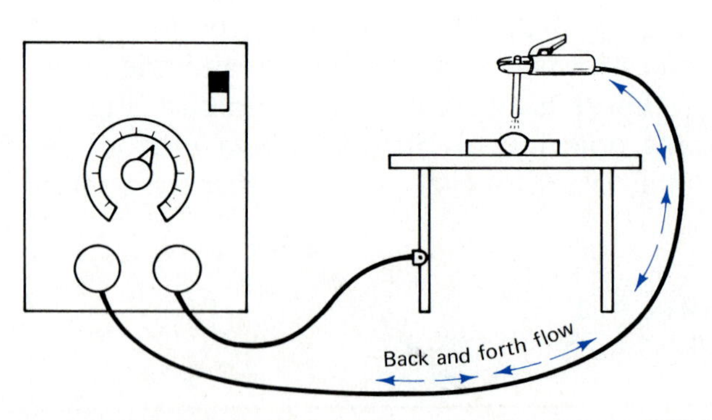

AC (alternating current): current is alternating back and forth between DCSP and DCRP at a rate of 60 complete cycles per second

FIGURE 6-3
SMAW polarities.

of welding circuits, some new terms have been added. DCSP is now also known as DCEN, for direct current electrode negative, and DCRP is now known as DCEP for direct current electrode positive. We will discuss what happens with each of these circuits, but let's first understand some basic electrical terms.

Voltage, also known as potential or pressure, is the electrical force that causes or pushes the flow of electrons. When arc welding, you will notice when holding a short arc length, the arc voltage is lower. When the arc length is increased, the voltage rises until the arc length exceeds the arc voltage capacity of the power source, and the arc is extinguished.

Amperage, also known as current or flow, is the actual electron movement through the conductor. Amperage is also referred to as heat, because the more amperes, the more heat in the arc.

Wattage is simply a total measure of electrical energy. When you purchase electricity from the electric company, you pay for the total wattage used because it is the amperage times the voltage (watts = volts × amperes). Whether we use 10 volts (V) and 5 amperes (A) or 5 V and 10 A, we still have a total of 50 watts (W).

Resistance is the opposition to the flow of electricity. It is measured in ohms (Ω). Resistance is measured by welders who are using long lengths of welding cable. They can calculate the reduction in energy that is available due to the long cable.

Understanding the "I^2R = loss" formula will help a welder calculate this reduction in energy due to lengthening the welding cable. For example, the arc voltage is 25 V and the amperage is 100 A at the welding terminals. The overall energy, or wattage, is 2500 W. The cable resistance is 0.00025 Ω per foot.

We now need to add welding cables of 100-ft lengths. What is the energy output at the end of the 100-ft cable? Here are the formulas:

watts = volts × amperes — (E) volts = 25
(I) amperes = 100
watts = 2500
(R) resistance = 0.000250 Ω (per foot)

$I^2 \times R$ = loss (in watts) — I^2 = 10,000 (100 squared)
R = 0.025 Ω (0.000250 × 100 ft)
10,000 × 0.025 = 250
2500 − 250 = 2250

That is, 2500 original watts at terminals minus 250 W loss in the 100 ft of cable equals 2250 W.

Now, this does not tell us exactly what loss in amperage we have, but it is obvious that we have energy (watts) loss due to the long cable. What the welders must realize is that they must raise the amperage as cables are lengthened. A good rule of thumb is to increase your amperage 5% for every 25 feet of welding cable added.

Heat needed for welding is produced and controlled by the amperage, but the polarity will also affect the location of the heat. With DCRP (DCEP) about two-thirds of the heat produced is at the electrode tip (Figure 6–4). This happens primarily because the electrons, which are traveling from negative to positive, are bombarding and decelerating on the electrode tip. This energy is what turns to heat energy.

On DCSP (DCEN) the situation is reversed. The electrode is negative and the plate is positive. This causes the electrons to flow from the electrode to the plate, where they bombard and decelerate, creating most of the heat at this part of the plate and in the weld (Figure 6–5).

As the term implies, alternating current (ac) alternates back and forth between straight and reverse polarity, a total of 60 complete cycles per second (Figure 6–6). The electrons actually change direction of flow 120 times per second. Heat balance with ac is one-half at the

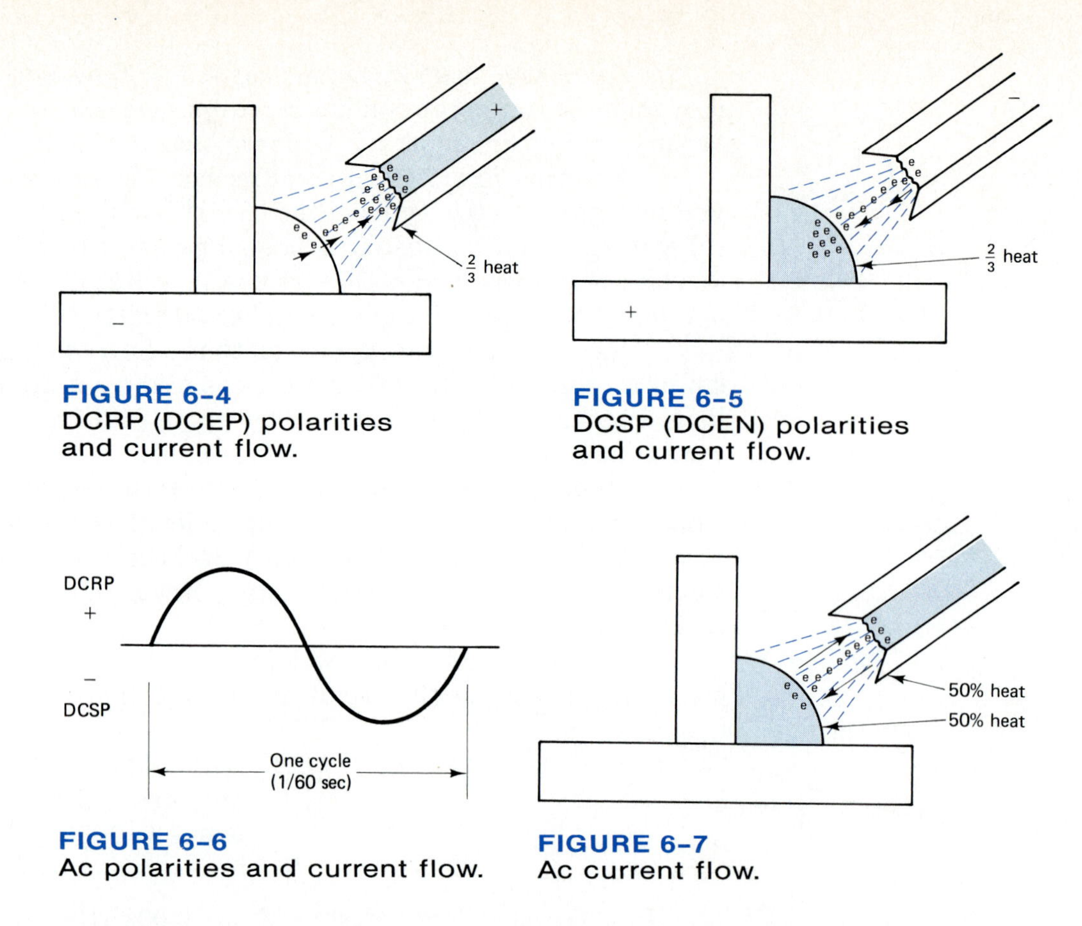

FIGURE 6-4
DCRP (DCEP) polarities and current flow.

FIGURE 6-5
DCSP (DCEN) polarities and current flow.

FIGURE 6-6
Ac polarities and current flow.

FIGURE 6-7
Ac current flow.

plate and one-half at the electrode (Figure 6-7). This is because 50% of the time electron flow is at the electrode and 50% at the plate.

Selecting the Polarity

Changing polarity may produce a pronounced effect on the penetration, the way the electrode operates, and even the way the arc sounds. For example, if we had been operating E6010 on DCRP, you would notice a good, deep penetration, with a "biting" arc. If we now change to DCSP, we will notice less penetration, a softer arc, and a different-sounding arc. With DCSP we have more heat at the plate but much less at the electrode. This results in less burnoff of the electrode and flux. The flux that burns off and creates the shield and ionization path (electrical path) for the electricity contains less hydrogen because of the slowed burn-off rate. Hydrogen is what causes the deep-penetrating, "biting" arc in E6010 electrodes; therefore, with SMAW on DCSP, we have a softer arc.

With SMAW, using DCSP (DCEN), there is usually less penetration but a good deposition rate. Running DCSP downhill makes thin material easy to arc weld.

If a dc power source is not available, some electrodes can be run on ac. The welding apparatus used is sometimes referred to as an ac "buzz box" because it uses only a transformer (no rectifier to change ac to dc). These machines are very portable and reliable. These electrodes usually have a flux coating with potassium to stabilize the arc. It would be difficult to strike and maintain an arc, an ac, without some type of stabilizer, because ac actually has electrical flow stop momentarily

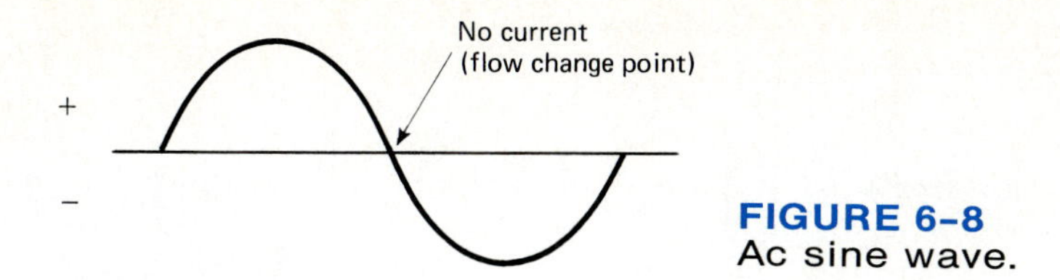

FIGURE 6–8
Ac sine wave.

while the electrons change flow direction (Figure 6–8). The zero-current point occurs 120 times per second in a 60-cycle frequency.

The arc must actually reestablish immediately after the zero-current point. This appears as arc sputtering to the welder who is trying to arc weld without a stabilized electrode. Let's look at which electrodes run best on which polarities.

Strictly DCRP: E6010
E7015

AC or DCRP: E6011
E7016
E7018
E7028

AC or DCSP: E6012

AC, DCSP, or DCRP: E6020
E6013
E7014
E7024
E7027

These electrodes are discussed further in Chapter 7.

SMAW EQUIPMENT AND CONTROLS

There are a number of different types of SMAW power sources available today, with a variety of capabilities and options.

Transformer Rectifier

The standard transformer rectifier machine uses the standard power coming into a building and adjusts it into electricity we can weld with. The power coming into the building (from the electric company) has much voltage and not enough amperage, so in your machine we run it through a step-down transformer, which steps down the voltage and increases the amperage. Remember, the amperage is heat and that is what we need to weld.

For the transformer shown in Figure 6–9, the primary side of the transformer is receiving 440 V at 45 A from the electric company. Because of the number of wire wraps from the primary to the secondary side around the iron core of the transformer, we have only 80 V but 250 A output on the secondary side. So now we have a desirable volt-ampere balance, but we still have alternating current (ac). For most

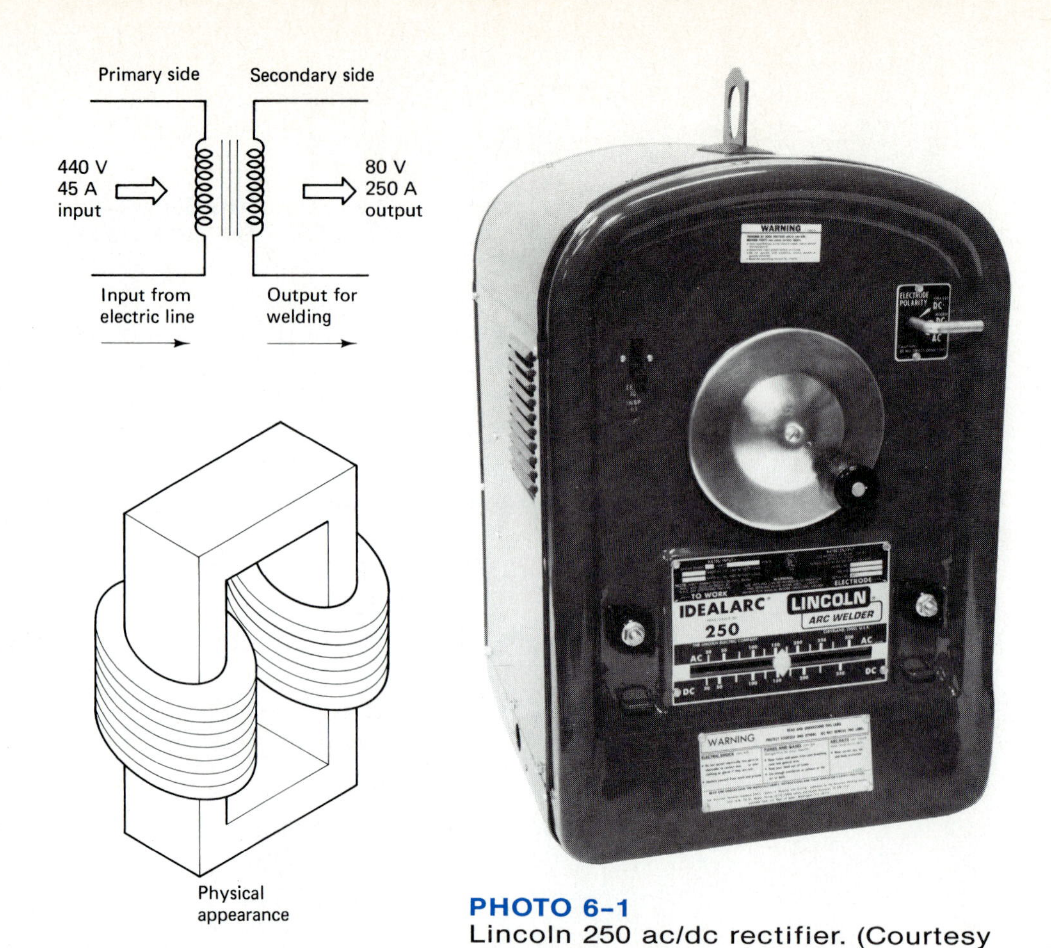

FIGURE 6–9
Transformer.

PHOTO 6–1
Lincoln 250 ac/dc rectifier. (Courtesy of The Lincoln Electric Co.)

SMAW applications we would prefer using direct current (dc). Changing ac to dc is accomplished with a rectifier (Figure 6–10). A rectifier is made up of diodes (solid-state components that allow electrical flow in one direction only). By allowing flow in only one direction, the diodes can resist out the return flow of the alternating current. This energy cannot just disappear, so it turns to heat, and the current flow is now in just one direction (dc). The advantage of the transformer rectifier machine is its low maintenance and its relative inexpensiveness.

Motor Generator

The motor generator power source simply uses an ac motor to turn a dc generator (Figure 6–11). Many welders prefer this type of power source because it is very smooth; that is, there is almost no fluctuation in the current. With the transformer rectifier power source, small changes in line voltage may be noted at the arc as hot or cold flashes. With the motor generator power source, most minor line voltage changes do not happen long enough to affect the ac motor rpm, so no fluctuation in dc generation is seen. This system does have more moving parts, such as the brush and commutator, to wear out.

Engine-Driven Generator

The engine-driven generator uses a gas or diesel engine to drive a dc generator. It is equipped with a special throttle and governor to adjust the rpm to the dc generator for smooth dc output. The advantage of

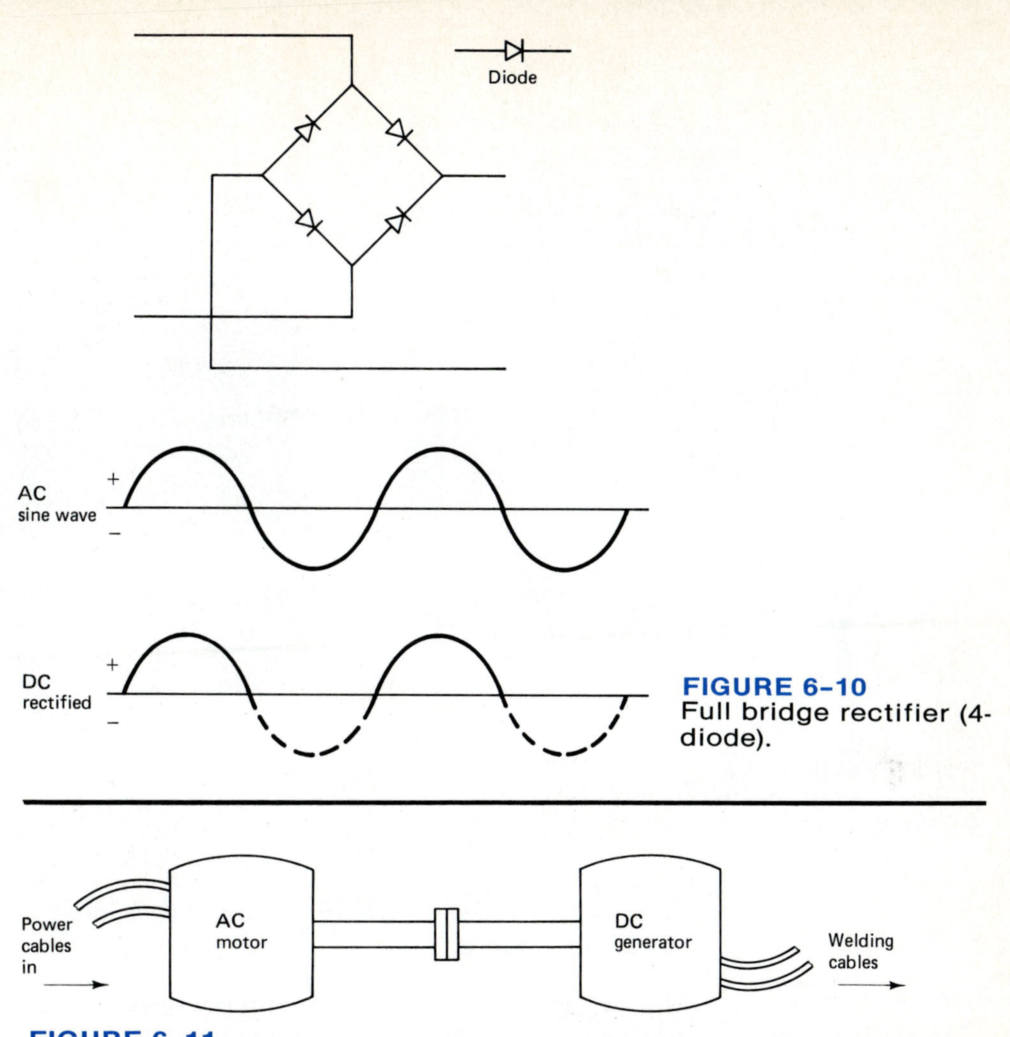

FIGURE 6–10
Full bridge rectifier (4-diode).

FIGURE 6–11
Motor generator.

PHOTO 6–2
Lincoln motor generator. (Courtesy of The Lincoln Electric Co.)

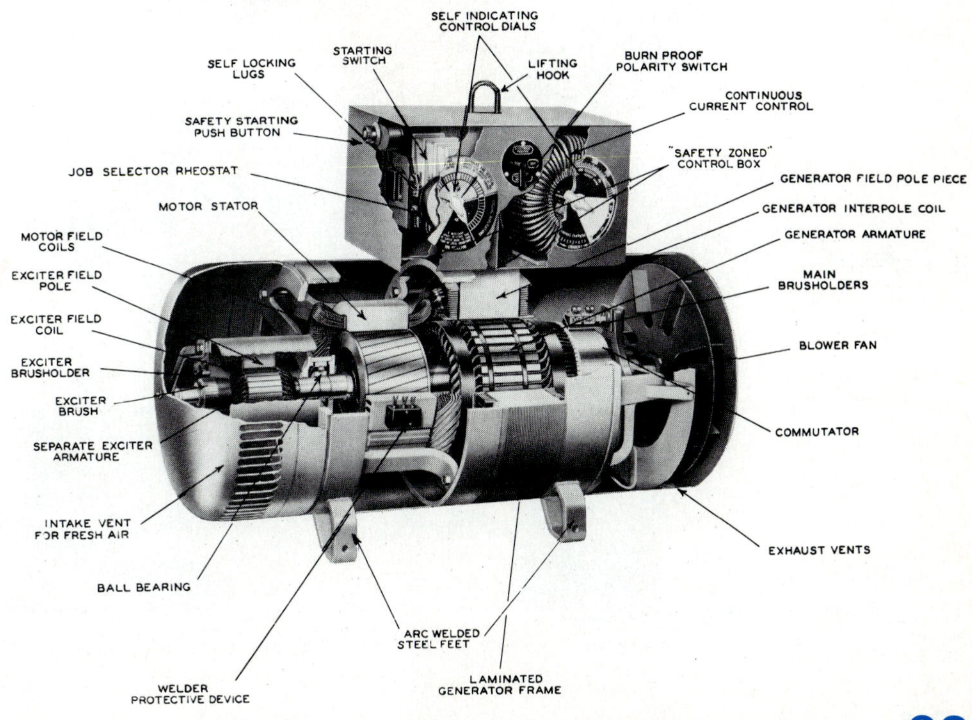

PHOTO 6–3
Lincoln engine-driven machine. (Courtesy of The Lincoln Electric Co.)

PHOTO 6–4
Lincoln heavy-duty engine-driven welder. (Courtesy of The Lincoln Electric Co.)

this system is its portability. Since it is engine driven, no external power is required.

AC Transformer

The ac transformer machine simply uses an ac transformer to step down the voltage current for welding. This system is one of the simplest; it contains only a transformer, cooling fan, and an on–off switch. They are also light and portable and can run off most 220-V power outlets. The limitations are that with only an ac power output, you must use only ac electrodes (see Chapter 7).

PHOTO 6–5
Lincoln 225–A ac machine. (Courtesy of The Lincoln Electric Co.)

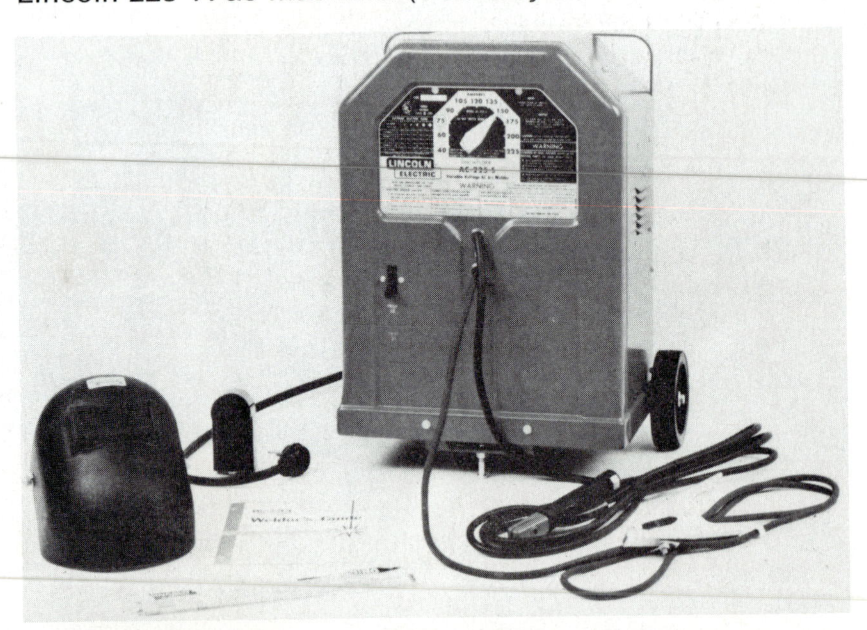

Inverter Power Sources

One of the newest types of welding power source is the inverter power source. Inverter power sources are smaller than conventional power sources, due to the application of a smaller transformer. Traditionally, the transformer is one of the largest and heaviest components of any welding power source. Inverter power sources use higher frequencies, which use fewer transformer windings, and smaller transformer cores; therefore they use a smaller, lighter transformer. These inverter machines have made possible portable and even hand-carry welding machines!

How Does It Work?

The inverter power source first takes the 60-Hz ac (alternating current) input to the machine and converts it to dc (direct current); this is done with a rectifier. The dc is then converted back to ac, but this time it is a higher frequency ac which now allows us to use a smaller transformer; this is accomplished with the *inverter.* The power is now rectified back to dc for welding.

The inverter, using solid-state, high-speed electronic switching components, takes the dc and alternates the direct current very quickly, or at high frequencies, back and forth into the line as alternating current (ac).

The power is now ready to go into the transformer to raise the amperage and lower the voltage, which is then rectified back to dc for welding. Remember all these steps are necessary because we need to obtain high frequency for small transformers, and we can only transform alternating current to obtain high amperages for welding. (Note steps in Figure 6–12.)

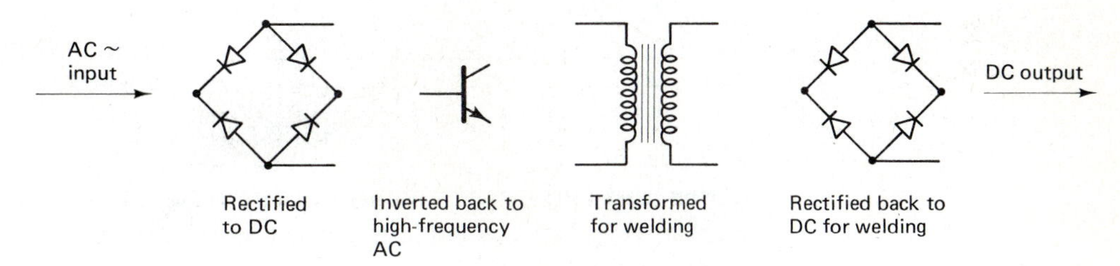

FIGURE 6–12

Controlling the Current

There are a number of ways to control the amperage output of a SMAW power source. Regardless of how it's done, infinite amperage control is always desirable. The **tapped reactor machine** uses "taps" from the control reactor to set the selected amperage (Figure 6–13). It must be set at one of the predetermined ampere settings or taps. There is no way to set amperages between tap settings. For example, if tap 4 is 85 A and tap 3 is 70 A, there is no way to set amperages between 85 and 70 A; you must select 85 or 70.

A **movable coil machine** (Figure 6–14) physically moves the primary or secondary side coil either closer or farther apart to raise or lower the amperage. The closer the primary side is to the secondary side, the more the magnetic flux can saturate the secondary side, resulting in higher amperage. Inversely, the farther apart, the less flux satu-

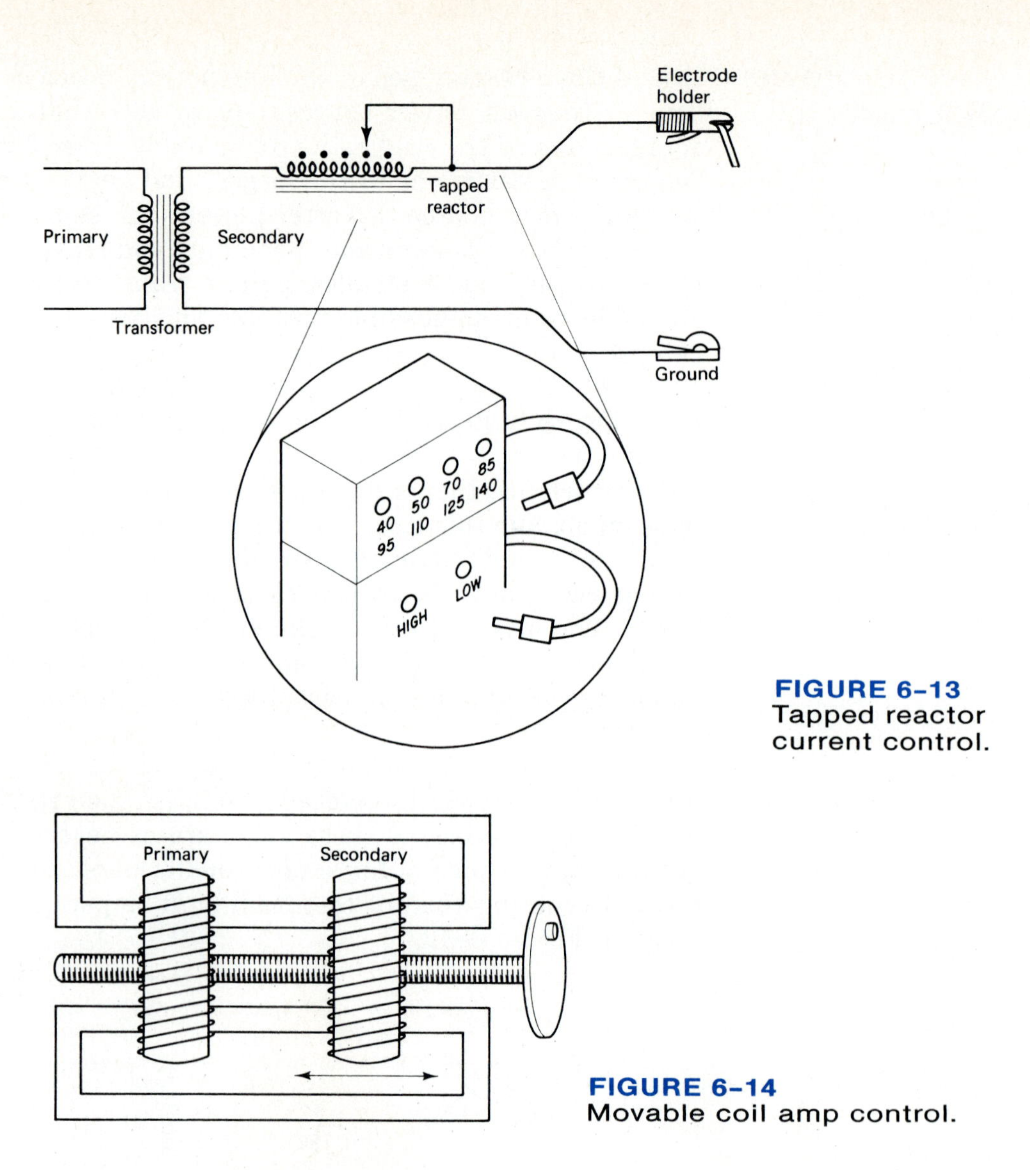

FIGURE 6–13
Tapped reactor current control.

FIGURE 6–14
Movable coil amp control.

ration and the lower the amperage output. This system is simple and troublefree, but its settings are easily affected by any atmospheric or humidity change.

The **movable shunt** uses a movable laminated iron shunt to come between the primary and secondary sides of the transformer. (Figure 6–15). When the shunt is fully retracted, we have maximum amperage. When the shunt is cranked in all the way, amperage is minimal because the shunt is diverting much of the flux away from the secondary side of the transformer. Infinite amperage control is obtained by cranking the shunt part way in, until the desired amperage is obtained.

The easy-adjusting, dial-controlled power sources usually use a rheostat to control the current to a saturable reactor circuit (Figure 6–16). Also known as a magnetic amplifier circuit, the saturable reactor circuit relies on the introduction of a dc current to reduce impedance in the control reactor, thus allowing a higher current flow through the reactor. The reactor acts like a control valve, allowing either high or low amounts of current to flow through, thus raising or lowering the amperage on the machine.

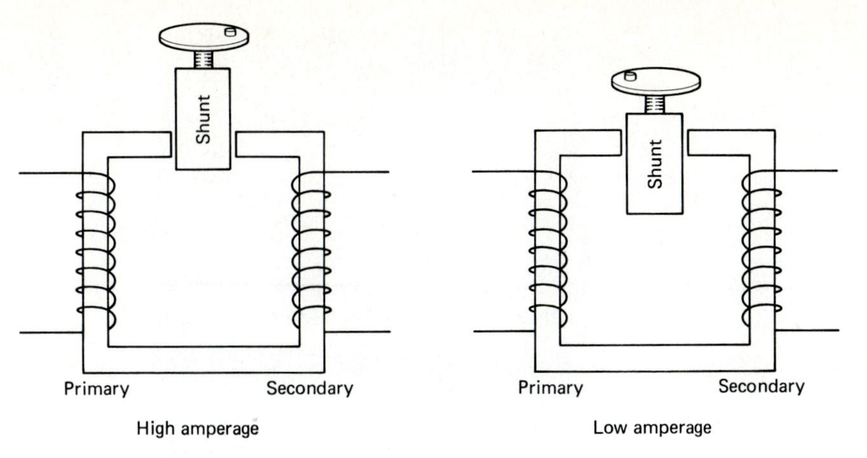

FIGURE 6–15
Movable shunt amp control.

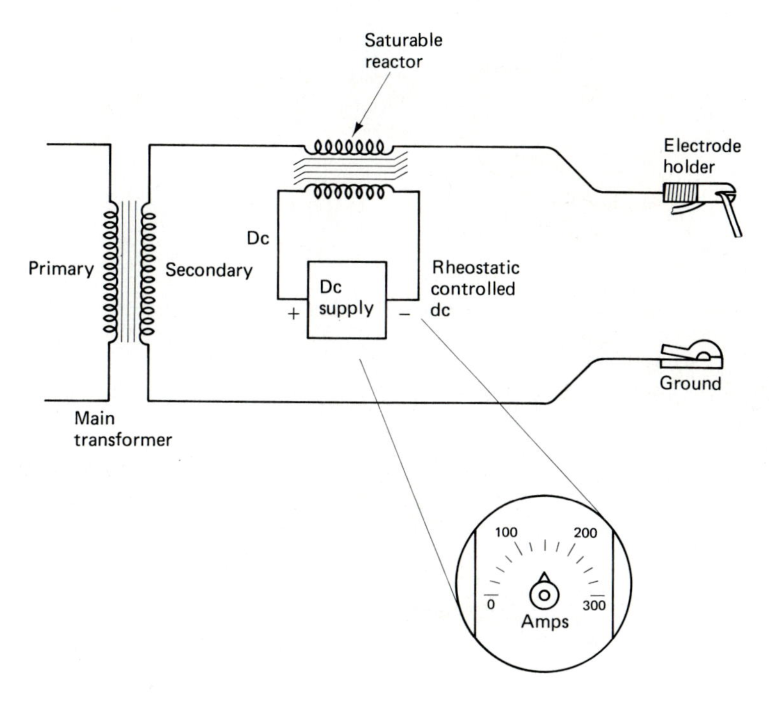

FIGURE 6–16
Saturable reactor amp control.

SMAW Machine Controls

The primary heat-controlling function with SMAW is the amperage control. It has the most effect of any machine control on the quality of the weld bead. We can control the size, deposition rate, and quality of the bead just by adjusting the amperage. Increasing the amperage will tend to speed the deposition rate and flatten the bead. If it gets too high, a flat drawn-out bead with spatter will occur. Lowering the amperage will produce a cooler, more convex bead. If it gets too low, electrode sticking and difficulty starting may occur. Electrode containers will have a recommended amperage range printed on the container. The range is usually quite wide; for example, the range for a $\frac{3}{32}$-in. E7016 is 60 to 100 A. If you experiment with high and low settings within this

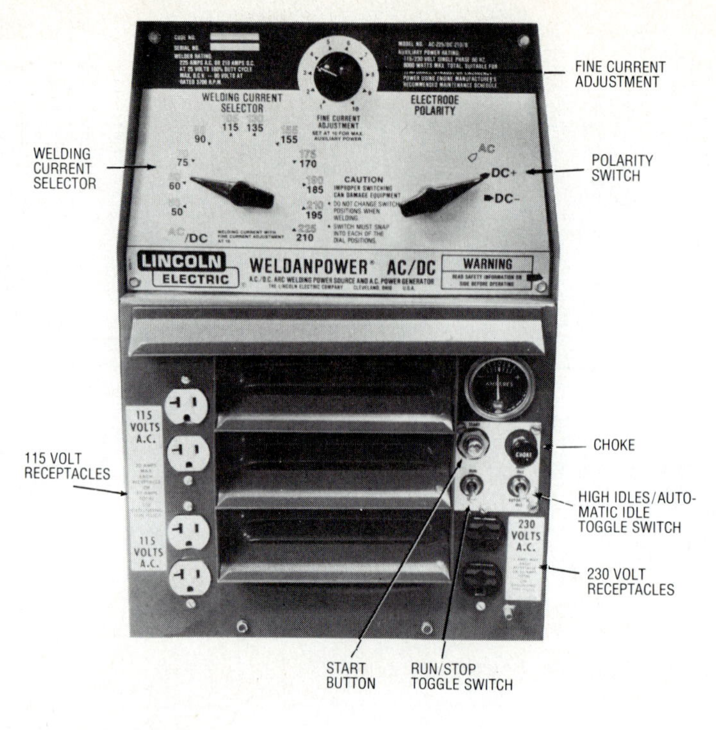

PHOTO 6-6 Typical machine controls for an engine-driven welder. (Courtesy of The Lincoln Electric Co.)

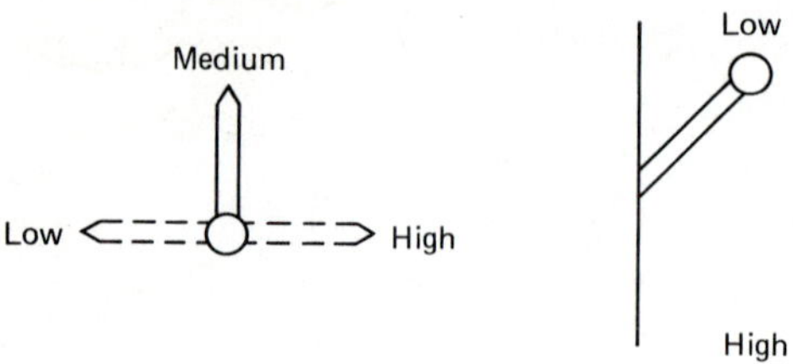

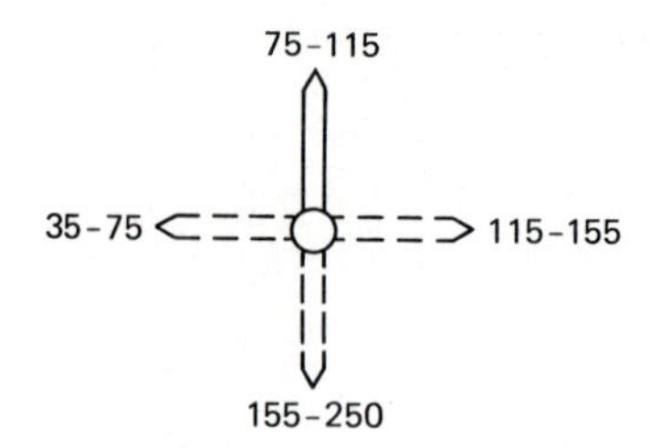

FIGURE 6-17 Range switches.

range, you will zero in on the correct setting for a given situation. Typical range switches are shown in Figure 6-17.

How To Correctly Set the Amperage

First, you must realize that there is no exact or correct setting that will work for everyone. We all weld slightly differently from one another. Some welders run hot and move fast; while others run cooler and travel slow with the electrode. Here are some basic steps for getting a setting that will work for you.

1. Check the recommended amperage range on the container, and set your machine at the median (about in the middle of the range). If, for example, the range is 60 to 100 A, start at 80 A.

2. Now, get some scrap plate together, check your polarity, and run a sample bead.
3. Examine what happened: Did the electrode strike easily? Did it run smoothly? Examine the bead: Is there spatter, or overlap? Is there a bead contour or fusion problem?
4. Correct the amperage setting up or down and run another bead to check the results of the corrective setting. If you need some help trouble shooting your bead, consult Table 7–2.

The four-step technique explained here is called the "set by feel method." You will probably have to reset your machine with this system for each new project.

The *polarity* allows you to select ac, DCRP, or DCSP simply by turning the switch. Without a polarity switch, switching electrode cables was necessary to change polarities on SMAW machines. Many SMAW power sources have an *amperage range switch*. It may have high, medium, and low settings, or may actually list the range in each position.

Meters for SMAW

Knowing the exact voltage and amperage status of the welding arc is not just convenient but necessary if doing code work. Welding procedures, which are required to be written by most welding codes, require exact voltage and amperage ranges to be listed for each pass on the procedure form (see Chapter 23). Voltage or amperage meters can have either analog or digital readout (Figure 6–18). Well-maintained meters are usually quite accurate. Digital readout meters are usually accurate to $\frac{1}{10}$ of a volt or ampere.

Machine Ratings

The *duty cycle* (Figure 6–19) is one way of rating the capability of a welding power source. It simply bases its rating on how long a machine can run, under load, in a 10-minute period. If a machine can run full power, under load, for a maximum of 6 minutes without internal dam-

FIGURE 6–18
Analog and digital readout.

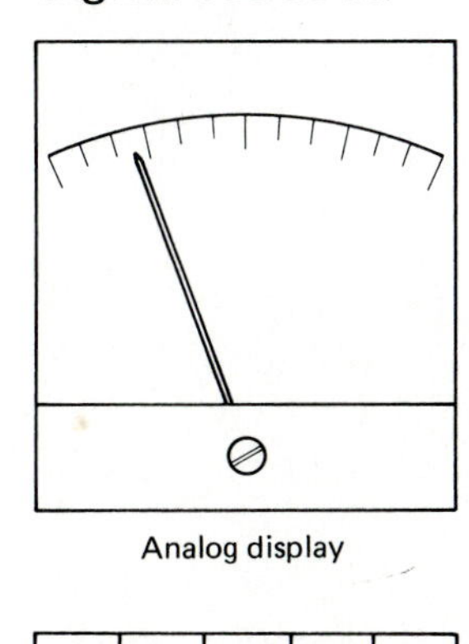

FIGURE 6–19
Duty cycle chart.

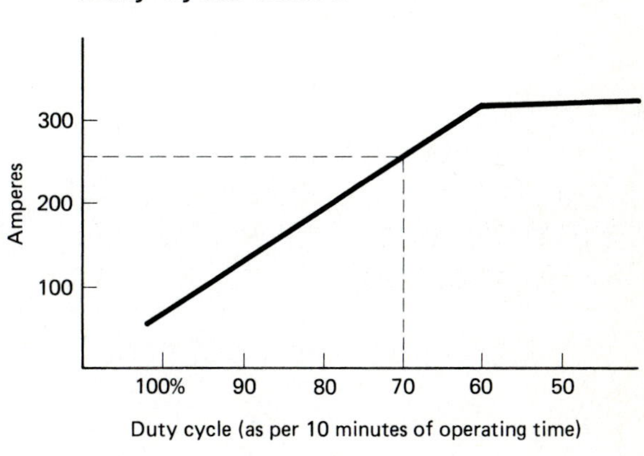

age or overheating, it would have a 60% duty cycle (6 minutes out of 10 minutes). Some manufacturers show a machine's duty cycle based on the amperage it is running at.

The power source *amperage rating* is also a good indicator of the durability of the machine. Some machines may be used for welding and air carbon arc cutting (AAC) and gouging. This requires high amperage settings (400 to 600 A is desirable for air carbon arc).

It is best to use a machine that has more than enough power for your specific application. You never know when you may need more power and your machine will not be in danger of overheating.

REVIEW QUESTIONS

1. What does SMAW signify?
2. What metals can we weld with the SMAW process?
3. What effect does your arc length have on the arc voltage?
4. You are welding with 50 feet of welding cable, on 120 A. You now need to extend your cable 25 feet (to 75 feet). What should your amperage be now?
5. On what polarity should E6010 be run?
6. On what polarity can E7018 be run?
7. What is the purpose of the welding transformer?
8. What is the purpose of the rectifier in the welding machine?
9. What is the advantage of the engine driven generator welding machine?
10. What is the advantage of the inverter welding power source?
11. Explain duty cycle.
12. Explain the effect of the amperage of your weld bead.

7

SMAW ELECTRODES

GENERAL CLASSIFICATIONS

We can classify SMAW electrodes by AWS classification, trade names, metal classifications, or general classes. Let's start with a broad look at mild steel electrodes in the general classifications. General classification breaks mild steel electrodes into four basic classes: (1) fast freeze, (2) fast fill, (3) fill freeze, and (4) low-hydrogen classes.

Fast Freeze Electrodes

Fast-freeze electrodes are the AWS class E6010 and E6011 cellulose coating electrodes. As the name states, these electrodes have a fast freezing action; that is, the puddle solidifies rapidly after being deposited from the electrode. When these electrodes burn, their decomposing shielding atmosphere is high in hydrogen; therefore, they produce intense heat and a deep biting action. This makes these electrodes excellent for burning through rust, grease, or used steel and is a popular electrode classification for many maintenance welders. Fast-freeze electrodes are well suited for out-of-position welding, due to their excellent puddle control and rapid solidification, which prevents the puddle from dripping when in vertical and overhead positions. Fast-freeze electrodes are considered to be one of the most difficult to oper-

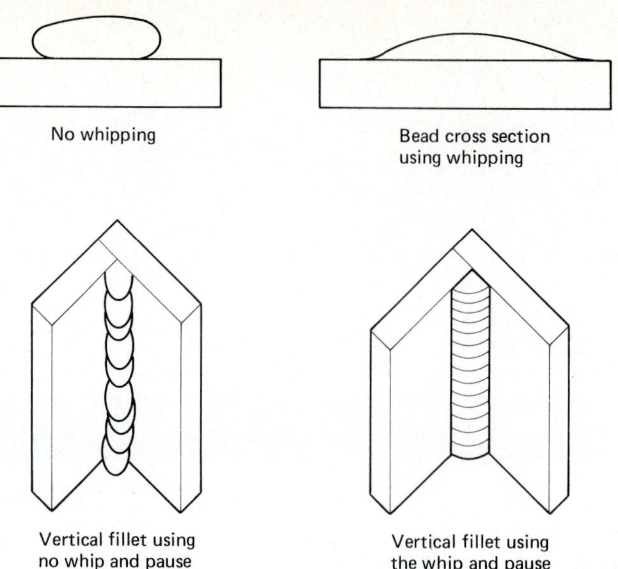

FIGURE 7–1
Whipping motion effects on beads.

ate. This is due in part to the requirement of a whip-and-pause technique or whipping motion with these electrodes (Figure 7–1). If this technique is not used, the weld beads will not flatten out or produce overlap while in the flat and horizontal positions, and will drip when in the vertical and overhead positions.

The whipping motion actually causes molten puddle stimulation (or waves), which wash out the puddle over a larger surface area and produce a flat bead with well-feathered edges. These electrodes can utilize the drag technique; however, overlap (or cold lap) will usually result.

Fast-Fill Electrodes

At the other end of the spectrum of general class electrodes, we have *fast-fill electrodes.* These are the AWS class E7024, E7027, and E6020 electrodes. These electrodes are very high deposition or just plain fast electrodes! Most contain iron powder in the flux coating, which helps create the high deposition rate. This classification is also known as production electrodes. Most fast-fill electrodes are recommended for flat and horizontal positions only because of their highly fluid puddle. Most welders consider these electrodes quite easy to operate. They utilize the drag technique, which simply requires the welder to set high amperages, lay the electrode in the joint at the correct angle, and apply a slight drag pressure to the electrode. The result is a smooth, extremely uniform weld.

Fill-Freeze Electrodes

The *fill-freeze* (also known as fast-follow) electrodes, E6012, E6013, and E7014, have the characteristics of both fast-fill and fast-freeze electrodes. They solidify faster than the fast-fill class, so they run well out of position. But they contain many of the same elements of the fast-fill class, so they have higher deposition rates than do the fast-freeze class. They are recommended for the occasional welder who may be called upon to do minor maintenance or repair work. It should be noted that fill-freeze class electrodes do not penetrate or "bite" as much as fast-freeze electrodes, so surfaces to be welded must be free of oil, grease, rust, and contamination.

ELECTRODE FLUXES AND WHAT THEY DO

As the electrode burns off, some of the flux forms a gas shield around the molten puddle. Other portions mix into the molten puddle and act as deoxidizers to keep porosity from the final weld deposit. As the flux deoxidizes it floats to the surface of the weld, where it forms a slag blanket. This blanket slow cools the weld, allowing the grains to refine slightly and giving better mechanical properties to the weldment. Elements such as iron, powder, and titania in fluxes increase the deposition rate (go faster) and improve bead appearance; cellulose and iron oxide improve penetration, while potassium stabilizes the arc for ac electrodes (see Table 7-1, page 94).

AWS ELECTRODE CLASSIFICATION SYSTEM

The American Welding Society (AWS) has standardized a system for classifying SMAW welding electrodes. You should use this system whenever you order or specify electrodes. Here's how it works. An E6010 electrode classification would indicate:

E = Electric (electrode)

60 = Tensile strength in thousands of pounds per square inch (in this case, 60,000)

1 = Position
- 1 = all-position electrode
- 2 = flat and horizontal only
- 3 = flat only

0 = Flux content (see Table 7-1)

Although using trade names for identifying electrodes is acceptable under some conditions, try to use the above standardized system as much as possible. This prevents confusion, as well as the accidental use of the wrong electrode for a given job.

TABLE 7-1
Electrode Characteristics

AWS CLASS	POLARITIES	POSITIONS	FLUX CONTENT	OPERATION TECHNIQUE	DEPOSITION RATE	GENERAL CLASS	COMMENTS
E6010	DCRP	ALL	Cellulose	Whip and pause	Average	Fast freeze	This is a deep-penetrating, good-biting electrode that freezes quickly, so it is excellent handling out-of-position and open-root welds. It is also a good choice for welding through rust and contaminated or galvanized surfaces.
E6011	DCRP or ac	All	Cellulose–potassium	Whip and pause	Average	Fast freeze	This electrode is the same as E6010 except that it will run on ac.
E6012	DCSP or ac	All	Sodium–titania	Drag or whip and pause	Good	Fill freeze	This electrode is a good choice for poorly fitted joints or large openings that must be bridged together. It runs excellently on vertical-down sections of sheet metal. It is primarily a single-pass electrode.
E6013	DCSP, DCRP, or ac	All	Titania–potassium	Drag or whip and pause	Good	Fill freeze	This electrode has many of the same qualities as E6012 but a slightly softer arc and runs ac or DCSP.
E7014	DCSP, DCRP, or ac	All	Iron powder–titania	Drag or side to side	High	Fill freeze	This high-deposition-rate electrode is one of the few high-speed electrodes that runs in all positions! Vertical-up may be somewhat difficult; so use small diameters when going uphill.
E7015	DCRP	All	Sodium	Drag or side to side	High	Low hydrogen	This is an excellent low-hydrogen electrode for running in all positions on DCRP.
E7016	DCRP or ac	All	Potassium	Drag or side to side	High	Low hydrogen	This is a super-smooth-flowing low-hydrogen electrode that will run ac or DCRP.
E7018	DCRP and ac	All	Iron powder	Drag or side to side	Very high	Low hydrogen	This is one of the most popular low-hydrogen electrodes. It has the highest deposition rate of the all-position low-hydrogen electrodes.
E7028	DCRP or ac	Flat and horizontal	High-iron powder	Drag	Very high	Low hydrogen	The E7028 has the highest deposition rate of all low-hydrogen-class electrodes. It is limited to flat and horizontal positions.
E6020	DCSP, DCRP, and ac	Flat and horizontal	Iron oxide	Drag	Very high	Fast fill	The E6020—or "quick stick" as it used to be called—is a deep-penetrating fast-fill electrode.
E7024	DCSP, DCRP, and ac	Flat and horizontal	Iron powder–titania	Drag	Very high	Fast fill	The E7024 combines a superhigh deposition rate with a very smooth wetting action.
E7027	DCSP, DCRP, and ac	Flat and horizontal	Iron powder–iron oxide	Drag	Very high	Fast fill	E7027 combines high deposition, fairly good wetting action, and good penetration.

LOW-ALLOY SUFFIX ELECTRODES

There are many applications of shielded metal arc welding on steels other than mild steels; for example, some boilers and boiler piping contain molybdenum. Some high-corrosion applications are made of steels that contain nickel.

These and other applications may be SMAW welded with special suffix electrodes. They are classified just like standard mild steel electrodes. But they are followed by a suffix, designating the particular additional alloy. An example is shown in Figure 7-2. Here is what the suffixes stand for:

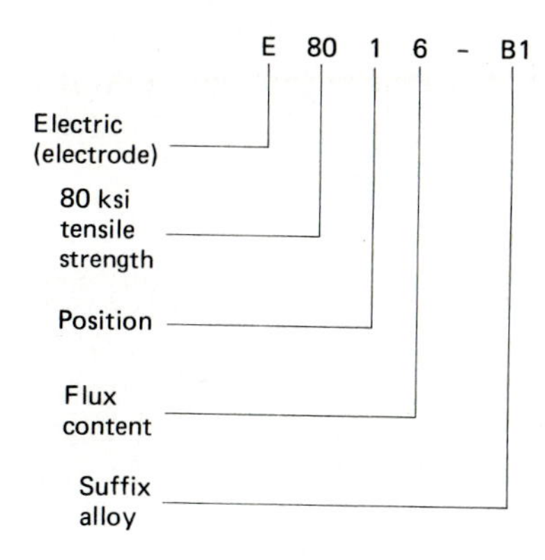

FIGURE 7-2
Classification system for SMAW electrodes.

Al	Carbon-molybdenum
Bl	Chromium-molybdenum
Cl	Nickel
Dl	Manganese-molybdenum
M	Nickel-manganese (military specifications)
G	All other low-alloy electrodes

STAINLESS STEEL ELECTRODES

Shielded metal arc welding of stainless steels produces excellent-quality welds on commonly-welded stainlesses. In most cases the weld properties exceed that of the base metal. It is suggested that you study the discussion of stainless steel in Chapter 19. This will give you an understanding of the basic classes of stainless steels.

For most stainless steel welds, we can use the "300" series stainless electrodes. These chromium-nickel grades produce a tough but ductile weld. As stated earlier, when selecting a filler metal, always match filler and base metal properties and elements as closely as possible. For example, if we are welding a 308 grade stainless, the best choice would be a E308-XX electrode.

Flux Coatings of Stainless SMAW Electrodes

There are two types of stainless steel electrode flux coatings: lime and titania. Lime coatings are represented by the number 15 after the stainless classification, and titania by the number 16 (EXXX-15 = lime, EXXX-16 = titania).

The lime-coated stainless electrodes are highly penetrating and work well for welding out of position because of the quicker freezing action. These should be run on dc polarities only.

The titania-coated stainless electrodes have a very smooth and stable arc with good wetting action. They are a little more difficult to use on vertical and overhead joints, but they will run on ac as well as dc polarities. They would be the best choice for production jobs because of the higher deposition rate. They also have a good welder appeal; that is, they look good.

AWS Classification System

An example of a stainless steel electrode classification is shown in Figure 7-3.

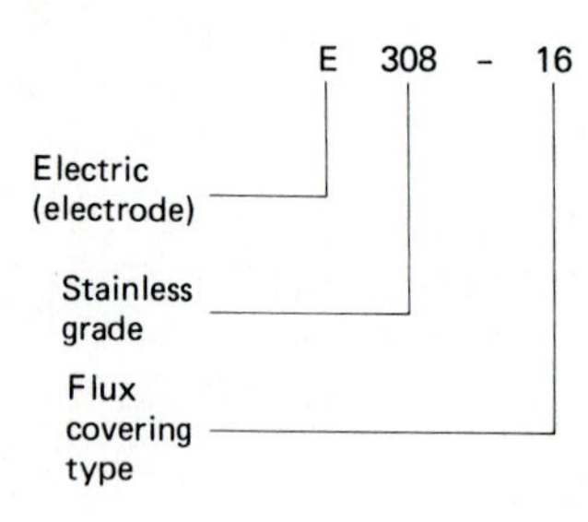

FIGURE 7-3
AWS classification system.

Precautions for Welding Stainless Steel

The welder, welding technician, and welding engineer must all be concerned with the following aspects of welding stainless steels:

1. Warpage and distortion
2. Carbide precipitation
3. Chemical properties

As you may know from studying chapter 19, stainless steels tend to have lower thermal conductivity than carbon steels. This results in an uneven distribution of heat and resulting distortion. The best solution for reducing distortion is to lower the heat input. This lowers the heat imbalance and expansion differences. Small weld beads, using stringers instead of weaves, or using intermittent or back-steeping beads will help lower the heat input.

Another area of concern is carbide precipitation. This is where carbon precipitates into the weld zone and combines with the chromium atom to form chromium carbide. This new substance will not shield with a protective oxide such as pure chromium does. The result is eventual oxidation (rust) in the chromium carbide areas.

Have you ever observed a stainless steel weld that has rusted or has some oxidation (rust) along the sides of the weld (probably in the heat-affected zone)?

There are a few ways of eliminating this problem. One of the easiest and cheapest ways is to use an L class electrode; for example, E308L. The "L" stands for low carbon. Less carbon in the weld area means less chance of chromium carbide forming. Carbon precipitates into the weld zone due to heat. Carbon loves heat and precipitates toward it; therefore, keeping the heat input low also helps to reduce carbide precipitation. Another method involves using "stabilized" electrodes. These electrodes have elements such as columbium and titanium that act as stabilizers. Here is how they work: As the heat precipitates carbon into the weld zone, it combines with the stabilizing element, forming columbium carbide, instead of with the chromium atoms, leaving the pure chromium oxide free to shield the metal from further oxidation.

Another concern involves matching up chemical elements. For example, if the weld has more chromium than the surrounding base metal, the weld would be harder (and may be more brittle). Since we usually want both weld and base metal to act in unison under load, this would be an undesirable property. The reverse could also happen, producing a "too soft" weld or a weld that would be more susceptible to oxidation and corrosion. The solution is quite simple. Use the same filler metal grade as the base metal to be welded. However, be aware that some engineers do require a higher-grade filler metal to be used when it may be beneficial. For example, a stainless container, where it is known that corrosion starts in welded corners and weld seams, may use a 308 base metal but require the use of E309-16 electrodes to overcome the problem.

HAND TOOLS FOR SMAW

The chipping hammer and wire brush are two basic tools used for SMAW jobs. The chipping hammer may look crude, but it is actually a carefully designed tool, and when used correctly will make your job much easier. Some slags are easier than others to remove. You will find that using the sharper, pointed end of the chipping hammer in the toes of the weld will loosen the slag (Figure 7–4). Use the flat side of the chipping hammer on the face of the weld (Figure 7–5). This will remove the loosened slag. Finally, brush the entire weld longitudinally to clean any remaining slag and slag dust. Repeat the entire cleaning procedure.

Other tools, such as grinders, pliers, wire wheels, and a flat file, may be used. A grinder and flat file are used for plate preparation in removing burrs and smoothing out bevels. Pliers should always be used to hold hot metal. Do not pick it up with your leather gloves. The high heat will burn out all the natural oils and harden the gloves. In addition to these basic tools, there are many tools to help you fit, align, and lay out your work.

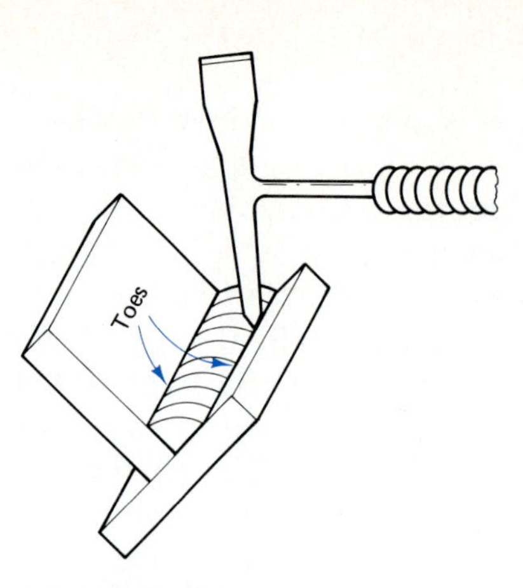

FIGURE 7-4
Chipping hammer cleaning (toes).

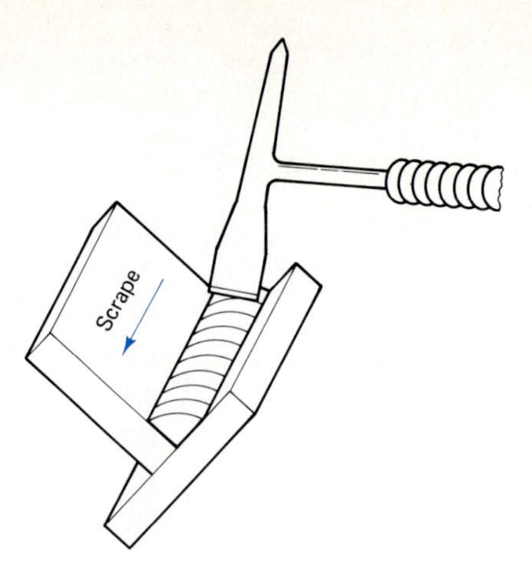

FIGURE 7-5
Chipping hammer cleaning (face).

STRIKING AND STARTING BEADS

There are two basic methods for striking and starting SMAW electrodes; (1) the scratch start and (2) the tap start. The scratch start works well because it produces a little light, so the welder can find his starting position when he or she is under the welding hood. It also allows the electrode to have time to develop the gas shield before it gets to the puddle. It also prevents electrode sticking (if done correctly) because the electrode is heated as it scratches into the starting position. The tap start works well because it leaves few or no arc strikes as the scratch start may, but the scratch start is recommended for most SMAW applications.

To use the scratch start, simply strike the electrode as you would strike a match. Use a quick, steady motion, holding the electrode still in a long arc once the arc has started. Once the puddle widens correctly, start traveling along the joint (Figure 7-6).

The tap start is done by tapping the electrode once at the starting point and then holding a long arc until the puddle widens (Figure 7-7). If the electrode should stick (fuse to the plate), try to restart it. If sticking continues, turn the amperage up, or use a smaller-diameter electrode. If the electrode sticks and does not easily break off the plate, quickly squeeze free the fused electrode out of the electrode holder. Use pliers to break off the fused electrode.

If you cannot consistently strike and establish the arc without sticking the electrode, you need more practice. A good exercise to practice involves simply scratching an electrode across a scrap plate. Continue scratching, then start holding an arc at the end of each scratch. Do this until you feel confident and comfortable with the striking technique.

Always consume electrodes down to about 1 to $1\frac{1}{2}$ in. (Figure 7-8). If shorter, heat may damage the electrode holder. Leaving more than $1\frac{1}{2}$ in. is going to be costly. The few cents wasted from each partially consumed electrode will add up quickly.

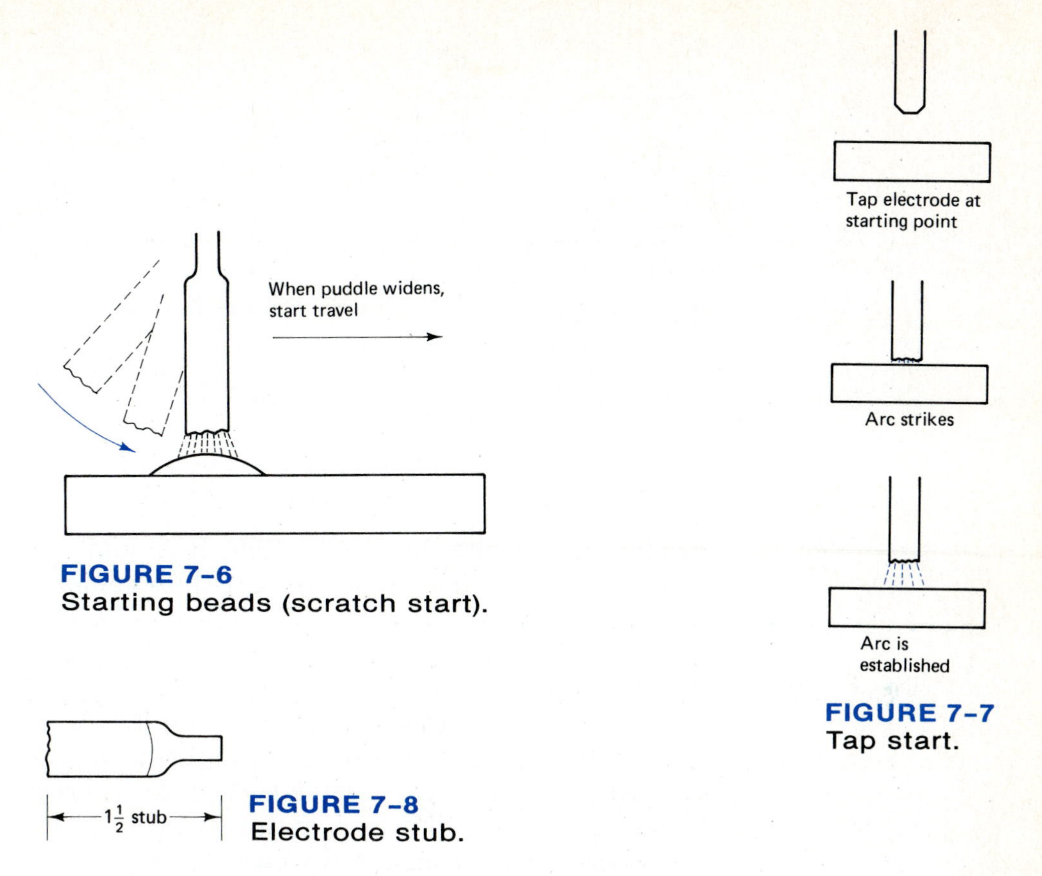

FIGURE 7–6
Starting beads (scratch start).

FIGURE 7–7
Tap start.

FIGURE 7–8
Electrode stub.

PROCEDURE FOR RUNNING MILD STEEL ELECTRODES

For the E6010 and E6011, use the whip-and-pause motion (Figure 7–9). This allows for a flat, better-fused bead. It also makes it much easier to run E6010 and E6011 vertically up and overhead. The whip-and-pause motion is done very quickly. The hold position time is about $\frac{1}{2}$ to 1 second; the whip takes only a fraction of a second. The arc is not

FIGURE 7–9
Whipping motion sequence.

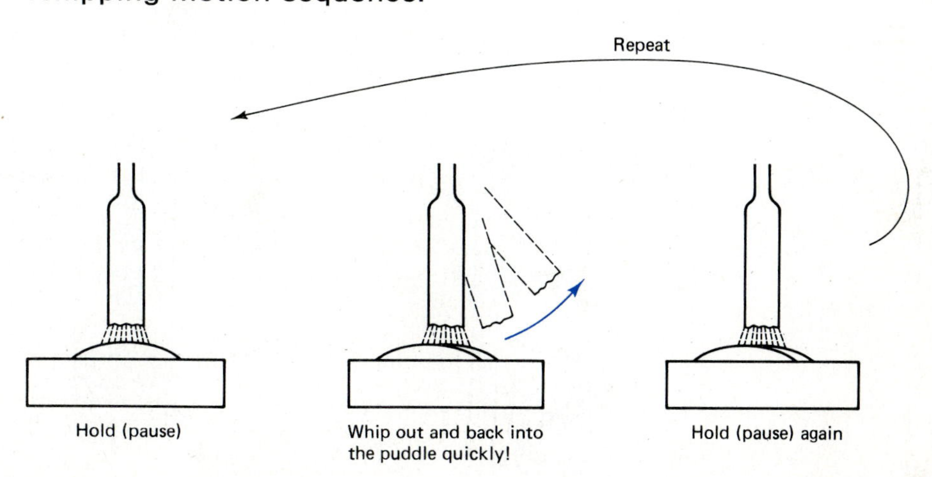

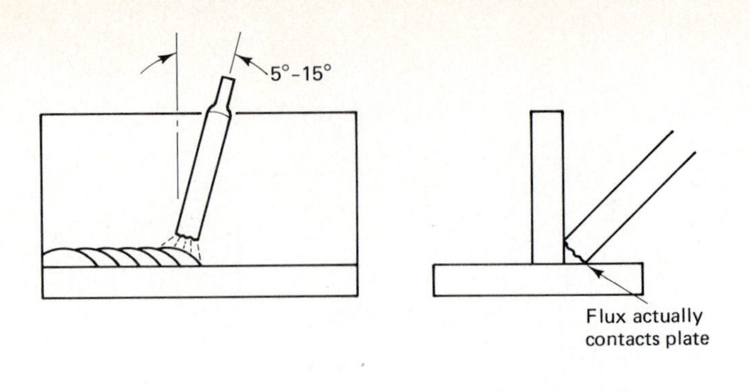

FIGURE 7-10
Drag technique.

broken during the whip-out motion. It should be done quickly and not too far out. The best way to learn this technique is to observe your instructor and practice. The final technique you develop may differ slightly from the one described above. That is okay if it attains the same results.

For electrodes such as E7024 and E7024 (fast fill) the drag technique is used (Figure 7-10). This simple technique requires actual plate contact after the arc is established. A slight pressure and slight angle (5 to 15°) in the direction of travel is all that it takes to produce excellent-quality beads. If the heat and angle are correct, the slag will neatly peak up when the weld starts to cool.

The *low-hydrogen* class electrodes (E7015, E7016, E7018, and E7028) require a very close arc length. For flat and horizontal positions, the drag technique can be used for the E7028 and E7018. But the most popular technique for low-hydrogen electrodes is the "close arc, side-to-side motion" (Figure 7-11). For this the scratch start technique is used and a very closely held arc with a slight side-to-side zigzag motion. This side-to-side technique requires a slight pause on each side. This allows fusion into the sides and prevents undercut. Some welders count "one-two left and one-two right" as they proceed up the joint; but if you use

FIGURE 7-11
Side-to-side motion.

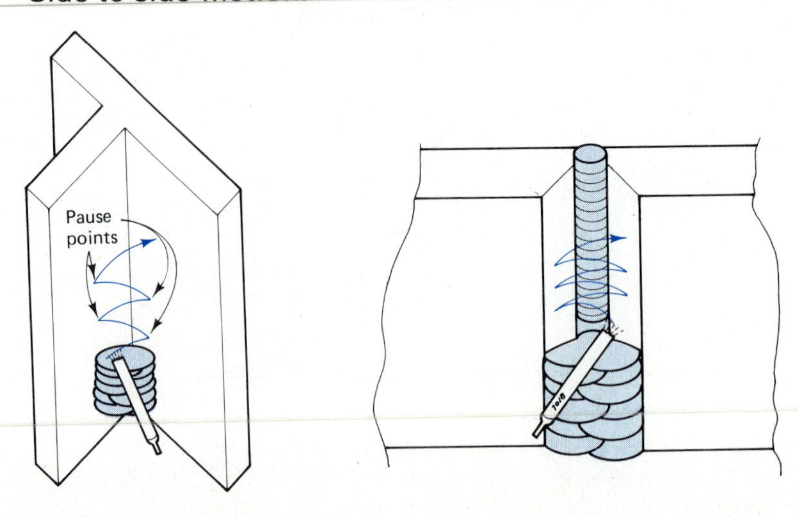

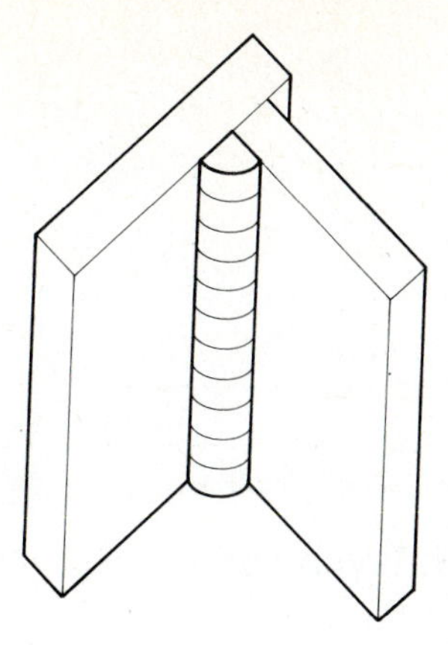

FIGURE 7–12
Bead sag.

the counting technique, make sure you are *reading the puddle* for the correct fill at the sides. This helps keep the bead uniform. If you do not hold on the corners enough, the bead may sag too much in the center (Figure 7–12). This technique, like the whip and pause, requires considerable practice—watch closely when your instructor demonstrates. There are other techniques used by some welders, but the methods described previously are the basics. Learn these first!

WELDING ADJUSTMENTS (THE FIVE VARIABLES)

The five variables are five basic changes or adjustments that can easily be made in the application of SMAW to correct the bead.

1. Electrode size
2. Amperage or heat
3. Electrode angle
4. Speed of travel
5. Arc length

Electrodes that are too large may stick or be hard to start. The final bead will usually be too convex and overlapped. Electrodes that are too small may spatter and may even catch fire. It will also be difficult to hold a steady arc length because the electrode is being consumed too fast.

If the *amperage* (heat) is too high, spatter, undercut, and overheating of the plate will occur. If the heat is too low, the arc will be difficult to strike and the bead will be overlapped.

There are two types of *electrode angles*: (1) travel angle and (2) angle of attack (Figure 7–13). If the travel angle is too great from perpendicular to the plate, a bead with poor penetration that is too convex will result. The correct travel angle is 5 to 15° in the direction of travel. The angle of attack should always be about one-half the joint angle. For example, for a 90° tee joint, the electrode angle should be 45°. An

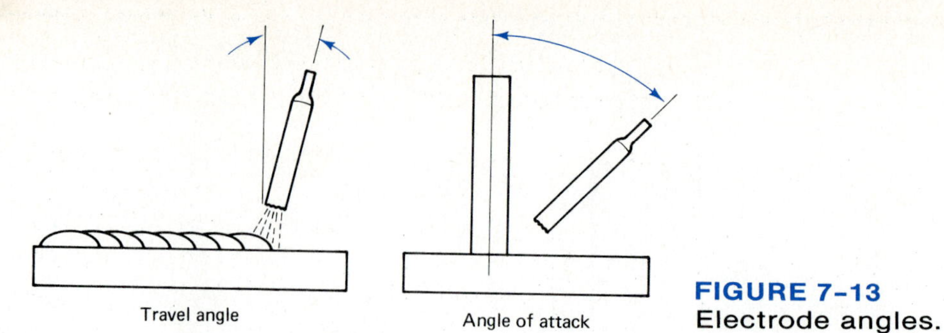

FIGURE 7–13
Electrode angles.

incorrect angle will result in uneven leg length, poor fusion, and possibly, undercut on the top leg.

If the *travel speed* is too slow, the bead will pile up and overlap will result. If it is too fast, the bead will be sparse and have poor fusion. Gaining skill in adjusting to just the right travel speed will come only from practice and watching the puddle flow out.

The *arc length* will be changing continually as the electrode is consumed. If it gets too long, an erratic arc and spatter will result. It is good to hold a close arc, especially with low-hydrogen electrodes. But if it gets much too close, the rod will not arc properly. Along with improper shielding, sticking and an erratic arc may result.

Any time beads are not coming out correctly, analyze these five variables first. Chances are that an adjustment of one of them will solve the problem (Table 7–2).

TABLE 7–2
Quick Reference Machine Setting Table for SMAW

Spatter	Decrease (shorten) arc length Decrease amperage
Overlap	Increase amperage Increase travel speed
Bead too convex	Increase amperage Increase travel speed
Undercut	Adjust electrode angle Decrease amperage Decrease travel speed Decrease (shorten) arch length
Lack of penetration	Increase amperage Increase root opening Decrease root face
Lack of fusion	Increase amperage Use slight side-to-side technique Adjust electrode angle
Porosity	Clean joint of any oil, grease, or moisture Keep close arc length Keep electrodes dry
Other variables to check	1. Correct polarity for electrodes used 2. Condition of electrodes 3. Size and type of electrodes 4. Angle and amperage of electrodes

SURFACING ELECTRODES

Surfacing or hardfacing is a welding procedure that builds up or resurfaces wear points on machinery, heavy equipment buckets and blades, mining equipment, and tools and shafts. There are many different types of electrodes for various surfacing applications. The correct choice of surfacing electrodes requires analyzing what type of wear conditions the surface must undergo while in service. There are surfacing electrodes for conditions of impact, abrasion, high-wear surfaces (such as bearing surfaces), and other conditions.

FIGURE 7–14
Surfacing symbol.

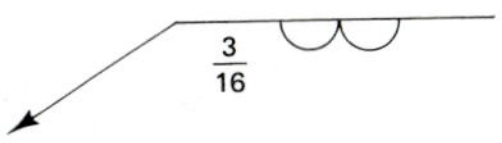

3/16

The AWS/ANSI welding symbol to indicate that surfacing is required is shown in Figure 7–14. The tail may contain information, such as the AWS electrode classification.

Electrodes designed for abrasion may contain very hard substances such as tungsten (Figure 7–15). When deposited, the tungsten carbide is held "in suspension" in the deposited weld. Tungsten carbide wears much longer than most materials. Eventually, the first layer of the tungsten carbide will wear down, and the next layer takes over to resist abrasion. These are known as composite surfacing electrodes.

The tungsten carbide electrode or filler rods are classified by the size of the tungsten carbide particles (Figure 7–16). It is based on the U.S. Standard mesh sizes; the lower the number, the larger the tungsten carbide size. An example classification is shown in Figure 7–17.

The electrode classes (SMAW) are:

EWC-12/30
EWC-20/30
EWC-30/40
EWC-40
EWC-40/120

FIGURE 7–15
Surfacing beads on dozer blade.

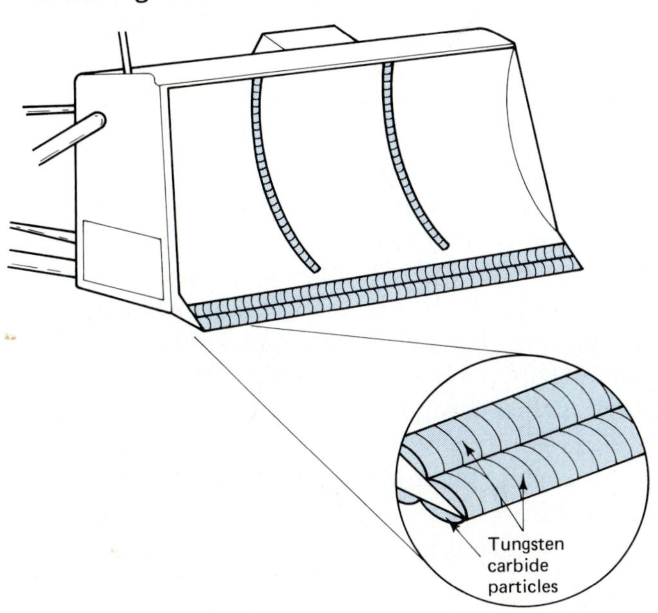

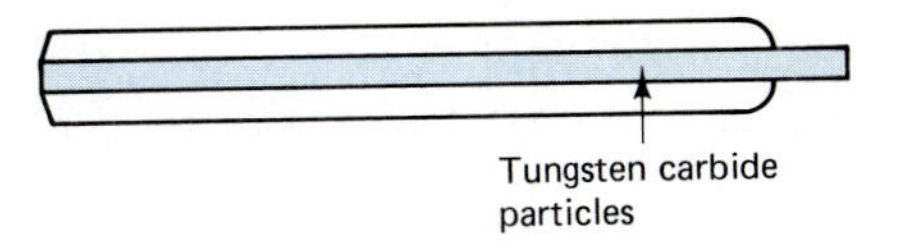

FIGURE 7–16
Surfacing electrode.

FIGURE 7–17
Classification system for tungsten carbide electrodes.

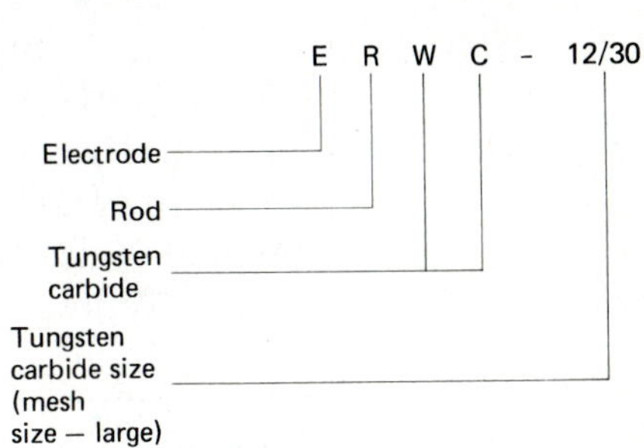

The filler rod classes (OFW) are:

RWC-5/8
RWC-8/12
RWC-12/20
RWC-20/30
RWC-30/40
RWC-30
RWC-40
RWC-40/120

The tungsten carbide makes up about 60% of the total composite material. Specific applications of tungsten carbide composite surfacing electrodes and rods include dozer and snowplow blades and cutting teeth on mining equipment or earthmoving shovels.

Surfacing electrodes and filler rods of the standard type (noncomposite type) vary in their applications. To decide what surfacing electrode or rod should be used for a given application, we must examine what type of wear it undergoes and what elements resist these types of wear stresses.

Copper–aluminum-based alloys: Good impact and wear properties for gears, cam lubes, and bearings

Nickel–chromium alloys: good corrosion resistance and wear properties for valves, seal rings, and cams

Iron–manganese alloys: good for high-impact and high-wear surfaces, such as railroad rails

Cobalt–chromium and iron alloys: good for high-temperature and high-corrosion surfaces such as engine valves, seats, and shafts

Iron–carbon and low alloys: good where toughness and hardness is needed, such as shear blades, cutting tools, and dies

Procedure for Surfacing

Surfaces should be free of any foreign material, including grease, oil, and dirt. Feather out any deep nicks or gouges on surfaces.

Preheating is recommended any time the carbon equivalent of the material's surface exceeds 0.40% (see Chapter 19). Before you start to do any welding, you should check the electrode manufacturer's recommended amperage setting. At this point you should also decide on a bead pattern. Weave and stringer beads are commonly used for surfacing. Higher-carbon-steel surfaces, such as shear blades, will be less likely to crack if stringer beads are used. Rapid buildups can be accomplished using weave and buttering techniques on other surfaces.

Where total coverage surfacing is required, sequencing is necessary to balance the heat input and reduce distortion. A typical sequencing technique is shown in Figure 7–18 and typical surfacing patterns are shown in Figure 7–19.

For many applications where toughness and hardness are required, such as shear blades, dies, and punches, a soft bedding layer is first deposited (Figure 7–20). This layer will act as a cushion for the impact surface, keeping it from cracking. Any compatible weld metal that is softer than the surfacing weld material will do. But some maintenance electrode manufacturers make special electrodes for this purpose.

Selecting the Surfacing Electrode

Applications such as bearing surfaces, gear teeth, and cam lubes (Figure 7–21) use electrode and fillers high in copper, aluminum, iron, and silicon (electrode classes ECu-Al-A, ECu-Si, ECu-Sn, and others). Materials requiring moderately hard surfaces, such as rollers, pulleys, wear pads, and cable drums (Figure 7–22), use electrodes and rods containing chromium, manganese, tungsten, vanadium, and molybdenum (electrode classes RFe 5-A, RFe 5-B, EF3 5-A, EFe 5-B, and others). Materials requiring very hard wear, high-resistance surfaces, such as shear blades, punch dies, augers, and stone crushers (Figure 7–23), contain chromium, carbon, cobalt, nickel, molybdenum, and iron (electrode classes RCo-Cr-A, RCo-Cr-B, ECo-Cr-A, ECo-Cr-B, and others).

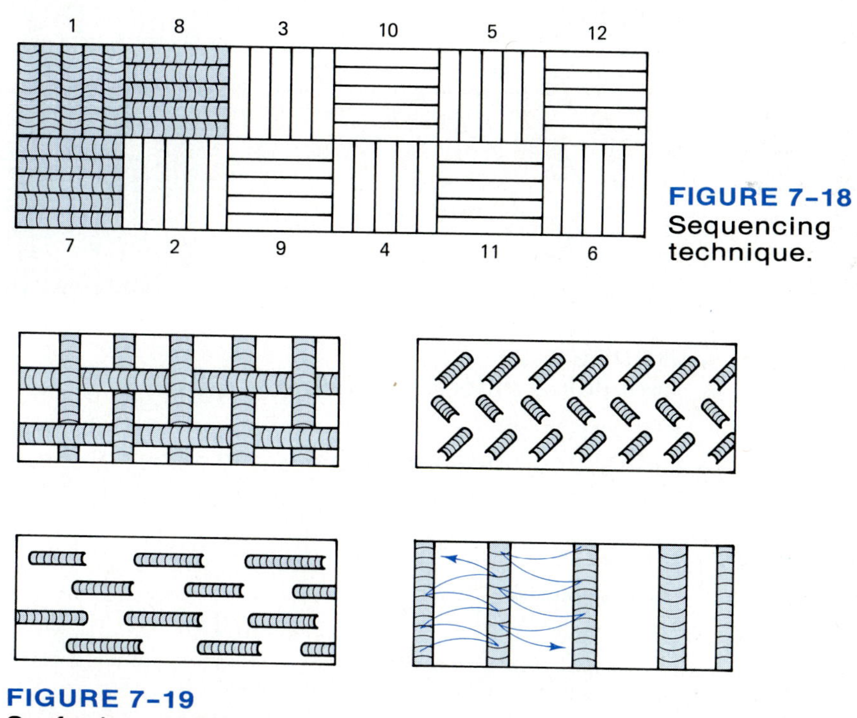

FIGURE 7–18
Sequencing technique.

FIGURE 7–19
Surfacing patterns.

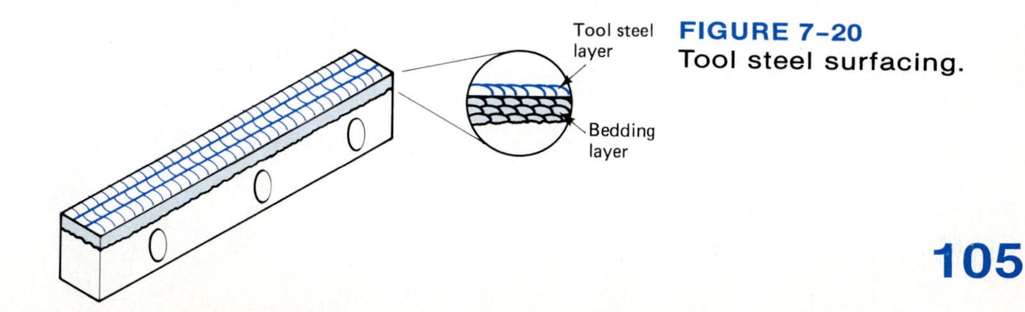

FIGURE 7–20
Tool steel surfacing.

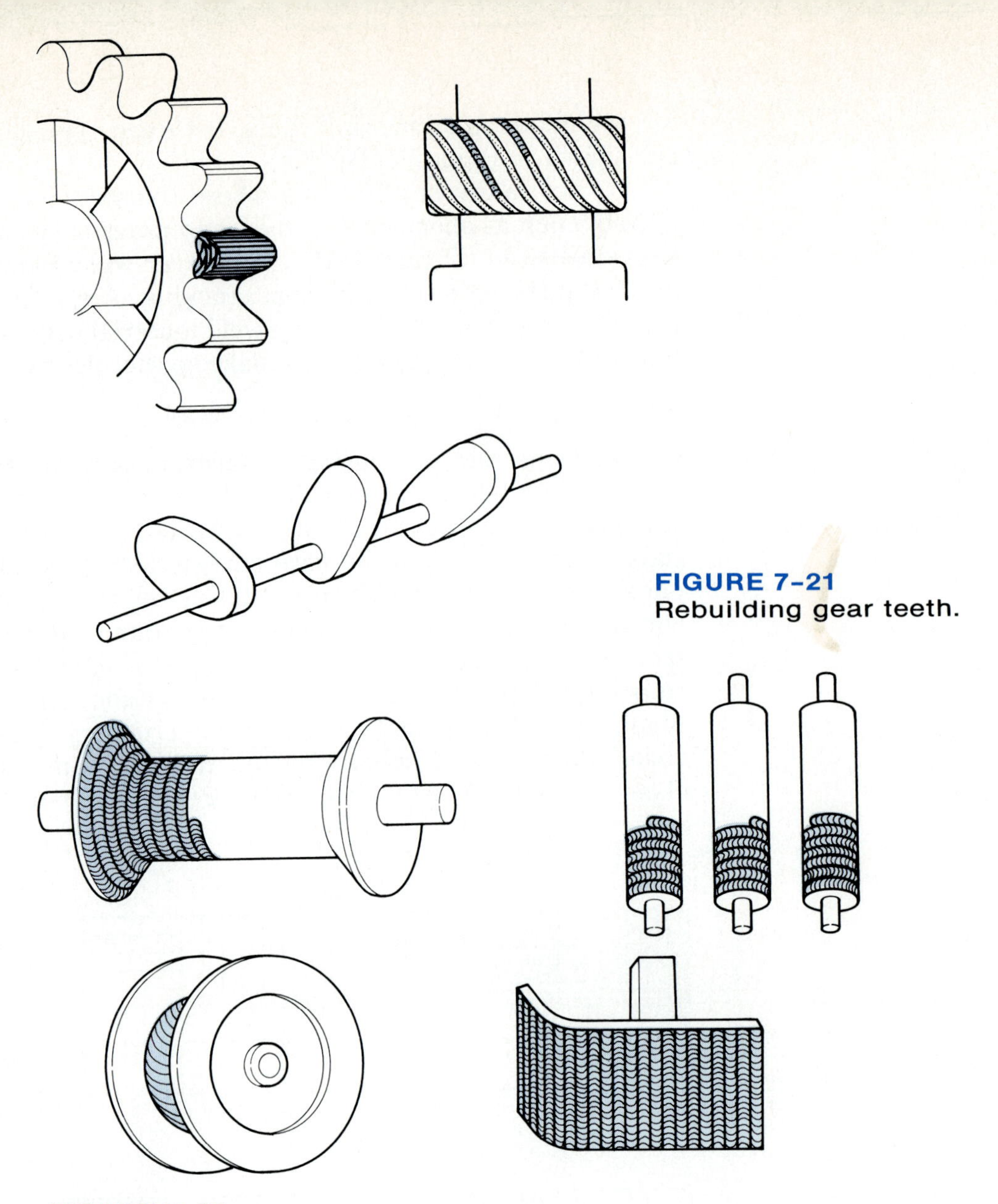

FIGURE 7-21
Rebuilding gear teeth.

FIGURE 7-22
Rebuilding rollers and wear surfaces.

FIGURE 7-23
Rebuilding punches, augers, and shear blades.

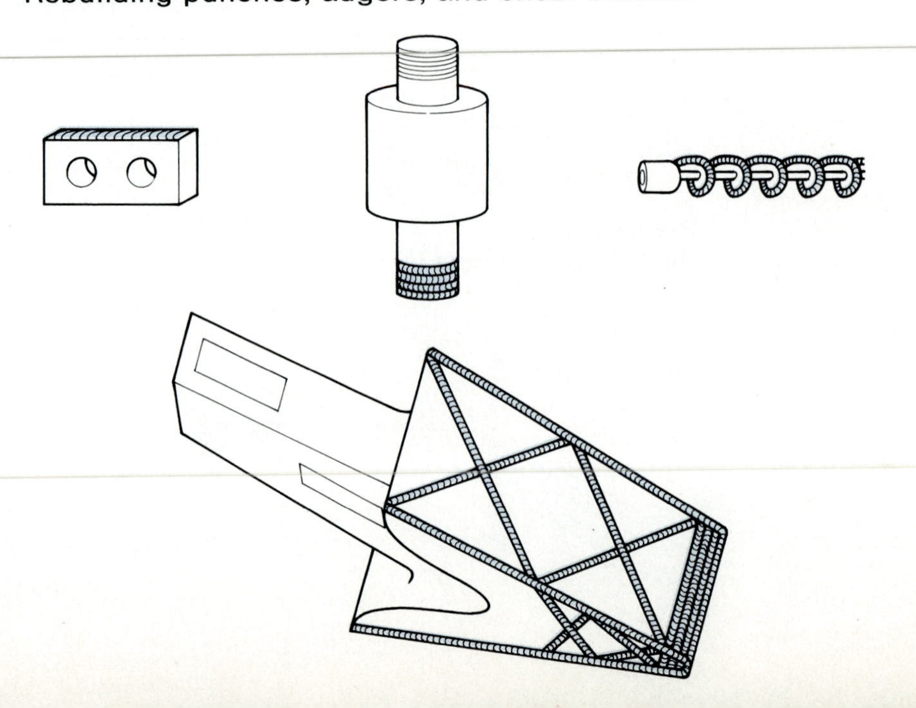

REVIEW QUESTIONS

1. What are typical applications of the fast fill electrode classifications?
2. Why are fast freeze classifications good for out-of-position welding?
3. What is the purpose of electrode flux?
4. What is the effect of iron powder in electrode fluxes?
5. What does the E, 70, 1, and 8 signify in an E7018 electrode?
6. What are the application differences between a 308-15 and a 308-16 stainless steel electrode?
7. What are the problems you may encounter when welding stainless steel?
8. How far down should electrodes be consumed?
9. List the 5 variables for shield metal arc welding.
10. What is the effect of travel speed weld beads.
11. What adjustments would you make to correct for weld spatter?
12. What is surfacing, or hardfacing, used for in industry?

SMAW PROJECT 1

Flat Fillet, E6010

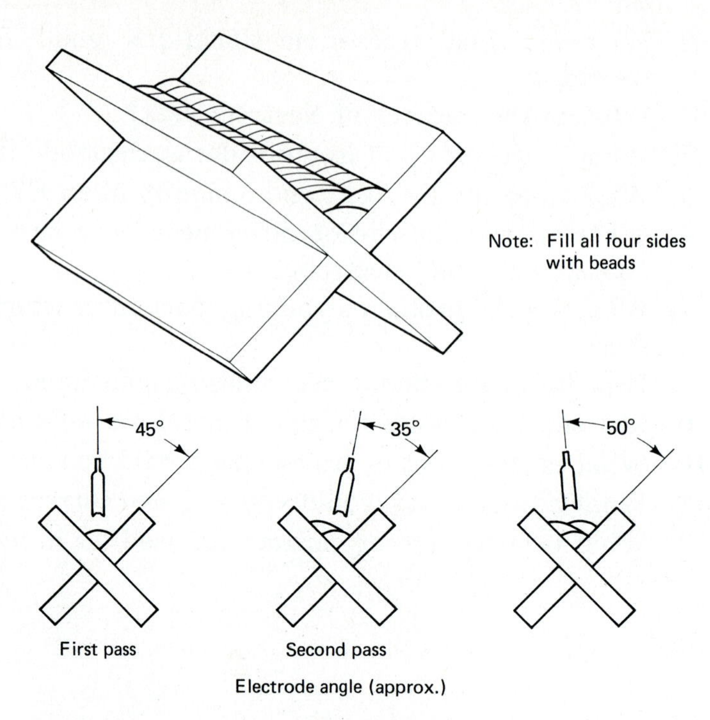

Material: A-36 (mild steel)

Size: Approximately $\frac{3}{8}$ in. × 3 in. × 6 in. plate

Joint Design: Multipass fillet

Position: Flat fillet (1F)

Technique: Stringers

Polarity: DCRP (DCEP)

Electrode: E6010, $\frac{1}{8}$-in. or $\frac{5}{32}$-in. diameter

Amperage: $\frac{1}{8}$-in. electrode, 80–120A
$\frac{5}{32}$-in. electrode, 95–135A

Initial and Interpass Cleaning: Chipping hammer and wire brush

Procedure: Set up and tack a double-sided tee joint. This gives four joints for welding. Deposit at least three beads in each of the four joints, rotating the plate into the 1F (flat fillet) position each time.

Visual Inspection: Beads should be smooth and clear of any visible porosity, slag, or cracks. Light spatter is acceptable, but large spatter should be removed with the chipping hammer and wire brush.

SMAW PROJECT 2

Horizontal Fillet, E6010

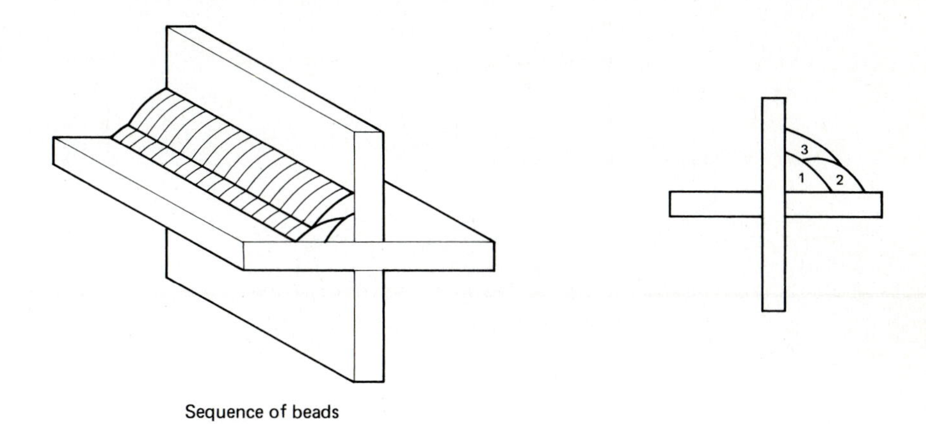

Sequence of beads

Material: A-36 (mild steel)

Size: Approximately $\frac{3}{8}$ in. × 3 in. × 6 in. plate

Joint Design: Multipass fillet

Position: Horizontal fillet (2F)

Technique: Stringers

Polarity: DCRP (DCEP)

Electrode: E6010, $\frac{1}{8}$ -in. or $\frac{5}{32}$ -in. diameter

Amperage: $\frac{1}{8}$ -in. electrode, 80–120A
$\frac{5}{32}$ -in. electrode, 95–135A

Cleaning: Chipping hammer and wire brush

Procedure: Set up and tack a double-sided tee joint. This will give you four joints for welding. Deposit at least three beads in each of the four joints, rotating the plate into the 2F (horizontal fillet) position each time.

Visual Inspection: Beads should be smooth and clear of any visible porosity, slag, or cracks.

SMAW PROJECT 3

Overhead Fillet, E6010

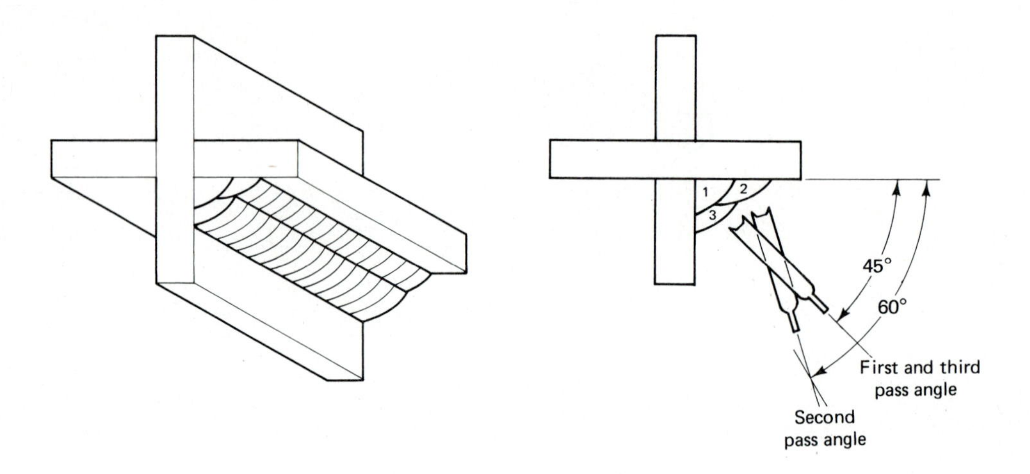

Material: A-36 (mild steel)

Size: $\frac{3}{8}$ in. × 3 in. × 6 in. plate

Joint Design: Multipass fillet

Position: Overhead fillet (4F)

Technique: Stringer beads

Polarity: DCRP (DCEP)

Electrode: E6010, $\frac{1}{8}$-in. or $\frac{5}{32}$-in. diameter

Amperage: $\frac{1}{8}$-in. electrode, 80–120A
$\frac{5}{32}$-in. electrode, 95–135A

Cleaning: Chipping hammer and wire brush

Procedure: Set up and tack a double-sided tee joint. This gives four joints for welding. Deposit at least three beads in each of the four joints, rotating the plate into the 4F (overhead-position fillet) each time.

Visual Inspection: Beads should be smooth and clear of visible discontinuities or porosity. Spatter should be very light in the 4F position.

SMAW PROJECT 4

Vertical Fillet, E6010

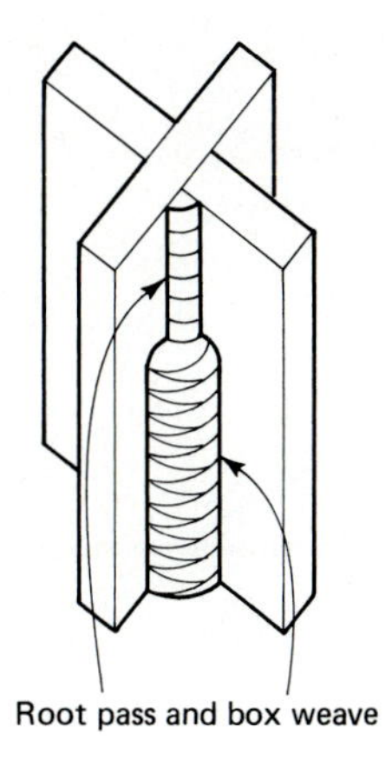

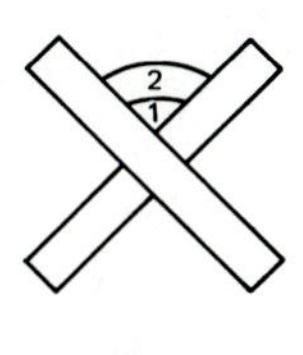

Material: A-36 (mild steel)

Size: $\frac{3}{8}$ in. × 3 in. × 6 in. plate

Joint Design: Multipass fillet

Position: Vertical fillet (3F)

Technique: Stringer (root pass) and weave (cover pass)

Polarity: DCRP (DCEP)

Electrode: E6010, $\frac{1}{8}$ -in. or $\frac{5}{32}$ -in. diameter

Amperage: $\frac{1}{8}$ -in. electrode, 75–120A
$\frac{5}{32}$-in. electrode, 90–130A

Cleaning: Chipping hammer and wire brush

Procedure: Set up and tack a double-sided tee joint. This gives four joints for welding. Deposit one stringer bead in the root pass. Then deposit a box weave three-fourths of the way up the joint for the cover pass, leaving one-fourth of the plate unwelded.

Visual Inspection: Beads should be examined for incomplete fusion at the sides. Beads should be overlapping, but not too convex. Restarts should be well tied in and free of voids.

SMAW PROJECT 5

Flat Groove, E6010

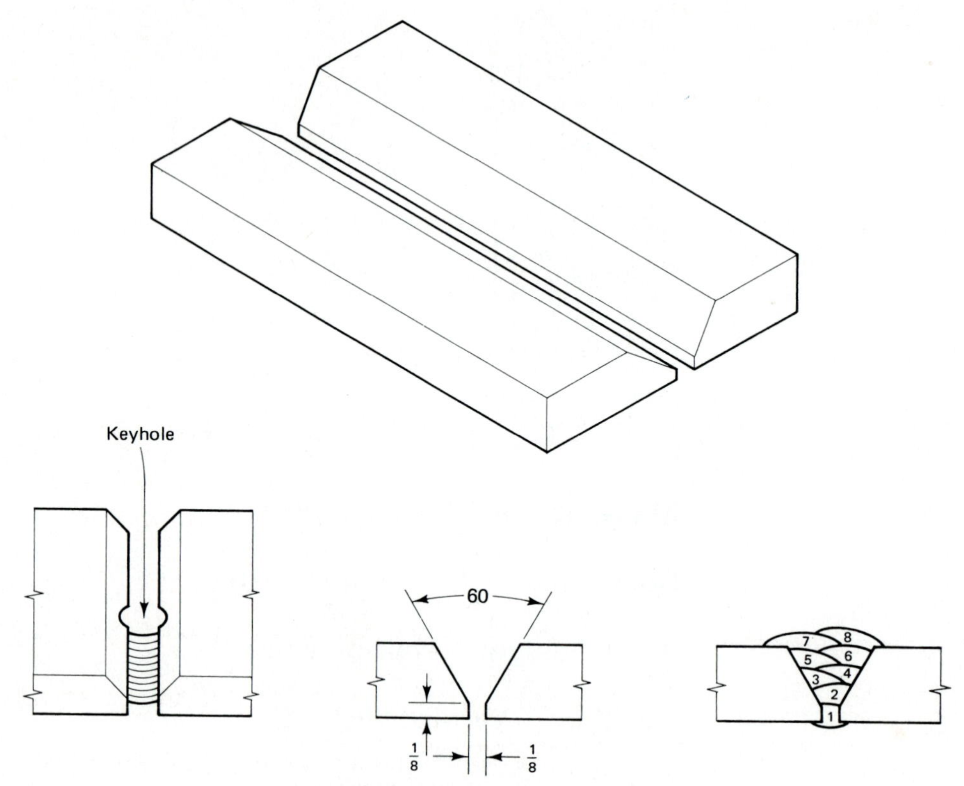

Three variables for a successful root pass:
$\frac{1}{8}$-in. electrode, $\frac{1}{8}$-in. root opening, $\frac{1}{8}$-in. root face

Material: A-36 (mild steel)

Size: $\frac{3}{8}$ in. × 3 in. × 6 in. plate

Joint Design: V-groove, 60° included angle

Position: Flat groove (1G)

Technique: Stringers

Polarity: DCRP (DCEP)

Electrode: E6010, $\frac{1}{8}$-in. or $\frac{5}{32}$-in. diameter

Amperage: Root pass, 60–75A
Fill passes and cover pass, 95–135A

Cleaning: Chipping hammer and wire brush

Procedure: Set up and tack both ends of the V-groove joint. Deposit a root pass using the keyhole technique to assure complete penetration. Use an $\frac{1}{8}$-in. root face, $\frac{1}{8}$-in. root opening, and a $\frac{1}{8}$-in. electrode E6010. Clean the root pass extremely well before depositing the "hot pass." The next pass, called the "hot pass," should be run about 15% hotter than the other fill passes. This pass will help burn out any slag that may remain trapped in the toes of the root pass. Continue to fill the groove with stringer beads until it is filled to its full cross section.

Visual Inspection: The back side of the plate at the root pass should be examined for unfused edges, suck-back, or burn-through. Convexity at the root should range from flush to $\frac{1}{16}$ in. The face of the weld should appear smooth, uniform, and free of cracks, porosity, or discontinuities. Light spatter is acceptable.

SMAW PROJECT 6

Horizontal Groove, E6010

Horizontal butt: fill joint completely

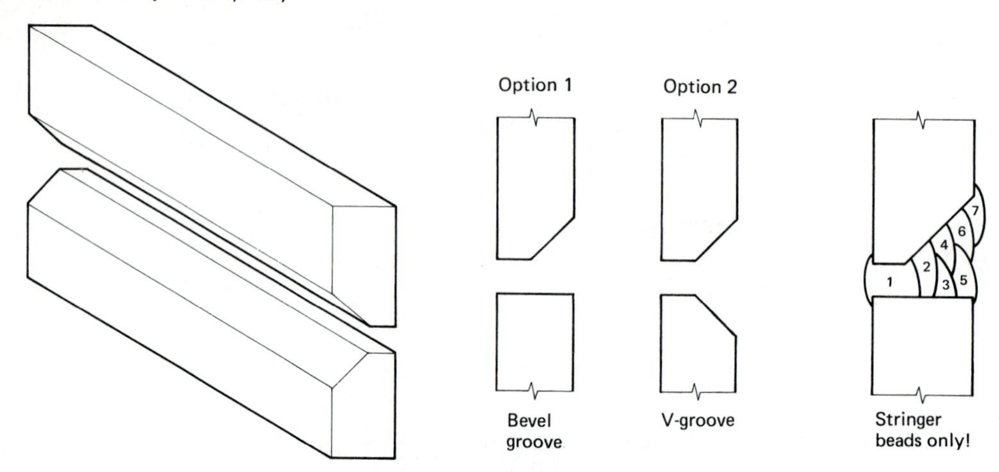

Material: A-36 (mild steel)

Size: $\frac{3}{8}$ in. × 3 in. × 6 in. plate

Joint Design: Bevel groove, 45° included angle

Position: Horizontal groove (2G)

Technique: Stringer

Polarity: DCRP (DCEP)

Electrode: E6010, $\frac{1}{8}$ -in. or $\frac{5}{32}$ -in. diameter

Amperage: Root pass, 60–75A
Filler pass, 95–135A

Cleaning: Chipping hammer and wire brush

Procedure: Set up and tack both ends of the bevel groove joint. Deposit a root pass using the keyhole technique to assure complete penetration. Clean the root pass extremely well before depositing the next pass. The next pass, called the "hot pass," should be run about 15% hotter than the other fill passes. This pass will try to burn out any slag that may remain trapped in the toes of the root pass. Continue to fill the groove with stringer beads, starting from the bottom and working up. Fill the groove to its full cross section.

Visual Inspection: The back side of the plate at the root pass should be examined for unfused edges, suck-back, or burn-through. Convexity at the face should range from flush to $\frac{1}{8}$ in. The face of the weld should be smooth, uniform, and free of cracks, porosity, or discontinuities. Light spatter on the lower plate is acceptable.

SMAW PROJECT 7

Vertical Groove, E6010

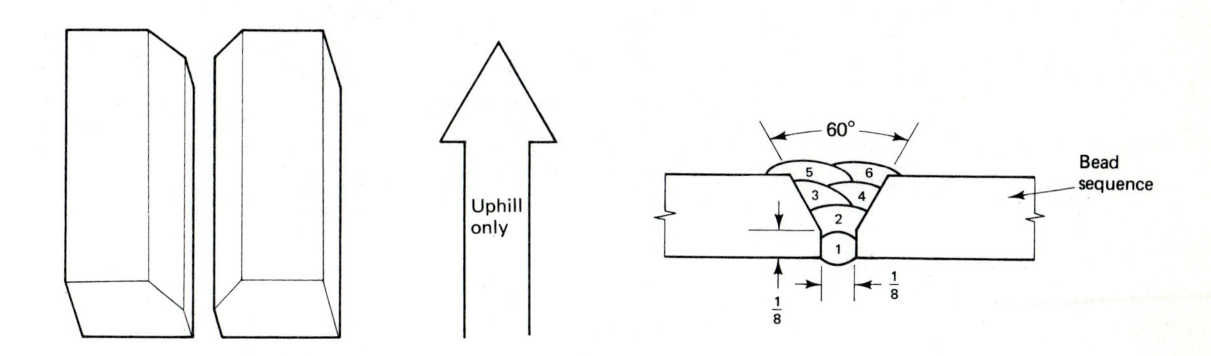

Material: A-36 (mild steel)

Size: $\frac{3}{8}$ in. × 3 in. × 6 in. plate

Joint Design: V-groove, 60° included angle

Position: Vertical groove (3G)

Technique: Stringer (root pass), weave (cover and filler passes)

Polarity: DCRP (DCEP)

Electrode: E6010, $\frac{1}{8}$-in. or $\frac{5}{32}$-in. diameter

Amperage: Root pass, 60–75A
Filler pass and cover, 95–135A

Cleaning: Chipping hammer and wire brush

Procedure: Set up and tack both ends of the V-groove joint. Deposit the root pass using the keyhole technique described earlier. Clean the root pass extremely well before depositing the next weld. The next pass, called the "hot pass," should be run about 15% hotter than the other fill passes. This pass will help burn out any trapped slag in the toes of the root pass. Continue to fill the groove using the box weave technique. Fill the groove to its full cross section.

Visual Inspection: The back side of the plate at the root pass should be examined for unfused edges, suck-back, or burn-through. Convexity at the face should range from flush to $\frac{1}{8}$ in. The face of the weld should be smooth, uniform, and free of cracks, porosity, or discontinuities. Light spatter is acceptable.

SMAW PROJECT 8

Flat Fillet, E7018

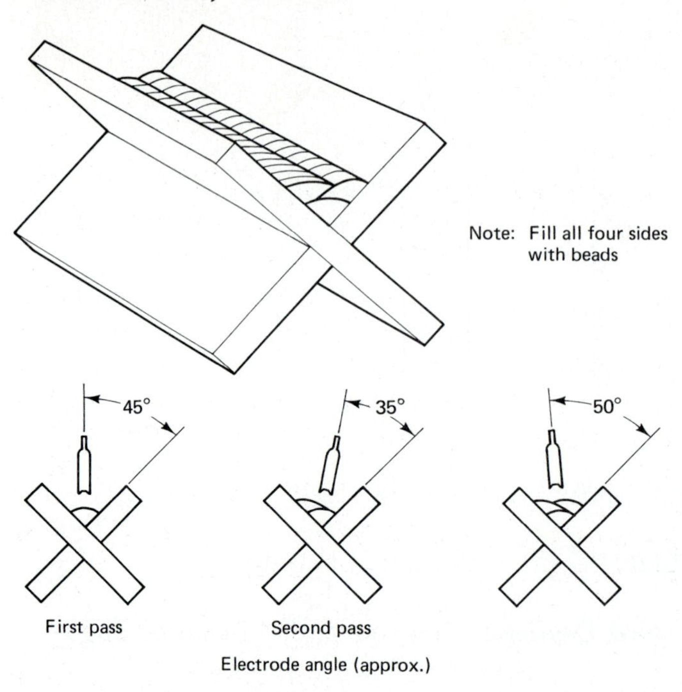

Material: A-36 (mild steel)

Size: $\frac{3}{8}$ in. × 3 in. × 6 in. plate

Joint Design: Multipass fillet

Position: Flat fillet (1F)

Technique: Stringers

Polarity: DCRP (DCEP)

Electrode: E7018, $\frac{1}{8}$ -in. or $\frac{5}{32}$ -in. diameter

Amperage: $\frac{1}{8}$ -in. electrode, 85–130A
$\frac{5}{32}$-in. electrode, 95–140A

Cleaning: Chipping hammer and wire brush

Procedure: Set up and tack a double-sided tee joint. This will give you four joints to weld in. Deposit at least three beads in each of the four joints, rotating the plate into the 1F position each time. Let the bead cool slightly before chipping E7018 slag, and make sure that you have safety glasses on and covering your eyes.

Visual Inspection: Beads should be uniform, smooth, and clear of slag, porosity, or open discontinuities. Light spatter is acceptable, but large spatter should be scraped off with the flat side of the chipping hammer.

SMAW PROJECT 9

Horizontal Fillet, E7018

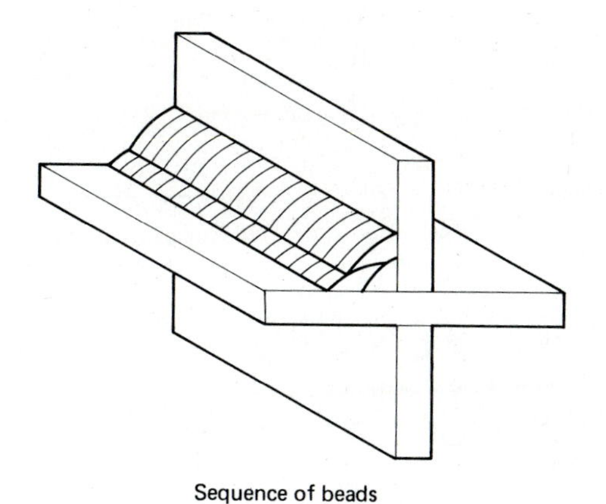

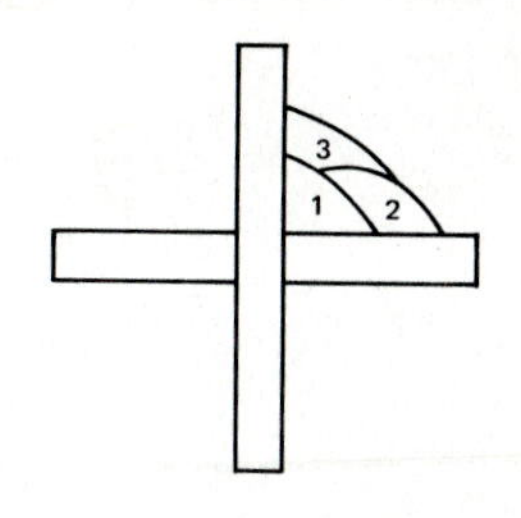

Sequence of beads

Material: A-36 (mild steel)

Size: $\frac{3}{8}$ in. × 3 in. × 6 in. plate

Joint Design: Multipass fillet

Position: Horizontal fillet (2F)

Technique: Stringers

Polarity: DCRP (DECP)

Electrode: E7018, $\frac{1}{8}$-in. or $\frac{5}{32}$-in. diameter

Amperage: $\frac{1}{8}$-in. electrode, 85–130A
$\frac{5}{32}$-in. electrode, 95–140A

Cleaning: Chipping hammer and wire brush

Procedure: Set up and tack a double-sided tee joint. This will give you four joints, rotating the plate into the 2F position each time. Let the bead cool slightly before chipping E7018 slag, and make sure that you are wearing safety glasses.

Visual Inspection: Beads should be uniform, smooth, and clear of slag, porosity, or open discontinuities. Light spatter is acceptable, but large spatter should be scraped off with the flat side of the chipping hammer.

SMAW PROJECT 10

Overhead Fillet, E7018

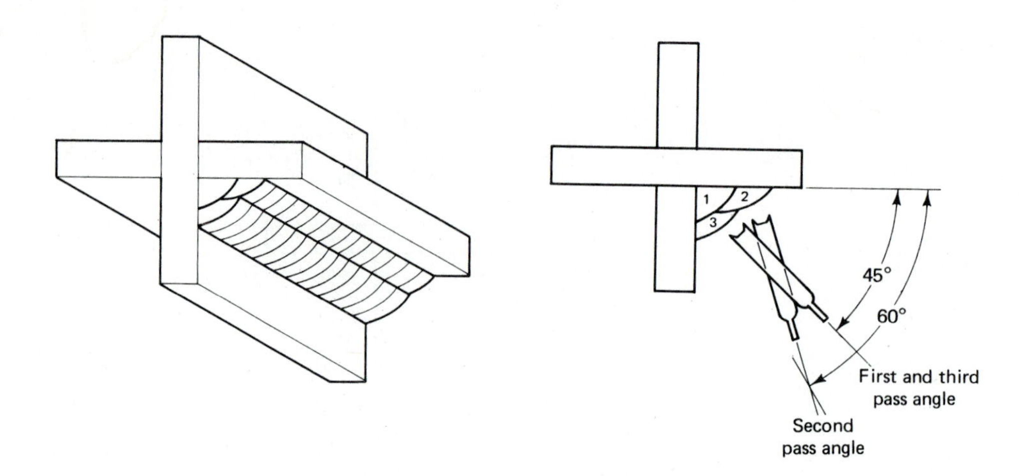

Material: A-36 (mild steel)

Size: $\frac{3}{8}$ in. × 3 in. × 6 in. plate

Joint Design: Multipass fillet

Position: Overhead fillet (4F)

Technique: Stringers

Polarity: DCRP (DCEP)

Electrode: E7018, $\frac{1}{8}$ -in. or $\frac{5}{32}$ -in. diameter

Amperage: $\frac{1}{8}$ -in. electrode, 85–130A
$\frac{5}{32}$-in. electrode, 95–140A

Cleaning: Chipping hammer and wire brush

Procedure: Set up and tack a double-sided tee joint. This will give you four joints to weld in. Deposit at least three beads in each of the four joints, rotating the plate into the 4F position each time. Let the bead cool slightly, before chipping E7018 slag, and make sure that you are wearing safety glasses.

Visual Inspection: Beads should be uniform, smooth, and clear of slag, porosity, or open discontinuities. Light spatter is acceptable, but heavy spatter should be scraped off with the flat side of the chipping hammer.

SMAW PROJECT 11

Vertical Fillet, E7018

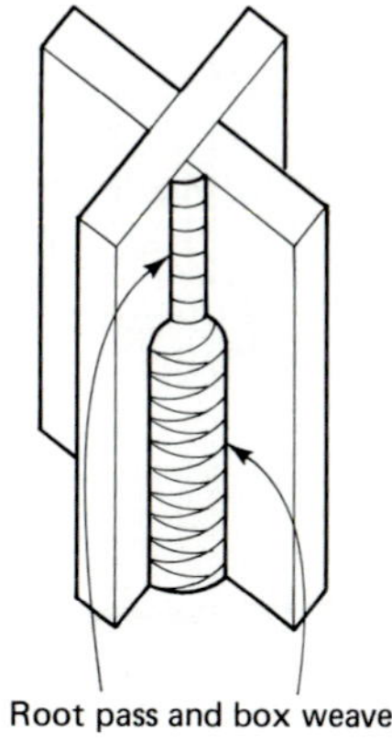

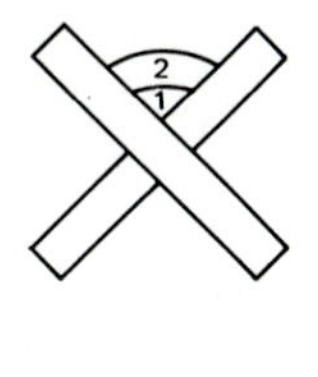

Material: A-36 (mild steel)

Size: $\frac{3}{8}$ in. × 3 in. × 6 in. plate

Joint Design: Multipass fillet

Position: Vertical fillet tee (3F)

Technique: Stringers and weave

Polarity: DCRP (DCEP)

Electrode: E7018, $\frac{1}{8}$ -in. or $\frac{5}{32}$ -in. diameter

Amperage: $\frac{1}{8}$ -in. electrode, 85–130A
$\frac{5}{32}$-in. electrode, 95–140A

Cleaning: Chipping hammer and wire brush

Procedure: Set up and tack a double-sided tee joint. This will give you four joints to weld in. Deposit one stringer in the root and one weave as the cover pass in each of the four joints, rotating the plate each time in the 3F position. Let the bead cool slightly, before chipping E7018 slag, and make sure that you are wearing safety glasses.

Visual Inspection: Beads should be uniform, smooth, and clear of slag, porosity, or open discontinuities. Light spatter is acceptable, but large spatter should be scraped off with the flat side of the chipping hammer.

SMAW PROJECT 12

Horizontal Groove, E6010 and E7018

Horizontal butt: Fill joint completely

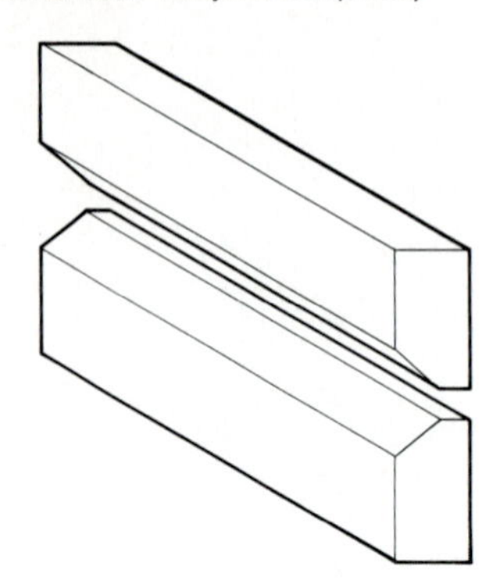

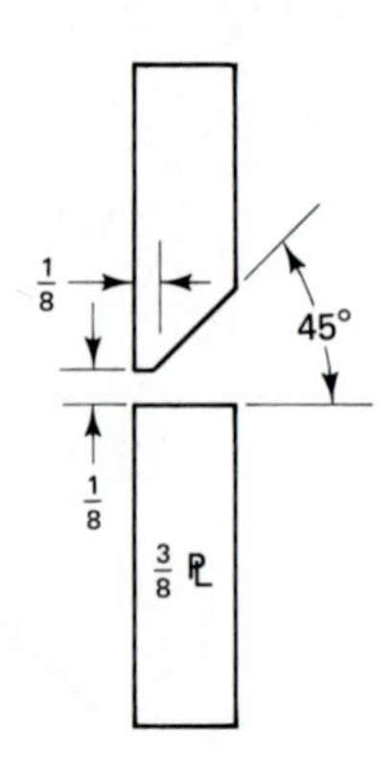

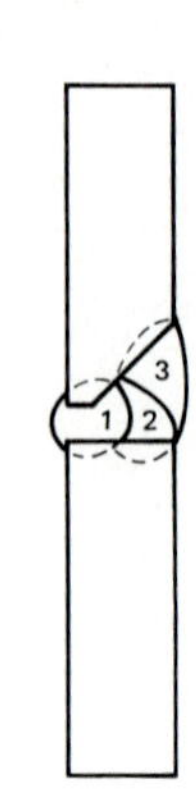

Material: A-36 (mild steel)

Size: $\frac{3}{8}$ in. × 3 in. × 6 in. plate

Joint Design: Bevel groove, 45° included angle

Position: Horizontal groove (2F)

Technique: Stringer only

Polarity: DCRP (DCEP)

Electrode: Root pass ($\frac{1}{8}$ in.), E6010
Filler passes and cover, E7018

Amperage: $\frac{1}{8}$-in. Root pass, 60–75A
Filler passes, 95–140A

Cleaning: Chipping hammer and wire brush

Procedure: Set up and tack both ends of the bevel groove joint. Deposit the root pass using the keyhole technique (this keyholing technique uses an $\frac{1}{8}$-in. root face, $\frac{1}{8}$-in. root opening, and an $\frac{1}{8}$-in. electrode). Clean the root pass extremely well before depositing the hot pass. The second pass, known as the "hot pass," is run about 15% hotter than the fill passes. This will burn out any slag trapped in the toes of the root pass. Continue to add fill passes until the groove is filled to its full cross section, working from the bottom up to the plate surface.

Visual Inspection: The back side of the plate at the root pass should be examined for unfused edges, suck-back, or burn-through. Convexity at the root should range from flush to $\frac{1}{16}$ in. The face of the weld should appear smooth, uniform, and free of cracks, porosity, or appear discontinuities. Light spatter is acceptable.

SMAW PROJECT 13

Vertical Groove, E6010 and E7018

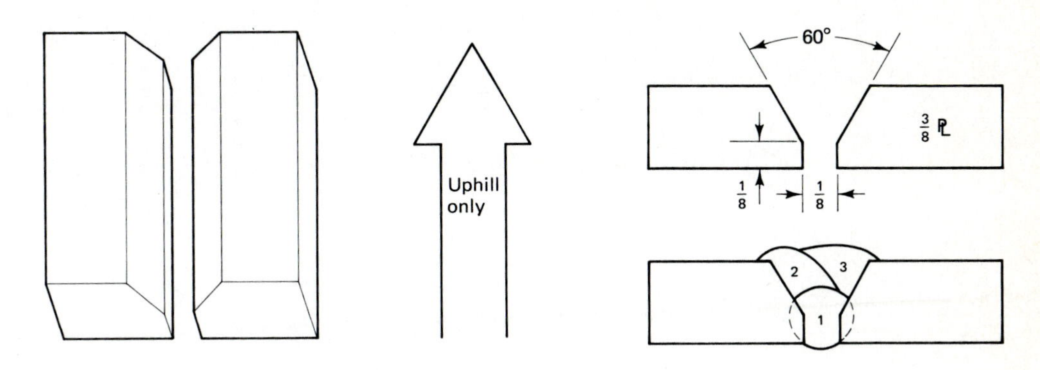

Material: A-36 (mild steel)

Size: $\frac{3}{8}$ in. × 3 in. × 6 in. plate

Joint Design: V-groove, 60° included angle

Position: Vertical groove (3G)

Technique: Stringer and weave

Polarity: DCRP (DCEP)

Electrode: Root pass, E6010
Filler passes, E7018

Amperage: $\frac{1}{8}$-in. Root pass, 60–75A
Filler passes, 95–140A

Cleaning: Chipping hammer and wire brush

Procedure: Set up and tack both ends of the Vee-groove joint. Deposit the root pass using the keyhole technique (this keyholing technique uses an $\frac{1}{8}$-in. root face, and $\frac{1}{8}$-in. root opening, and an $\frac{1}{8}$-in. electrode). Clean the root pass extremely well before depositing the hot pass. The second pass, known as the "hot pass," is run about 15% hotter than the fill passes. This will burn out any slag trapped in the toes of the root pass. Continue to add filler passes until the groove is filled to its full cross section.

Visual Inspection: The back side of the plate at the root pass should be examined for unfused edges, suck-back, or burn-through. Convexity at the root should range from flush to $\frac{1}{16}$ in. The face of the weld should appear smooth, uniform, and free of cracks, porosity, or open discontinuities. Light spatter is acceptable.

SMAW PROJECT 14

Vertical Groove 1" Plate, E6010 and E7018

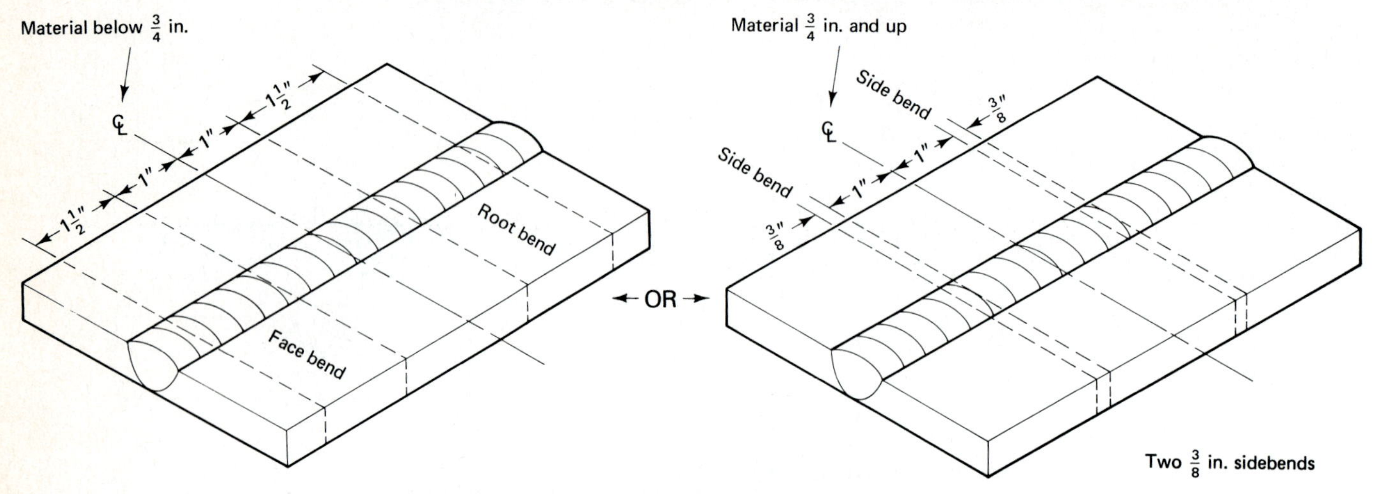

Material: A-36 (mild steel)

Size: 1 in. × 3 in. × 6 in. plate

Joint Design: V-groove, 60° included angle

Position: Vertical groove (3G)

Technique: Stringers and weave

Polarity: DCRP (DCEP)

Electrode: Root pass, E6010
Filler passes, E7018

Amperage: $\frac{1}{8}$-in. Root pass, 60–75A
Filler passes, 95–130A

Cleaning: Chipping hammer and wire brush

Procedure: Set up and tack both ends of the V-groove joint. Deposit the root pass using the keyhole technique (this keyholing technique uses an $\frac{1}{8}$-in. root face, an $\frac{1}{8}$-in. root opening, and an $\frac{1}{8}$-in. electrode). Clean the root pass extremely well before depositing the hot pass. The second pass, known as the "hot pass," is run about 15% hotter than the fill passes. This will burn out any slag trapped in the toes of the root pass. Continue to add fill passes until the groove is filled to its full cross section.

Visual Inspection: The back side of the plate at the root pass should be examined for unfused edges, suck-back, or burn-through. Convexity at the root should range from flush to $\frac{1}{16}$ in. The face of the weld should appear smooth, uniform, and free of cracks, porosity, or discontinuities. Light spatter is acceptable. Perform bend test as shown above.

GAS TUNGSTEN ARC WELDING (GTAW)

Gas tungsten arc welding (GTAW), also known as TIG (tungsten inert gas) welding, is an excellent-quality manual welding process that utilizes a nonconsumable tungsten electrode to heat and melt the workpiece. If required, filler metal can be fed into the puddle, similar to that of OFW. The molten puddle is shielded from the atmosphere with an inert gas supply feeding from the torch cup (Figure 8–1). Figure 8–2 shows a typical GTAW work cell. GTAW produces one of the highest-quality weld deposits of any manual welding process.

In this chapter we not only discuss how GTAW works, but why

FIGURE 8–1
GTAW welding.

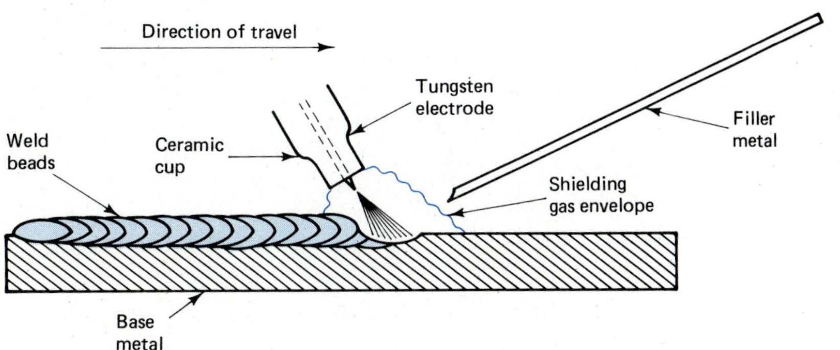

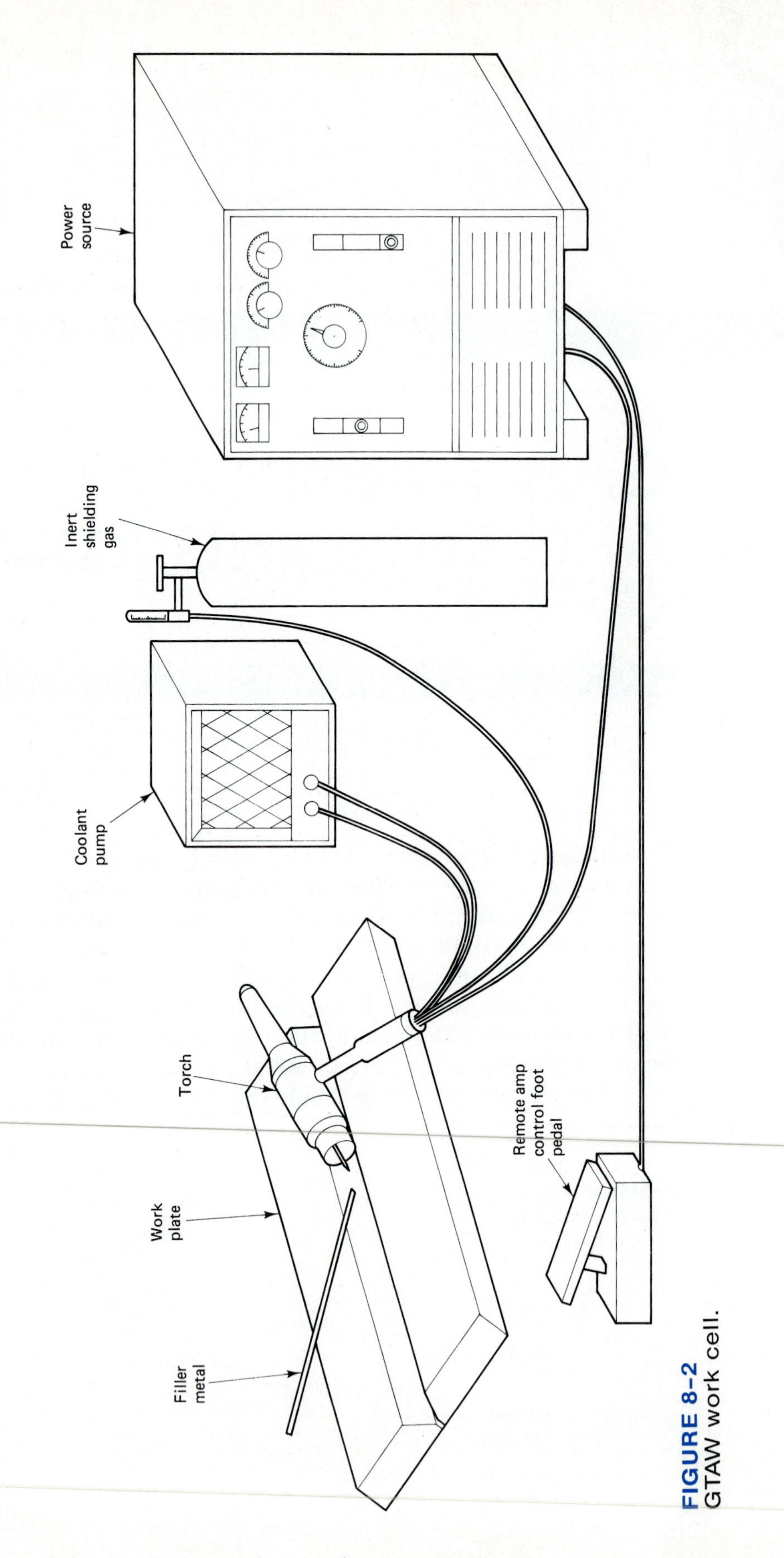

FIGURE 8-2
GTAW work cell.

it works like it does. This helps the welder, technician, or welding inspector to troubleshoot and apply the welding process more efficiently.

One of the common trade names for the GTAW process is TIG, which was the official AWS definition for the process during the 1950s and 1960s. Prior to that it was usually referred to as heliarc or heliweld. Today the official AWS definition for the process is GTAW (gas tungsten arc welding).

ADVANTAGES AND APPLICATIONS OF GTAW

The GTAW process will weld more different types of metals and alloys than will any other manual process. With GTAW we can weld most carbon steels and stainless steels, nickel steels (monel, inconel, etc.) aluminum, magnesium, copper, brass and bronze, titanium, and many others. We can also weld many unlike metals to each other: for example, mild steel to stainless steel, or brass to copper.

Other advantages of this process are the very low heat-affected zone produced by these processes (Figure 8-3). The heat-affected zone is the weak link in any weld produced by fusion welding. It is the area where a properly made weld should break. This is due to the enlarged grain structure, which produces a slightly brittle microstructure in this superheated zone. The area is actually superheated by the neighboring molten weld metal, but does not melt itself. At the same time it receives a rush or quench of cold base metal from the other side (Figure 8-4). Remember, a properly made weld should break in the heat-affected zone.

If a weld fails in any other zone it is usually the result of im-

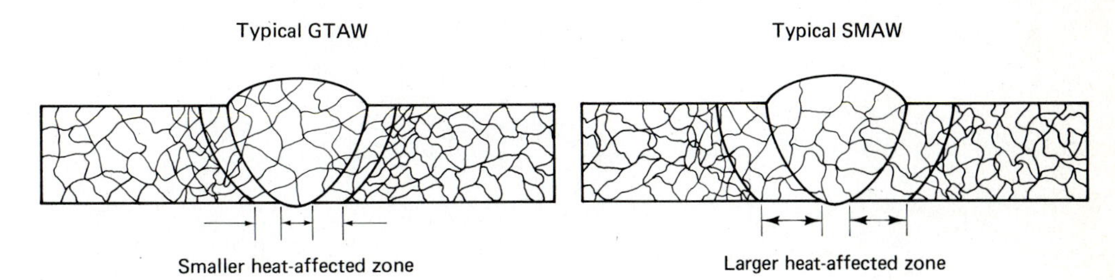

FIGURE 8-3
SMAW versus GTAW welds.

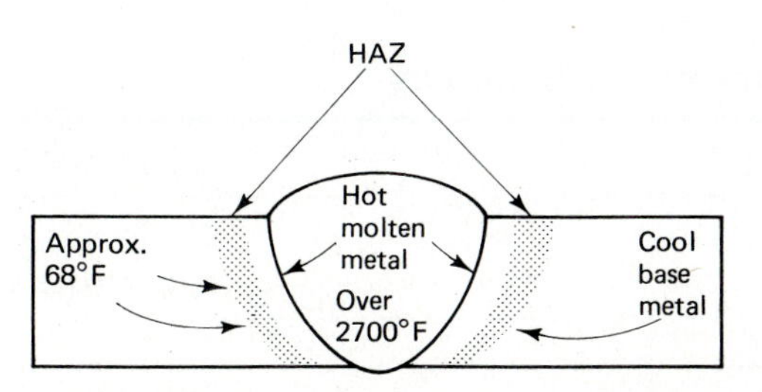

FIGURE 8-4
Temperature extremes on the heat-affected zone.

proper welding techniques (i.e., preparation or application of the process). For example, a break in the fusion zone usually indicates a lack of fusion. A failure in the weld can indicate any number of problems, such as porosity, slag, or tungsten inclusions. A failure in the base metal could be the result of poor steel from the mill (these defects would include laminations, seams, laps, etc.)

The GTAW process does not require filler metal to pass through the superheated electric arc, because filler is dipped directly into the puddle. In other electric arc processes filler is passed through the arc. This can vaporize small quantities of alloying elements out of the filler and result in a lower grade of weld deposit.

The GTAW process requires almost no cleanup; there is no slag or spatter. Ventilation requirements are minor due to the slight amount of smoke produced.

There are many cases when GTAW are applied to a new or experimental metal or alloy; in this case there would be no manufacturing filler metal available. Scraps or shavings from these materials can be used as filler; in the case of plate or sheet, just shear a thin strip and use as filler metal. The decision to use GTAW is usually based on the quality and flexibility required for the application. Another consideration with GTAW is the need for back-purging or some method of shielding the back side on an open butt weld (Figure 8-5).

Root shield can be accomplished in a number of effective ways. The simplest method consists of only a nonconsumable backup which will trap the shielding gas around the groove face and root (Figure 8-6). The nonconsumable backing can be any material with different enough properties that it will not fuse into the base metal (Table 8-1). Another

FIGURE 8-5
Back-shielding GTAW welds.

Shielding envelope
Back shield

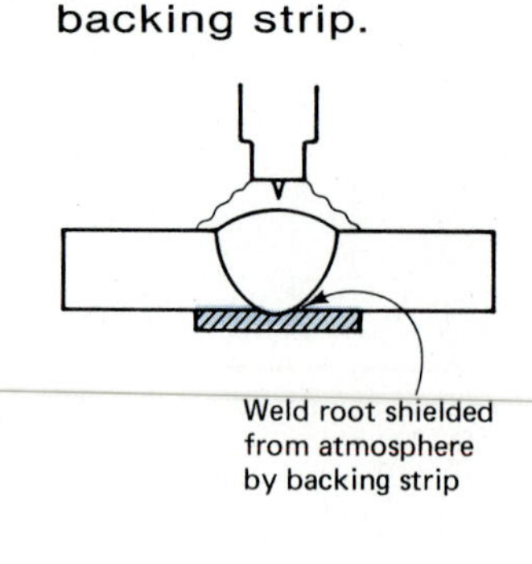

FIGURE 8-6
GTAW weld with backing strip.

TABLE 8-1
Suggested Backing for Materials

MATERIAL	SUITABLE BACKING MATERIAL
Carbon and stainless steels	Copper, aluminum, or ceramic
Aluminum	Stainless steel or ceramic

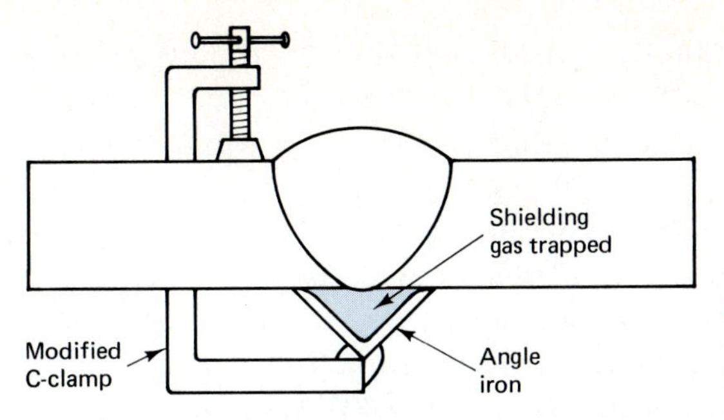

FIGURE 8-7
GTAW weld with angle iron backing.

method of back-shielding uses an angle iron to flow a stream of argon to the root of the groove (Figure 8-7). This method does not provide any help in root contour but does provide adequate shield for the root bead.

Typical applications of the GTAW process include aircraft welding of the structure, landing gear, and engine components, root pass on most nuclear reactor coolant pipes, and 100% GTAW on many nuclear applications. GTAW is used almost exclusively for stainless steel fusion welding in the food and chemical industries.

The GTAW process uses a nonconsumable tungsten electrode. These tungstens are one of the highest-melting-temperature metals known.

GTAW THEORY

The electric arc used for heating in GTAW (or any electric arc process) is known as a plasma gas. Some refer to plasma as the fourth state of matter (liquids, solids, gases, and plasma gases). A plasma gas is simply a gas that is conducting electricity. Many people talk of electricity as "jumping the gap" between the electrode and work. However, the electrons do not actually jump this distance from the electrode to the work; they travel through the ions of the gas.

Let us actually trace the path of electricity from the electrode to the work. First we must understand some basic electrical principals. *Voltage* is the pressure or potential that initially causes the dislodge of the electron in the outer orbit of the conductor (Figure 8-8). In our case the conductor will be copper, tungsten, argon, and finally steel.

FIGURE 8-8
Electron activity in the conductor.

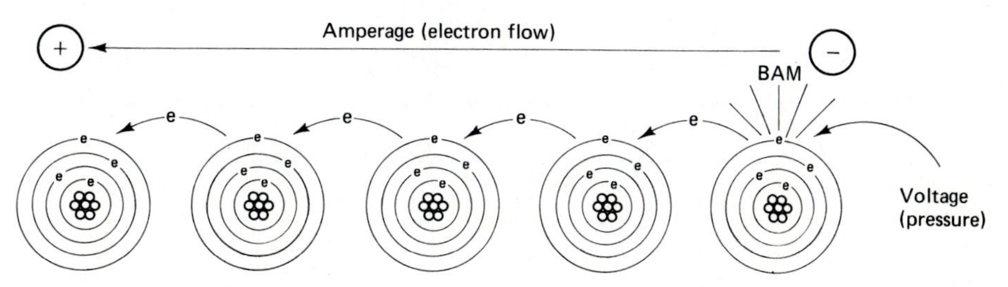

Amperage is the actual flow of the electrons from orbit to orbit. Do not confuse electricity with amperage. Electricity is the phenomenon that is produced when these electrons change orbit. Electricity travels at about the speed of light, 186,000 miles per second. The electrons do not travel that fast.

Once volts free an electron, that atom momentarily is deficient one electron and is known as an ion. This process is known as ionization. In arc welding we are ionizing the gases in the arc, and therefore creating a plasma gas. In GTAW welding we are ionizing the argon gas.

Electricity comes into the welding power source in the high voltage–low amperage state. This is relatively useless for welding; for welding we need high amperage and lower voltage. High amperage produces heat where high voltage is pressure or potential. The transformer turns the amperage up and the voltage down. It may then go to the rectifier to convert it to direct current (dc) if we intend to weld with dc.

The electricity will then travel through the copper wire conductor to the torch, where it transfers through the collet and into the tungsten electrode. From the tungsten electrode down to the work, the electricity must have sufficient voltage and ionize the gas between the tungsten and the work. Once the gas is ionized the electricity will return to the circuit through the ground cable.

Polarities for GTAW

GTAW uses all three polarities for various applications of the process (Figure 8-9):

1. DCSP (DCEN)
2. DCRP (DCEP)
3. ACHF

Direct-current straight polarity is also referred to as DCEN, direct current electrode negative. As the name implies, this polarity uses the tungsten electrode negative and the work positive (Figure 8-10). Electricity travels from negative to positive, so electrons will flow from the tungsten to the work. With DCSP, about two-thirds of the heat is at the work. This is due to the electrons decelerating at the work. DCSP is used for most GTAW applications where deep penetration and a narrow bead is required (Figure 8-11), such as mild steel, stainless steel,

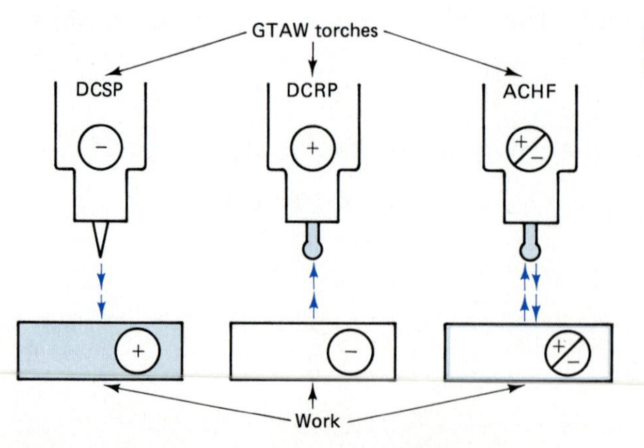

FIGURE 8-9
GTAW polarities.

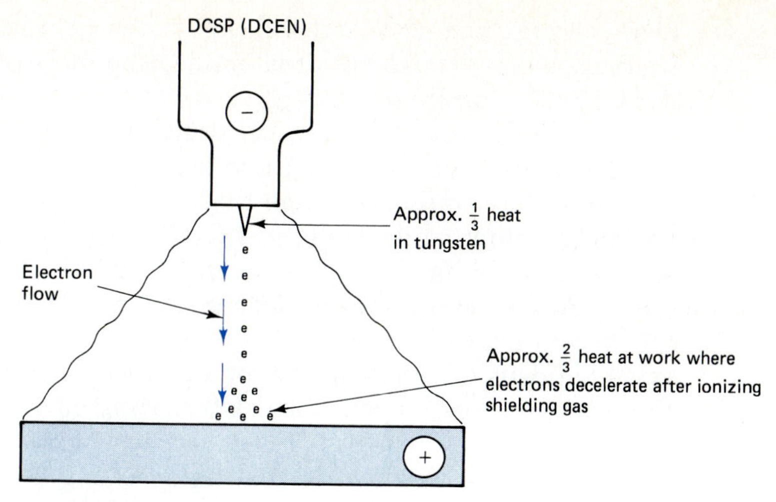

FIGURE 8–10

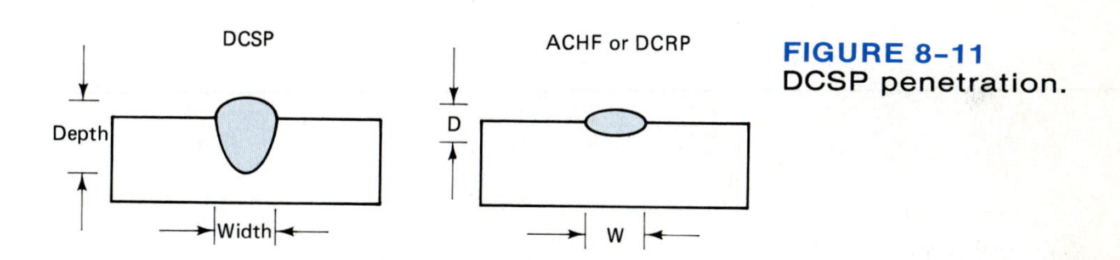

FIGURE 8–11
DCSP penetration.

copper, and titanium. DCSP should be used on the following metals: carbon steels, stainless steels, copper and copper alloys, nickel steels, brass and bronzes, titanium, and reactive and refractory metals.

Direct-current reverse polarity (DCRP) is also referred to as DCEP, direct-current electrode positive. As its name indicates, in DCRP (DCEP) the tungsten is positive and the work is negative. This puts about two-thirds of the heat at the tungsten and about one-third of the heat at the work. The electrons decelerate at the tungsten electrode, which is the positive side of the arc (Figure 8–12). This polarity is generally limited to preparing the tungsten. When welding aluminum the tungsten tip should be rounded or balled. This gives the tip much more

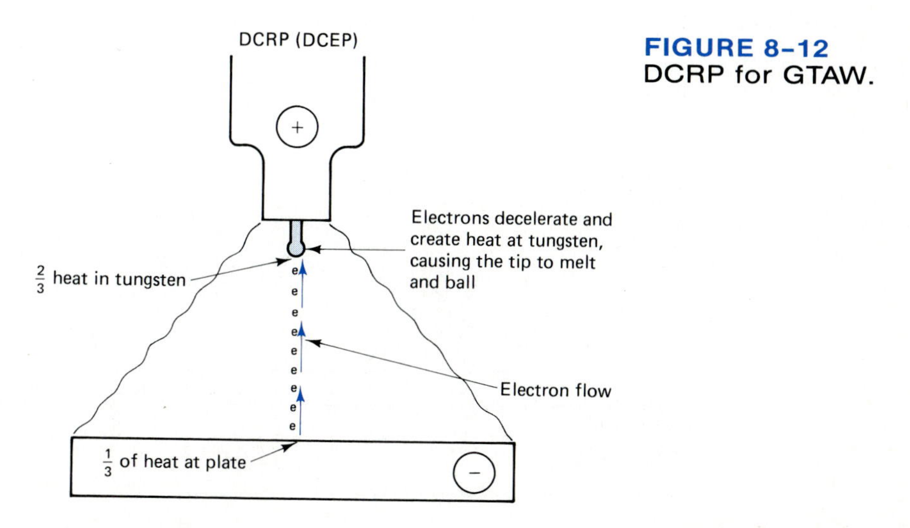

FIGURE 8–12
DCRP for GTAW.

heat-carrying capacity and will prevent the tip from melting off when welding on ac. Another limited application of DCRP would be on heavily oxidized aluminum castings.

Alternating-current high frequency (ACHF) is used primarily for the welding of aluminum and magnesium. Aluminum and magnesium both have oxide coatings that contaminate the molten aluminum and magnesium while welding. ACHF actually helps remove these oxides while welding. Let's first examine ACHF.

Ac, as the name indicates, is an actual alternating between DCSP and DCRP (Figure 8–13). In the United States this alternation cycle happens 60 times per second or 60 cycles. We use ac because in the reverse polarity cycle of the alternating current (when the electrons leave the plate; Figure 8–14) they actually cut away the aluminum (Al) and magnesium (Mg) oxide coatings, leaving a clean uncontaminated weld. DCRP cannot be used because the concentration of heat into the tungsten would melt the tungsten.

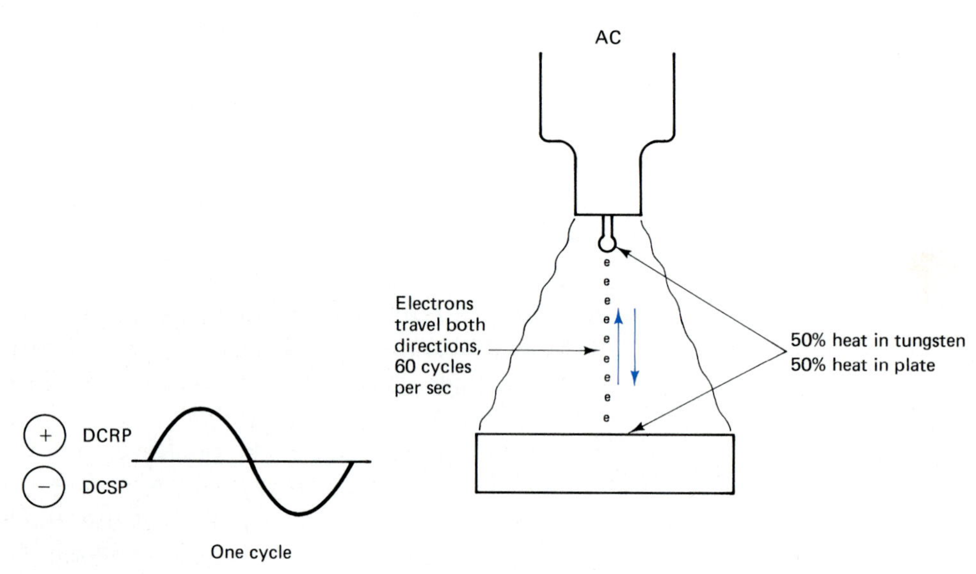

FIGURE 8–13
Ac for GTAW.

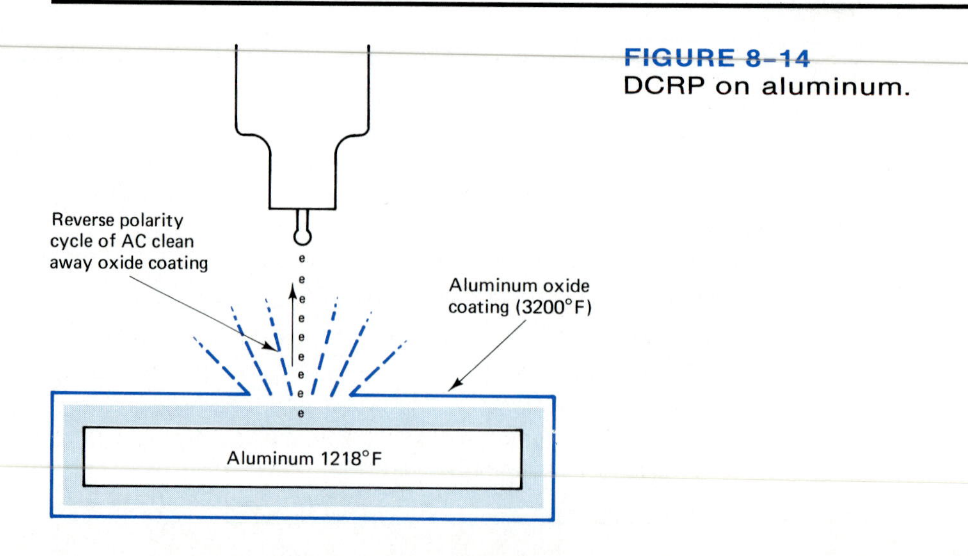

FIGURE 8–14
DCRP on aluminum.

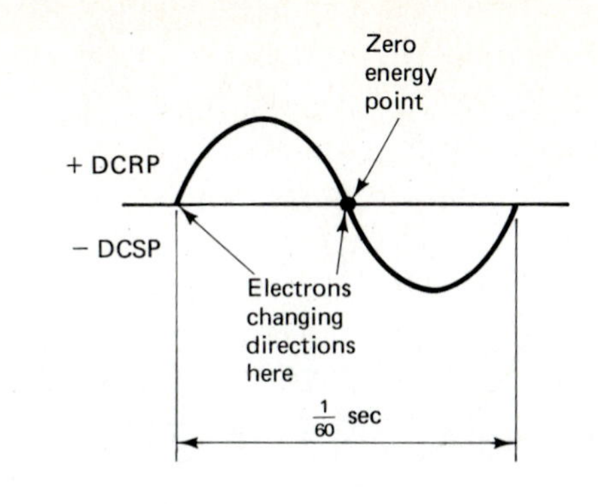

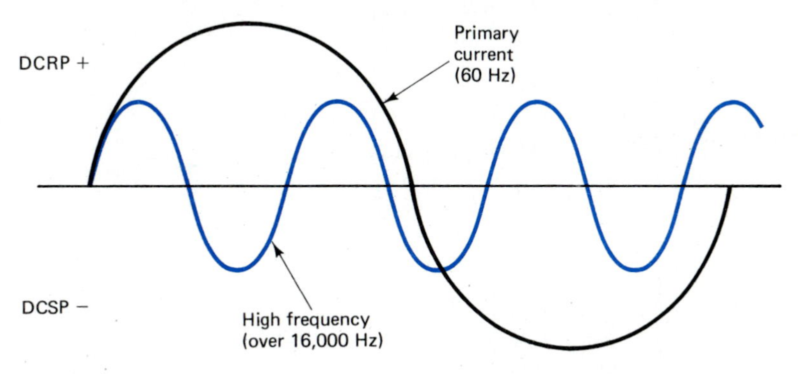

FIGURE 8–15
Ac high frequency.

Ac is, however, not used alone. It must utilize a superimposed (which means to add over top of) high-frequency circuit (Figure 8–15). This high frequency (HF) is used to eliminate the effects of the zero point in ac welding. As ac alternates between DCRP and DCSP, it reaches a point of no energy during this change. Attempting to weld without HF on ac will make arc starting and maintaining an arc very difficult. HF may be 15,000 to 22,000 cycles per second. Therefore, when the primary 60-cycle current is going through its zero point, the HF may go through many cycles, thus preventing the arc from stopping. These extra HF pulses in the reverse-polarity mode are also helping the oxide cleaning action.

GTAW TUNGSTEN ELECTRODES

There are five basic types of GTAW tungstens. The proper selection of these tungstens will make for the best quality and economic application of the GTAW process. Tungsten is utilized as the nonconsumable electrode in the GTAW process because it is one of the highest-melting-temperature metals known. It also has an abundance of electrons, which makes it an excellent conducting electrode. Electron boil-off temperature for pure tungsten is 5400°F. Boil-off is the point at which electrons are easily emitted from the tungsten for ionization. The melting temperature is 6160°F (3392° C). Current ranges are given in Table 8–2.

TABLE 8-2
Typical Current Ranges for Tungsten Electrodes[a]

ELECTRODE DIAMETER		DCSP (A): EWP, EWTH-1, EWTH-2, EWTH-3	DCRP (A): EWP, EWTH-1, EWTH-2, EWTH-3	HIGH-FREQUENCY UNBALANCED WAVE, AC (A)			HIGH-FREQUENCY BALANCED WAVE, AC (A)		
IN.	MM			EWP	EWTH-1, EWTH-2, EWZR	EWTH-3	EWP	EWTH-1, EWTH-2, EWZR	EWTH-3
0.010	0.25	Up to 15	[b]	Up to 15	Up to 15	[b]	Up to 15	Up to 15	[b]
0.020	0.5	5–20	[b]	5–15	5–20	[b]	10–20	5–20	10–20
0.040	0.1	15–80	[b]	10–60	15–80	10–80	20–30	20–60	20–60
$\frac{1}{16}$	1.6	70–150	10–20	50–100	70–150	50–150	30–80	60–120	30–120
$\frac{3}{32}$	2.4	150–250	15–30	100–160	140–235	100–235	60–130	100–180	60–180
$\frac{1}{8}$	3.2	250–400	25–40	150–210	225–325	150–325	100–180	160–250	100–250
$\frac{5}{32}$	4.0	400–500	40–55	200–275	300–400	200–400	160–240	200–320	160–320
$\frac{3}{16}$	4.8	500–750	55–80	250–350	400–500	250–500	190–300	290–390	190–390
$\frac{1}{4}$	6.4	750–1000	80–125	325–450	500–630	325–630	250–400	340–525	250–525

[a] All values are based on the use of argon as the shielding gas. Other current values may be used, depending on the shielding gas, type of equipment, and application.

[b] These particular combinations are not commonly used.

Source: American Welding Society.

Tungsten Electrode Classifications

Tungsten electrodes have classifications just as to SMAW or other types of electrodes. It would be difficult to print these AWS classifications on the tungsten electrodes themselves because of the small diameters commonly used for GTAW. For this reason color codes are used on these electrodes; the AWS class is shown on the end of the electrode box. The system works as follows: E stands for electric and W stands for tungsten (W is the chemical symbol for tungsten) these are followed by P, TH1, TH2, TH3, or ZR (Figure 8–16a).

1. Pure tungsten, EWP (green)
2. 1% Thoriated EWTH1 (yellow)
3. 2% Thoriated EWTH2 (red)
4. Striped EWTH3 (blue)
5. Zirconium EWZR (brown)

In *pure tungsten* no additional elements are alloyed into the tungsten. It is identified by a green color-coded stripe on the end of the tungsten (Figure 8–16b). It will operate effectively only on ACHF and is designed for aluminum and magnesium welding only. The reason for these limitations is pure tungsten's melting point, 6160°F. Attempting to use it on DCSP with other metals will round the sharpened point and an unstable arc will result. When preparing to weld aluminum or magnesium, the pure tungsten tip should be rounded. This will give it more heat-carrying capability and the tungsten will not melt.

The 2% *thoriated tungsten* (EWTH2) can be used for almost any metal or tungsten preparation (sharpened or rounded). However, it is more expensive than pure tungsten and thus should be used on metals that require a higher melting temperature. The addition of thorium oxide (thoria) elevates the tungsten's melting point to 6290°F. It also lowers electron boil-off temperature. The higher melting temperature allows the sharpened point to remain on the tip of the tungsten while welding. The 2% thoriated tungstens have a red color code (Figure 8–16c). It is designed primarily for DCSP but will run in ACHF applications. This tungsten is used for welding all-carbon steels, stainless

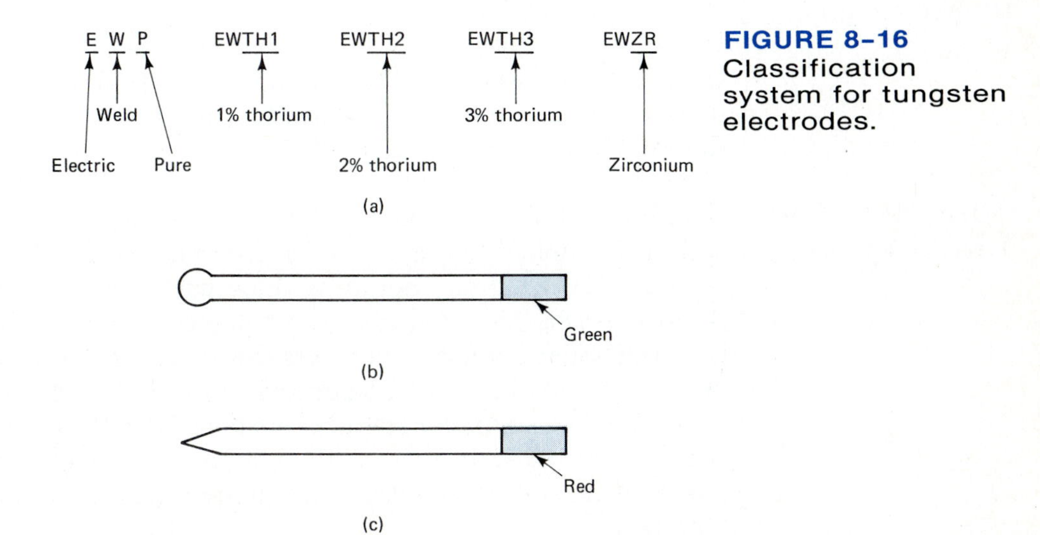

FIGURE 8–16 Classification system for tungsten electrodes.

steels, nickel alloys, titanium, and all refractory and reactive metals. The 2% thoriated tungstens will work on aluminum, magnesium, and copper. However, they are not economical because these metals will run on the cheaper pure and 1% thorium tungstens.

The 1% *thoriated tungstens* (EWTH1) were designed exclusively for copper and copper alloys; this is because the melting temperature of copper is moderate (1990°F). The 1% thorium is enough thorium to elevate the melting temperature of this tungsten to hold the sharpened point for welding copper. It is slightly cheaper than the 2% thoriated and slightly more expensive than pure tungsten. The color code is yellow.

Zirconium (EWZR) tungstens are designed for the same applications as those for pure tungstens, aluminum, and magnesium. The difference between pure and zirconium tungstens are $\frac{1}{2}$% zirconium added to the zirconium tungsten. This zirconium reduces the contamination effects of dipping the tungsten into the molten puddle while welding aluminum and magnesium. Tungsten reacts rapidly with any oxide when it is hot. Dipping it into molten aluminum or magnesium puts these oxides into the tungsten rapidly, causing the tungsten to conduct poorly and contaminate the molten weld metal. Zirconium is an excellent deoxidizer and has characteristics similar to those of tungsten, therefore alloys well with it. If a zirconium tungsten is dipped, the zirconium will absorb much of the oxide, preventing contamination. These tungstens are more expensive than pure tungstens but should be used on very critical aluminum and magnesium work. Zirconium tungstens also reduce the chances of tungsten inclusions, due to tungsten dipping; therefore, x-ray-quality welds can be achieved. EWZR tungstens have a brown color code.

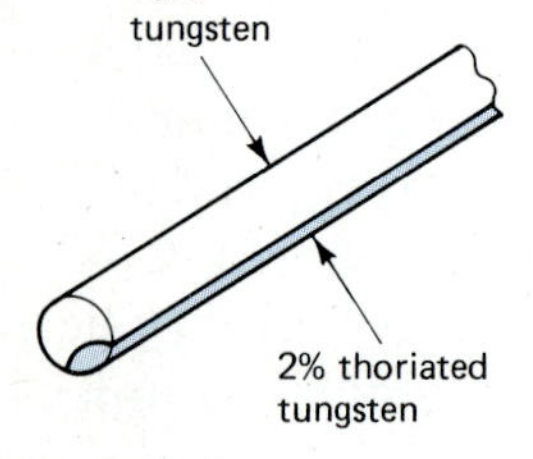

FIGURE 8–17
Striped tungsten.

Striped tungsten (EWTH3) is the newest tungsten classification. This tungsten combines pure tungsten and a strip of 2% thoriated tungsten (Figure 8–17). The thoria content is only about $\frac{1}{2}$ of 1% (0.35 to 0.55%). This additional thoria percentage helps keep a stabilized arc and increases the melting temperature over that of pure tungsten. This results in slightly better characteristics for ac welding. However, it is also acceptable for DCSP welding. The EWTH3 is considered a utility-type tungsten; it will work extremely well on aluminum and magnesium, and fairly well on moderate to thin sections of all other metals. Under all situations the EWTH3 tungsten should be melted and balled before ac or dc welding. This will mix the thoria with the pure tungsten; now the tungsten can be ground to a point (for DCSP) or left in a ball (for ACHF).

Tungsten End Preparation

Grinding or balling the end of a tungsten should be done to the opposite end of the identifying color. This will allow you to identify it at a later date. ACHF welding should be done with a ball or rounded tungsten (Figure 8–18). This will give the tungsten more surface area to carry the additional tungsten heat created with ac. For thick sections and high heat settings, use a large ball. For thin metal and a more directional arc, use a smaller rounded surface. DCSP welding should be done with a sharpened tungsten. Remember, with DCSP, electrons will be emitting from the tungsten. Sharpening gives a highly directional arc stream.

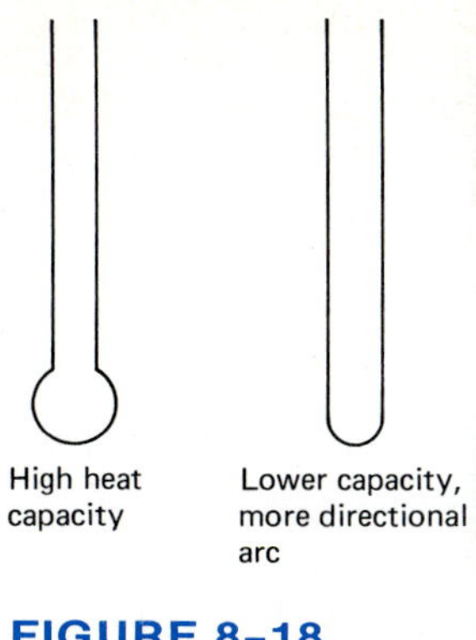

FIGURE 8–18
Tungsten end for ACHF.

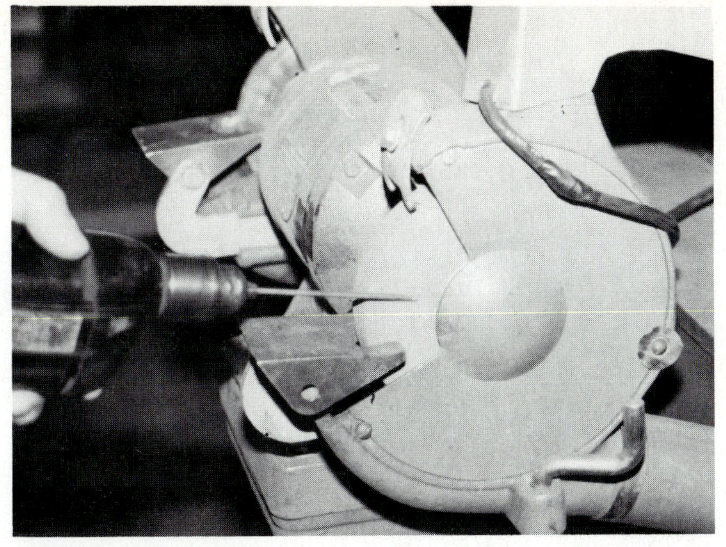

PHOTO 8–1
Using the side of the grinding wheel and a hand drill, you can produce very uniform tungsten turns.

Tungsten Tapers

There are some advantages to varying degrees of tungsten tapers. For open root butt welds a sharp tungsten with a slight flat spot on the tip gives the best results (Figure 8–19a). For fillet welds, a long sharp tungsten with plenty of stickout will give a puddle that is easy to reach and easy to control (Figure 8–19b). For beads on plate or other flat surfaces, a blunt tungsten tends to give deeper penetration (Figure 8–19c).

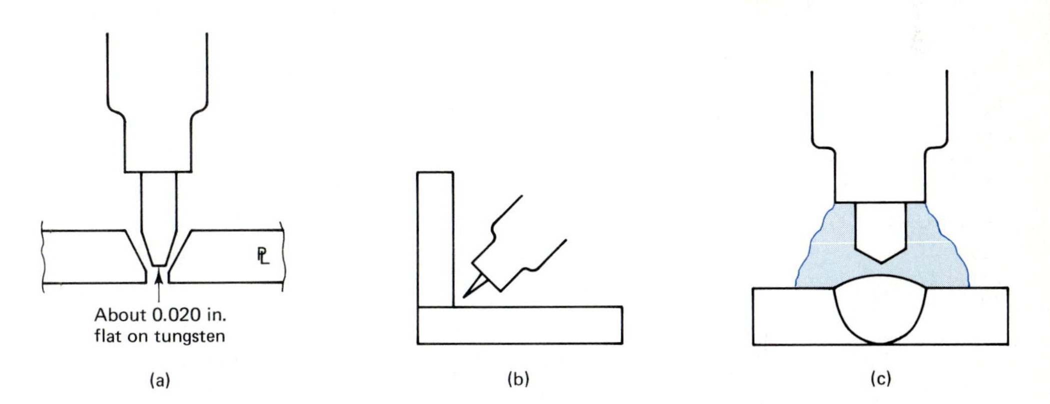

FIGURE 8–19
Tungsten tapers.

Preventing Tungsten Contamination

The key to preventing tungsten contamination is shielding it from the atmosphere and not touching the weld puddle. Remember, tungsten will react and absorb oxygen from the air when it is heated (it does not have to be molten). Points to remember:

1. Inert gas flow must be sufficient for the amount of tungsten stickout (Figure 8–20).

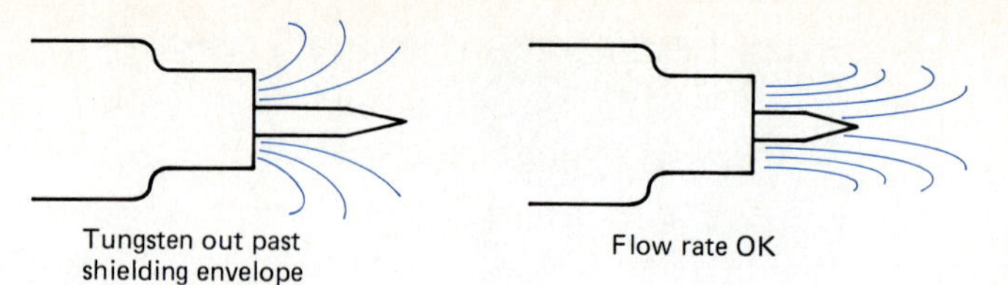

FIGURE 8–20 Gas flow rates.

2. Achieve laminar flow. *Caution:* Too much shielding gas will cause turbulent flow and pick up pockets of oxygen which will contaminate both tungsten and puddle. Ideal gas flow is the laminar or straight-line rate (Figure 8–21). Actual gas flow rates will depend on cup size.

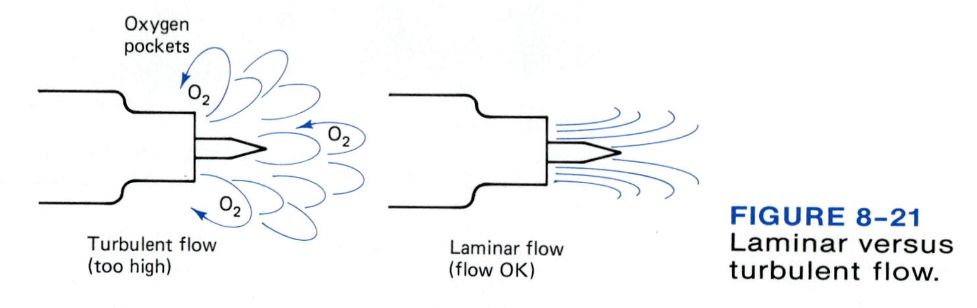

FIGURE 8–21 Laminar versus turbulent flow.

3. Keep the postpurge on long enough to allow tungsten to cool to a nonreactive state (about 1000°F).
4. Keeping the tungsten stickout low will also reduce the chance of accidentally dipping into the molten metal and the resulting tungsten contamination.

Preventing Puddle and Weld Contamination

Puddle or weld contamination refers to any contaminates that end up in the molten puddle or finished weld. These usually occur while welding when the puddle is molten. When inspecting welds these contaminates are called discontinuities.

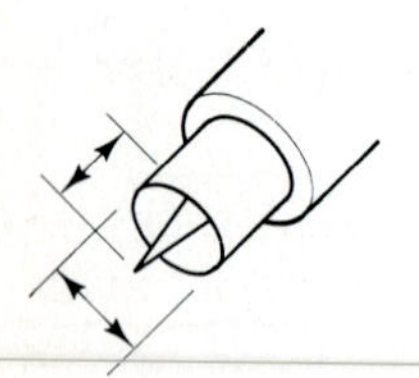

FIGURE 8–22 Tungsten stickout.

1. Tungsten inclusions result when the tungsten is accidentally dipped or short-circuited into the puddle. Carefully watching the arc length and keeping electrode stickout to a minimum will reduce the chances of tungsten inclusions. The absolute maximum stickout should be no more than the diameter of the cup being used (Figure 8–22). If applicable, use thoriated or zirconium tungstens, which are less likely to deposit tungsten if dipped.
2. Dirty puddle results when dirty metal or a dirty filler rod is used. You will observe these contaminates in and around your puddle while welding. Prevention requires adequate cleaning and removal of oxide coatings on the base metal. A wire brush or chemical deoxidizers or degreasers are usually sufficient. Filler metal can be cleaned in the same way, and fine sandpaper works well to remove both dirt and the thin layer of copper or rust that may be on some filler rods (Figure 8–23).
3. Silicon puddle results from using mild steel fillers that use silicon as the deoxidizer, such as AWS class E70S-3 or E70S-1. This

FIGURE 8–23 Cleaning filler metal.

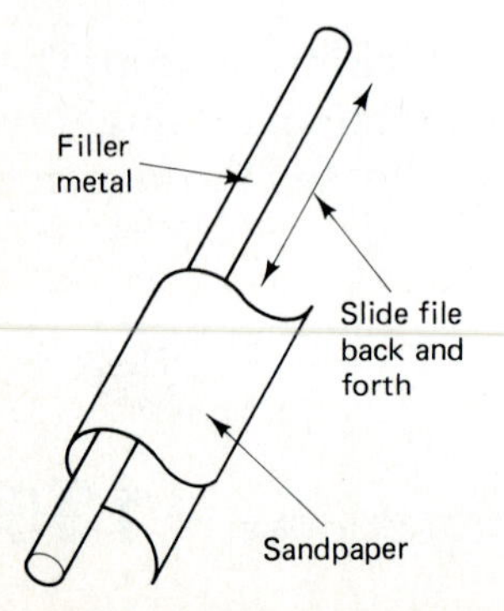

is not a problem for welding most steel, except on pipe welds, where complete puddle visibility is critical. The silicon floats in with the steel and can block the view of the puddle. Using ER70S-2 or ER70S-5 will eliminate this problem, as this AWS classification use triples deoxidation (aluminum, titanium, zirconium).

4. Porosity can result from any of the following: improper shielding, gas flow; dirty filler metal or tungstens; dirty, oily, or rusty base metal; or improper cup size for the application (shielding). Remember, steel is usually shipped with a thin film of oil to prevent surface rust in shipping. This can be removed by heating or chemically, but should definitely be removed before attempting a gas tungsten arc weld.

SHIELDING GASES FOR GTAW

There are two general classes of shielding gases used in the welding industry today, the active gases and the inert gases. The inert gases are most commonly used for GTAW. An inert gas such as argon or helium is chemically inactive; in other words, it will not break down into other substances. In welding this means that we can weld with it and pass electricity through it, and it will remain the same. Argon, helium, neon, krypton, xenon, and radon are all considered inert gases. For GTAW argon and helium are by far the most popular for industrial welding.

Argon

The atomic weight is 39.9481. Today, argon has surfaced as the most popular gas used for GTAW. The reasons for this are twofold.

1. Argon is heavier than air and therefore produces a superior shield from the atmosphere.
2. Argon ionizes (conducts electricity) at only 17 V; this low resistance to electricity makes for an extremely smooth, quiet arc.

Argon has also been known as an isolating gas. This means that the gas ionizes in a very narrow stream, which contributes to the many advantages of the GTAW process, such as the high depth-to-width ratio (Figure 8–24) and low heat input.

Helium

The atomic weight is 4.0026. Helium is an extremely light gas that provides a good shield for molten metal. Helium ionizes at 24.2 V, which make it produce a higher volume of heat in the arc stream and therefore a wider, deeper-penetration puddle (Figure 8–25). Helium should be used for thicker sections of steel and for metals such as copper and aluminum that have high thermal conductivity. The helium will allow

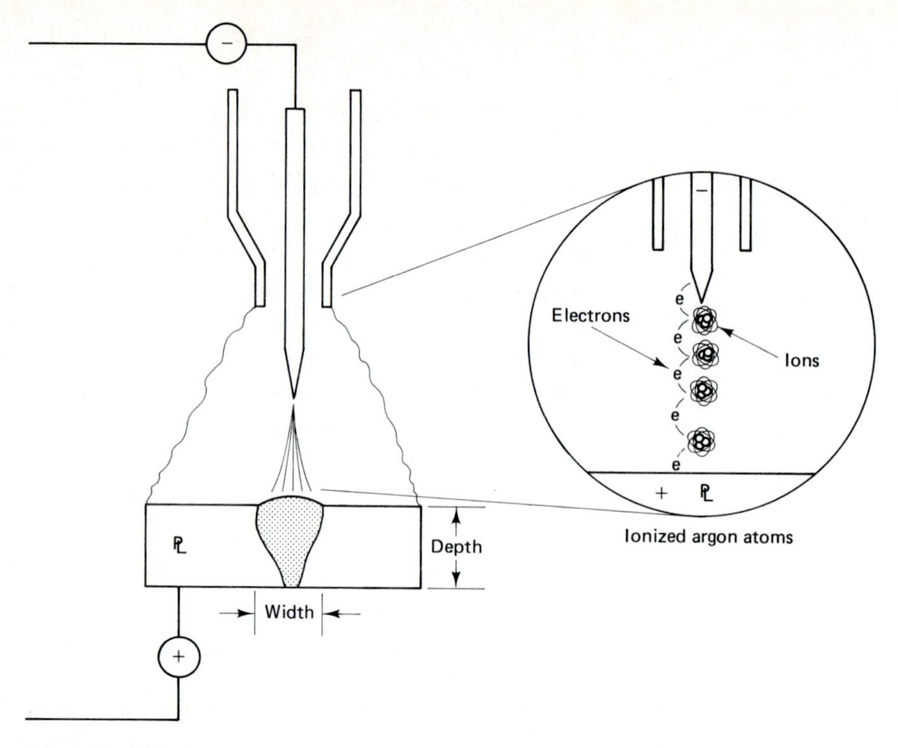

FIGURE 8–24
GTAW circuit.

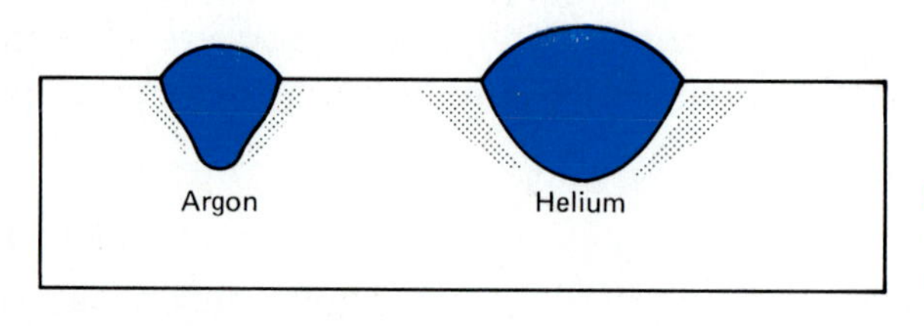

FIGURE 8–25
GTAW heat input from gas used.

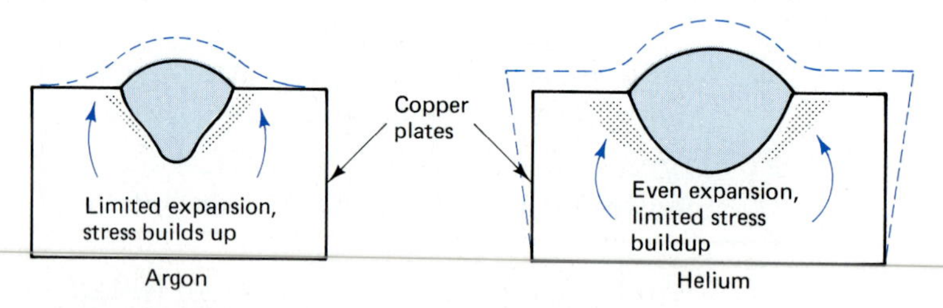

FIGURE 8–26
GTAW heat input and stress induced.

fast puddle formation and broad, even heat input. Helium actually produces a slight preheating effect. These characteristics are necessary to prevent cracking in high-thermal-conducting metals, such as copper (Figure 8–26).

Gas Mixtures

Argon–Helium This mixture can be used in any proportion, 50%–50%, 25% Ar–75% He, 25% He–75% Ar, or others. This combines the characteristics of the excellent shielding of argon and the broader,

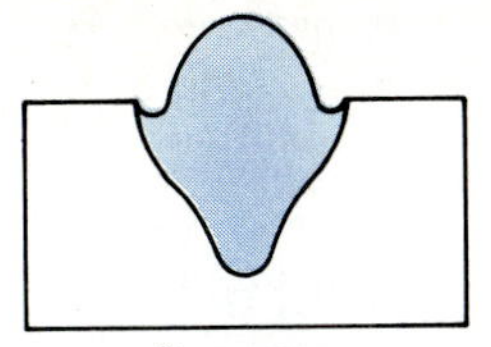

Pure argon

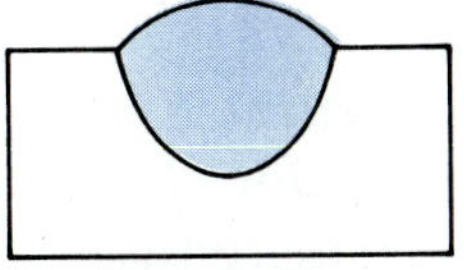

Argon-hydrogen

FIGURE 8-27
GTAW effects of gas used on weld contour.

wider arc stream and resultant wider, deeper puddle of helium. Remember, the more helium, the higher the volume of heat in the arc. Thick sections of aluminum and copper will weld much faster and easier with this mixture than with pure argon. This mixture is also good for automatic GTAW processes, because it will maintain a steady puddle size regardless of minor arc-length changes.

Argon–Hydrogen Some metals have ultralow thermal (heat) conductivity; austentic stainless steels and nickel steels are typical examples. These metals tend to "ball up" and not flatten out when welding. Other terms for this action are "poor wetting action" or "poor fluidity." The visible result of poor fluidity is undercutting. The addition of small amounts of hydrogen to the shielding mixture (5% maximum for stainless and for nickel) will increase the wetting action considerably (Figure 8-27). The addition of hydrogen also makes these metals easier and faster to weld. Do not use any gas mixture with hydrogen on carbon steels, copper, aluminum, or titanium, as they will absorb hydrogen and can eventually crack.

Argon–CO_2 This mixture can be used with carbon steels only. Contamination will occur when welding metals such as aluminum or magnesium with Argon/CO_2. The only advantage of this mixture over straight argon is a slight increase in wetting action and it is a more economical shielding gas.

Remember, shielding gases do more than shield molten metal from the atmosphere. Utilizing the correct gas can make difficult-to-weld metals much easier to weld and of higher quality.

PHOTO 8-2
Airco Heliwelder 250. (Courtesy of ESAB Welding Products, Inc.)

GTAW MACHINE CONTROLS

Understanding all machine controls on a GTAW machine will help you get the most out of your machine and give you maximum capability and flexibility for any given welding job.

Polarity Switch This control will allow you to select DCRP (DCEP), DCSP (DCEN), or ac for the given GTAW application (Figure 8-28a). For which polarity is applicable, see the appropriate metal section below.

Course Range Switch This is usually a high-, medium-, or low-amperage setting: for example, 20-60 A, 60-120 A, or 120-200 A (Figure 8-28b). Some manufacturers have four ranges.

Fine Current Adjustment This control will fine tune the course range amperage (Figure 8-28c). It usually has a 0-10 or 0-100 selection on the dial face. This is a percentage of the course range setting. Do not make the mistake of thinking this is the actual amperage. Example:

course adjustment = 60-120
fine current adjustment = 70%

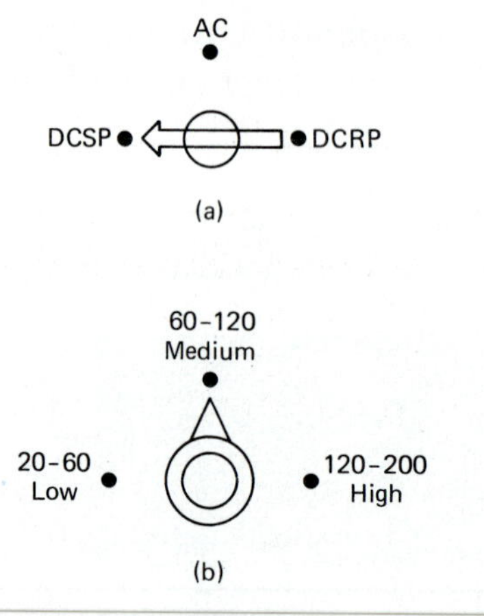

FIGURE 8-28 GTAW machine switches.

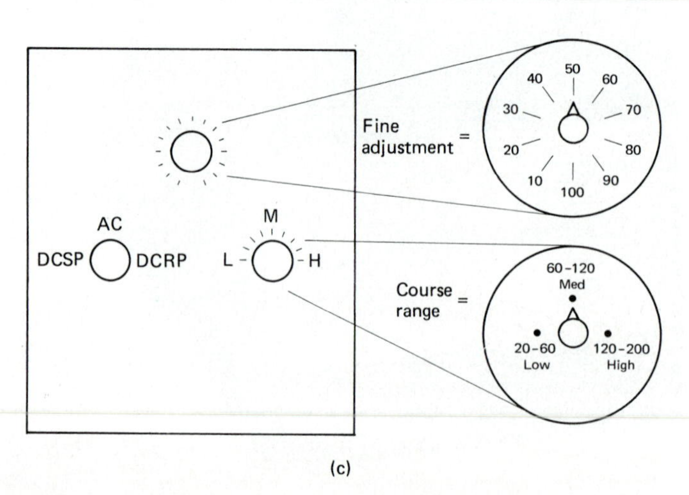

1. Find the difference in the range: = 60 (subtract the highest setting from the lowest setting).
2. Take the percentage of fine adjustment. Find the percentage of that in the course range: 60 × 70%. Percentage to decimal: 60 × .70 = 42.
3. Now add the lowest setting to the percentage found in step 2: 42 + 60 = 102. Your Amperage = 102.

The only truly accurate way to measure arc amperage is with an analog or digital meter, so use only machine dial settings to get you in the "ballpark."

Slow Start or Soft Start Adjustment This will control the speed at which the primary current will come in, replacing over the high-frequency start current. For example, a low soft start setting will give only a fraction of a second of high frequency before the primary current sets in; a high soft start will give 4 to 5 seconds of only high frequency before the primary current phases in (Figure 8–29).

High-Frequency Intensity This will control the actual intensity or volume of the HF. A high setting of HF intensity can increase the oxide cleaning action when welding aluminum and result in a more stable arc for any ac application. Since high frequency tends to erode the tungsten, use only the level of intensity needed to produce a stable arc. If your machine is equipped with this control, you will need to experiment with a variety of settings to achieve the smooth arc, and cleanse the puddle. Generally speaking, thin sections and oxide metals require higher high-frequency intensity.

Postpurge When the arc is extinguished, the puddle may still be molten for a short period, or the solidified metal and tungsten electrode may still be hot enough to be reactive and may contaminate. Shielding gas must continue flowing to prevent oxidizing of these areas. The postpurge controls the time of this gas after flow. A high current setting and reactive metals (metals that will easily react with the atmosphere) require longer postpurge times. Postpurge also helps cool the ceramic cup. Five to 15 seconds is usually sufficient postpurge; however, high current settings and reactive metals may require up to 45 seconds.

FIGURE 8–29
Effect of start adjustment.

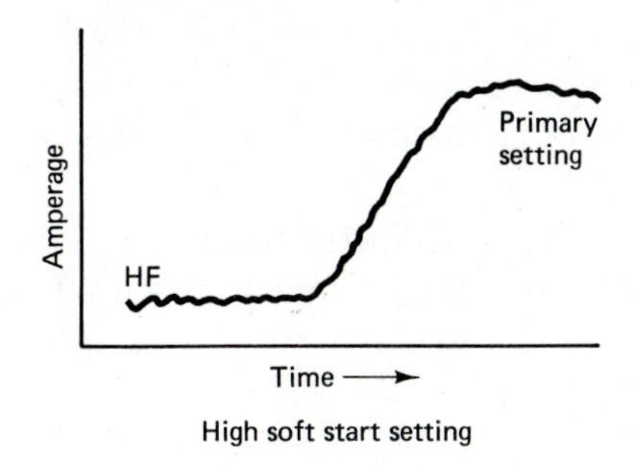

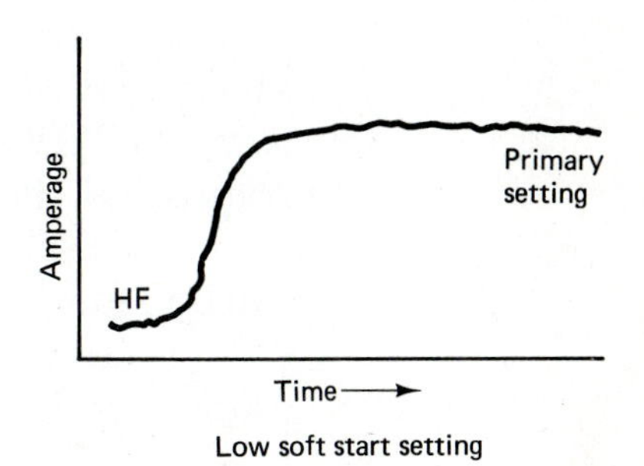

High-Frequency Continuous/Start/Off Control This is usually a double-pole, double-throw (three-position) toggle switch. In the continuous position, high frequency will be on continuously during welding; this is primarily for ac welding for arc stabilization. In the start position the high frequency will shut off a few seconds after the arc is started. This short term of high frequency is just enough to help start the arc without having to torch the tungsten to the work. This setting is used for DCSP applications. In the off position, no high frequency will be superimposed into the circuit. This would be for non-GTAW applications such as SMAW. Remember, a GTAW machine is a constant current machine, so it can double as an arc welder. *Note:* Some GTAW may use the HF off position, but the tungsten electrode must be struck to start the arc. This can be done with minimum tungsten contamination if a copper starting block is used to strike the tungsten on. Preferred arc starting is always with HF on (start or continuous position). In this way the tungsten simply needs to be brought close to the work and the arc will initiate.

Standard/Remote Switch This control will activate or deactivate any remote amperage or arc-start control, such as the foot puddle. For most applications the remote will allow for rheostatic amperage control (foot pedal on), allowing the welder complete puddle control at the beginning and end of the weld bead. However, in conditions where dragging the foot pedal to the welding area would be difficult, flipping to the standard position will deactivate the foot pedal (or whatever is being used as remote amperage control), and current will be ready to flow once the tungsten is struck. Remember, copper starting blocks can be used here if the switch is in the standard position.

Flowmeter Settings The flowmeter is calibrated in cubic feet per hour (cfh) or SCFH standard cubic feet per hour (scfh). This is because cfh allows much more practical range of controllability for the volumes used in GTAW when compared to psi volumes. When reading a flowmeter setting, read the top of the ball. Make sure that the gas solenoid is open by activating the remote amperage control switch (probably the foot pedal in this case). Now adjust the flow rate. Most GTAW flow rates should be set between 12 and 30 cfh. Cup size, amperage, the metal to be welded, and the shielding gas used are factors to be considered when setting shielding gas flow rate. Use large cups higher flow rates (18 to 30 cfh) for metals such as aluminum and copper. These metals have relatively large molten puddles. When GTAW welding, reactive and refractory metals should also use large cups and higher flow rates for better shielding due to their highly reactive nature when exposed to the atmosphere (see the GTAW metals, page 155). Stainless steels and nickel steels usually use moderate to small cups and flow rates. GTAW welding of these metals is usually done at lower amperage to prevent distortion.

Therefore, smaller cups and lower gas flow rates can be used. Carbon steels should use small or large cups with moderate flow rates (20–28 cfh). Set the flow rate for steel based on the amperage: high amperage, higher flow rate.

The shielding gas should also be considered when setting the flow rate. Argon is heavier than air and can effectively shield at lower flow rates, whereas helium is lighter than air and requires higher flow rates for shielding. Shielding gas mixtures should use high or low flow rates based on their proportional mix.

GTAW TORCH

GTAW torches (Figure 8–30) come either water or air cooled; of course, heavy-duty uses should use water-cooled torches. There are some other advantages to water-cooled torches, such as a smaller, lighter torch and generally an easier-to-handle, cooler torch. Air-cooled torches are acceptable for light-duty, lower-amperage work.

Caps come in long, short, and flat types. Flat and short caps are designed for getting into tight spots, and are lighter. Long caps, usually 7 in. in length, are the most common for standard bench jobs. The cap must seal tight to the torch body to prevent air from drawing into, and mixing with, the shielding gas. The cap also serves to tighten in the tungsten by pushing on the collet.

Cups are made of ceramic to withstand the high temperature of the arc. Cups also serve to direct the shielding gas to the work. Standard sizes are the 4, 6, 8 and 10. These numbers represent the number of sixteenths; for example, no. 4 = $\frac{4}{16}$- or $\frac{1}{4}$-in.-diameter cup; no. 6 = $\frac{6}{16}$- or $\frac{3}{8}$-in.-diameter cup; and so on.

The gas lense cup is a special cup that directs a longer stream of laminar gas flow (see the earlier "Preventing Tungsten Contamination" section). This means that the effective shielding area is increased considerably over that of standard cups. Gas lenses are more expensive than standard cups but are necessary when more tungsten stickout is required to reach into tight spots.

The collet is what holds the tungsten in the torch; it must match the tungsten size. Some manufacturers' torches also require a matching collet body.

FIGURE 8–30
GTAW torch.

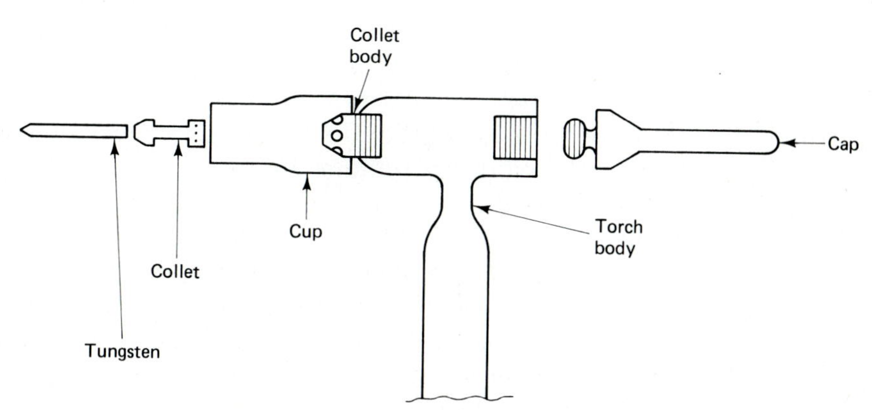

GTAW OF ALUMINUM

One of the highest-quality methods of welding aluminum is the GTAW process. To effectively weld, inspect, and troubleshoot GTAW of aluminum, we should understand some of the properties of this metal.

Nonferrous Aluminum is nonferrous; this means that it contains little or no iron (Fe). The magnet test will quickly identify ferrous and nonferrous metals.

Oxide Coating Aluminum has a heavy oxide coating that melts at 3700°F. Aluminum itself melts at 1218°F. The problem with welding aluminum is that the arc must first penetrate this oxide coating before it can melt the actual aluminum. If the oxide gets into the molten puddle, it causes puddle contamination. The welder cannot see the puddle once it is contaminated, and welding must stop until the oxide is wire-brushed off. Aluminum oxide is actually a protective shield that prevents further corrosion and oxidation of aluminum in the solid state. Removing the oxide coating can be done by wire brushing or with chemical deoxidizers. Wire-brushing aluminum should be done with a stainless steel wire brush. A standard mild steel wire brush will put small particles into the pores of the aluminum and can cause contamination, which may show up in radiographs (x-rays) of the aluminum. Always remove the oxide coating before welding. Heating aluminum slightly makes the oxide coating much easier to wire-brush off.

Weight Aluminum is light in weight, approximately one-third of the weight of steel, which makes it easy to identify.

Color Aluminum is light silver in color; some heavily oxidized aluminum castings may be a dull silver. When welding aluminum you will notice no red color when molten. This sometimes makes the puddle a little harder to spot. Always look for a silverish-color puddle.

Thermal Conductivity Aluminum has high thermal (heat) conductivity, which means that it conducts heat rapidly across the plate (Figure 8–31). You will find when welding any metal with high thermal conductivity, such as aluminum, that the high rate of heat conduction away from the arc causes a slow-to-develop puddle. With aluminum, once the puddle is formed, it rapidly grows larger, which requires the welder to increase his or her travel speed gradually to maintain a uniform puddle. A slow uniform travel speed will result in a nonuniform, gradually widening puddle (Figure 8–32).

Hot Shortness This property means the lack of metal strength at high temperatures. The result of hot shortness in aluminum is usually a throat crack down the center of an overheated aluminum weld. This results because with hot-shortness metals, the weakest area is the hottest area; this of course would be the center of the weld. Preventing hot-shortness cracking can be done simply by controlling the volume

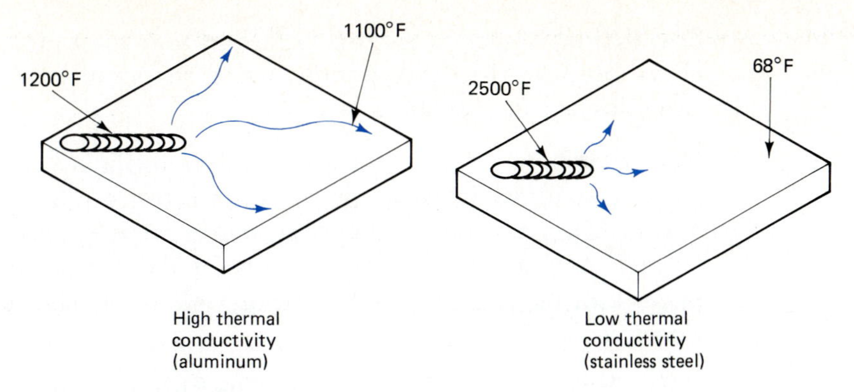

FIGURE 8-31
Thermal conductivity.

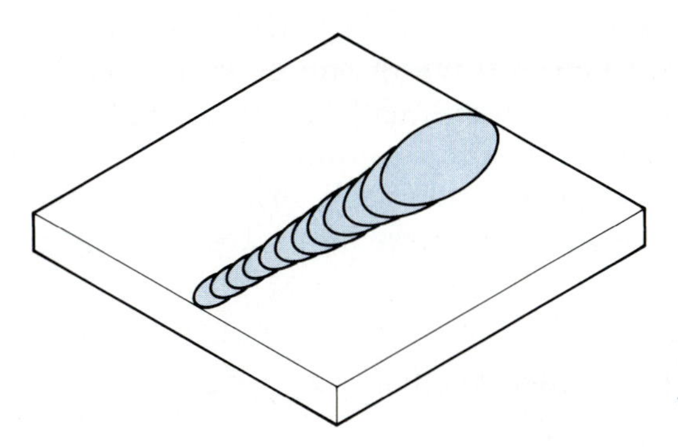

FIGURE 8-32
Widening puddle in aluminum.

of heat input into the metal. Use higher travel speeds and do not use rewash (or wash-in-pass) on aluminum welds. Use filler metal whenever possible. Remember, the filler metal is in the solid state and much cooler than the molten aluminum puddle. Filler metal gives a quenching effect to the puddle and quickly cools it when filler is dipped into the puddle.

GTAW Setup for Aluminum

Tungsten Electrodes

As stated previously, pure or zirconium tungstens are suggested for GTAW of aluminum. The pure tungsten (EWP) is a cheaper tungsten but will easily contaminate if dipped into the puddle. The zirconium tungsten (EWZR) will absorb a number of dips without contaminating, therefore can be a better choice for super-high-quality jobs.

FIGURE 8-33
Tungsten ball tapers.

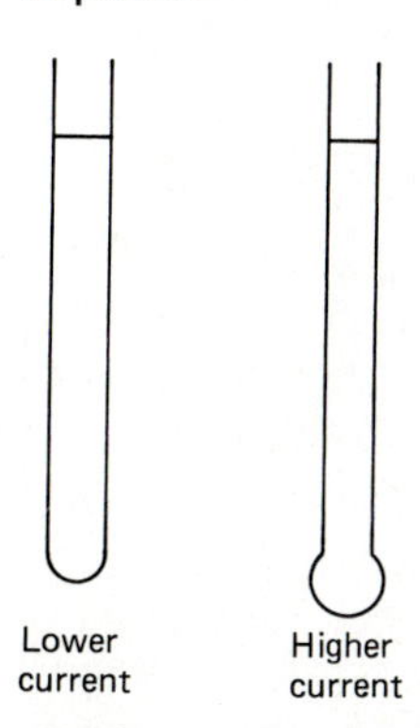

Tungsten Preparation

The tungsten electrode will be operating on ACHF (alternating current high frequency). This will put 50% of the heat generated into the tungsten; this fact, along with the erosion characteristics of high frequency, put a great deal of strain on the end of the tungsten. The key to preventing tungsten deterioration is rounding or balling the tip of the tungsten (Figure 8-33). To accomplish this, switch to DCRP; use high frequency on "start." Once your machine is started, start the arc and slowly increase the amperage until a molten ball develops on the end of the tungsten. Now slowly decrease the amperage until the tungsten has solidified. The more heat that is applied, the larger the ball will be, and the more current-carrying capacity it will have.

Shielding Gas Argon is acceptable for most aluminum applications. Helium or argon helium mixtures should be used for thick sections (over $\frac{1}{2}$ in. thick).

Polarities ACHF is used for most applications. Occasionally, very heavily oxidized aluminum castings require additional oxide cleaning action. This can be achieved by using DCRP; care must be taken not to melt over heat the tungsten on DCRP. Use larger tungstens and a large ball on the end of the tungsten. Keep the amperage as low as possible.

High Frequency Setting The high frequency should be on "continuous" for AC, and "start" for DC applications.

GTAW Techniques for Aluminum

When making any GTAW weld with aluminum, always allow the puddle to develop before dipping the filler metal. The "forward back and dip" motion is quite successful in producing high-quality GTAW welds in aluminum. This technique allows the puddle to form fully. Bring the torch forward slightly, then back, bringing the filler metal into the leading edge of the puddle. These steps are steadily repeated to produce a uniform bead.

Aluminum Filler Metals

A guide to the choice of filler metals is given in Table 8-3.

TABLE 8-3
Guide to the Choice of Filler Metal for General-Purpose Welding of Aluminum

BASE METAL	319, 333, 354, 355, C355	13, 43, 344, 356, A356, A357, 359	214, A214, B214, F214	7039 A612, C612, D612, 7005[k]	6070	6061, 6063, 6101, 6151, 6201, 6951	5456	5454
1060, EC	ER4145[c,i]	ER4043[i,f]	ER4043[e,i]	ER4043[i]	ER4043[i]	ER4043[i]	ER5356[c]	ER4043[e,i]
1100, 3003, Alclad 3003	ER4145[c,i]	ER4043[i,f]	ER4043[e,i]	ER4043[i]	ER4043[i]	ER4043[i]	ER5356[c]	ER4043[e,i]
2014, 2024	ER4145[g]	ER4145	...	...	ER4145	ER4145	...	...
2219	ER4145[g,c,i]	ER4145[c,i]	ER4043[i]	ER4043[i]	ER4043[f,i]	ER4043[f,i]	ER4043	ER4043[i]
3004, Alclad 3004	ER4043[i]	ER4043[i]	ER5654[b]	ER5356[e]	ER4043[e]	ER4043[b]	ER5356[e]	ER5654[b]
5005, 5050	ER4043[i]	ER4043[i]	ER5654[b]	ER5356[e]	ER4043[e]	ER4043[b]	ER5356[e]	ER5654[b]
5052, 5652[a]	ER4043[i]	ER4043[b,i]	ER5654[b]	ER5356[e,h]	ER5356[b,c]	ER5356[b,c]	ER5356[b]	ER5654[b]
5083	...	ER5356[c,e,i]	ER5356[e]	ER5183[e,h]	ER5356[e]	ER5356[e]	ER5183[e]	ER5356[e]
5086	...	ER5356[c,e,i]	ER5356[e]	ER5356[e,h]	ER5356[e]	ER5356[e]	ER5356[e]	ER5356[b]
5154, 5254[a]	...	ER4043[b,i]	ER5654[b]	ER5356[b,h]	ER5356[b,c]	ER5356[b,c]	ER5356[b]	ER5654[b]
5454	ER4043[i]	ER4043[b,i]	ER5654[b]	ER5356[b,h]	ER5356[b,c]	ER5356[b,c]	ER5356[b]	ER5654[c,e]
5456	...	ER5356[c,e,i]	ER5356[e]	ER5556[e,h]	ER5356[e]	ER5356[e]	ER5556[e]	
6061, 6063, 6101, 6201, 6151, 6951	ER4145[c,i]	ER4043[b,i]	ER5356[b,c]	ER5356[b,c,h,i]	ER4043[b,i]	ER4043[b,i]		
6070	ER4145[c,i]	ER4043[e,i]	ER5356[c,e]	ER5356[c,e,h,i]	ER4043[e,i]			
7039, A612, C612, D612, 7005[k]	ER4043[i]	ER4043[b,h,i]	ER5356[b,h]	ER5039[e]				
214, A214, B214, F214	...	ER4043[b,i]	ER5654[b,d]					
13, 43, 344, 356, A356, A357, 359	ER4145[c,i]	ER4043[d,i]						
319, 333, 354, 355, C355	ER4145[d,c,i]							

TABLE 8–3 (continued)

5154 5254[a]	5086	5083	5052, 5652[a]	5005, 5050	3004, ALC. 3004	2219	2014, 2024	1100, 3003, ALC. 3003	1060 EC
ER4043[e,i]	ER5356[c]	ER5356[c]	ER4043[i]	ER1100[c]	ER4043	ER4145	ER4145	ER1100[c]	ER1260[c,j]
ER4043[e,i]	ER5356[c]	ER5356[c]	ER4043[e,i]	ER4043[e]	ER4043[e]	ER4145	ER4145	ER1100[c]	
. . .	. . .	. . .	. . .	. . .	. . .	ER4145[g]	ER4145[g]		
ER4043[i]	ER4043	ER4043	ER4043[i]	ER4043	ER4043	ER2319[c,f,i]			
ER5654[b]	ER5356[e]	ER5356[e]	ER4043[e,i]	ER4043[e]	ER4043[e]				
ER5654[b]	ER5356[e]	ER5356[e]	ER4043[e,i]	ER4043[d,e]					
ER5654[b]	ER5356[e]	ER5356[e]	ER5654[a,b,c]						
ER5356[e]	ER5356[e]	ER5183[e]							
ER5356[b]	ER5356[e]								
ER5654[a,b]									

Notes:

1. Service conditions such as immersion in fresh or salt water, exposure to specific chemicals, or exposure to a sustained high temperature (over 150°F) may limit the choice of filler metals.
2. Recommendations in this table apply to gas shielded-arc welding processes. For gas welding, only R1100, R1260, and R4043 filler metals are ordinarily used.
3. Filler metals designated with ER prefix are listed in AWS specification A5.10.
   [a] Base metal alloys 5652 and 5254 are used for hydrogen peroxide service. ER5654 filler metal is used for welding both alloys for low-temperature service (150°F and below).
   [b] ER5183, ER5356, ER5554, ER5556, and ER5654 may be used. In some cases they provide: (1) improved color match after anodizing treatment, (2) highest weld ductility, and (3) higher weld strength. ER5554 is suitable for elevated temperature service.
   [c] ER4043 may be used for some applications.
   [d] Filler metal with the same analysis as the base metal is sometimes used.
   [e] ER5183, ER5356, or ER5556 may be used.
   [f] ER4145 may be used for some applications.
   [g] ER2319 may be used for some applications.
   [h] ER5039 may be used for some applications.
   [i] ER4047 may be used for some applications.
   [j] ER1100 may be used for some applications.
   [k] This refers to 7005 extrusions only.
4. Where no filler metal is listed, the base metal combination is not recommended for welding.

GTAW of Stainless Steel

Stainless steels are welded similarly to mild steels, but require some special care and understanding of their properties for high-quality stainless welds. Stainless steels are available in a number of grades; most contain chromium and many contain nickel and chrome. These elements give stainless excellent corrosion resistance (see Chapter 19).

Thermal Conductivity A property you will observe when welding stainless steels is that of low heat (thermal) conductivity. Unlike metals such as aluminum and copper, stainless steels transfer heat very slowly throughout the stainless plate. This property contributes to the high amount of warpage in stainless. This property also makes for rapid forming of the fluid stainless puddle. This occurs because the heat does not dissipate rapidly into the surrounding material. The metal will turn red just before the material reaches its melting point.

GTAW Setup for Stainless Steel

Tungsten Electrode The best tungsten for stainless is the 2% thoriated (red color code EWTH2). This tungsten has sufficient heat-carrying capacity.

Tungsten Preparation Sharpen the tungsten to a sharp point for the best puddle control. Always use DCSP (DCEN) for stainless.

Shielding Gas The shielding gas can be pure argon, a mixture of argon–helium, or argon–hydrogen (95 argon, 5% hydrogen) for chrome–nickel grades. The addition of helium or hydrogen to argon shielding gas will increase the fluidity and wetting action of the stainless steel puddle. Of course, too much hydrogen in the atmosphere may result in absorption into the puddle and cause hydrogen cracks. Therefore, hydrogen should be limited to 5% maximum. The argon–helium mix can be in any combination, but remember, the more helium, the deeper the penetration and the more fluid the puddle. Most popular for this mixture is 25% argon–75% helium.

High-Frequency Setting Set the high frequency on the start position; this will give just enough high frequency to establish the arc.

GTAW Techniques for Stainless Steel

The key to high-quality GTAW weld deposits in stainless steel is low heat and small beads. The first reason for this is the distortion factor. Due to the low heat conductivity of most stainless, the heat and therefore the expansion are limited to the immediate area surrounding the puddle and arc. The remaining metal is relatively cool, and therefore has not expanded. It is this difference in expansion areas (on nonexpansion areas) that creates the distortion in stainless. The second consideration for holding the heat input down is carbide precipitation.

This is the migration of carbon atoms into the weld and heat-affected zone at high temperatures. If too much carbon is absorbed into any one area of stainless, it will combine with the chromium atoms and form chromium carbide. Unlike pure chromium, chromium carbide has no shielding oxide, and therefore this area will rust.

Limiting the overall heat input into stainless will prevent or at least reduce both carbide precipitation and distortion. This can be accomplished by methods such as backstepping, intermittent welding, and small narrow beads (Figure 8–34). Obviously, beads must be sufficiently large for good fusion and penetration, but keeping beads narrow but deeply penetrating will accomplish both good fusion and lower heat input.

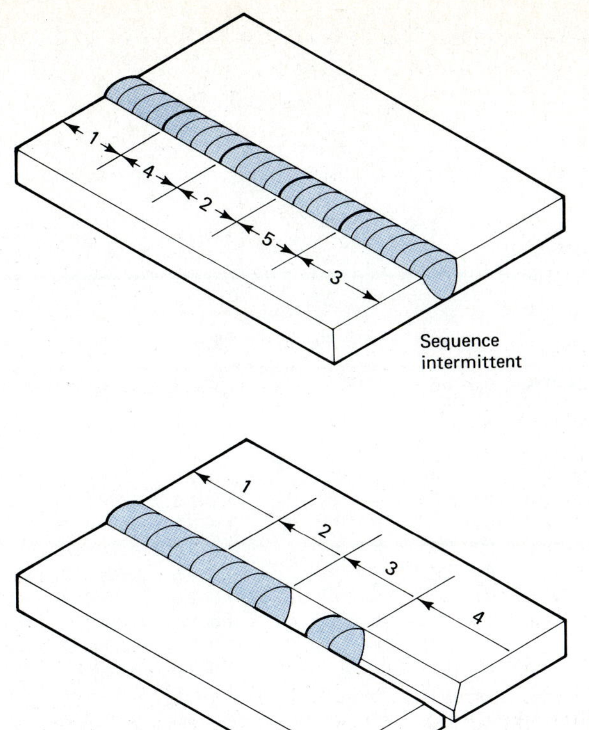

FIGURE 8–34
Sequence and intermittent steps.

Stainless Steel Filler Metals

When selecting stainless steel filler metals, always match up the base metal with the filler metal (Table 8–4). Example, when using 309 base metal, use an ER 309 filler metal. Some job specification or procedures may require a filler metal alloy different from that of the base metal, but this is the exception, not the rule. An example of the filler metal classification system as shown in Figure 8–35.

Using stainless filler metals with the suffix "L" in the last digit or ER321 or ER347 indicates a stabilized-grade filler metal. Suffix "L" indicates low carbon. This will allow the weld to absorb carbon when welding without oversaturating the chromium atoms and causing carbide precipitation (chromium carbide). Columbium (ER347) and titanium (ER321) alloy fillers stabilize stainless by forming columbium carbides or titanium carbides, thus leaving the chromium atoms free to shield and keep the stainless corrosion free.

Stainless steel GTAW can be done with or without filler metal. The standard forward, back, and dip technique is quite successful for most plate applications, (Figure 8–36). On pipe root passes the walk the cup technique is popular (Figure 8–37).

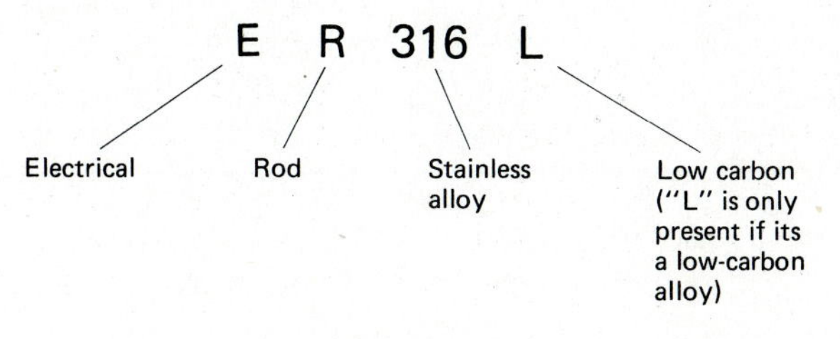

FIGURE 8–35
Classification system for stainless steel filler metals.

TABLE 8-4
Chemical Composition Requirements (%)

AWS CLASSIFICATION	C	Cr	Ni	Mo	Cb + Ta	Mn	Si	P	S	Fe	W	Cu
ER307	0.04–0.14	19.5–22.2	8.0–10.5	0.5–1.5	—	3.75–4.75	0.03–0.65	0.03	0.03	Rem.	—	0.5
ER308[a]	0.08	19.5–22.0	9.0–11.0	0.5	—	1.0–2.5	0.30–0.65	0.03	0.03	Rem.	—	0.5
ER308L[a]	0.03	19.5–22.0	9.0–11.0	0.5	—	1.0–2.5	0.30–0.65	0.03	0.03	Rem.	—	0.5
ER308Mo	0.08	18.0–21.0	9.0–12.0	2.0–3.0	—	1.0–2.5	0.30–0.65	0.03	0.03	Rem.	—	0.5
ER308MoL	0.04	18.0–21.0	9.0–12.0	2.0–3.0	—	1.0–2.5	0.30–0.65	0.03	0.03	Rem.	—	0.5
ER309[a]	0.12	23.0–25.0	12.0–14.0	0.5	—	1.0–2.5	0.30–0.65	0.03	0.03	Rem.	—	0.5
ER309L	0.03	23.0–25.0	12.0–14.0	0.5	—	1.0–2.5	0.30–0.65	0.03	0.03	Rem.	—	0.5
ER310	0.08–0.15	25.0–28.0	20.0–22.5	0.5	—	1.0–2.5	0.30–0.65	0.03	0.03	Rem.	—	0.5
ER312	0.15	28.0–32.0	8.0–10.5	0.5	—	1.0–2.5	0.30–0.65	0.03	0.03	Rem.	—	0.5
ER16-8-2	0.10	14.5–16.5	7.5– 9.5	1.0–2.0	—	1.0–2.5	0.30–0.65	0.03	0.03	Rem.	—	0.5
ER316[a]	0.08	18.0–20.0	11.0–14.0	2.0–3.0	—	1.0–2.5	0.30–0.65	0.03	0.03	Rem.	—	0.5
ER316L[a]	0.03	18.0–20.0	11.0–14.0	2.0–3.0	—	1.0–2.5	0.30–0.65	0.03	0.03	Rem.	—	0.5
ER317	0.08	18.5–20.5	13.0–15.0	3.0–4.0	—	1.0–2.5	0.30–0.65	0.03	0.03	Rem.	—	0.5
ER317L	0.03	18.5–20.5	13.0–15.0	3.0–4.0	—	1.0–2.5	0.30–0.65	0.03	0.03	Rem.	—	0.5
ER318	0.08	18.0–20.0	11.0–14.0	2.0–3.0	8 × C min. to 1.0 max.	1.0–2.5	0.30–0.65	0.03	0.03	Rem.	—	0.5
ER320	0.07	19.0–21.0	32.0–36.0	2.0–3.0	8 × C min. to 1.0 max.	2.5	0.60	0.04	0.03	Rem.	—	3.0–4.0
ER321[b]	0.08	18.5–20.5	9.0–10.5	0.5	—	1.0–2.5	0.30–0.65	0.03	0.03	Rem.	—	0.5
ER330	0.18–0.25	15.0–17.0	34.0–37.0	0.5	—	1.0–2.5	0.30–0.65	0.03	0.03	Rem.	—	0.5
ER347[a]	0.08	19.0–21.5	9.0–11.0	0.5	10 × C min. to 1.0 max.	1.0–2.5	0.30–0.65	0.03	0.03	Rem.	—	0.5
ER348	0.08	19.0–21.5	9.0–11.0	0.5	10 × C min. to 1.0 max.[c]	1.0–2.5	0.30–0.65	0.03	0.03	Rem.	—	0.5
ER349[d]	0.07–0.13	19.0–21.5	8.0– 9.5	0.35–0.65	1.0–1.4	1.0–2.5	0.30–0.65	0.03	0.03	Rem.	1.25–1.75	0.5
ER410	0.12	11.5–13.5	0.6	0.6	—	0.6	0.50	0.03	0.03	Rem.	—	0.5
ER410 NiMo	0.06	11.0–12.5	4.0– 5.0	0.4–0.7	—	0.6	0.50	0.03	0.03	Rem.	—	0.5
ER420	0.25–0.40	12.0–14.0	0.6	0.5	—	0.6	0.50	0.03	0.03	Rem.	—	0.5
ER430	0.10	15.5–17.0	0.6	0.5	—	0.6	0.50	0.03	0.03	Rem.	—	0.5
ER26-1[e]	0.01	25.0–27.5	[f]	0.75–1.50	—	0.40	0.40	0.02	0.02	Rem.	—	0.20[f]
ER502	0.10	4.5– 6.0	0.6	0.45–0.65	—	0.6	0.50	0.03	0.03	Rem.	—	0.5
ER505	0.10	8.0–10.5	0.5	0.8–1.2	—	0.6	0.50	0.04	0.03	Rem.	—	0.5
ER630	0.05	16.0–16.75	4.5– 5.0	0.75	0.15–0.30	0.25–0.75	0.75	0.04	0.03	Rem.	—	3.25–4.00

Notes:

1. Analysis should be made for the elements for which specific values are shown in this table. If, however, the presence of other elements is indicated in the course of routine analysis, further analysis shall be made to determine that the total of these other elements, except iron, is not present in excess of 0.50%.
2. Single values shown are maximum percentages except where otherwise specified.

[a] These grades are available in high-silicon classifications which shall have the same chemical composition requirements as given below with the exception that the silicon content should be 0.65 to 1.00%. These high-silicon classifications will be designated by the addition "Si" to the standard classification designations indicated below. The fabricator should consider carefully the use of high-silicon filler metals in highly restrained fully austenitic welds.

[b] Titanium, 9 × C min. to 1.0 max.

[c] Tantalum, max. 0.10%.

[d] Titanium, 0.10 to 0.30.

[e] Nitrogen, 0.015 max.

[f] Nickel, max., 0.5 minus the copper content.

Source: American Welding Society.

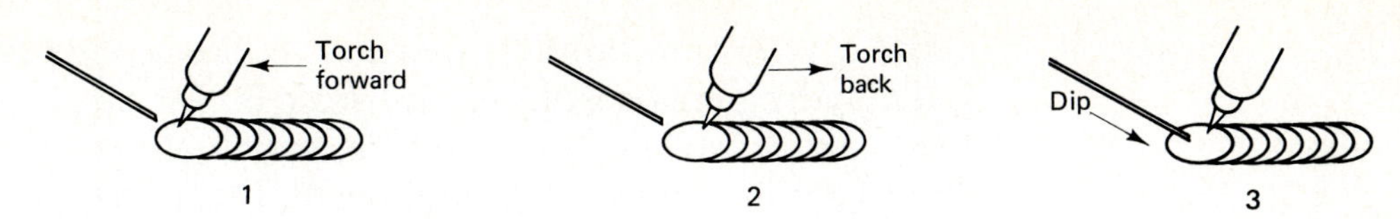

FIGURE 8–36
Forward, back, and dip technique.

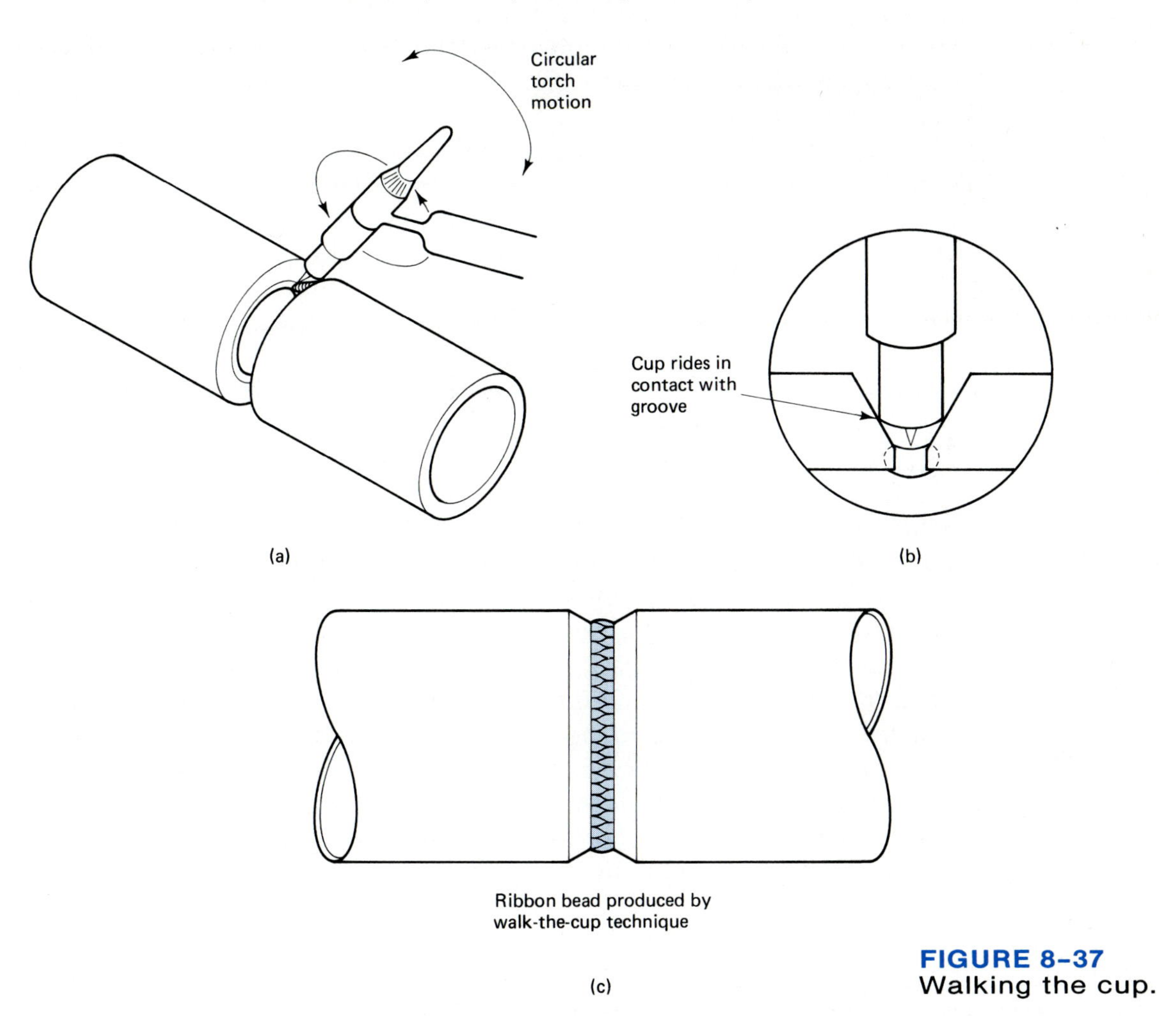

FIGURE 8–37
Walking the cup.

GTAW of Carbon Steels

Plain carbon steels use just about the same setup and procedure as for stainless steels; however, slightly higher current (heat) levels are acceptable. Use tungsten 2% thoriated (EWTH2) red color code. The tungsten should be sharpened. Use DCSP (DCEN) for a stable arc and deep penetration. The most common shielding gas for carbon steel is pure argon; occasionally, a little helium might be used for thicker sections, but carbon steels have better wetting action and generally weld easier than stainless steels.

One precaution that should be observed when welding carbon steels is that of base metal cleanliness. First, many carbon steels are shipped with a thin oil film to prevent oxidation (rusting). Remember, GTAW is a cooler-running welding process and thus this light coating of oil may not burn off. The result is porosity in the weld, and perhaps even hydrogen absorption. Mill scale is also heavy on hot-rolled carbon steels. This scale traps moisture, which also may cause porosity, and hydrogen absorption contamination can result in delayed cracks (see Chapter 19).

Proper material preparation will reduce the carbon steel contamination problem. Passing the oxyfuel torch over the steel will burn off any oil or moisture (use adequate ventilation); next, wire brush all surfaces to be welded. A good clean brush is essential (stainless steel brushes work best).

If, despite these efforts, porosity and contamination are still occurring, a filler metal with higher levels or deoxidizers will be necessary. ER70S-2 or ER70S-5 are good choices. These wires are triple deoxidized and will even weld through some moderately dirty surfaces.

GTAW OF NICKEL STEELS

Nickel steels are similar to stainless steels; however, they are nickel based whereas stainless steels are iron based. Typical nickel steels are monels (nickel–copper), inconel (nickel–chromium), and hastalloy (nickel–molybdenum). They are welded much the same as with stainless steels except for a few minor adjustments, using slightly lower amperages and a little more shielding gas. Because of the poor wetting action of nickel steels, argon hydrogen (95% argon–5% hydrogen) is a good shielding gas mixture. Argon–helium also works well, and pure argon is acceptable. Extreme cleanliness is also a necessity for successful GTAW on nickel steels. Degrease any oily surfaces and wire-brush to remove the surface oxide before welding.

GTAW OF COPPER

GTAW produces the highest-quality copper welds of any manual process. Practice is necessary to make successful welds on copper—not that it requires any special manipulation, but it has properties that are considerably different from that of carbon steel and its alloys. As with any metal, understanding these properties will enhance your knowledge and ability to weld them.

Copper is a nonferrous metal, therefore is nonmagnetic. Do not forget to use your magnet to verify low iron (Fe) content in metals. Copper is red in color and remains red when it reaches its melting point (1990°F). Copper also has a protective oxide coating (copper oxide) that makes it very resistant to oxidation and corrosion. Copper is also very

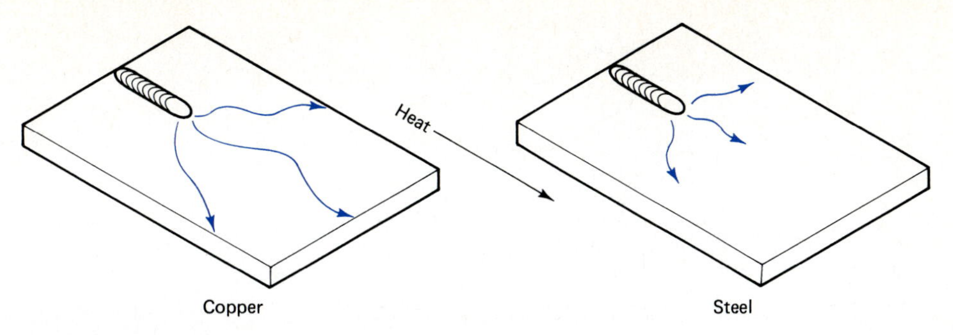

FIGURE 8-38
Heat input in copper and steel.

tough but ductile, which makes it a useful metal where these properties are needed.

The most important properties that affect the welding of copper is the high thermal (heat) conductivity and thermal expansion (Figure 8-38). Copper acts much like aluminum in that the whole plate will get hot very fast. We call this property "hot shortness." As with aluminum, copper conducts heat much faster than steel.

GTAW Setup for Copper

Always wire-brush all copper surfaces to be welded; this will remove much of the copper oxide coating. Note that this oxide will not contaminate quite like aluminum oxide does to aluminum, but it should be removed for clean, high-quality copper welds.

Polarities DCSP (DCEN) should be used to produce the best penetration and puddle control. ACHF can be used to produce a shallower, penetrating bead, but it is not necessary for oxide cleaning, as with aluminum or magnesium.

Shielding Gases One of the best shielding gas mixtures for copper is helium and argon. Any mixture will work; however, the thicker the copper, the more helium percentage in the mix is desirable. The helium produces a "preheating effect" because of its wider ionization pattern. This produces faster puddle. The argon in the mix, of course, stabilizes the arc and helps shield the puddle. Pure argon can be used for moderate and thin sections ($\frac{3}{8}$ in. and under); pure helium can also be used for very thick sections ($\frac{1}{2}$ in. and up). A popular mixture is 25% argon–75% helium.

Preheating Preheating is another way to improve the weldability of copper. Preheating will improve the fluidity of the puddle and help prevent cracking (Figure 8-39). Remember, the high thermal expansion of copper can cause a stressful situation in the weld area. Preheating will reduce the temperature difference, thus reduce the expansion difference and take some of the stress out of the weldment. Use approximately 400°F preheat. Use higher temperatures for thick sections and lower temperatures for thin sections.

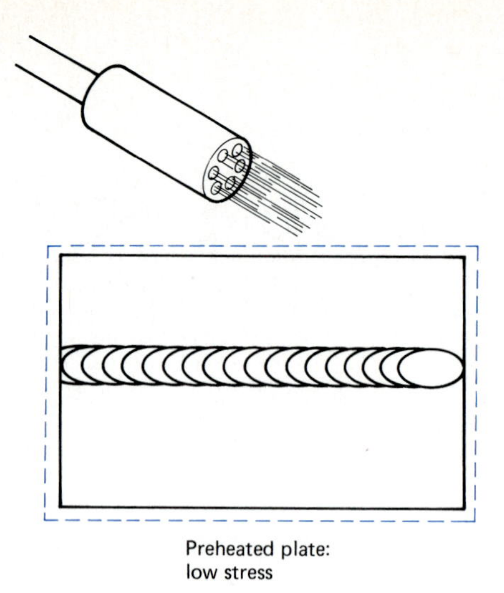

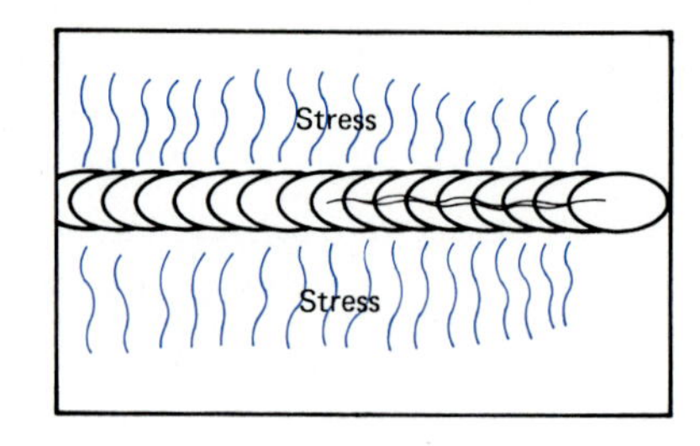

FIGURE 8–39
Heat stress reduction by preheating.

GTAW Techniques for Copper

Copper acts somewhat like aluminum and somewhat like steel. The puddle is heavy like steel, but it widens out like that of aluminum. Be aware of this high thermal conductivity and resulting pedal widening, and use corrective action by slowly letting up on the foot pedal as the puddle progresses across the plate. For more information on copper, see Chapter 19.

GTAW OF TITANIUM

Titanium is an amazing metal; it has a very high strength-to-weight ratio. This means that it is very light in weight for the level of strength it has. Some of titanium's uses include high-speed jet engine turbines and some aircraft and space vehicle skins. It has a very high melting temperature of 3270°F.

GTAW Setup for Titanium

Polarities DCSP should be used for most GTAW applications of titanium; this will give optimum penetration. Titanium does have a light titanium oxide coating, but it can be removed by wire brushing and will not require ACHF for oxide cleaning.

Shielding Gases Argon is by far the best shielding gas for titanium, due to it being heavier than air. An argon–helium mix can be used for very thick sections of titanium. Shielding is the most important consideration for successful welds on titanium. This is because titanium will react and contaminate if exposed, when molten, to the air (hydrogen is the primary contaminate). If possible, weld in a dry box (Figure 8–40). This device is a chamber that is completely filled with inert shielding gas; therefore, no atmospheric contaminates can reach the molten weld metal.

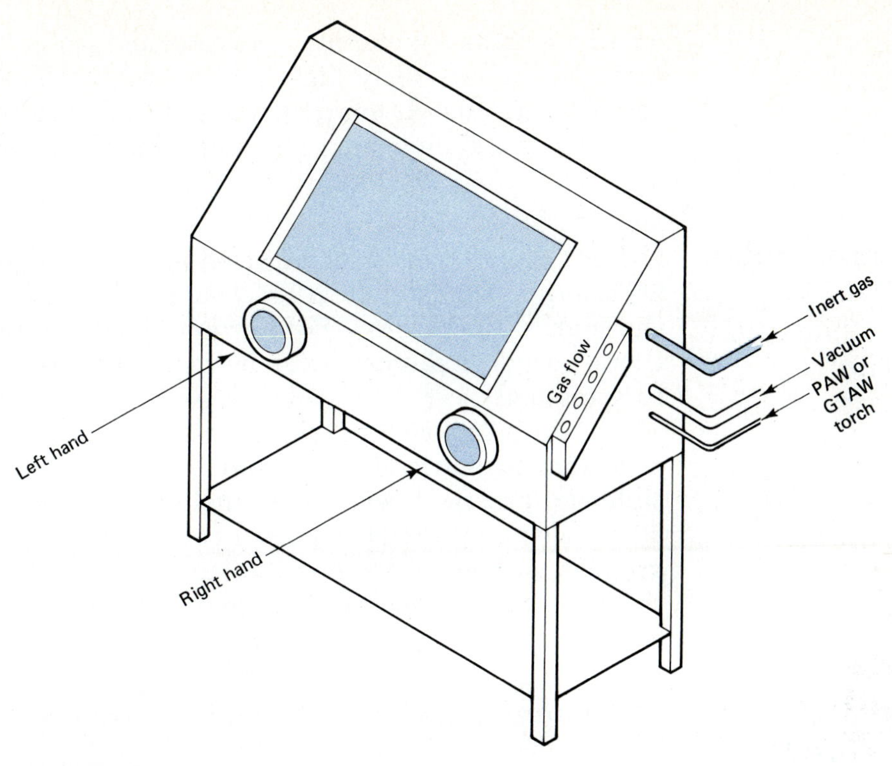

FIGURE 8–40
Dry box.

Other methods of "extra" shielding include using a back purge for open butt groove welds or trailing purges placed behind the torch as it travels along the joint.

The 2% thoriated (EWTH2) tungsten should be used for most all applications of GTAW of titanium. As mentioned earlier, the EWTH2 tungsten has a lower electron boil-off point and a higher melting temperature, thus making it excellent for welding this high-melting-point metal.

Additional Considerations In addition to extra shielding precautions, titanium requires extreme cleanliness. Degreasing with acetone will help. Any contaminants in the final titanium weld will cause a brittle weld.

GTAW OF REACTIVE AND REFRACTORY METALS

Reactive refractory metals are two separate classes of metals, but each has similar "reactive and refractory" characteristics: reactive metals, such as beryllium, titanium, and zirconium, absorb atmospheric gases (oxygen, hydrogen, and nitrogen) when heated or when in the molten state. The higher the metal temperature, the more rapid is this chemical reaction. Absolute shielding is required for these metals. Refractory metals such as columbium, molybdenum, tantalum, and tungsten are metals with extremely high melting points. They also share the high atmospheric absorbtion properties of the reactive properties. Absolute cleanliness and shield are also required when welding these metals.

General Rules for Welding Reactive and Refractory Metals

DCSP should be used for most all application on reactive metals. The 2% thorium tungsten (EWTH2) tungsten should be used to withstand the high temperatures. Pure argon shielding gas should be used, although argon-helium mix can be used for very high temperature refractory metals.

As mentioned previously with titanium, extreme cleanliness is necessary for successful welds on refractory and reactive metals. If possible and practical, weld these metals in a dry box; this will eliminate all outside contamination from the atmosphere. When welding open grooves without a dry box, be sure to use a back purge. Remember, these metals will absorb hydrogen, oxygen, and nitrogen at high temperatures.

Properties and Notes for Welding Refractory and Reactive Metals

Molybdenum (Mo) gas turbines, ramjets, and rocket motor parts

1. Color: silver, white (very hard)
2. Melting point = 4730°F
3. Density = 638
4. Welds can be brittle; preheat and postweld heat treating help

Zirconium (Zr) high-temperature nuclear components, chemical

1. Some alloys are Zircaloy 2 and Zircaloy 3
2. Melting point = 3360°F
3. Density = 402 lb/ft^3
4. Very weldable, but must have full shielding

Columbium (Cb) very corrosion resistant, high-temperature pipe for nuclear

1. Melting point = 4379°F
2. Density = 524 lb/ft^3
3. Will start to oxidize at 750°F
4. Extreme shielding required

Tantalum (Ta) high resistance to acids, used for pipes; looks like tungsten with blue tint

1. Melting point = 5425°F
2. Density = 218 lb/ft^3
3. Low thermal conductivity, can use helium to promote fluid
4. Welds best with no filler metal
5. Reacts with all gases except inert gases

Tungsten (W)

1. Melting point = 6170°F
2. Density = 1190 lb/ft^3
3. Very reactive with O_2
4. Should preheat

GTAW FILLER METALS

GTAW filler metals for carbon steel are classified under the American Welding Society's Section A5.18. Both GTAW and GMAW are covered in this section since they both use the same wire chemical compositions for welding carbon steels. The only difference is that the filler wire for GTAW will be cut into a 36-in. stick, whereas GMAW wires will be on a 25-, 50-, or 60-lb spool.

Some GTAW systems have a power feed, filler wire feed. These systems use the spool instead of the 36-in. stick because they continuously feed when the feed button is depressed. These systems make it easier on the welder to feed filler into the puddle. It is quite important that you understand what each digit in the classification system means. For example for ER70S-3:

E means "electric" or "electrode" and that the wire can conduct electricity when welding (for GMAW).

R means that the wire is also suitable for nonconductive filler rod and can be dipped into the puddle, as with GTAW and PAW.

70 indicates the minimum tensile strength of the deposited weld metal in ksi or thousand pounds per square inch (example: 70 = 70,000 psi).

S indicates a solid filler wire.

3 indicates the chemical content range of the wire. To find the range of any wire, check the ANSI/AWS specification.

For example, the specification for ER70S-3 is 0.06 to 0.15 carbon, 0.9 to 1.4 manganese, and 0.45 to 0.70 silicon. Each digit will vary and there may be more elements in the filler rod.

Here is a brief description of the steel AWS-classified GTAW wire.

- *ER70S-2:* Al, Zr, Ti triple deoxidized, good for welding dirty or rusty surfaces.
- *ER70S-3:* silicon content, most popular solid wire, similar to E70S-1 but contains more silicon.
- *ER70S-4:* super high silicon, good for large puddles and large welds.

- *ER70S-5:* triple deoxidized, similar to E70S-2, but available only in large diameters.
- *ER70S-6:* similar to E70S-3 but better deoxidized. Good for large welds; premium filler wire.
- *ER70S-7:* high manganese content; gives super deoxidation, good wetting action, and a smooth appearance. Like the 4 and 6 classes, it is good for large welds.
- *ER70S-G:* all-purpose wire; open-class filler can have any contents but must meet all requirements of ANSI/AWS A5.18 specification. E70S-G usually contains many deoxidizers, x-ray-quality welds, mild steel, and high-strength low-alloy. Some contain moly for pipe welding.

GTAW MACHINE SETTING SUMMARY

Use Table 8–5 as a quick reference for GTAW machine settings.

TABLE 8–5
GTAW Setup Chart

METAL	POLARITY	SHIELDING GAS[a]	TUNGSTEN[b]
Aluminum and aluminum–magnesium alloys	ACHF: DCSP can be used on dirty castings; use continuous high frequency (HF)	Argon or helium; use helium only on heavy sections	Pure zirconium (EWP, EWZR); zirconium is the preferred tungsten if available
Carbon steels	DCSP	Argon helium	2% thorium (EWTH2)
Copper and deoxidized copper alloys	DCSP	Helium or argon; Helium is the suggested shielding gas for copper	1% thorium (EWTH1); 2% thorium is also an acceptable tungsten
Brass and Bronze	DCSP/ACHF	Argon	1 or 2% thorium (EWTH1/EWTH2)
Stainless and NI alloys	DCSP	Argon	2% thorium (EWTH2)
NI alloys	DCSP	Argon/hydrogen	2% thorium (EWTH2)
Titanium and refractory metals	DCSP	Argon; always use additional shielding gas and back-purge, if necessary, for titanium	

[a] Use slightly high flow rates for helium.

[b] EWP, pure tungsten, green color code; EWTH1, 1% thorium tungsten, yellow color code; EWTH2, 2% thorium tungsten, red color code; EWZR, zirconium, brown color code.

REVIEW QUESTIONS

1. List some of the advantages of the GTAW process.
2. How is the weld puddle shielded with GTAW?
3. What is the purpose of high frequency used in GTAW?
4. What causes tungsten contamination?
5. What tungstens can be used for GTAW of aluminum?
6. What tungsten is used for mild steel and stainless steel?
7. What tungsten tip preparations are used for welding various metals?
8. What polarity is used for GTAW of mild steel?
9. What polarities are used for GTAW of aluminum?
10. What metals can be welded with GTAW?
11. What surface preparation must aluminum have before gas tungsten arc welding?
12. What does "EWTH2" stand for?
13. What effect does thorium have on tungsten?
14. Why is postpurge needed with GTAW?
15. Name the parts of the GTAW torch.
16. What are some of the problems encountered when GTAW-welding aluminum?
17. What are some of the problems encountered when GTAW-welding stainless steel?
18. What shielding gas works well when GTAW-welding copper?
19. What problems are encountered when GTAW-welding titanium?
20. What might ER70S-2 filler rod be used for?

GTAW PROJECT 1

Aluminum Flat Lap Joint

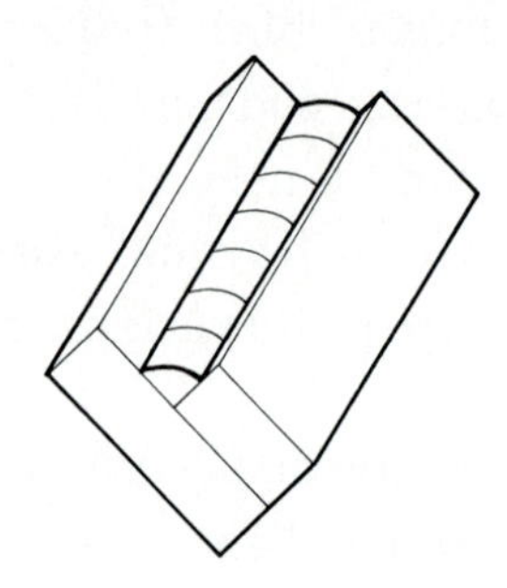

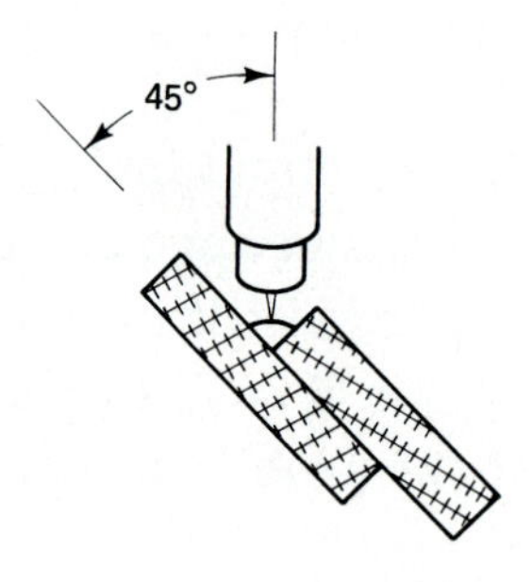

Material: Aluminum

Size: $\frac{1}{4}$ in. × 3 in. × 6 in. plate

Joint Design: Lap joint

Position: Flat fillet (1F)

Cleaning: Wire brush or chemical deoxidizer

Technique: Stringer beads

Polarity: ACHF or DCRP

Shield Gas: Argon

Gas Flow: 18–25 cfh

Filler Rod: 4043, $\frac{3}{32}$ in. diameter

Amperage: 15–145 A.

Procedure: Wire-brush or use a chemical deoxidizer on the plate surfaces. Using dc reverse polarity, ball the tungsten electrode by slowly increasing the amperage with the foot pedal (or other remote amperage control device, such as a thumb switch or thumb wheel). Once the tungsten end starts to round or ball, slowly remove the heat by coming off the foot pedal. Now switch to ac and turn the high frequency to the continuous position. Position and tack your plates into a double lap joint configuration. Set the amperage so that the puddle forms in a few seconds with the pedal all the way down. This will reduce the need for excessive pedal adjusting during welding. Deposit a bead into each side of the lap joint as you approach the end of the plate and back off the foot pedal to reduce the heat input into the bead, thus controlling the bead size and excessive cratering.

Visual Inspection: Beads should be uniform through their full length and should appear shiny on the surface. This indicates good deoxidation of the puddle. The bead should be well fused into the base metal and free of overlap. The end of the bead should be free of crater cracks and hot cracks.

GTAW PROJECT 2

Aluminum Flat Tee Joint

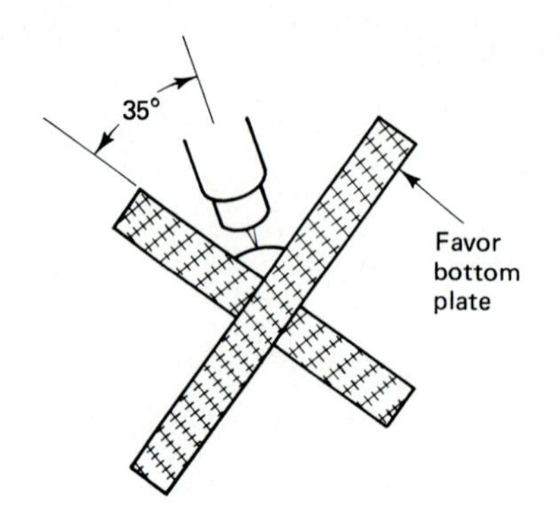

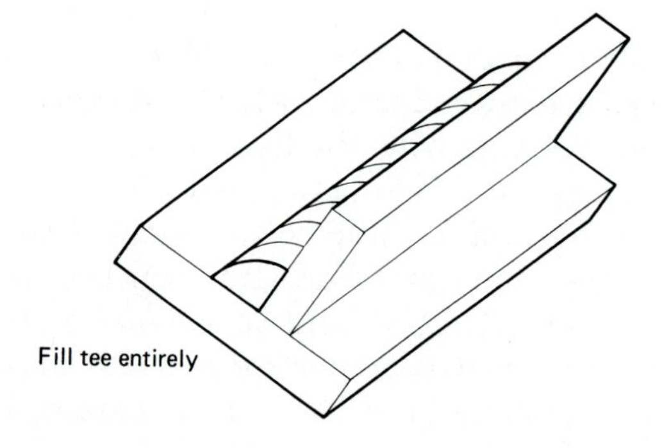

Material: Aluminum

Size: $\frac{1}{4}$ in. × 3 in. × 6 in. plate

Joint Design: Tee joint

Position: Flat fillet (1F)

Technique: Stringer beads

Polarity: ACHF or DCRP

Shield Gas: Argon

Gas Flow: 18–25 cfh

Filler Rod: 4043, $\frac{3}{32}$ in. diameter

Amperage: 120–155 A.

Cleaning: Wire brush or chemical deoxidizer

Procedure: Wire-brush or use a chemical deoxidizer on the plate surfaces. Using dc reverse polarity, ball the tungsten by slowly increasing the amperage with the foot pedal (or other remote amperage control device, such as a thumb switch or thumb wheel). Once the tungsten end starts to round or ball, slowly remove the heat by coming off the foot pedal. Now switch to ac and turn the high frequency to continuous operation. Position and tack your plates into a double-sided tee joint. Set the amperage so that the puddle forms in a few seconds with the pedal all the way down. This will reduce the need for excessive pedal adjustment during welding. Deposit a bead into each of the four tee joints. As you approach the end of the plate, back off the foot pedal to reduce the heat input into the bead, thus controlling the bead size and excessive cratering.

Visual Inspection: Beads should be uniform through their full length and should appear shiny on the surface. This indicates good deoxidation of the puddle. The bead should be well fused into the base metal and free of overlap. The end of the bead should be free of crater cracks and hot cracks.

GTAW PROJECT 3

Aluminum Horizontal Tee Joint

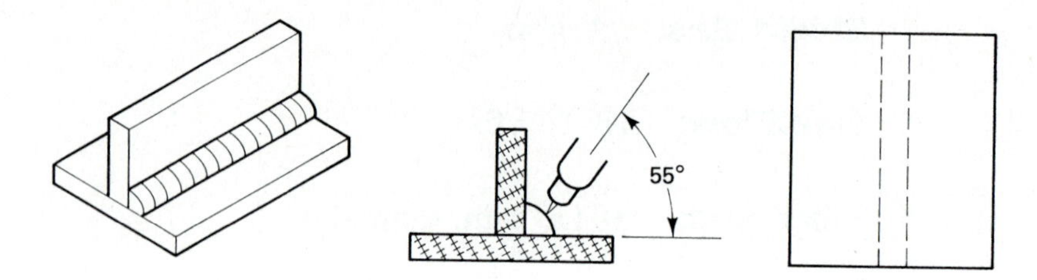

Material: Aluminum

Size: $\frac{1}{4}$ in. × 3 in. × 6 in. plate

Joint Design: Tee joint

Position: Horizontal fillet (2F)

Technique: Stringer beads

Polarity: ACHF and DCRP

Shield Gas: Argon

Gas Flow: 18–25 cfh

Filler Rod: 4043, $\frac{3}{32}$ in. diameter

Amperage: 120–155 A.

Cleaning: Wire brush or chemical deoxidizer

Procedure: Wire-brush or use a chemical deoxidizer on the plate surfaces. Using dc reverse polarity, ball the tungsten by slowly increasing the amperage with the foot pedal (or other remote amperage control device, such as a thumb switch or thumb wheel). Once the tungsten end starts to round or ball, slowly remove the heat by coming off the foot pedal. Now switch to ac and turn the high frequency to continuous operation. Position and tack your plates into a double-sided tee joint. Set the amperage so that the puddle forms in a few seconds with the pedal all the way down. This will reduce the need for excessive pedal adjustment during welding.

Visual Inspection: Beads should be uniform through their full length and should appear shiny on the surface. This indicates good deoxidation of the puddle. The bead should be well fused into the base metal and free of overlap. The end of the bead should be free of crater cracks and hot cracks.

GTAW PROJECT 4

Aluminum Flat Groove Joint

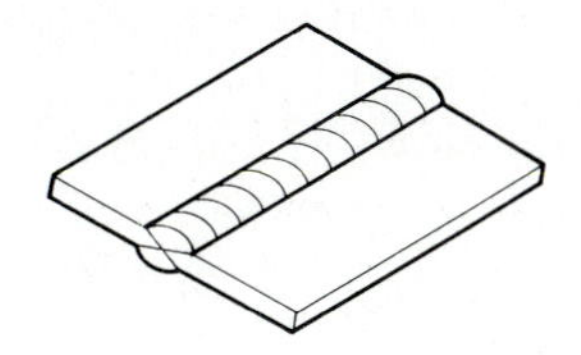

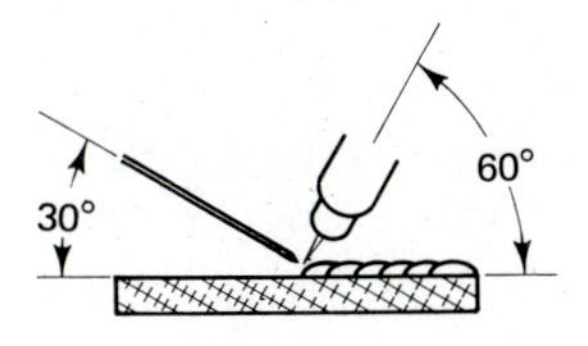

Material: Aluminum

Size: $\frac{1}{4}$ in. × 4 in. × 6 in. plate

Joint Design: Double Vee-groove

Position: Flat groove (1G), 60°–75° included angle

Technique: Stringer beads

Polarity: ACHF and DCRP

Shield Gas: Argon

Gas Flow: 18–28 CFH

Filler Rod: 4043, $\frac{3}{32}$ in. diameter

Amperage: 110–155 A.

Cleaning: Wire brush or chemical Deoxidizer

Procedure: Using the dc reverse polarity, ball the tungsten by slowly increasing the amperage with the foot pedal. Once the tungsten end starts to round or ball slowly, remove the heat by letting off the foot pedal. Now switch to ac and turn the high frequency to continuous operation. Bevel, position, and tack both ends of your butt joint. Set the amperage so that the puddle forms in a few seconds with the pedal all the way down. This will reduce the need for excessive pedal adjustment during welding. Wire-brush or use a chemical deoxidizer on the edges to be welded. Deposit a stringer bead on the first side. Examine the root side of the joint for penetration, then deposit a bead into the other side of the groove joint. As you approach the end of the plate and back off the foot pedal to reduce the heat input into the bead, thus controlling the bead size and excessive cratering.

Visual Inspection: Beads should be uniform through their full length, and should be well fused into the base metal and free of overlap. The end of the bead should be free of crater cracks and hot cracks.

GTAW PROJECT 5

Aluminum Vertical Lap; Tee Joint

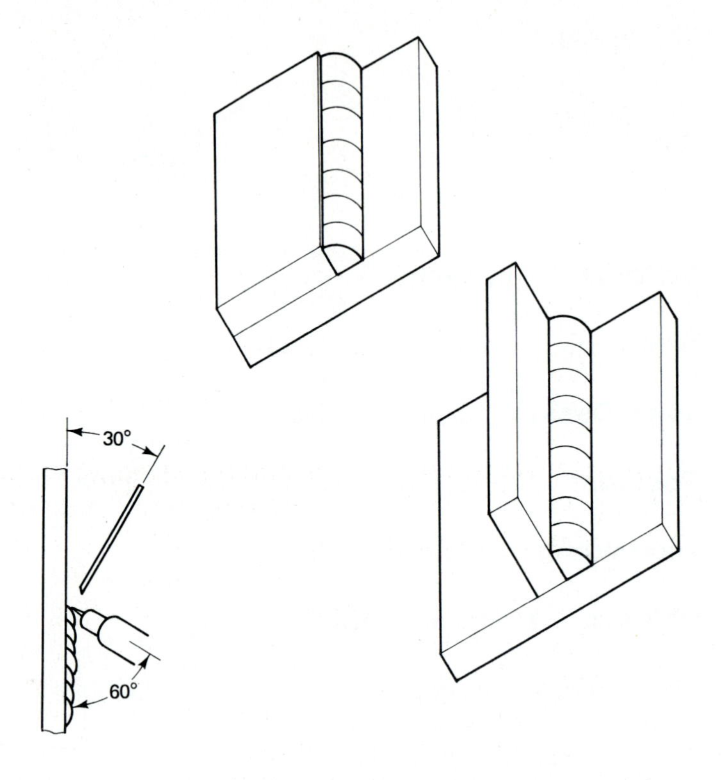

Material: Aluminum

Size: $\frac{1}{4}$ in. × 3 in. × 6 in. plate

Joint Design: Tee joint

Position: Vertical fillet (4F)

Technique: Stringer beads

Polarity: ACRP and DCRP

Shield Gas: Argon

Gas Flow: 18–25 cfh

Filler Rod: 4043, $\frac{3}{32}$ in. diameter

Amperage: 115–155 A.

Cleaning: Wire brush or chemical deoxidizer

Procedure: Wire-brush or use a chemical deoxidizer on the plate surfaces. Using dc reverse polarity, ball the tungsten by slowly increasing the amperage with the foot pedal (or other remote amperage control device, such as a thumb switch or thumb wheel). Once the tungsten end starts to round or ball, slowly remove the heat by coming off the foot pedal. Now switch to ac and turn the high frequency to continuous operation. Position and tack your plates into a double-sided tee joint. Set the amperage so that the puddle forms in a few seconds with the pedal all the way down. This will reduce the need for excessive pedal adjustment during welding. Deposit a vertical up bead into each side of the four tee joints. As you approach the end of the plate, back off the foot pedal to reduce the heat input into the bead, thus controlling the heat input into the bead. This will continuously control the bead size and prevent excessive cratering.

Visual Inspection: Beads should be uniform through their full length and should appear shiny on the surface. This indicates good deoxidation of the puddle. The bead should be well fused into the base metal and free of overlap. The end of the bead should be free of crater cracks and hot cracks.

GTAW PROJECT 6

Stainless Steel Flat Lap Joint

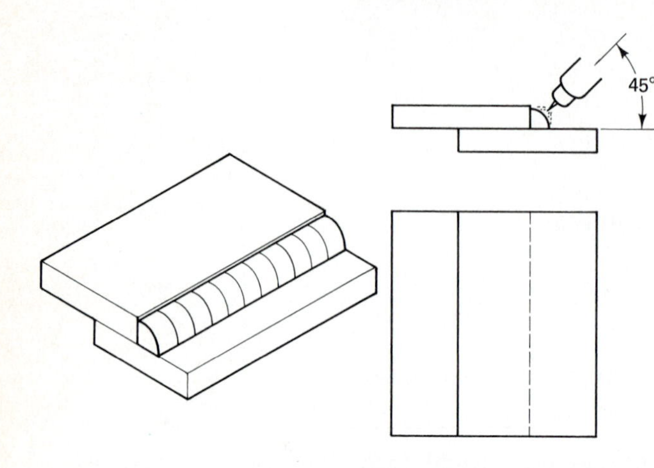

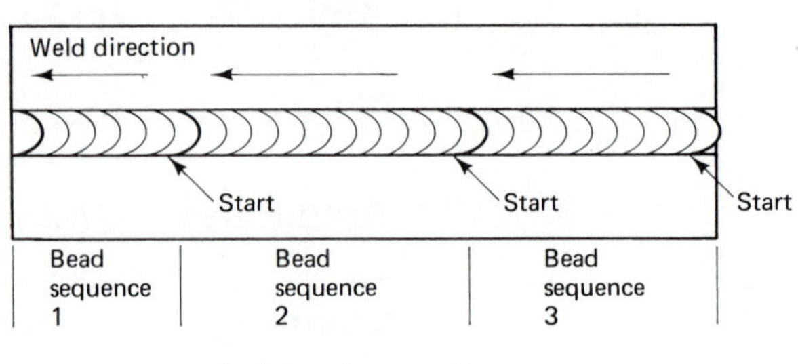

Back-stepping procedure

Material: Stainless steel

Size: $\frac{3}{16}$ in. × 3 in. × 5 in. plate

Joint Design: Lap joint

Position: Horizontal fillet (2F)

Technique: Small stringer

Polarity: DCSP (DCEN)

Shield Gas: Argon

Gas Flow: 12–22 cfh

Filler Rod: ER308, $\frac{3}{32}$ in. diameter

Amperage: 110–145 A.

Cleaning: Wire brush

Procedure: Clean both the surfaces to be welded and the filler rod. Grind the tungsten to a sharp point. Position and tack plates into a lap joint configuration. Deposit small beads into each side of the lap joint, positioning the plate into the 2F position each time. (The fillet weld leg length should not exceed the thickness of the base metal.) If distortion occurs, use the backstepping technique. Wire-brush the completed welds.

Visual Inspection: Welds should show complete fusion at the toes and should be a bright silver or light blue color. Dark purple beads indicate excessive heat. Absolutely no cracks, inclusions, or porosity should be visible.

GTAW PROJECT 7

Stainless Steel Flat Tee Joint

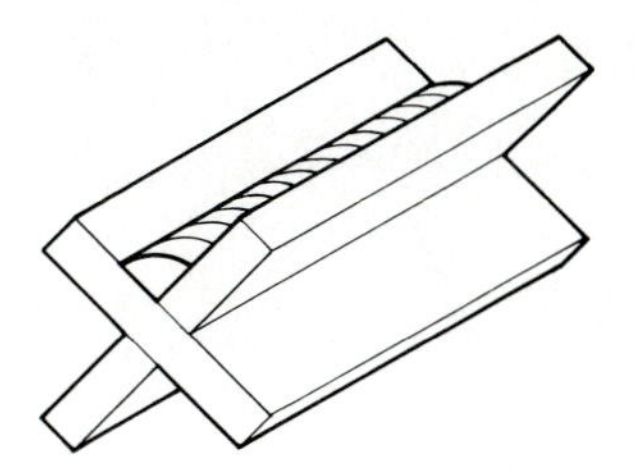

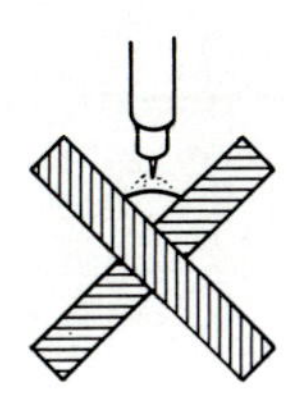

Material: Stainless steel

Size: $\frac{3}{16}$ in. × 3 in. × 5 in. plate

Joint Design: Tee joint

Position: Flat fillet (1F)

Technique: Small stringers

Polarity: DCSP (DCEN)

Filler Rod: ER308, $\frac{3}{32}$ in. diameter

Shield Gas: Argon

Gas Flow: 12–20 cfh

Amperage: 75–105 A.

Cleaning: Wire brush

Procedure: Clean both the surfaces to be welded and the filler rod. Grind the tungsten to a sharp point. Position and tack plates into a lap configuration. Deposit small beads into each of the four sides of the tee joint, positioning the plate into the 1F position each time. (The fillet weld leg length should not exceed the thickness of the base metal.) If distortion occurs, use the backstepping technique. Wire-brush the completed welds.

Visual Inspection: Welds should show complete fusion at the toes and should be a bright silver or light blue color. Dark purple indicates excessive heat. Absolutely no cracks, inclusions, or porosity should be visible.

GTAW PROJECT 8

Stainless Steel Vertical Tee Joint

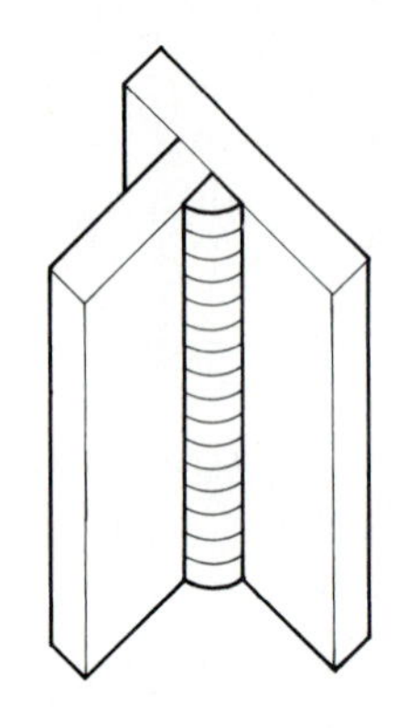

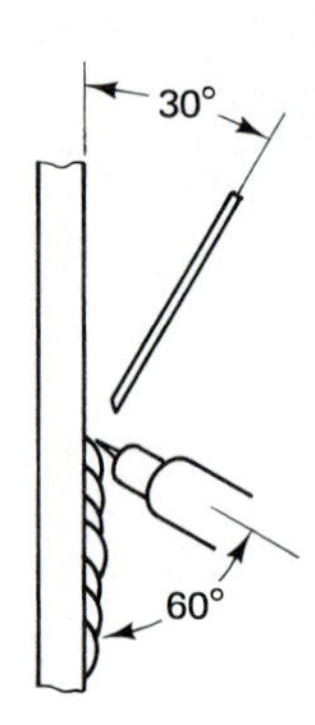

Material: Stainless steel

Size: $\frac{3}{16}$ in. × 3 in. × 5 in. plate

Joint Design: Tee joint

Position: Vertical fillet (3F)

Technique: Small stringers

Polarity: DCSP (DCEN)

Filler Rod: ER308, $\frac{3}{32}$ in. diameter

Shield Gas: Argon

Gas Flow: 12–20 cfh

Amperage: 75–105 A.

Cleaning: Wire brush

Procedure: Clean both the surfaces to be welded and the filler rod. Grind the tungsten to a sharp point. Position and tack plates into a vertical tee-joint configuration. Deposit small beads into each side of the tee joint, positioning the plate into the 3F position each time. (The fillet weld leg length should not exceed the thickness of the base metal.) If distortion occurs, use the backstepping technique. Wire-brush the completed welds.

Visual Inspection: Welds should show complete fusion at the toes and should be a bright silver or light blue color. Dark purple indicates excessive heat. Absolutely no cracks, inclusions, or porosity should be visible.

GTAW PROJECT 9

Stainless Steel Flat Groove Joint

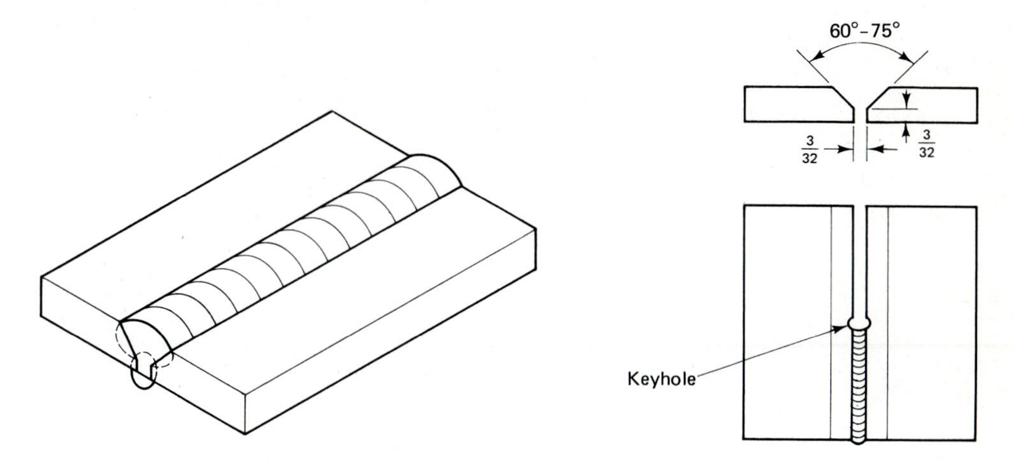

Material: Stainless steel

Size: $\frac{1}{4}$ in. × 3 in. × 5 in. plate

Joint Design: Vee-groove

Position: Flat groove (1G), 60–75° included angle

Technique: Small stringers

Polarity: DCSP (DCEN)

Shield Gas: Argon

Gas Flow: 15–24 cfh

Filler Rod: ER308, $\frac{3}{32}$ in. diameter

Amperage: 75–135 A.

Cleaning: Wire brush

Procedure: Clean both the surfaces to be welded and the filler rod. Grind the tungsten to a sharp point. Position and tack plates into a butt-joint configuration, leaving about $\frac{1}{16}$ in. of root opening. Deposit the root pass using the keyholing technique. Clean the joint. Now deposit filler beads until the joint is filled to its full cross section. Wire-brush the completed welds.

Visual Inspection: Welds should show complete penetration at the root and no unfused edges. The bead color should be a light silver-blue. A dark purple bead indicates excessive heat. No cracks, inclusions, or porosity should be visible.

GTAW PROJECT 10

Stainless Steel Vertical Groove Joint

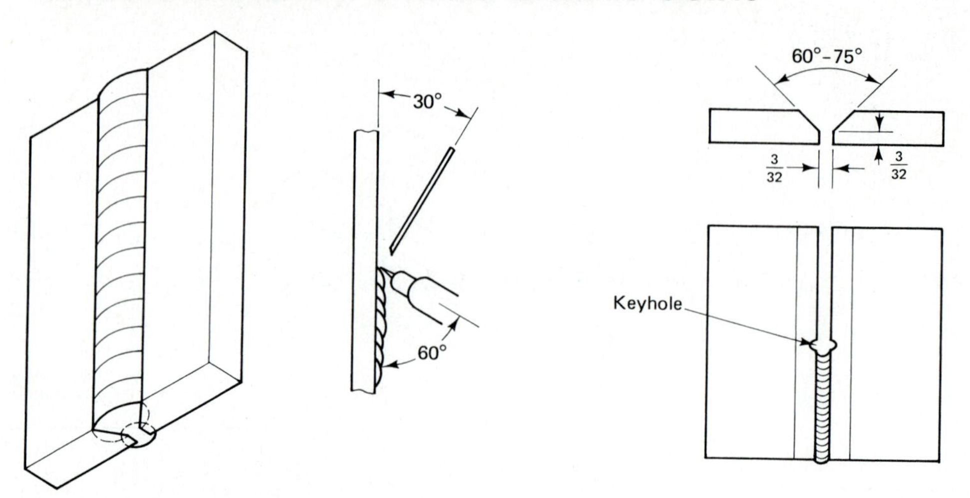

Material: Stainless steel

Size: $\frac{1}{4}$ in. × 3 in. × 5 in. plate

Joint Design: V-groove

Position: Vertical groove (4G), 60–75° included angle

Technique: Small stringers

Polarity: DCSP (DCEN)

Shield Gas: Argon

Gas Flow: 15–24 cfh

Filler Rod: ER308, $\frac{3}{32}$ in. diameter

Amperage: 75–135 A.

Cleaning: Wire brush

Procedure: Clean both the surfaces to be welded and the filler rod. Grind the tungsten to a sharp point. Position and tack plates into a butt-joint configuration, leaving about a $\frac{1}{16}$-in. root opening. Deposit the root pass using the keyholing technique. Clean the joint. Now deposit the filler beads until the joint is filled to its full cross section. Wire-brush the completed welds.

Visual Inspection: Welds should show complete penetration at the root and no unfused edges. The bead color should be a light silver-blue. A dark purple bead indicates excessive heat. No cracks, inclusions, or porosity should be visible.

PULSED GTAW

The pulsed GTAW process is similar to the simple GTAW process; however, it utilizes a high–low current pulsation. This would be similar to the welder pumping the remote amperage control (foot pedal) up and down. However, the high–low amperage pulse is instant or what we call a square-wave pulse (Figure 9–1a). Pumping the foot pedal would *not* produce a square-wave pulse, but a graduated pulse (Figure 9–1b).

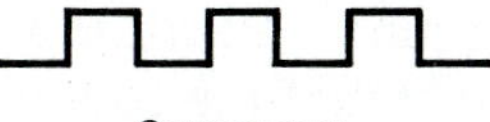

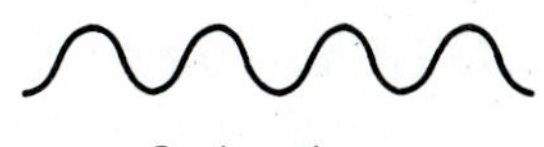

FIGURE 9–1 Square wave and graduated wave pulses.

ADVANTAGES OF PULSED GTAW

The advantages of pulsed GTAW are:

1. Less heat input and deep penetration
2. Less distortion
3. Easy to weld thick-to-thin sections
4. Less foot pedal control manipulation
5. Excellent puddle control for out-of-position welding

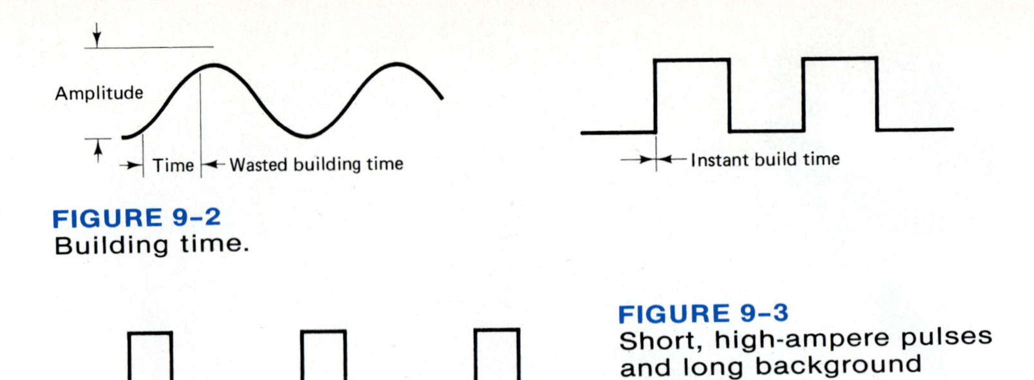

FIGURE 9–2
Building time.

FIGURE 9–3
Short, high-ampere pulses and long background current.

The most important advantage of the pulsed GTAW process is the fact that it actually produces the same weld as standard GTAW but with much less heat input into the metal. This actually happens due to the pulsating action. As the heat (amp) level is reached, penetration and heat are very quickly put into the plate. However, before the plate is saturated with excess heat, the pulsed GTAW drops off to the background current, where the level of heat input is much less but the arc is maintained. Another factor that reduces the overall heat input is the square-wave pulse. The square-wave pulse produces an instant "on"–instant "off" cycle of pulsing current. There is no wasted time where amperage was building and failing (Figure 9–2). The results of this lower heat input are smaller heat-affected zones and a less brittle, more ductile weld. Distortion is, of course, a direct result of heat expansion and is therefore reduced.

Thick-to-thin sections are a problem on most metals. Usually by the time the thicker plate starts to melt, the thinner plate has long since melted through. Pulsed GTAW reduces this problem. With pulsed GTAW, as the melting temperature is reached, both plates start to melt. The pulse frequency is set for short, high-amperage pulses and long background current settings (Figure 9–3). This allows just enough heat to surge and melt both plates, but drops off before the thin plate melts through.

Pulsed GTAW requires less foot pedal control because of the lower heat input. For example, when welding aluminum with conventional GTAW, more heat must be used at the start of the bead and much less heat at the end of the bead (due to aluminum's high-thermal-conductivity characteristics). This requires a lot of foot pedal movement to control the heat and puddle. With pulsed GTAW, once the puddle size is set with the foot pedal position, very little movement is required to maintain that same puddle size. This keeps out-of-position beads uniform.

CONTROLS FOR PULSED GTAW

Pulsed GTAW gives the welder total arc and heat control. The size and shape of the pulses can be controlled with the following main pulse controls (Figure 9–4):

PHOTO 9–1
Airco square-wave pulsed GTAW welder. (Courtesy of ESAB Welding Products, Inc.)

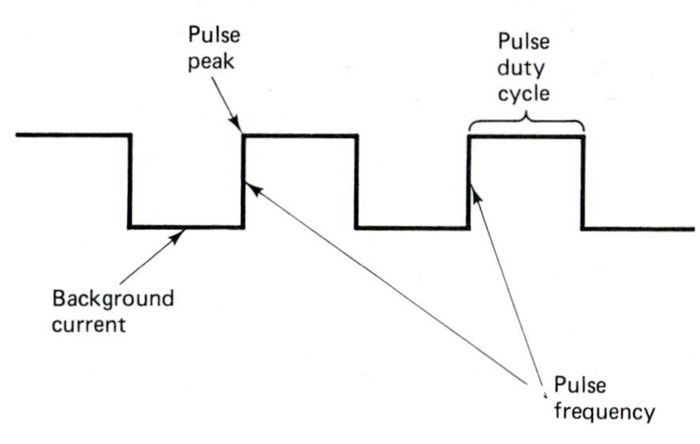

FIGURE 9–4
GTAW pulse controls.

1. Pulse peak current
2. Background current
3. Pulse frequency or pulses per second
4. Pulse duty cycle

The *pulse peak* current is the *high heat level* of the pulse cycle. For the most part this control is set fairly high on an infinite amperage control. The *background* current sets the low heat level. This should, of course, be set lower than the pulse peak current. The *pulses per second* (pps), or pulse frequency as some machine manufacturers call this control, is the pulse repetition rate, or how often the current pulses to the high heat level. The *pulse duty cycle* is the pulse peak duration as a percentage of total time. This actually controls how long the high heat level is maintained before it drops off to the background.

The degree of current high–low, or pulses per second, and fast–

slow is totally dependent on the given conditions for each situation. The variables for machine settings are:

1. The type of metal and the properties that metal possesses (thermal conductivity, hot shortness, etc.)
2. The metal's thickness (its ability to handle heat based on its volume)

Additional Controls for Pulsed TIG GTAW

Some additional controls are appearing on new pulsed TIG machines for increased application and flexibility. They include:

DCRP/DCSP Balance Control This control can adjust the balance of DCRP and DCSP in the ac sine wave. For example, if you need additional oxide cleaning action, increase the reverse polarity balance; for more penetration, increase the straight polarity balance (Figure 9–5). This control is ac only.

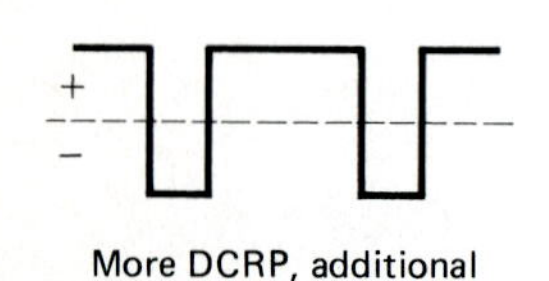

More DCRP, additional oxide cleaning

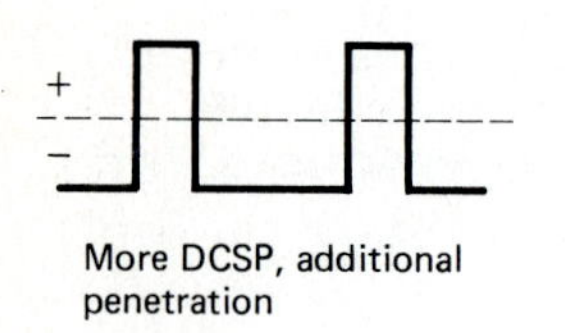

More DCSP, additional penetration

FIGURE 9–5 Ac balance control.

Start Current Control (Slow Start). This controls the speed at which the current will reach its set amperage. With this control you can produce a rapid start and puddle, or a soft start, and slowly develop a puddle.

Crater Fill This control allows the welder to reduce craters by slowly dropping off current at the end of the weld. This also helps to reduce crater cracks.

Preflow This allows the shielding gas to start flowing before the arc is started, assuring shielding gas coverage.

Amp Preset This allows the welder to preset the amperage digitally without striking an arc.

Spot Time Control This allows for spot or intermittent welding with automatic timing. Every time an arc is started, the spot timer starts and will stop the arc at a preset time. This keeps intermittent and spot welds uniform as well as giving all welds the same arc time.

PULSED GTAW MACHINE SETTINGS

Do not make the mistake of thinking that pulsed GTAW is infallible; you can burn through, overheat, and crack the weld or experience any of the problems you would with conventional GTAW if you do not set the machine correctly for the given condition. The following conditions are possible with pulsed GTAW (Figure 9–6):

	Pulse peak	Background	pps	Pulse duty cycle	
High-pulse duty cycle	80-150	50%	1	80%	
High-pulse peak/low-pulse duty cycle	100-150	40%	0.5	30%	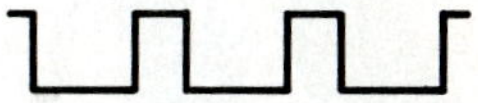
Rapid pps	100-150	60%	3	50%	
Slow pps/low-pulse duty cycle	100-150	40%	0.5	20%	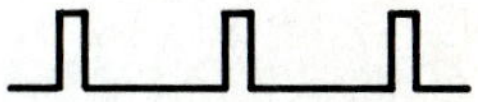

Exact settings may vary among machines

FIGURE 9–6
Pulsed GTAW machine settings.

- *High pulse duty cycle (pulse peak duration).* This setting produces considerable heat with only occasional drops to background. This would be good for thicker materials but would still produce a slightly lower heat input over conventional GTAW.
- *High pulse peak–low pulse duty cycle.* This setting would be excellent for thick-to-thin or moderately thin sections. Again, the pulse peaks are very high but the durations are very short, making this setting good for thick-to-thin sections.
- *Fast pps or high pulse frequency.* This setting produces rapid, fast pulsing and in turn produces puddle stimulation. A very uniform bead is the result.
- *Slow pps or low pulse frequency and low pulse duty cycle.* This setting would be primarily for extremely thin or low-melting temperature metals.

Additional settings and combinations of these basic settings are possible for each unique situation. Experiment with your machine, trying different settings and combinations. This is the best way to get the most out of pulsed GTAW.

REVIEW QUESTIONS

1. What are the advantages of pulsed GTAW?
2. What does the background control do?
3. What does the pulse peak control do?
4. What effect does the start current control have on the pulsed GTAW process?
5. How might you set the pulsed GTAW machine for thick to thin metal?

PLASMA ARC WELDING (PAW)

Plasma arc welding (PAW) is a process that stems from the GTAW process. It creates heat for welding in much the same way that GTAW does. It ionizes the gases between the tungsten and work, but with plasma we restrict and condense this plasma gas into a high-velocity plasma jet (Figure 10–1).

ADVANTAGES OF PAW

By controlling the oriface gas flow rate, PAW can actually control the amount of penetration. High oriface gas pressure will produce a stiff arc and deep penetration; lower-pressure oriface gas will produce a soft arc and less penetration (Figure 10–2). This gives welders a controllability they have not experienced with other processes.

The penetration with PAW can also be narrow and deeper than GTAW. This is due to the denser plasma gas. This also results in a lower heat-affected zone. There is no problem of touching the tungsten in the puddle as with GTAW. This is because the PAW tungsten is placed back into the torch, and does not stick out.

The torch stand-off distance can be much greater with PAW than with GTAW. This allows the welder to reach into deep, hard-to-

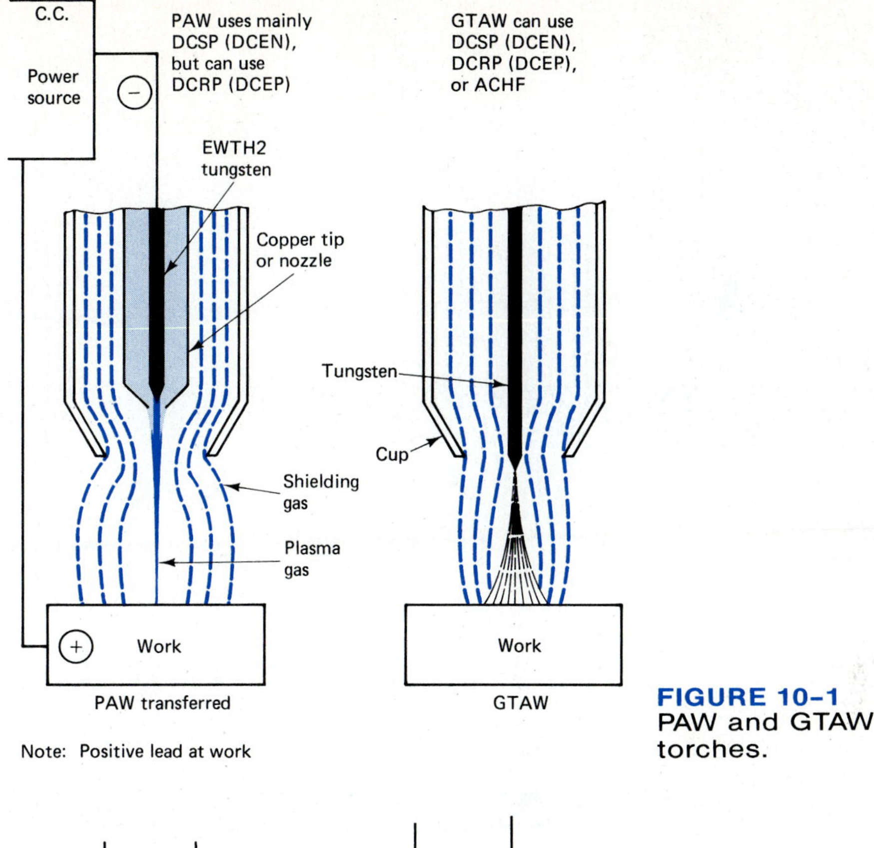

FIGURE 10–1 PAW and GTAW torches.

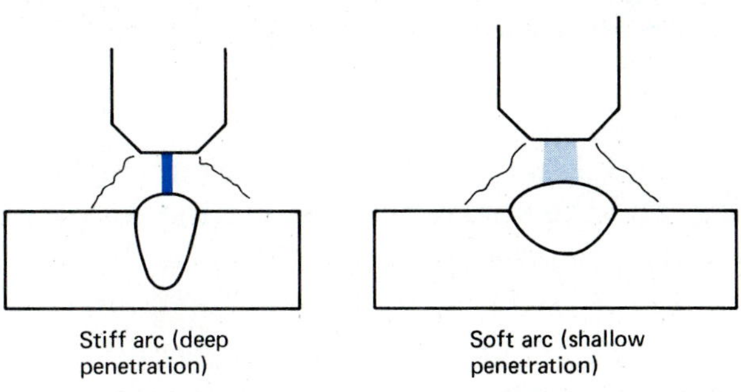

FIGURE 10–2 Stiff arc and soft arc.

get-at areas. PAW can also perform keyhole welding techniques. More about keyholing later. Along with these advantages, PAW shares all the advantages mentioned for GTAW in Chapter 8.

PLASMA THEORY

The basic states of matter are solids, liquids, and gases. Plasma gas is known as the fourth state of matter because it is a gas that is being ionized or is conducting electricity. Plasma gas acts differently than standard gas; for example, researchers have found that if the surrounding shield gas is cooled, the primary plasma gas will condense and become denser. In Chapter 8 we discussed plasma gas in some detail.

There are two types of plasma applications: (1) transferred and (2) nontransferred (Figure 10–3). The transferred system uses the work to complete the circuit, whereas with nontransferred the circuit is com-

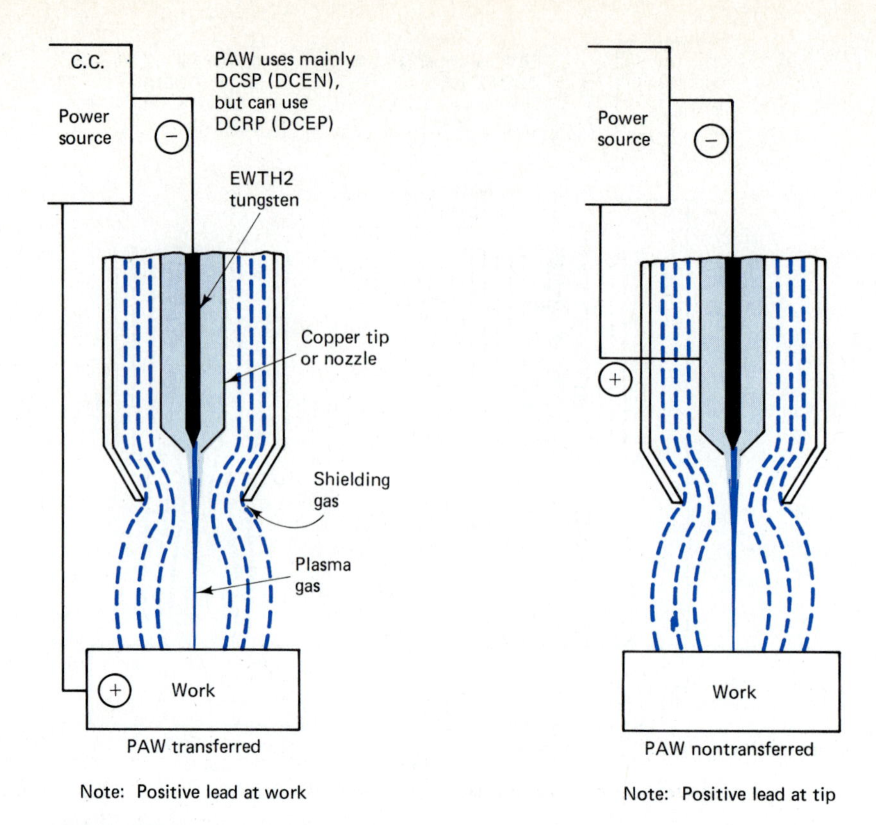

FIGURE 10-3
PAW transferred and PAW nontransferred.

pleted through the tip. The transferred method is used for welding and cutting, and the nontransferred is used for plasma spray surfacing and some cutting applications (see Chapter 18).

PAW GASES

Two basic types of gases are used with PAW: (1) the primary plasma gas (also known as the oriface gas), and (2) the secondary shielding gas. Primary plasma gas is what actually conducts electricity and creates the heat for welding. Therefore, if we control the primary gas, we can control the size, shape, and penetration of the weld. The primary gas can be set "soft" for low penetration on thin material or set higher to create a "stiff" arc for deeper penetration. Pure argon is almost always used as the plasma gas (oriface gas) for PAW.

The secondary gas or shielding gas produces an additional, outer shielded zone. This keeps any foreign gases away from the plasma gas and weld. If any foreign gases were to mix with the plasma gas, it would cause it to become erratic and turbulent, in addition to contaminating the weld puddle. The secondary gas is usually argon, but other gases, such as CO_2, argon-hydrogen, or argon helium, can be used. For example, hydrogen or helium may be added to the secondary gas to increase fluidity for nickel-steel plasma arc welding.

POLARITIES USED WITH PAW

For most PAW applications DCSP (DCEN) is used. Remember, this polarity directs most of the heat at the plate. It also causes the electrons to flow in the same direction as the plasma gas flow, contributing to the effectiveness of the process.

The use of DCRP (direct current reverse polarity) is limited almost exclusively to the plasma arc welding of aluminum and magnesium. Special high-heat oriface tips are used to absorb the heat in the tip created by DCRP. DCRP is used on aluminum and magnesium for the same reason that it is used occasionally with GTAW. DCRP helps to remove the oxides on aluminum and magnesium, which will contaminate the weld if not removed.

PAW TORCH

The key to the PAW process is its unique torch. Even though PAW uses the same constant-current power source as GTAW, it has a special plasma control unit and torch. It is a bit bulkier than a GTAW torch because it contains additional parts, such as the gas distributor and tip (Figure 10-4). These parts are responsible for directing the plasma gas. The plasma gas is created inside the torch tip and creates a large amount of heat; therefore, torches are almost always water cooled.

It is very important that PAW torches be assembled exactly as the manufacturer has designed. If any parts are left out when reassembling, improper arcing may overheat the torch and destroy it.

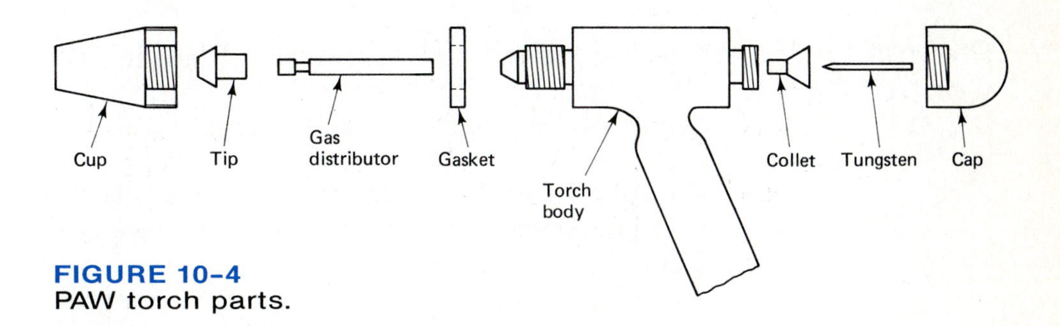

FIGURE 10-4
PAW torch parts.

PAW TUNGSTEN ELECTRODES

The 2% thoriated tungsten (EWTH2—red color code) is the best tungsten for PAW. It should be sharpened to a point with a vertex angle of 20 to 60° (Figure 10-5). Tungstens must also be "gaged." Gaging the

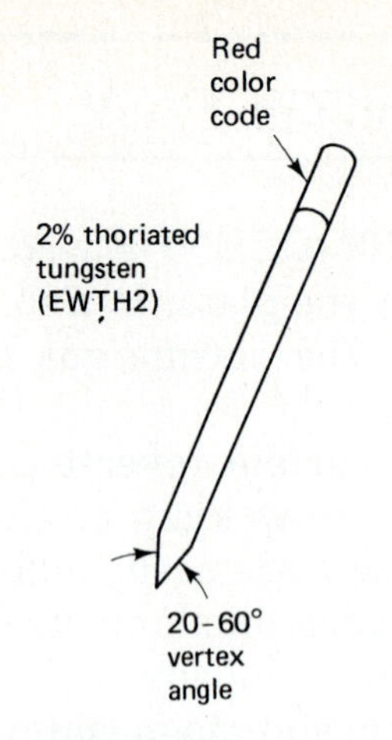

FIGURE 10-5
Thoriated tungsten angles.

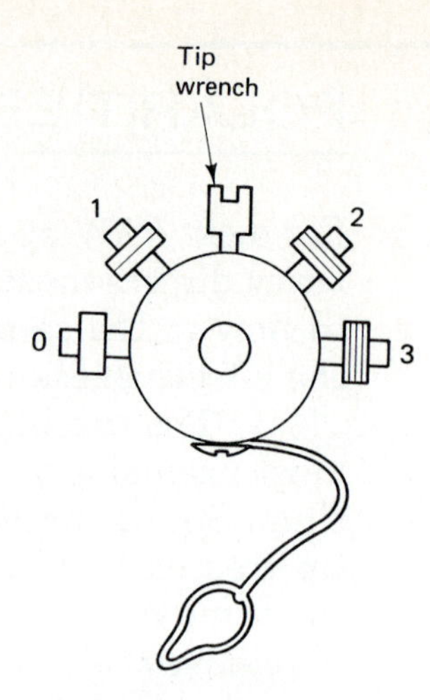

FIGURE 10-6
PAW tip gage.

tungsten is adjusting how far the tungsten extends out into the oriface tip. The exact distance is determined by the tip that is used. Each plasma torch manufacturer gives suggested gaging distances. The gap is set by a special PAW tip gage (Figure 10-6). This gage is pushed up into the torch before the tungsten is fully tightened. The tungsten is then tightened, locking it into its gaged positions.

PAW MACHINE CONTROLS

As stated previously, there are two main parts to the PAW machine: the constant-current power source and the plasma control panel (Figure 10-7).

Plasma Control Panel

Plasma Flow 1 This control adjusts low flow rates in the primary plasma gas. Its adjustment ranges from 0 to 10 cubic feet per hour (cfh) and is set for standard bead welding.

Plasma Flow 2 This control adjusts high flow rates and is activated for keyhole technique welding. Its range is usually between 0 and 5 cfh.

Shielding Flow This sets the rate of shielding gas flow. It is usually set between 20 and 30 cfh. Its range is usually 0 to 60 cfh.

High/Low Plasma Switch This two-position switch sets either plasma flow 1 or both plasma flow 1 and 2 simultaneously. The plasma flow 1 position is for standard bead welding, and the plasma flow 1 and 2 setting is for keyhole welding.

Pilot Control Switch The pilot arc is a high-frequency, low-current arc that takes place in the tip. The pilot helps start the arc,

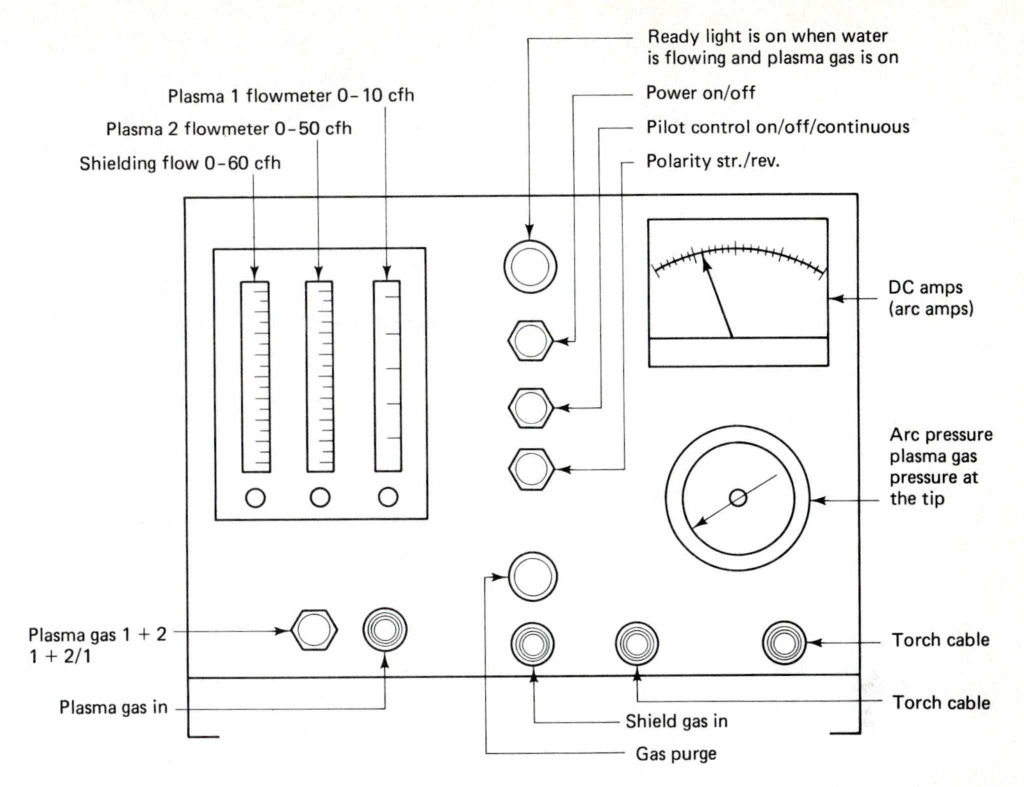

FIGURE 10-7
PAW control panel.

much as the high frequency does with GTAW. The pilot arc has a stabilizing effect on low-current settings and can actually be used all by itself to weld extremely thin materials. The pilot arc is usually controlled by a three-position "on-off continuous" switch. The "on" position keeps the pilot on until the primary current comes in; then it shuts off. This position should be used for amperages above 50. The continuous setting keeps the pilot on continuously while welding and will not shut off. It will maintain the arc even at low amperages; therefore, use the continuous setting for very thin material, low-amperage applications.

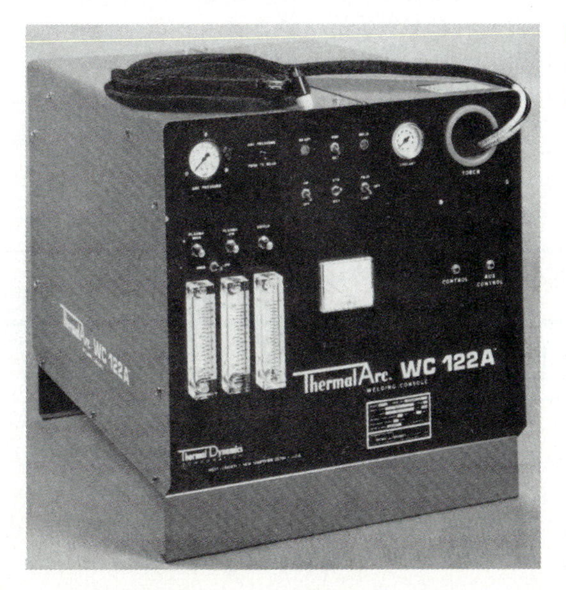

PHOTO 10-1
PAW power source and control panel. (Courtesy of Thermal Dynamics Corp.)

Straight/Reverse Polarity Switch This switch simply allows the welder to select between direct current straight polarity or direct current reverse polarity. DCSP will be used for almost all applications of PAW, except that DCRP will be used for aluminum and magnesium.

Ready Light Before attempting to weld, always make sure that the ready light is on. The ready light will come on only if coolant is flowing and gas pressure is available at the torch. Most PAW panels have a gas purge button to set the gas pressures on all flowmeters.

Power Source

The *power source* is set for PAW much the same as it is for GTAW. Amperage adjustments will be made at the power source, but most other adjustments are made at the plasma control unit. Always make sure the polarities of the power source and plasma control unit match.

Machine Settings for Standard PAW

With the possible combinations of settings, how do we get just the right setting for a particular job? If some controls, such as plasma flow 1, are not correctly set, the torch tip may melt. Each manufacturer of PAW equipment has a precise setup chart. Always consult this chart for the best results. We have provided Table 10-1 to help you with approximate settings for varying situations.

TABLE 10-1
PAW Setup Chart

POLARITY	MAXIMUM CURRENT (A)	PRIMARY PLASMA GAS TYPE	PLASMA GAS FLOWS (CFH)		SHIELD GAS (FLOW RATE 20-30 CFH)	RECOMMENDED ORIFICE (IN.) SIZE	TIP GAGE SET
			SOFT ARC	STIFF ARC			
Straight	30	Argon	0.5-1.0	1.0-4.5	Argon	0.031	0
	50	Argon	0.5-1.5	1.5-4.5	Argon	0.046	
	75	Argon	0.5-2.0	2.0-4.5	Argon	0.062	
	100	Argon	0.5-2.5	2.5-4.5	Argon	0.081	
	10	Argon	0.5-1.0	1.0-4.5	Argon	0.031	1
	20	Argon	0.5-1.5	1.5-4.5	Argon	0.046	
	55	Argon	0.5-2.0	2.0-4.5	Argon	0.062	
	75	Argon	0.5-2.0	2.0-4.5	Argon	0.081	
	100	Argon	0.5-2.5	2.5-4.5	Argon	0.093	
Reverse	30	Argon	0.4-0.5	0.5-4.5	Helium	0.046	2
	45	Argon	0.8-1.0	1.0-4.5	Helium	0.062	
	50	Argon	1.1-1.5	1.5-4.5	Helium	0.081	

PAW SETUP AND TECHNIQUES

If you are already familiar with GTAW, PAW will be an easy conversion. Fortunately, almost all metals that can be plasma welded use the same technique as used with GTAW. Please consult Chapter 8 for information on welding techniques for the various metals. Only minor adjustments will be required to adapt to plasma arc welding. Here are some special notes for PAW setup on commonly welded metals.

PAW of Aluminum and Magnesium

1. Gas: helium
2. Polarity: DCRP
3. Pilot arc: continuous
4. Use only DCRP PAW tips (heavy-duty tips)

Be sure to remove oxide coating (a stainless steel wire brush works best).

PAW of Stainless Steel

1. Preparation: sharpened to a point (use a chart for gaging)
2. Polarity: DCSP
3. High frequency, pilot continuous
4. Gas: argon plasma, argon, or argon–helium mixture for shielding gas

Technique

1. Small beads only
2. Filler must match base metal
3. Filler or no filler

PAW of Carbon Steel

1. Same settings as for stainless steel, but slightly higher current settings
2. E70S-3 provides good economical filler (on clean steel); E70S-2 provides excellent filler; Al, Zr, Ti deoxidizers
3. Plasma gas: argon; shielding gas: argon

Technique The procedure is similar to that for stainless steel, but:

1. Amperage slightly lower
2. More shielding gas required
3. Can use Ar/H_2 gas (5% maximum H_2 shielding gas)
4. Polarity: DCSP

PAW of Copper

1. Wire-brush to remove oxides
2. Polarity: DCSP deep penetration
3. Shielding gas; pure helium (best), 25% argon–75% helium (good); plasma gas: argon
4. Preheat to approximately 400°F

Technique Very fluid puddle (run similar to aluminum GTAW).

PAW of Titanium

1. Polarity: DCSP
2. Shielding gas: argon or argon–helium; plasma gas: argon
3. Tungsten 2% Th (EWTH2)
4. Degrease with acetone

Additional shielding gas and postpurge should always be used for titanium; hydrogen is the primary contaminant.

PAW Keyholing

Another unique capability of plasma is its keyholing capability. In keyholing a small keyhole is actually pierced into the weld joint. The walls of the keyhole are molten (Figure 10–8). As the keyhole is moved forward, the molten walls flow together at the rear, forming the weld. Little or no beveling is required for most butt joints. Keyholing is faster, uses less filler metal, and puts less heat into the weld. It is quite difficult to use keyholing manually, so it is usually done automatically with automatic travel speed equipment. Always refer to the manufacturer's suggested machine and gas flow settings.

Setup for Keyholing

The keyholing mode is usually fully automatic travel; however, it can be done manually. It is usually limited to thickness ranging from $\frac{1}{8}$ to $\frac{3}{8}$ (0.125 to 0.375) in. The plasma gas setting must be high enough to penetrate and create the keyhole, but not so high as to cause collapse of the molten metal (PAC fall-through). Plasma arc cutting fall-through is the result of too much plasma gas, causing a cutting action instead of welding.

The travel speed limitations are as follows:

1. If travel is too fast, the keyhole will disappear and penetration will be lost.
2. If the travel is too slow, the puddle will be too large and fall through will result.

Plate preparation is similar to GTAW but more land (root face) and a less included angle are used (Figure 10–9).

FIGURE 10–8
PAW keyhole.

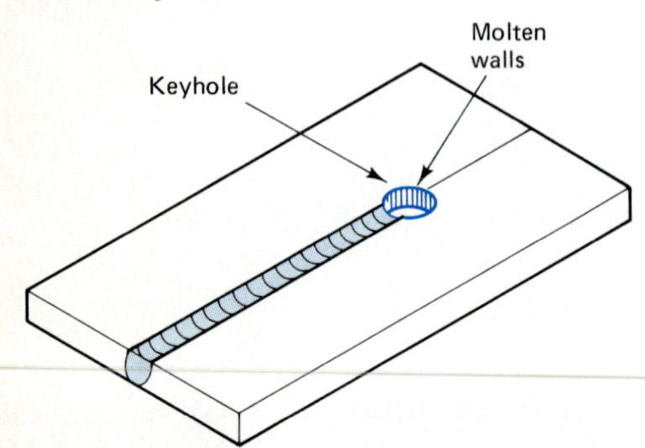

FIGURE 10–9
PAW keyhole setup angles.

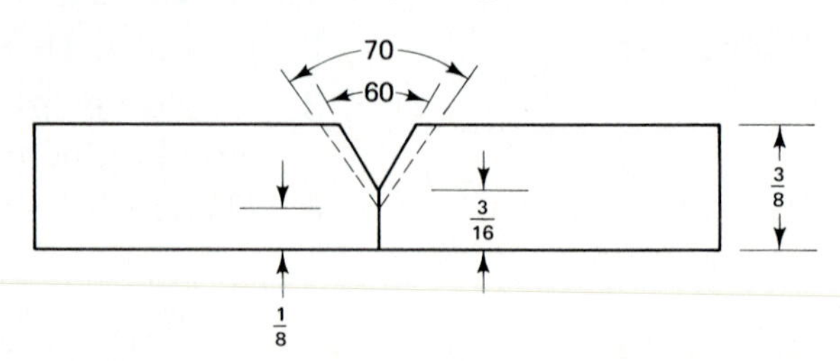

REVIEW QUESTIONS

1. What are the advantages of PAW over the GTAW process?
2. What is a plasma gas?
3. What types of gases are used with PAW?
4. What polarities are used with PAW, and why?
5. What tungsten is used for PAW?
6. What machine controls would you adjust to produce a stiff arc?
7. What effect does the secondary gas have on a plasma arc weld?
8. What are the two types of PAW welding techniques?

11 GAS METAL ARC WELDING (GMAW)

The gas metal arc welding (GMAW) process, also popularly known as MIG (metal inert gas), has been a sleeping giant of production welding. This fast and economical welding process is just now replacing some SMAW applications. This is *not* to say that it can replace all SMAW applications, but it is clearly faster in many production applications. Since its introduction in the 1940s, steady improvements in power sources, shielding gases, and variations of the basic GMAW process,

PHOTO 11-1
Modern GMAW system. (Courtesy of Miller Electric Mfg. Co.)

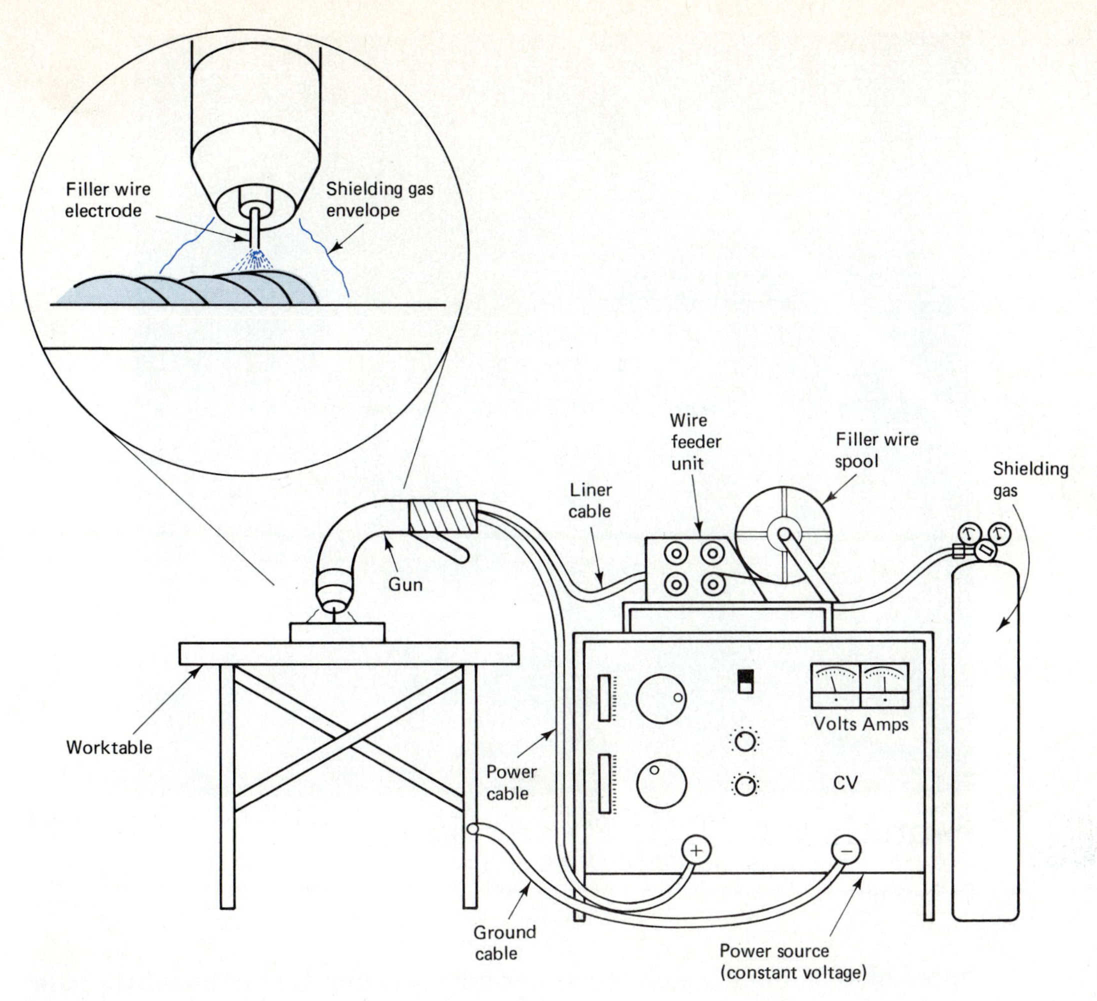

FIGURE 11–1
GMAW system and equipment.

such as FCAW (flux-cored arc welding), has resulted in good to very high quality welds. Proper application of this process is the key to fast, high-quality welds.

Here is how it works. Electricity travels through the cable to a contact tip, where it makes contact with the filler wire (Figure 11–1). The energized wire makes contact with the plate, starting the arc. The arc creates the heat needed for welding, and the filler wire feeds the puddle with weld metal. The wire, the arc—the entire operation—is shielded from the atmosphere with the shielding gas, which is flowing from the gun. Drive wheels in the wire feeder unit feed filler wire from the spool through a liner to the gun.

ADVANTAGES OF GMAW

As stated previously, the primary advantage is its high welding speed for production welding, but other time savings are realized in that there is no need to stop and change electrodes in the middle of a pass. The

PHOTO 11-2
Welders performing GMAW. Note the low smoke produced by the process.

spool of wire must be changed periodically, but this is usually accomplished quickly and is done only at about 8-hour intervals when in maximum use. Like GTAW, GMAW can have a small heat-effective zone if applied properly. This advantage carries the same advantages as with GTAW (see Chapter 8).

The GMAW process also produces very little smoke and a very light silicone dioxide slag. This slag is also known as glass slag. Many metals that are badly distorted by some welding processes will distort very little with GMAW. Another major advantage is GMAW's ease of operation. In other words, you can quickly learn to make good welds in a relatively short time compared to other welding processes.

The actual gun manipulation is quite easy with GMAW. This is due to the automatic arc length control. The welder need only be concerned with gun angles and speed of travel. However, the real skill in GMAW is setting the machine for the variable welding situations. Tying in the volts, amps (wire feed) and slope controls is a skill in itself. Welders should concern themselves with learning these machine controls. We discuss the machine adjusting procedure later in this chapter.

The GMAW process is now one of the more flexible welding processes. We can weld heavy plate or switch to very thin plate with only a few machine adjustments. This is due to the different modes of metal transfer (spray, globular, short circuit, etc.). We can also weld many different types of metals: mild steel, medium-carbon steels, many tool steel grades, stainless steels, coppers, aluminum, and magnesium. In the future, other metals and alloys will be "GMAW" weldable.

Some of the other forms of GMAW that have been developed as a result of GMAW are FCAW (flux-cored arc welding), SAW (submerged arc welding), and ESW (electroslag welding). These processes are discussed in Chapters 13 to 15.

GMAW THEORY

The key to GMAW's successful operation is the constant-voltage (CV) power source. Constant voltage means that the voltage is held constant even though varying tip-to-work distance changes. In contrast, a constant-current (CC) power source, used for GTAW and SMAW, varies the arc voltage as the operator physically changes the arc length. This helps hold the current steady.

To observe the characteristic differences between the CC and CV power sources, perform the following procedure. Have a fellow welder run a bead while you observe the volt/ammeter. Have your partner change the arc length (move the gun up and down) as the bead is run. Do this with both GTAW or SMAW and a GMAW machine. You should observe a varying voltmeter and steady ammeter with the constant-current (SMAW and GTAW) power source, and a varying ammeter and steady voltmeter with the constant-voltage (GMAW) power source (Figure 11-2). With the CV, GMAW power source, as the arc length trys to shorten, the amperage will increase to make up for the energy difference. And when the arc tries to lengthen, the amperage will drop, leaving a steady voltage and resulting in a *steady arc length*. Simply stated, the constant-voltage power source maintains a constant arc length regardless of stickout (contact tip-to-work distance).

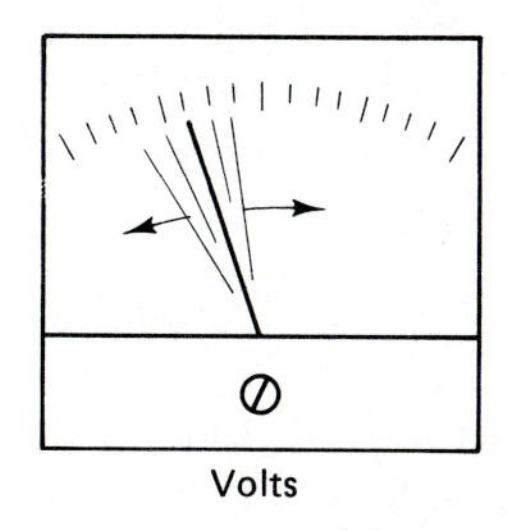

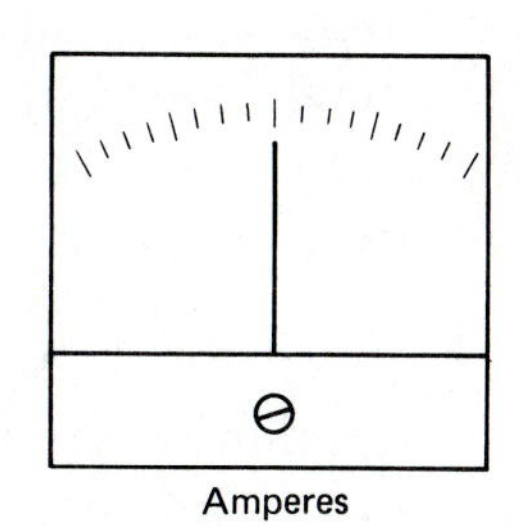

CC, constant current (SMAW or GTAW power sources)

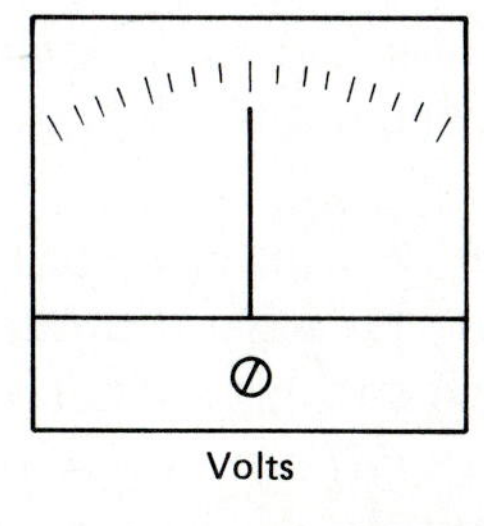

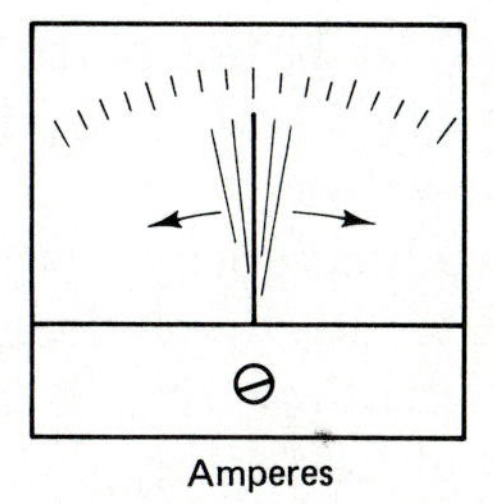

CV, constant voltage (GMAW, FCAW, or SAW power sources)

FIGURE 11-2
Constant-current and constant-voltage readings.

Polarities for GMAW

Most GMAW uses DCRP (DCEP). However, some limited applications of DCSP (DCEN) and AC are used with some filler wires. The use of DCRP puts about two-thirds of the heat of the arc into the filler wire (Figure 11-3). This produces a smooth flat bead. The wire is continuously feeding from the GMAW gun. DCRP is usually necessary to burn off this rapidly feeding wire and create a hot arc. When DCSP is used, it results in a cool running bead.

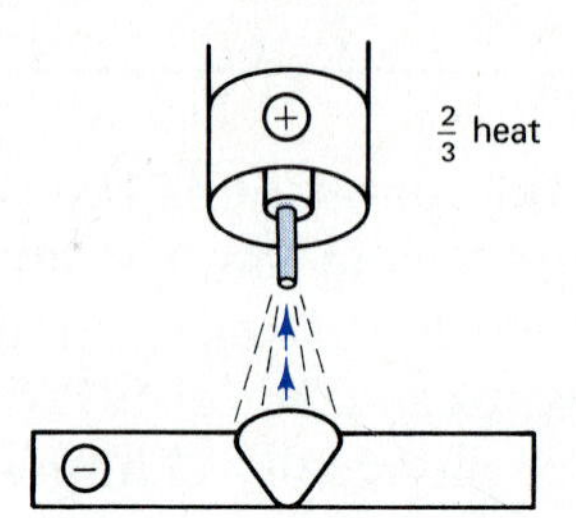

FIGURE 11-3
Polarity (DCRP) and heat concentration.

MODES OF METAL TRANSFER

The method or mode in which the filler metal transfers from solid wire to molten weld bead through the arc is known as the transfer mode. There are five basic modes of metal transfer:

1. spray
2. short circuit
3. globular
4. buried arc
5. pulsed arc

These modes are what makes the GMAW process so versatile. We change the heat, speed of travel, penetration, and other variations just by changing modes. The key to successful GMAW is knowing how to change and properly apply these modes. Let's examine each mode.

Spray Transfer

In the spray transfer mode small droplets of molten filler metal are sprayed through the arc and into the puddle (Figure 11-4). Machine settings are quite high. Therefore, it gives very deep penetrations and a hot wide puddle. If run properly, spray should deposit a flat, uniform, well-tapered-in bead. The best application for spray would include relatively thick ($\frac{3}{8}$ to $1\frac{1}{2}$ in.) plate in the flat and horizontal positions. The puddle is so fluid, vertical and overhead work is extremely difficult and may be impossible.

FIGURE 11-4
Spray transfer: high voltage, high amperage (wire feed).

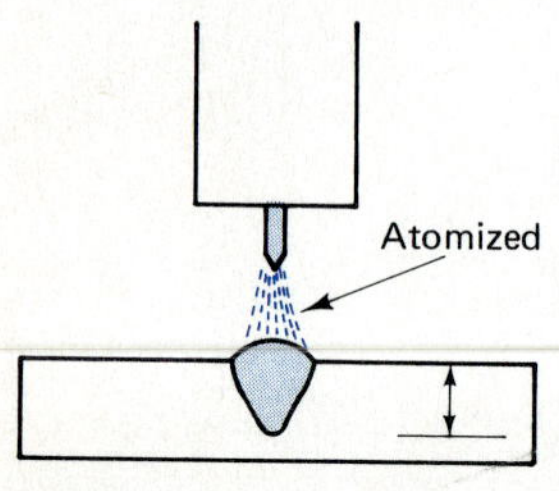

To produce spray transfer, the power source voltage should be set *high* (28 to 36, depending on the wire diameter and shielding gas used). This will allow for a long arc and high electron impact on the wire to produce the fine spray droplets. Of course, wire feed (current) must also be fairly high to keep the supply of filler wire coming and to prevent burnback. Spray transfer used direct-current reverse polarity

(DCRP), which puts most of the heat of the arc in the positively charged wire. It also requires a gas that will ionize a narrow enough stream of electrons to cause spray. Argon does this extremely well. An argon and oxygen (98% argon and 2% oxygen) mixture is very popular. Other mixtures are available for spray, but they usually have a high percentage of argon.

Short-Circuit Transfer

Short-circuit transfer is the coolest-running mode and the least penetrating, so caution must be exercised when applying or designing for welding in this mode of transfer. DCRP is still employed for most short-circuit transfer. Very low voltages and moderately high amperage (wire feed) settings should also be used (Figure 11-5). Short circuit actually relies on short circuiting of the filler wire onto the plate and the resulting heat generation to melt and fuse the plate. Let's examine the arc when using short-circuit transfer (Figure 11-6).

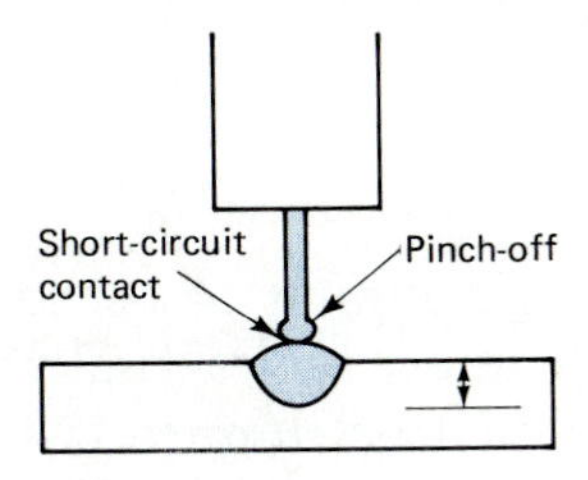

FIGURE 11-5 Short-circuit transfer: low voltage, moderately high wire feed.

Short circuit, because of its cool running bead, is excellent for vertical and overhead welds on thin plate and sheet metal. It also makes a nice root pass on open butt groove welds, but be sure to use a very small root face (landing); $\frac{1}{16}$ to $\frac{3}{32}$ in. is usually plenty (Figure 11-7).

Some codes will not allow use of short-circuit transfer GMAW unless it is first qualified (tested). Be aware that some very good-looking short-circuit welded beads may be poorly penetrated if welded on heavy plate. You will enjoy running short-circuit transfer because the puddle is very easily controlled and stays right where the welder puts it.

Almost any gas can be used for short circuit; but straight CO_2, and argon 75%-CO_2 25% mix are the most popular. Spatter may be higher with straight CO_2.

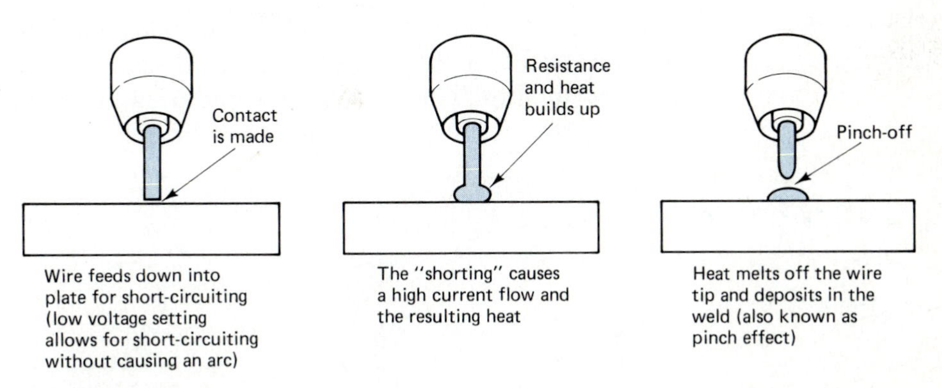

FIGURE 11-6 Short-circuit transfer.

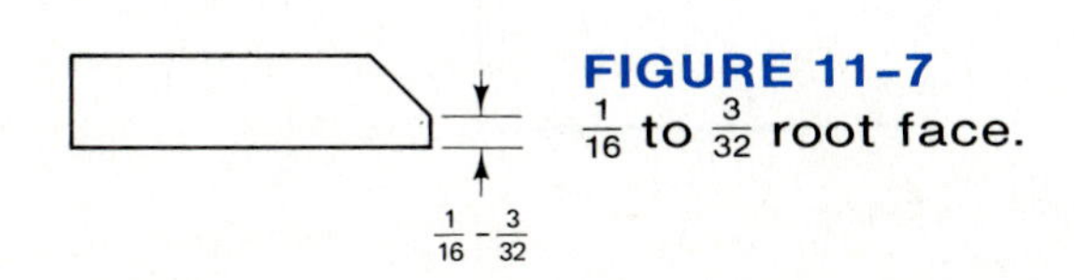

FIGURE 11-7 $\frac{1}{16}$ to $\frac{3}{32}$ root face.

Globular Transfer

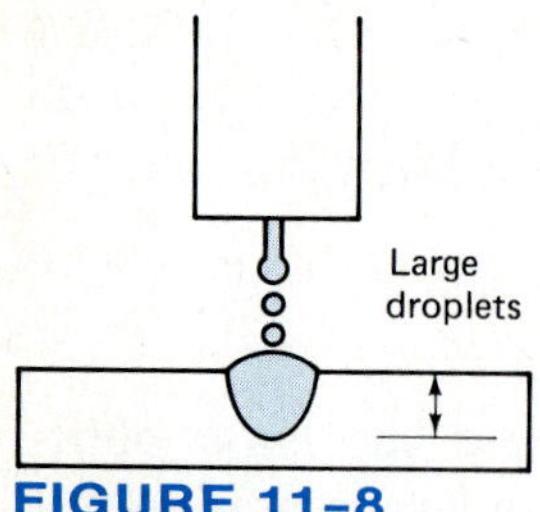

FIGURE 11–8 Globular transfer: moderately high voltage, moderately low amperage wire feed).

If you average what happens between spray transfer and short-circuit transfer, you now have globular (Figure 11–8). The energy in the arc is not high enough to vaporize the filler into fine droplets as in spray, but it is not low enough to allow the wire to contact the plate. The result is a large ball of molten metal forming on the tip of the wire as it feeds. The ball finally pinches off and is magnetically drawn across the arc [remember, opposites attract, and the wire is positive (+) and the plate is negative (−)]. Globular is handy for those in-between jobs where more penetration is required than short circuit will provide but the high heat input of spray is not wanted. The commonly welded carbon steel thickness for globular is $\frac{1}{4}$ to $\frac{1}{2}$ in. Globular can also have the flattest, smoothest, and best bead contour. This is due to the long open arc that exists before the filler drops off the wire.

Buried Arc Transfer

This unique method of GMAW transfer uses settings similar to those for spray, but higher wire feed settings are used. The arc will actually be below the surface of the plate if enough wire feed is used (Figure 11–9). The very fluid puddle restricts buried arc to the flat groove or flat and horizontal fillet weld positions. Being almost buried below the surface of the plate, the arc leaves almost no spatter, even when using straight CO_2 shielding gas. CO_2 is the most commonly used gas for buried arc.

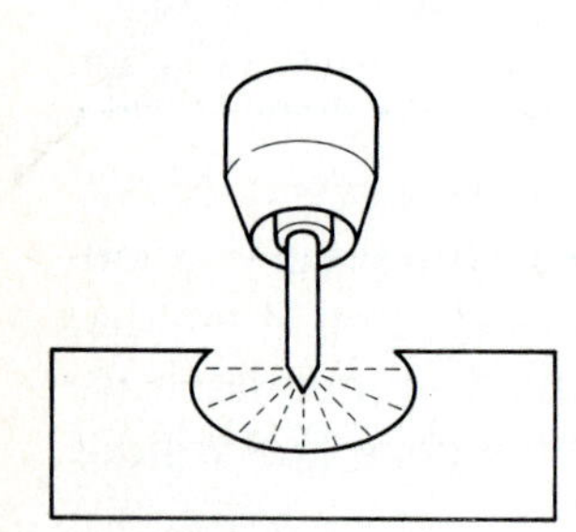

FIGURE 11–9 Buried arc transfer.

The primary advantages of this mode of transfer are its high speed, deep penetration, and low spatter. It is usually used on mild steel and stainless steels; however, on stainless steels, a shield gas mix of argon, CO_2 and helium works best (90% argon–7.5% CO_2–2.5% helium). To achieve buried arc transfer, increase the voltage and wire feed (amperage) until spray is reached. Now increase wire feed until buried arc is reached.

Pulsed Arc Transfer

Pulsed arc transfer (Figure 11–10) is also called discrete pulsed GMAW. This is because the pulse is so rapid that the welder cannot even see the arc pulsing. The extremely rapid pulsing keeps the mode in a continuous spray mode. The advantages of discrete pulses are:

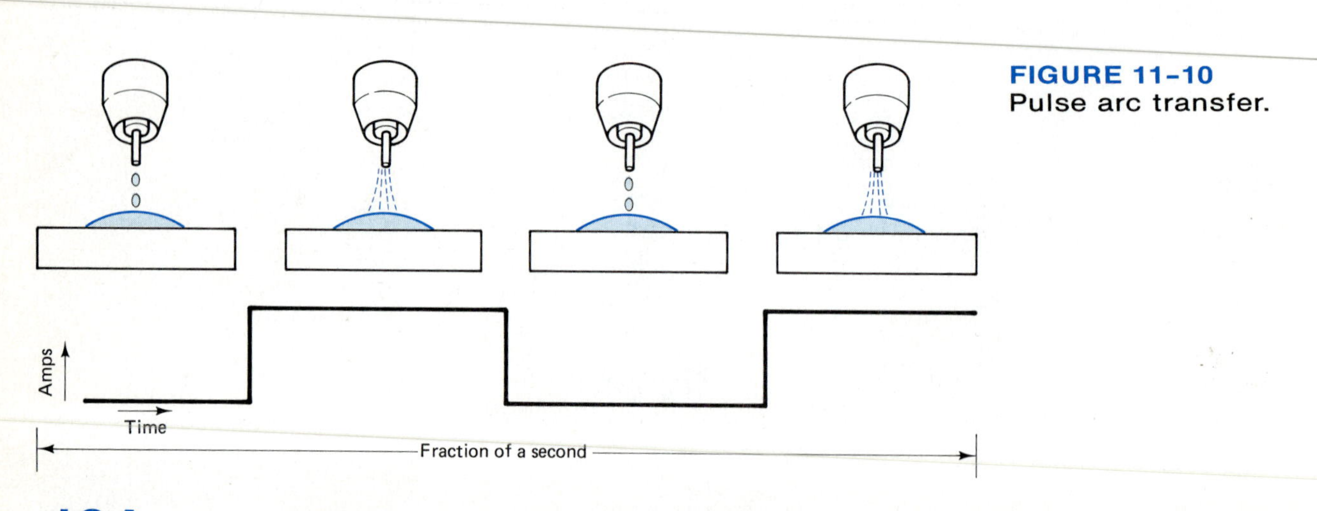

FIGURE 11–10 Pulse arc transfer.

1. Lower head input but maintained penetration
2. Less material distortion due to the lower heat input
3. Easy-to-weld thick-to-thin sections
4. No gun occillation to produce a smooth, flat bead

The hot-cold action of the arc keeps the heat input down, not allowing the plate to saturate with heat, while maintaining penetration due to the impact of the hot surge cycle of the pulsing action.

Some pulsed GMAW power sources have only one control for heat. This control automatically coordinates voltage, amperage, and wire feed simultaneously.

Shielding gases for pulsed transfer almost always include a high percentage of argon to reach the spray mode. Argon–helium, in a range of about 90% argon–10% helium, works quite well. You will notice a distinct sound and one of the smoothest, flattest, well-tapered beads of any mode of GMAW transfer. More on pulsed GMAW can be found in Chapter 12.

GMAW GASES

One of the most important variables in GMAW is the shielding gas. A change in shielding gases can completely change the characteristics of the process. The gas serves two purposes:

1. It shields the molten puddle from atmospheric contaminates.
2. It creates a smooth electrical conduction path for the electrons in the arc.

Some gases create a smooth electrical path, while others create a more resistant path for the current. An example of this could be observed with CO_2 compared to argon. Straight argon produces a smooth arc and spatter-free weld but a rather convex, built-up bead. Straight CO_2 produces rougher arc with moderate spatter.

Most gases for GMAW used today are mixes. These mixes combine the desirable characteristics in just the right proportions of two or more gases. Let's look first at the characteristics of the pure gases used in GMAW.

Argon

Argon is seldom used straight, except for GMAW aluminum. It is a super shield gas and a very good gas conductor of electricity. Argon produces one of the smoothest, quietest arcs. As mentioned in Chapter 8, argon ionizes a very narrow stream of electrons in the arc. This results in a well-penetrated but undercut bead (Figure 11–11).

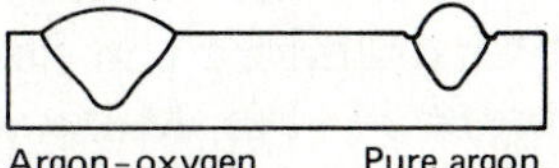

FIGURE 11–11
Argon–oxygen versus pure argon.

Argon–Oxygen

Argon-oxygen is a superior mix for running the spray mode of transfer. Oxygen can be added to argon in the range 2 to 5% for welding steel (95% Ar–5% O_2 or 98% Ar–2% O_2). The small amount of oxygen tends to flatten and taper the bead toes into the base metal (Figure 11–11). This is also known as good wetting action (good fluidity). The higher the percentage of oxygen, the better the wetting action. *Caution:* Using more than 5% oxygen may cause porosity in the weld.

Helium

Helium and helium mixtures (usually, argon–helium) are used with coppers and thick sections of aluminum. Helium is also used for automatic and robotic applications because of its good arc length control. Pure helium is very effective on copper, due to its broader arc stream. This tends to preheat the surrounding copper and reduce the chance of cracking. When using helium–argon mixtures, remember that the more helium, the higher the volume of heat and the larger the bead. Therefore, for a thick section more helium should be used in the mixture. Helium and argon can be used in almost any proportions to suit the material and thickness.

Carbon Dioxide

CO_2 (carbon dioxide) is the most economical gas for GMAW. It is an active gas, which means that it will break down into additional substances when ionized in the arc. Small amounts of CO_2 break down into carbon monoxide. These different gases have different ionization potentials. This tends to cause a slightly unstable arc and spatter. CO_2 works well on short-circuit and globular transfer modes, but cannot usually localize enough energy at the wire tip to go into the spray transfer mode.

Argon–CO_2

Argon–CO_2 is used in proportions of 75% argon–25% CO_2, 80% argon–20% CO_2, 92% argon–8% CO_2, and others. Argon–CO_2 mixtures combine the excellent shield and localized heat that argon provides with just enough CO_2 to flatten the bead and provide good wetting action. These mixes are also fairly economical, due to the CO_2 content.

Argon–Helium–CO_2

Argon-helium–CO_2 is usually proportioned 90% argon, 7.5% helium, and 2.5% CO_2. This mixture was developed primarily for stainless steels to give better wetting action while providing the good shielding required by stainlesses.

GMAW TORCH MANIPULATIONS

Pushing and pulling the gun are common GMAW torch manipulation terms. *Pulling* the torch involves dragging the gun, allowing the beads to build up behind the arc (Figure 11–12). Most welders find that it is easier to see the puddle when pulling the gun, but the bead is usually

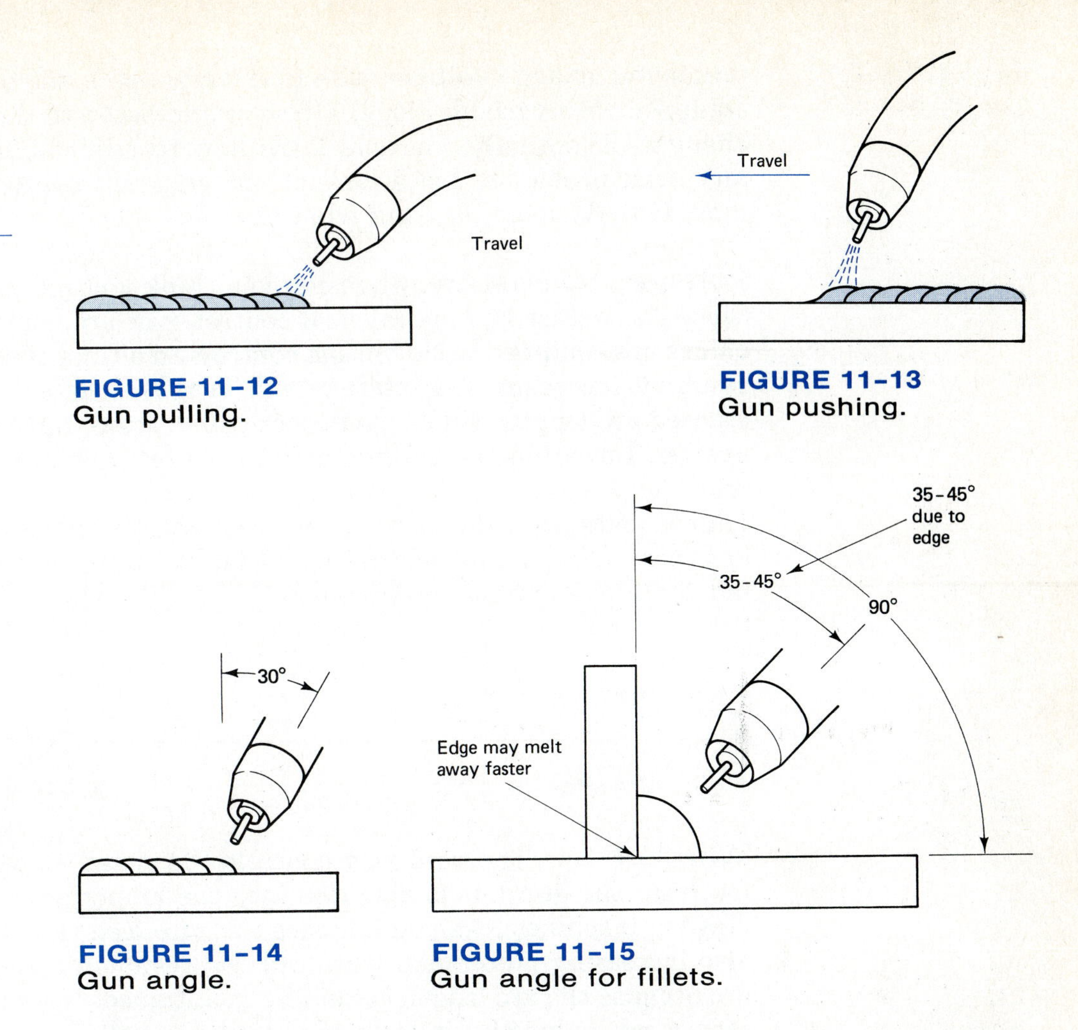

FIGURE 11-12
Gun pulling.

FIGURE 11-13
Gun pushing.

FIGURE 11-14
Gun angle.

FIGURE 11-15
Gun angle for fillets.

built up more (more convex) than with the pushing technique. *Pushing* the gun produces a flatter bead contour, because the arc force is heating and flattening the puddle (Figure 11-13). The gun tends to obstruct the welder's view of the back of the puddle; therefore, experience must be gained in achieving the proper travel speed and manipulations.

The *gun travel angle* is the angle at which the gun leans in the direction of travel. Every situation can require variation in gun travel angle, but a standard 30° angle is common (Figure 11-14).

The *transverse angle* (angle of attack) is the gun angle in relation to the base metal. It is especially critical with fillers to keep the leg lengths equal. A general rule for transverse angle is to split the angle of the joint. For example, in a 90° fillet tee joint, the transverse angle should be about 45° (Figure 11-15). In some cases you may want to angle slightly more in the direction of the thicker plate, or the flat surface versus an edge surface.

GMAW MACHINE CONTROLS

Your ability to properly set up and adjust your GMAW machine to variable conditions and situations is by far the most important variable that you will learn. Practice is the only way to perfect this skill. While practicing, change voltage, wire feed (amperage), and slope controls (if equipped with variable slope). Observe and listen to the result of these changes. Eventually, you will learn how to adjust quickly to correct any bead problems. Let's look at the common controls available on most GMAW machines and what they do.

Voltage In GMAW, whenever you think voltage, think arc length. Since the welder has no physical control over arc length, GMAW machines are equipped with voltage controls to adjust the arc length. The result of increasing the voltage will increase the arc length, and increased arc length will cause deeper penetration, but may cause some spatter. Lowering the voltage will do the reverse; penetration will decrease and spatter will be reduced. Voltage will also affect bead contour. Higher voltages will give a flatter bead, and lower voltages may result in a very convex and possibly cold lapped bead (Figure 11-16). Too much voltage may cause burnback.

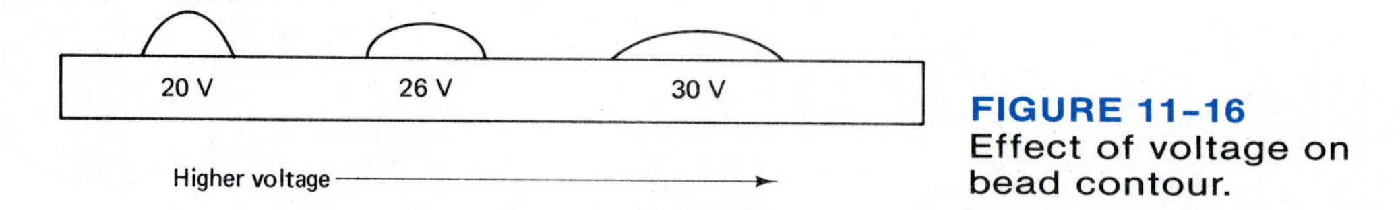

FIGURE 11-16
Effect of voltage on bead contour.

Wire Feed The wire feed controls the speed at which the wire is fed from the gun but is also tied into the amperage control (on most GMAW machines). As you increase the wire feed rate, the amperage is also increasing. Too much wire feed will result in a sporadic, popping arc because the arc length cannot be maintained. Not enough wire feed may result in burnback. Both wire feed and voltage must be increased and decreased together and balanced. As long as wire feed and voltage are increased and decreased within a balance of each other, the bead will remain acceptable. The wire diameter will also limit the maximum and minimum settings. Each wire diameter has a range in which it will run properly. Experience will help you learn the limits of these ranges. The voltage and wire feed (amperage) controls will have the most effect on the integrity and appearance of your weld bead. Figure 11-17 shows a few examples of improper machine settings and their causes.

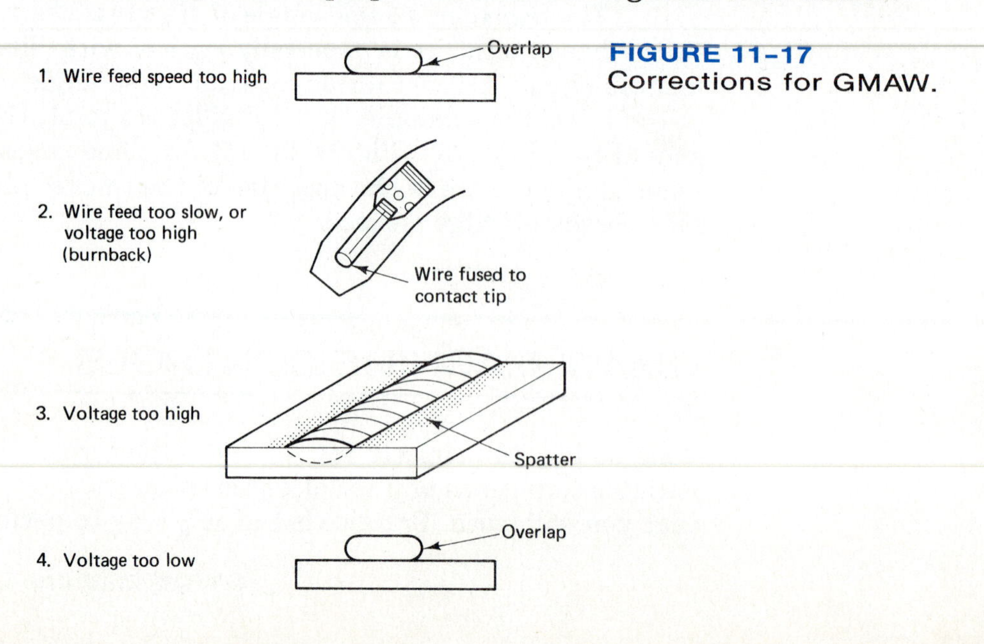

FIGURE 11-17
Corrections for GMAW.

PHOTO 11–3
Typical GMAW wire feeder unit. (Courtesy of ESAB Welding Products, Inc.)

PHOTO 11–4
Typical GMAW wire feeder unit. (Courtesy of Miller Electric Mfg. Co.)

Voltage Range Switch

This switch is used to set the approximate voltage range. It is usually a three-position—high, medium, or low range—switch.

Inch Switch

This switch is for feeding wire only. When respooling new filler wire, there is no need to purge expensive shielding gas or kick in the contactor (by pulling the trigger) when all that is needed is to feed wire through the liner and to the gun. *Note:* When using the inch switch to respool the wire, remove the contact tip to prevent wire jamming as the wire reaches the end of the gun.

Gas Purge Switch

This switch is for setting the flowmeter. Using the trigger switch on the gun to set gas flow will waste filler wire. The gas purge switch only activates the gas solenoid.

Adjusting the Slope

By adding slope turns you slow the rate of response and limit the maximum current of the power source to changing arc lengths (controls the amount of CV or CC characteristics of the power source). Most power sources range from 6- to 14-turn settings (Figure 11–18). The response time is how quickly it takes for the machine to go from a short circuit

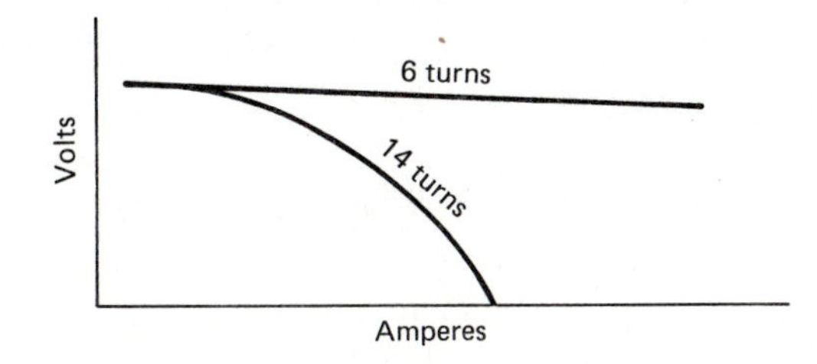

FIGURE 11–18
Turn settings.

PHOTO 11-5
GMAW power source equipped with slope and inductor. (Courtesy of Miller Electric Mfg. Co.)

at the arc back to its welding voltage or arc length. Things to be considered include mode of transfer, diameter of wire, and metal type and size.

1. At 0 turns the machine produces instant arc length correction.
2. At 14 turns the machine produces a very slow arc length correction. (This power allows for short circuiting.)
3. Higher slope settings (8 to 14) reduce spatter but also tend to unstabilize the arc (the arc length wanders).
4. Low slope settings (6 to 9) produce a very stable arc (arc length), but will cause small-diameter (0.030 to 0.035 in.) wires to spatter. This is because the high amperage the machine uses for arc length correction literally causes the wire tip to explode.
5. For short-circuiting transfer, use *higher slope settings* (9 to 14) and less heat input.
6. For spray transfer use lower slope settings (6 to 8); these settings also produce higher heat input.
7. Small-diameter wires use medium to high settings (8 to 14).
8. Large-diameter wires use medium to low settings (6 to 9).
9. Metals that are thick or dissipate heat rapidly (copper or aluminum) use a low slope setting.
10. Metals that are thin or have a slow dissipation of heat (stainless or nickel steels) use higher slope settings.

Setting the Inductor or Stabilizer

The inductor and stabilizer controls produce only a slight change in the rate of response of the power source without changing the volt–ampere curve (Figure 11–19).

1. The inductor regulates the amount of dc stabilization in the power source circuitry.

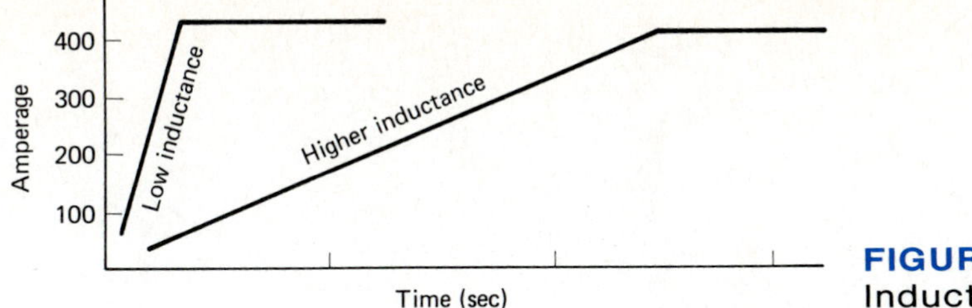

FIGURE 11-19
Inductor settings.

2. Increasing inductance also reduces spatter (the more inductance, the slower the response time).
3. Increasing the inductance also allows the arc to remain open longer (short circuit transfer) because of the slowed response, allowing the weld to flatten out (more fluid puddle).
4. Too much inductance will cause an erratic start.

Machine Setting Summary

Use Table 11-1 for quick reference when adjusting your machine.

TABLE 11-1
GMAW Machine Adjustments

Voltage
Long arc length ↑ short arc length ↓
Flatten bead ↑ (increase heat)
Reduce spatter ↓
Amperage (or wire feed)
More heat ↑ (voltage ↑ too)
Faster travel ↑
More filler ↑
Slope
Reduce spatter ↑
Stabilize arc length ↓
More heat ↓
Small-diameter wires ↑ or thin metal ↑
Large-diameter wires ↓ or thick metal ↓
Stainless or nickel steels ↑
Aluminum or coppers ↓
Inductor
Flatten bead ↑
Reduce spatter ↑
Stabilize arc ↓
Reduce erratic starts ↓

Changing Filler Wire Spools

1. Turn off the power source and feeder units.
2. Remove the cup and contact tip. Note that the contact tip may have to be removed by snipping off the balled end (Figure 11-20).

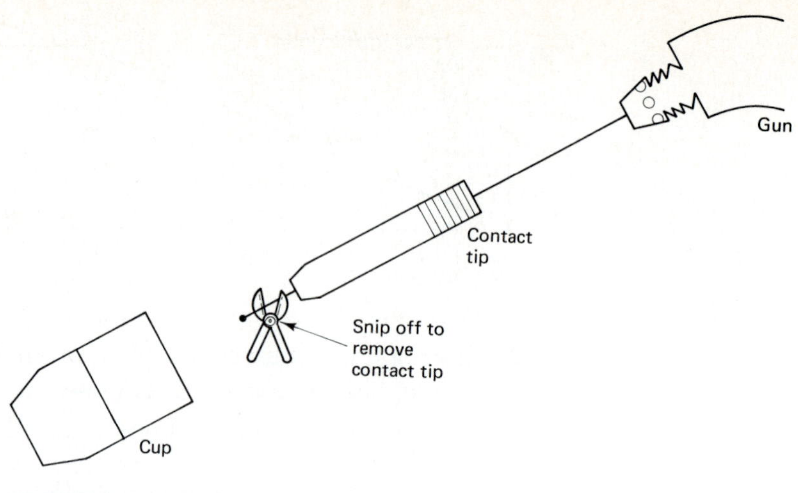

FIGURE 11-20
Removing contact tip.

3. Loosen the tension on the drive wheels.
4. Grip the wire with pliers or locking pliers at the gun end and pull the remaining wire from the liner and feeder unit.
5. Install a new filler wire spool.
6. For easier rethreading, cut off the new lead wire at an angle.
7. Pull the new lead wire through the drive wheels (Figure 11-21) and push partway down the liner tube.

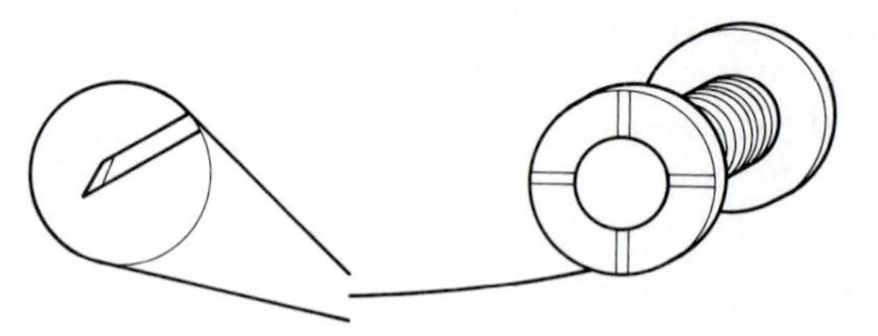

FIGURE 11-21
Pulling new lead wire.

8. Tighten the drive wheels and turn on the feeder unit and power source.
9. Push the "inch switch" to feed wire to the gun. If the machine is not equipped with an inch switch, turn off the shielding gas and push the gun trigger switch.
10. When the wire has fed out of the gun about 4 to 6 in., stop and replace the contact tip and cup.
11. Cut off the excess wire an inch or so past the cup. If you shut off the shield gas, turn it back on and adjust to the proper flow rate. You are now ready to weld again.

Preventing Burnback

Burnback can be very frustrating. So let's learn how to handle it and what causes it. Burnback is the result of the arc trying to take place back into the cup and actually past the contact tip. When this happens, the result is that the molten filler metal is fused into the copper contact tip. If the voltage is too high or the wire feed is much too low, burnback will result. It is more likely to result when using argon-oxygen, or argon-CO_2 shielding gas mixtures. It may also happen as a result of a wire jam in the liner or feeder unit. The solution is simply to turn down the voltage or increase the wire feed to the point where the arc length is well out of the cup. If the wire is feeding erratically, blow out the liner and blow any metal particles off the drive wheels with air pressure. Also check the contact tip and cup for cleanliness. *Note:* Make sure that any air hose system you use for cleaning equipment is equipped with a 15-psi safety tip.

Repairing a Burned-Back Contact Tip

Most burned contact tips can be repaired (Figure 11–22).

1. Remove the tip. This may require cutting the filler wire off at the spool and pulling off the tip with a section of wire attached.
2. Now, carefully grind the end of the tip until it is beyond the fused portion. Using pliers or locking pliers, pull the wire free of the contact tip.
3. Burrs left from grinding the tip can be removed with oxyfuel tip cleaners.

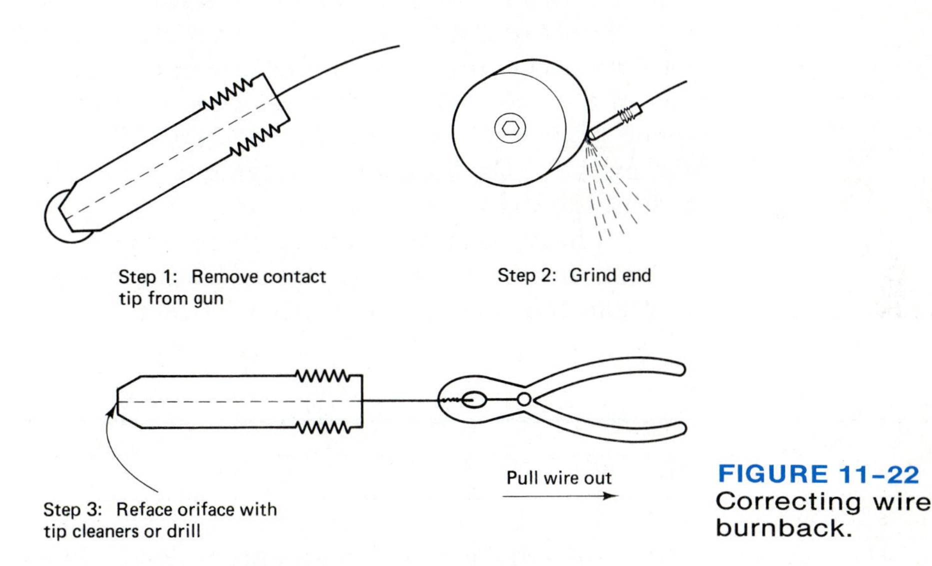

FIGURE 11–22 Correcting wire burnback.

GMAW OF ALUMINUM

Aluminum can easily be gas metal arc welded; however, a few simple procedures must be followed:

1. Clean and deoxidize the aluminum surface to be welded.
2. Preheat medium and heavy sections.
3. Push the torch.
4. Use DC Reverse Polarity.

Aluminum has a heavy oxide coating that must be at least

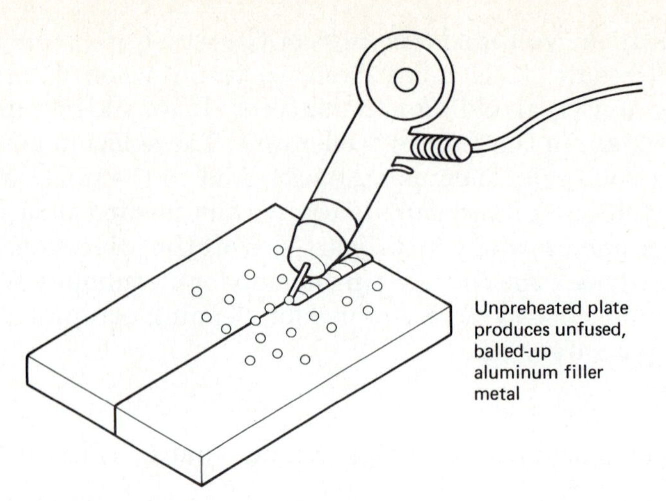

FIGURE 11–23
GMAW of aluminum.

partly removed for good-quality welds. A stainless steel wire brush works well for breaking down this oxide coating, but liquid chemical deoxidizers work faster when you have many feet of welding to do.

Preheating will help to give you a smooth, uniformly penetrated bead. Without preheating, aluminum welds will ball up and not penetrate until the aluminum plate warms up (Figure 11–23). The average preheat range for most aluminums is 150 to 300°F.

Pushing the torch is the next requirement for good-quality gas metal arc welds on aluminum. Pushing tends to deoxidize the area in front of the weld, allowing for a deposit on good, clean, deoxidized aluminum (Figure 11–24). Pulling the gun will simply lay the aluminum filler metal on the base metal, relying on deoxidizers in the filler metal to clean the weld.

FIGURE 11–24
Pushing the gun in GMAW of aluminum.

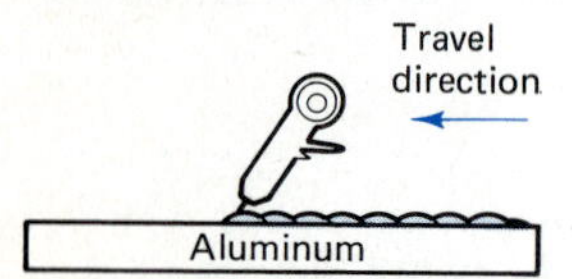

A heavy section of aluminum may benefit from the use of helium or argon–helium shielding gases. Helium will increase both the heat input and the penetration of these thicker aluminum sections.

GMAW OF MILD STEEL

Mild steels are easily gas metal arc welded. As mentioned earlier, the shielding gas can greatly affect weld qualities, such as puddle fluidity, which will affect your ability to weld in the vertical and overhead positions. Using CO_2 as a shielding gas will reduce puddle fluidity but may increase spatter. Also, remember that CO_2 tends to run short-circuit transfer if the voltage is not set high enough, and short circuit tends to result in poor penetration if applied improperly on heavier sections of steel. Using argon–O_2 will greatly increase penetration and will easily produce spray transfer. The only problem with argon–O_2 shielding gas is its highly fluid puddle. In many cases, GMAW with this gas mixture may be limited to flat and horizontal positions. Using argon–CO_2 (75% argon–25% CO_2 or 80% argon–20% CO_2) may be a good choice where all position welding of mild steel is required. The 75%–25% mixture allows for good penetration with globular transfer, yet still runs cool enough with short-circuit transfer for vertical and overhead welding.

PHOTO 11-6
GMAW of carbon steel. (Courtesy of Miller Electric Mfg. Co.)

Pushing the MIG gun tends to produce the flattest, deepest-penetrated bead. Pulling the MIG gun works well on thinner sections because it is less penetrating and leaves a more convex bead.

GMAW OF STAINLESS STEEL

Stainless steel runs much the same as mild steel, but stainless steel tends to run a bit more fluid. This puddle fluidity tends to make welding in the flat and horizontal positions smoother and easier, but vertical and overhead welding more difficult. Whenever possible, position stainless steel gas metal arc welds in the flat or horizontal positions. Many gas mixtures have been experimented with, but the mixture of argon 90%, helium 7.5%, and CO_2 2.5% has surfaced as one of the most popu-

PHOTO 11-7
GMAW of stainless steel. (Courtesy of Miller Electric Mfg. Co.)

lar for stainless steels. This triple-gas mixture gives a rapid deposition rate and a smooth fluid puddle.

The effect of pushing or pulling the MIG gun has much the same effect as it does on steel. Pushing gives a flatter, deeper-penetrating bead, and pulling leaves a more convex, shallower-penetrating bead.

GMAW OF COPPER

Copper is also easily gas metal arc welded, but a few precautions must be followed. Copper is quite crack sensitive when welded. This is due to its "hot shortness" characteristic. Preheating reduces the chances of these hot cracks by reducing the stress induced by uneven heat distribution. An average preheat temperature is 400°F. Thicker sections will require a slightly higher preheat temperature, and thinner sections, a slightly lower temperature. It is also helpful to use helium or at least a helium–argon mixture to increase the heat input. Remember, copper has high thermal conductivity, which means that the heat dissipates rapidly from the weld area, causing slow-developing, very convex beads. The addition of helium will cause heating of a wider area around the puddle and result in a flatter, more uniform copper weld.

REVIEW QUESTIONS

1. How does the GMAW power source differ from other arc welding power sources?
2. Most MGAW uses what polarity?
3. How does GMAW compare to other arc welding processes with regard to speed?
4. How is the GMAW process shielded from the atmosphere?
5. List some advantages of the GMAW over that of other arc welding processes.
6. How is the arc length controlled with GMAW?
7. List the modes of metal transfer used with GMAW.
8. What mode of metal transfer works well for sheet metal?
9. What mode (or modes) of metal transfer works well for heavy plate?
10. What gas or gas mixture is required for spray transfer GMAW?
11. Explain what effect adjusting the amperage and the voltage has on the arc and weld puddle.
12. What machine adjustments would be needed to correct a bead that is too convex?
13. What adjustments would you make to correct excessive spatter?
14. What effect does the slope control have on arc control?
15. What is burnback, and how can it be controlled?
16. List some important factors for welding aluminum.

PULSED GMAW

The pulsed GMAW process is similar to standard GMAW, but it utilizes a very rapid high-low pulsation while welding. Pulsed GMAW is quite similar to pulsed GTAW, but GMAW pulses discretely or so rapidly that you cannot see the actual pulse. However, a distinct high-pitched tone is produced by the rapid pulsing.

PHOTO 12-1
Pulsed GMAW system. (Courtesy of ESAB Welding Products, Inc.)

ADVANTAGES OF PULSED GMAW

The process is quite impressive, with the following list of advantages:

1. Less heat input into base material (rapid puddle solidification)
2. Less material distortion
3. Easy to weld out of position
4. Good penetration and wetting action
5. Much less gun occlusion required to produce a smooth, flat bead

These advantages are quite helpful for welding carbon steels, but their greatest potential is realized when welding metals such as aluminum and stainless. The low heat input keeps aluminum from saturating with heat (which can cause "hot shortness") while still providing good penetration and fusion. Low heat input also reduces distortion in stainless steels and can prevent carbide precipitation (see Chapter 19).

PULSED GMAW THEORY

The arc energy is rapidly pulsing high and low (Figure 12-1). On the high side (peak), penetration and fusion are achieved. On the low side (background), the work is allowed to cool slightly while keeping the arc established. The pulsing action causes a puddle stimulation that gives superior wetting action. This is all happening very rapidly and the welder notices little more than he or she is getting good puddle control. But there is actually a lower volume of heat being put into the work. The number of pulses per second (pps) can usually be adjusted for varying situations. Most pulsed GMAW remains in the spray mode of transfer during the pulsing cycle.

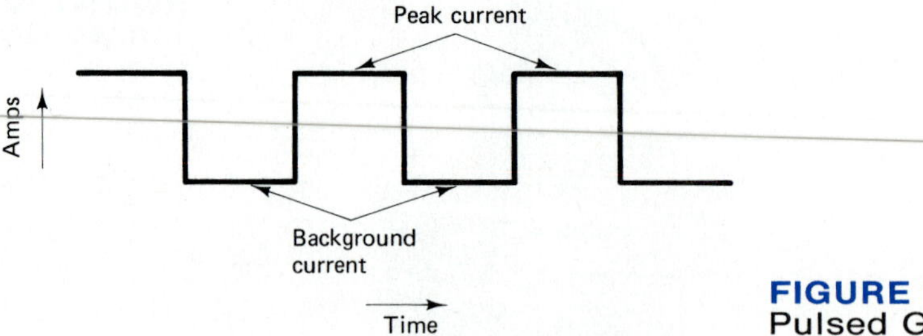

FIGURE 12-1
Pulsed GMAW waveform.

MACHINE CONTROLS FOR PULSED GMAW

Pulsed GMAW has three additional machine controls beyond that of standard GMAW.

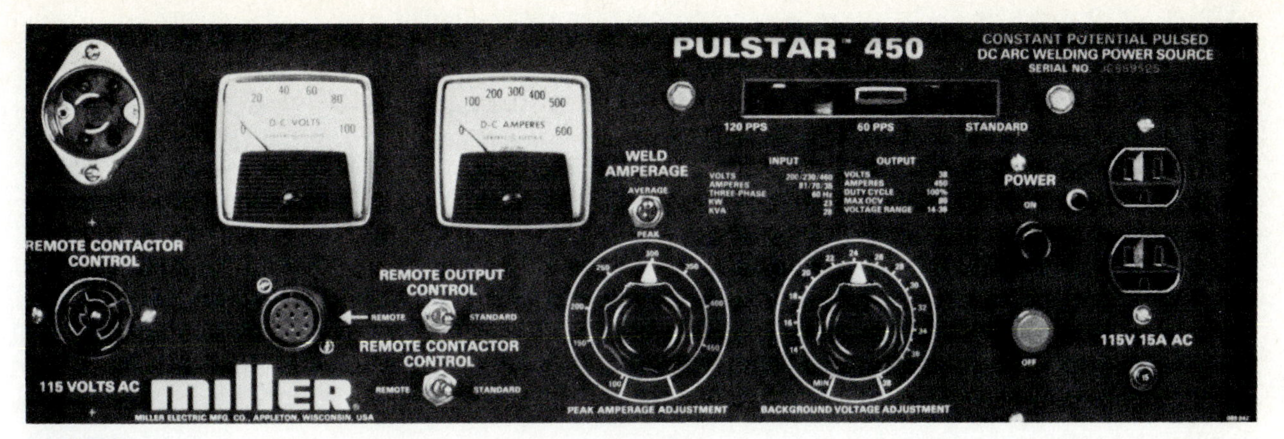

PHOTO 12-2
Pulstar 450 control panel. (Courtesy of Miller Electric Mfg. Co.)

PHOTO 12-3
Pulstar pulsed GMAW welder. (Courtesy of Miller Electric Mfg. Co.)

1. *Peak current.* This control sets the high side of the pulse cycle. It is also considered the main current control.
2. *Background current.* This sets the low side of the pulse cycle. It is set as a percentage of the peak current.
3. *Pulses per second (PPS).* This sets the number of pulses that will occur per second. It may only be a 2-position, high/low switch.

Machine Settings for Various Conditions

The pps usually has a high or low setting (Figure 12-2). On high (about 120 pps), there is more puddle stimulation and better wetting action for a flatter bead. On low, there is slightly less heat input and better puddle control for out-of-position welding. For very thin or heat-sensitive metals, use a low background and fairly low peak setting; pps should also be on low. For thicker sections or high-speed production jobs, use higher background with a fairly high peak current. PPS should also be in the high position.

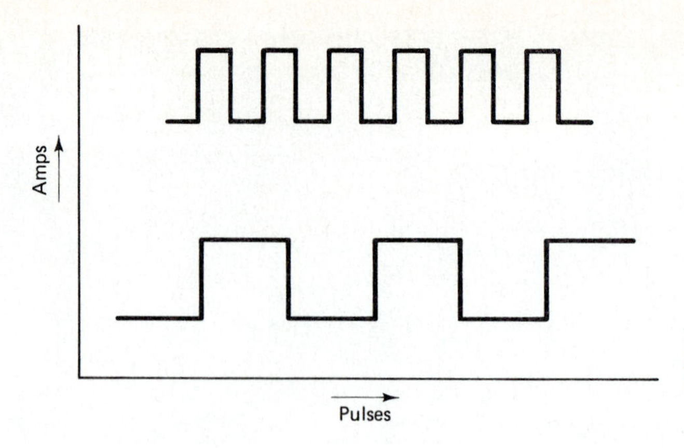

FIGURE 12–2 Waveforms for high- and low-pps settings.

SYNERGIC GMAW (SYNERGIC GUN)

A new system has been developed to simplify GMAW machine settings and improve weld quality (patented under number 4510373 U.S. and number 185329 Canada, by Controlled Systems division of A.M.H. Canada Ltd). This system is known as synergic GMAW (Gas Metal Arc Welding).

Synergic GMAW combines volt, amp, and pulse settings into one simple adjustment of gun handle or front panel control.

Most synergic systems are microprocessor controlled. This allows ultra-fine tuning and gives quick, accurate machine adjustments for varying conditions.

Some synergic systems still have voltage and amperage trim controls to fine tune the transfer mode, but once it is set, one knob does all the heat controlling.

Synergic also lessens problems such as burnback and excessive spatter.

PHOTO 12–4 Simplified control panel of the SAM system. (Courtesy of Controlled Systems division of A.M.H. Canada Ltd.)

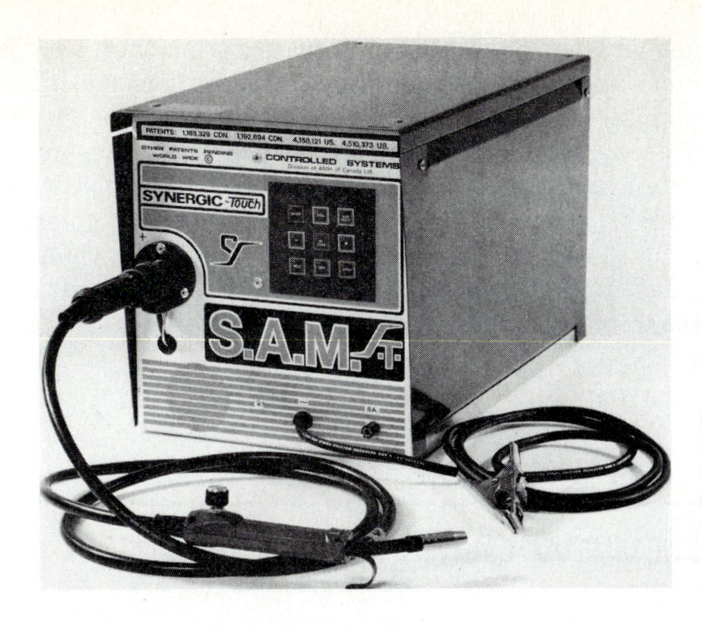

PHOTO 12-5
Lightweight simple-to-operate SAM Jr. (Courtesy of Controlled Systems division of A.M.H. Canada Ltd.)

COMPUTERS IN WELDING MACHINES

Yes, computers (microprocessors) are in televisions, radios, and cars, and are now in welding machines and equipment. With their introduction we can make extremely precise settings with incredible repeatability and very high weld quality. Did you ever have a day where you couldn't seem to get the correct machine setting and wished the machine could set itself? Or how about that day when the machine settings were perfect. How would you like to record those settings into the machine's computer to be called up any time in the future. Well, now you can!

Some of these new welding machines with on-board computers perform functions that weren't even imaginable a few years ago. Here are a few advantages of these machines:

1. Accurate recall of previously-recorded machine settings
2. More powerful machines with quite a bit of on-board memory
3. Pre-programmed machine settings for a variety of welding conditions
4. Self-diagnosing features for monitoring machine components
5. Extremely accurate readout of voltages and amperage

One of the most popular places to use computers is in GMAW wire feeder units. This is where the volts and amps (wire feed) are controlled. As you probably know, making machine settings for GMAW and FCAW is a bit trickier than with SMAW. This is because volts and wire feed must be carefully balanced to produce the correct bead. The computer memory can store your selected settings for your future recall, or better yet, make the selection for you! All that may be required of you is to specify what type of material you are welding, the shielding gas, and the thickness of your material; then let the machine make the volts and amp settings. Some microprocessor-based machines still let you trim the volts and amps, but they make the primary settings. By the way, for those who like to rely on their own on-board computer (your own brain) there is usually a *manual* toggle switch which will flip you back to standard machine settings without computer assistance.

Another big advantage is realized when using different filler wires. Different wires usually require different machine settings. These machine settings can be quickly recalled for the various wires being used.

REVIEW QUESTIONS

1. What are the advantages of pulsed GMAW?
2. What does the peak current control in pulsed GMAW?
3. What does the PPS control in pulsed GMAW?
4. What is the advantage of synergic GMAW?
5. What is the reason for putting computers (microprocessors) into welding machines?

GMAW PROJECT 1

Lap Joint

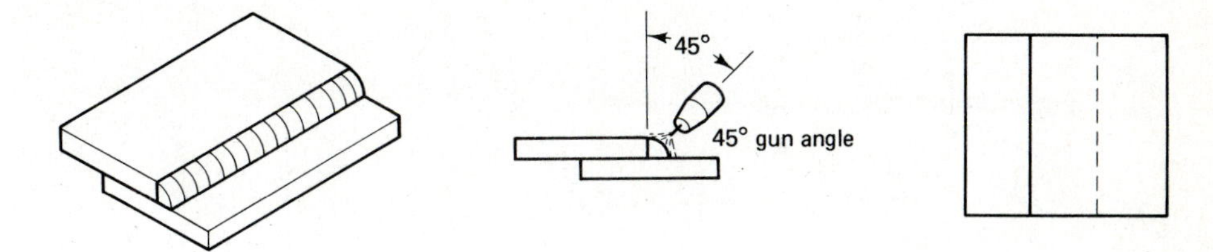

Material: A-36 (mild steel)

Size: $\frac{3}{8}$ in. × 3 in. × 6 in. plate

Joint Design: Lap joint

Position: Horizontal fillet (2F)

Technique: Stringer beads

Polarity: DCRP (DCEP)

Shield Gas: 75% Argon/25% CO_2

Gas Flow: 25–30 cfh

Filler Wire: E70S-3

Wire Feed: 180 ipm

Voltage: 25–30 V

Amperage: 180 A

Cleaning: Chipping hammer and wire brush

Procedure: Set up and tack a lap joint so that both sides can be welded. Use a 45° angle of attack. Deposit a bead on each side of the plate, using the pushing technique on one side and the pulling technique on the other.

Visual Inspection: The beads should be smooth, uniform, and free of any visible porosity, slag inclusion, cracks, or large craters at the end of the weld beads. A light-black glass type of substance is common on gas metal arc welds; this is a silicon dioxide formation from the filler wire deoxidizers. It should be removed when producing multipass welds.

GMAW PROJECT 2

Tee Joint

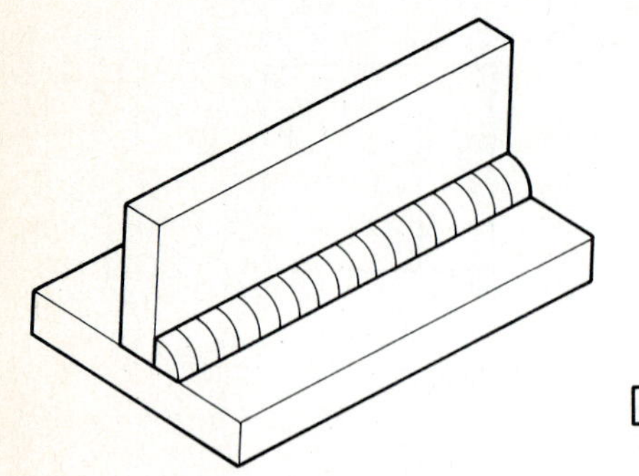

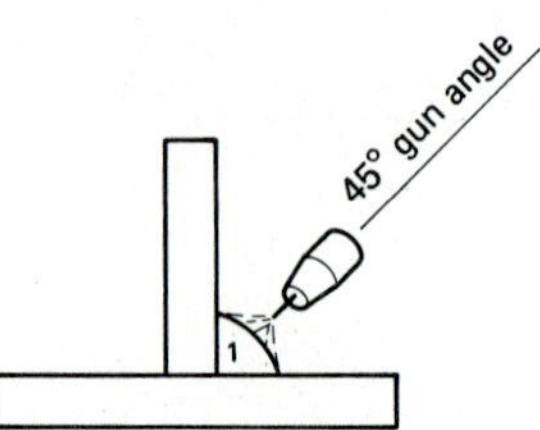

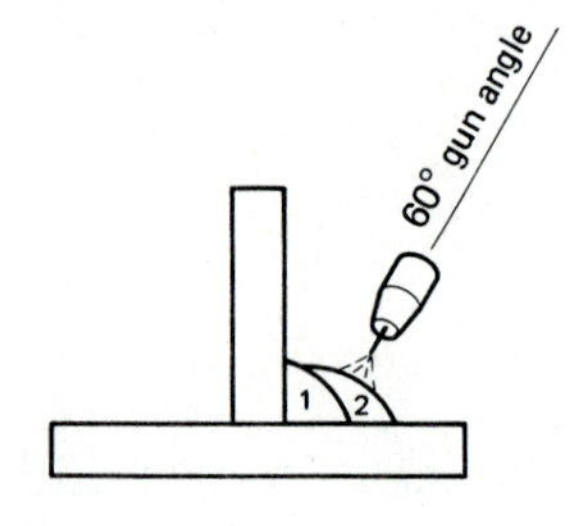

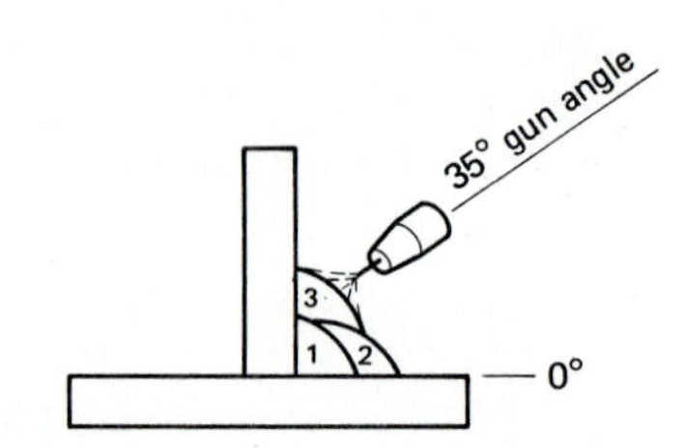

Material: A-36 (mild steel)

Size: $\frac{3}{8}$ in. × 3 in. × 6 in. plate

Joint Design: Tee joint

Position: Horizontal fillet (2F)

Technique: Stringer beads

Polarity: DCRP (DCEP)

Shield Gas: 75% Argon/25% CO_2

Gas Flow: 25–30 cfh

Filler Wire: E70S-3

Wire Feed: 180 ipm

Voltage: 25–30 V

Amperage: 180 A

Cleaning: Chipping hammer and wire brush

Procedure: Set up and tack a double-sided tee-joint configuration so that four sides can be welded. Use a 45° angle of attack. Deposit at least three passes in each of the four sides, rotating the plates into the 2F position each time. Try both the pushing and pulling techniques.

Visual Inspection: The leg lengths should be within $\frac{1}{8}$ in. of each other in the length. The face of the weld should be smooth, uniform, and free of visible discontinuities or craters. Undercut should be less than 0.010 in.

GMAW PROJECT 3

Vertical Tee Joint

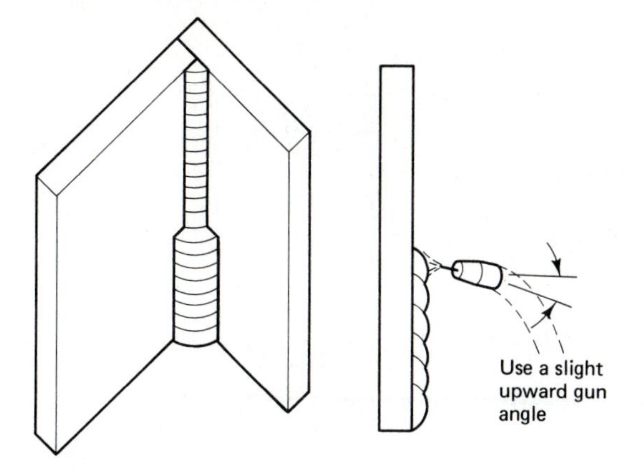

Material: A-36 (mild steel)

Size: $\frac{3}{8}$ in. × 3 in. × 6 in. plate

Joint Design: Tee joint

Position: Vertical fillet (3F)

Technique: Stringer and weave beads

Polarity: DCRP (DCEP)

Shield Gas: 75% Argon/25% CO_2

Gas Flow: 25–30 cfh

Filler Wire: E70S-3

Wire Feed: 165 ipm

Voltage: 24 V

Amperage: 165 A

Cleaning: Chipping hammer and wire brush

Procedure: Set up and tack a double-sided tee-joint configuration so that four sides can be welded. Using a 45° angle of attack, deposit a root pass using a slight side-to-side motion. After cleaning the root pass, deposit a cover pass using a side-to-side weave motion. *Note:* This project can also be done using a root pass, downhill, and cover pass weave vertical up.

Visual Inspection: The leg lengths should be examined and be within $\frac{1}{8}$ in. of each other. The weld face should be smooth, uniform, and free of visible discontinuities. The undercut should be less than 0.010 in.

GMAW PROJECT 4

Flat Groove Joint

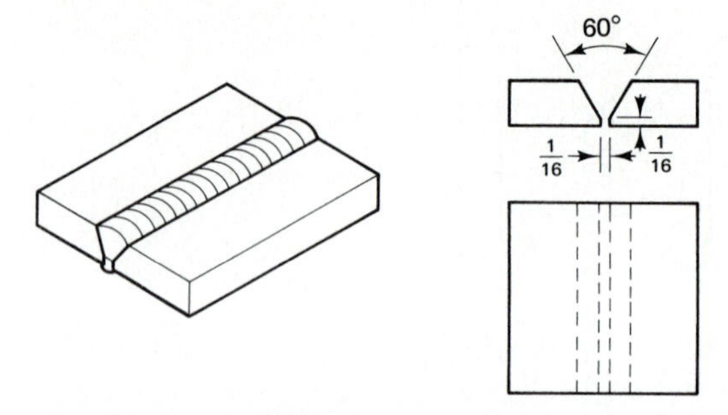

Material: A-36 (mild steel)

Size: $\frac{3}{8}$ in. × 3 in. × 6 in. plate

Joint Design: V-groove, butt joint, 60° included angle

Position: Flat groove (1G)

Technique: Stringer beads

Polarity: DCRP (DCEP)

Shield Gas: 75% Argon/25% CO_2

Gas Flow: 25–30 cfh

Filler Wire: E70S-3

Wire Feed: 150–180 ipm

Voltage: 23–25 V

Amperage: 150–180 A

Cleaning: Chipping hammer and wire brush

Procedure: Cut a 30° bevel angle in two plates, and grind a $\frac{1}{16}$- to $\frac{3}{32}$-in. root face on each plate. Position and tack your plate with a $\frac{3}{32}$- to $\frac{1}{8}$-in. root opening and a 60° included angle. Deposit a root pass using a side-to-side motion together with a bit of "U" motion. Adjust your amperage (wire feed), voltage, and travel speed to produce smooth, uniform penetration at the back side of the root. Increase your machine settings and continue to fill the groove.

Visual Inspection: The back side of the root pass should be free of any unfused edges and should not penetrate beyond $\frac{1}{16}$ in. or have any internal concavity. The face of the weld should be smooth, uniform, and free of visual discontinuities.

GMAW PROJECT 5

Vertical Groove Joint

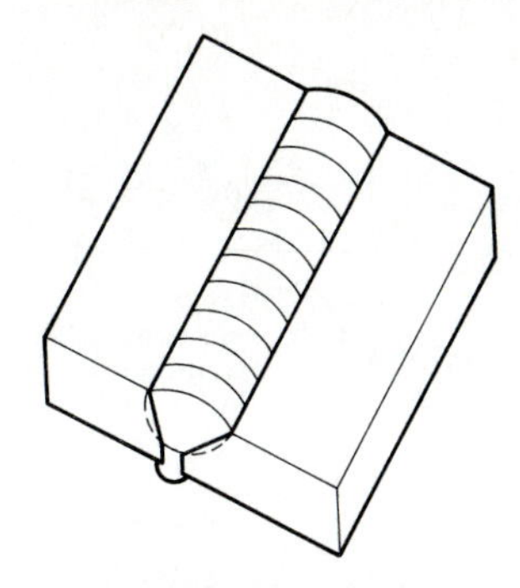

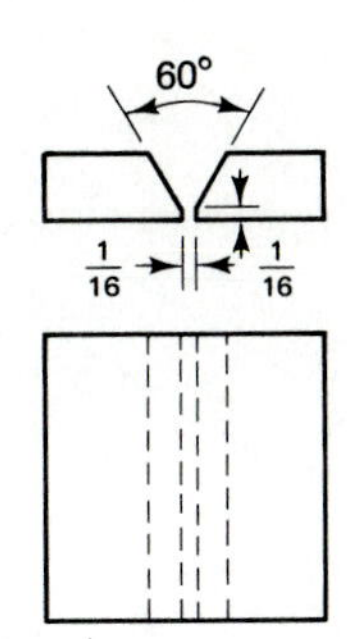

Material: A-36 (mild steel)

Size: $\frac{3}{8}$ in. × 3 in. × 6 in. plate

Joint Design: V-groove, butt joint, 60° included angle

Position: Vertical groove (3G)

Technique: Stringer beads

Polarity: DCRP (DCEP)

Shield Gas: 75% Argon/25% CO_2

Gas Flow: 23–30 cfh

Filler Wire: E70S-3

Wire Feed: 150–180 ipm

Voltage: 23–25 V

Amperage: 150–180 A

Cleaning: Chipping hammer and wire brush

Procedure: Cut a 30° bevel angle in two plates, and grind a $\frac{1}{16}$- to $\frac{3}{32}$-in. root face on each plate. Position and tack your plate with a $\frac{3}{32}$- to $\frac{1}{8}$-in. root opening and a 60° included angle. Deposit a root pass using a side-to-side motion together with a bit of "U" motion. Adjust your amperage (wire feed), voltage, and travel speed to produce smooth, uniform penetration at the back side of the root. Increase your machine settings and continue to fill the groove, using a side-to-side weave motion.

Visual Inspection: The back side of the root pass should be free of any unfused edges and should not penetrate beyond $\frac{1}{16}$ in. or have any internal concavity. The face of the weld should be smooth, uniform, and free of visible discontinuities.

GMAW PROJECT 6

Aluminum Lap Joint

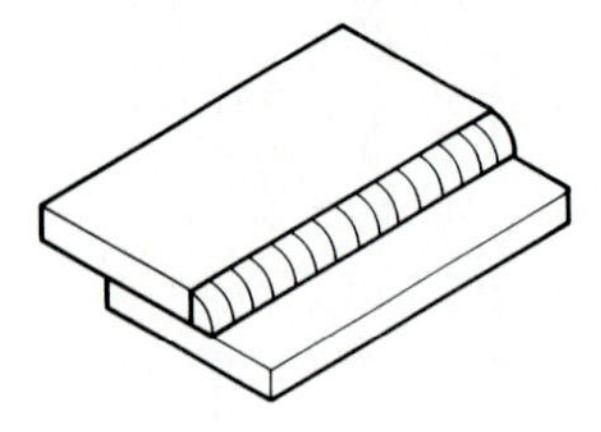

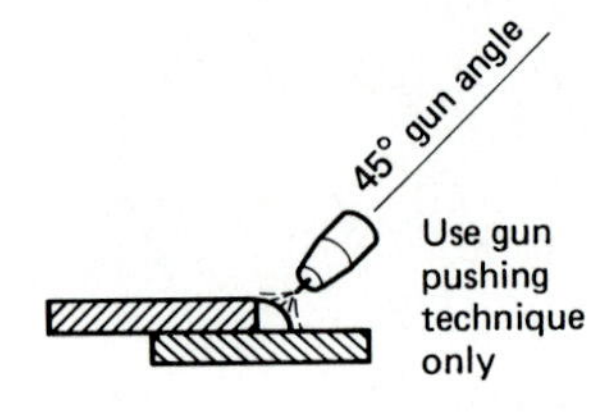

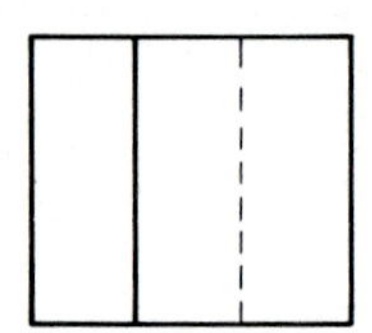

Material: Aluminum

Size: $\frac{1}{4}$ in. × 3 in. × 6 in. plate

Joint Design: Lap joint

Position: Horizontal (2F)

Technique: Stringers

Polarity: DCRP (DCEP)

Shield Gas: Argon

Gas Flow: 18–30 cfh

Filler Wire: ER4043, $\frac{3}{32}$-in. diameter

Wire Feed: 180 ipm

Voltage: 24 V

Amperage: 160 A

Cleaning: Wire brush *before* welding joint

Procedure: Wire-brush the plate with a stainless steel wire brush to remove the aluminum oxide coating. Tack plates into a lap joint configuration. With a 45° angle of attack, use the pushing technique to deposit at least three passes. If the beads tend to "ball-up" or do not fuse into the base metal, try preheating the base metal to about 150 to 200°F; this will improve weldability.

Visual Inspection: The face should show smooth, uniform beads with no undercut. Bead convexity should be uniform throughout the weld length, with limited cratering at the end of the beads.
Note: The key to producing high-quality aluminum GMAW welds is using the pushing technique.

GMAW PROJECT 7

Aluminum Tee Joint

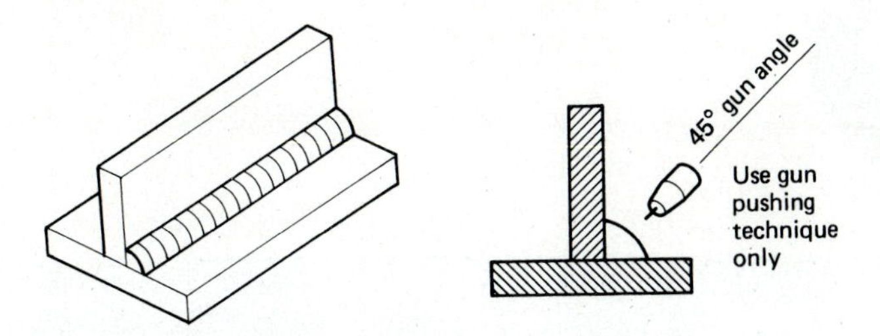

Material: Aluminum

Size: $\frac{1}{4}$ in. × 3 in. × 6 in. plate

Joint Design: Tee joint

Position: Horizontal fillet (2F)

Technique: Stringers

Polarity: DCRP (DCEP)

Shield Gas: Pure argon

Gas Flow: 18–30 cfh

Filler Wire: 4043, $\frac{3}{32}$-in. diameter.

Wire Feed: 180 ipm

Voltage: 24 V

Amperage: 165 A

Cleaning: Wire brush *before* welding joint

Procedure: Wire-brush the aluminum plate with a stainless steel wire brush to remove the aluminum oxide coating. Tack the plates into a horizontal tee-joint configuration with a 45° angle of attack. Use the pushing technique to deposit at least three passes. If the beads tend to "ball-up" or do not fuse into the base metal, try preheating to about 150 to 200°F. This will improve the weldability.

Visual Inspection: The face should exhibit smooth, uniform beads with no undercut on the top leg. The bead convexity should be uniform throughout the weld length, with limited cratering at the end of the weld bead.

Note: Always use the pushing technique.

FLUX-CORED ARC WELDING (FCAW)

Flux-cored arc welding (FCAW) is closely related to GMAW, but it uses tubular wire instead of solid wire (Figure 13–1). It also requires a power source with a higher amperage capacity. It may or may not require external shielding like GMAW; some wires are self-shielding and some require dual shielding of the flux in the core of the wire and external gas flow.

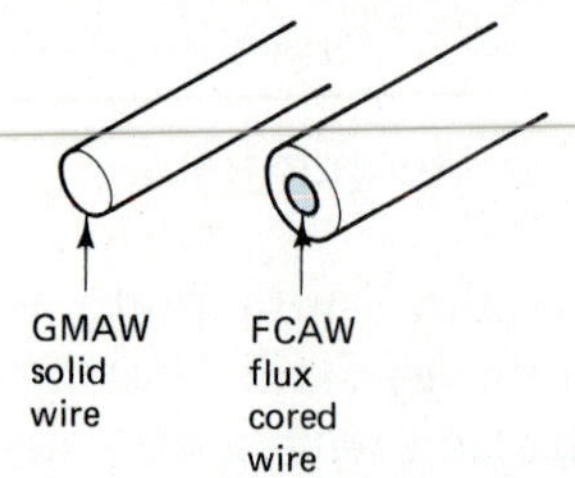

FIGURE 13–1
FCAW and GMAW wires.

ADVANTAGES OF FCAW

Like GMAW, FCAW is extremely fast, and actually has a higher deposition rate than GMAW. The flux in the core of the wire shields the long, molten, trailing puddle (Figure 13–2). GMAW cannot produce as large welds because of its lack of this additional shielding action.

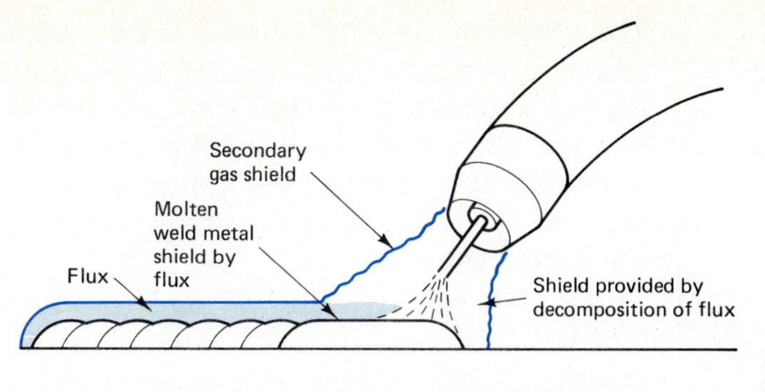

FIGURE 13–2
FCAW process.

Additionally, the flux chemically cleans and deoxidizes the weld metal. It also slowly cools the weld metal, giving the weld good ductile and mechanical properties. The dual shielded FCAW method using additional gas shielding is one of the highest quality of all production welding processes, and is considered "x-ray quality." This means that this process would meet or exceed x-ray (radiograph) quality requirements as spelled out by applicable welding codes (see Chapter 27).

FCAW uses primarily the spray and globular mode of metal transfer (see Chapter 11 for types of metal transfers). Spray FCAW is used for flat and horizontal work and some limited vertical applications. Globular is used for some horizontal and vertical welding. Only a few years ago, vertical welding was not possible with FCAW, but today's special vertical cored wires make all positions practical.

FCAW requires about the same amount of time to master as does GMAW. It is best to master GMAW first, then move on to FCAW. If you have already mastered GMAW, FCAW will be quite a simple conversion. Here are a few items to look for when FCAW welding.

PHOTO 13–1
Large FCAW weldments. (Courtesy of Miller Electric Mfg. Co.)

1. The process uses larger-diameter filler wires than most GMAW; this means more amperage and more welding rays. Many welders use a darker lense then they use for other processes. I use a number 12, but remember, high amperages need darker lenses and low amperages can use lighter ones. People's eyes are different, too, so you will need to find a shade with which your eyes are comfortable.
2. The FCAW weld puddle will be larger and more fluid than the GMAW puddle; keep this in mind when welding in the vertical position. A little side-to-side motion will help keep this fluid puddle flat while running in the vertical position.
3. A little more smoke is produced with FCAW. Make sure that there is good ventilation or a good working welding exhaust system. CO_2 has been the standard shielding gas for dual shielded FCAW. However, with the emerging vertical running FCAW wires, the 75% argon–25% CO_2 mix is quite popular. It gives a smooth, controllable arc wire, as CO_2 produces a little rougher arc, but it is cheaper. Both gases will run spray or globular transfer with FCAW.

PREPARING FOR FCAW

As when starting any welding process, the welder should be wearing all safety equipment and observe all the safety rules in Chapter 2. If you have not read Chapter 2, read it before going any further.

1. If all equipment is properly installed, turn on the power source, feeder unit, and open shielding gas bottle.
2. Push the purge switch if there is one; if not, turn the wire feed down, pull the gun trigger switch, and set the shield gas flow rate. (Gas flow rates for FCAW are between 20 and 35 cfh.)
3. Set the voltage at the rate suggested in Table 13-1.

TABLE 13-1
Typical Machine Setting Ranges for FCAW

WIRE DIAMETER (IN.)	CURRENT (A)	VOLTAGE	TRAVEL SPEED (IPM)	SHIELD GAS FLOW (CFH)
0.045	150–250	22–26	140–300	30–40
$\frac{1}{16}$	230–350	25–32	180–300	30–45
$\frac{5}{64}$	280–450	28–35	180–300	35–50
$\frac{3}{32}$	330–580	30–36	150–250	40–50
$\frac{7}{64}$	400–650	32–38	120–200	45–55
$\frac{1}{8}$	420–700	34–40	100–200	50–60

4. Set the amperage wire feed at the rate suggested in Table 13-1.
5. Position the gun in either the pushing or pulling direction, with a 1-in. stickout and the wire about $\frac{1}{4}$ to $\frac{1}{2}$ in. off the plate to be welded (Figure 13-3).
6. Pull the trigger and adjust the travel speed to produce a bead width that gives good fusion.

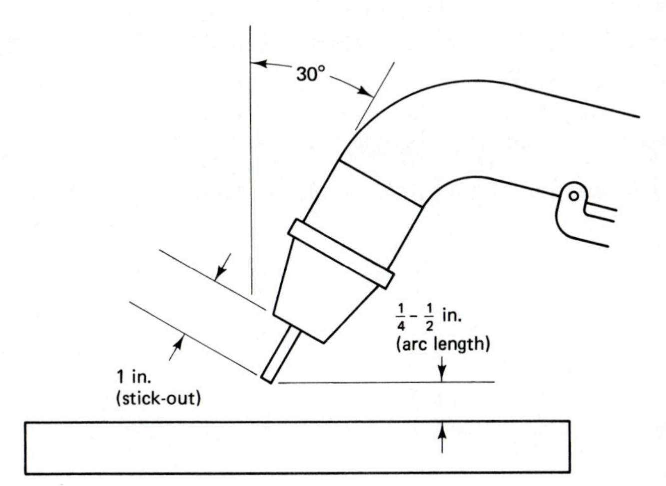

FIGURE 13-3
Stickout.

Table 13-1 will get you "in the ballpark" for machine settings. For best results, you will probably have to make some additional adjustments. Be sure to consult the wire manufacturer's recommendations for exact settings on the wire you are using.

REVIEW QUESTIONS

1. What are the advantages of FCAW over GMAW?
2. Are there any disadvantages of FCAW?
3. What modes of metal transfer are used with FCAW?
4. What is the purpose of the flux in the core of FCAW filler wire?
5. What shielding gases are used most commonly with FCAW?
6. What type of power source is used with FCAW?

SUBMERGED ARC WELDING (SAW)

Submerged arc welding (SAW) is a high-production welding process similar to GMAW (gas metal arc welding) or FCAW (flux-cored arc welding). What makes SAW unique is that the arc takes place below a blanket of granular flux. The heat for welding is obtained in the same way that it is in standard arc welding processes.

SAW THEORY

Here's how it works. A feed tube of flux first lays down a blanket of flux into the groove or fillet joint to be welded. The energized electrode wire feeding from the contact tip feeds through the flux and starts the arc on the plate. During the heat of welding, the flux forms a molten protective slag which shields the molten weld pool and slows the cooling rate of the weld (Figure 14–1).

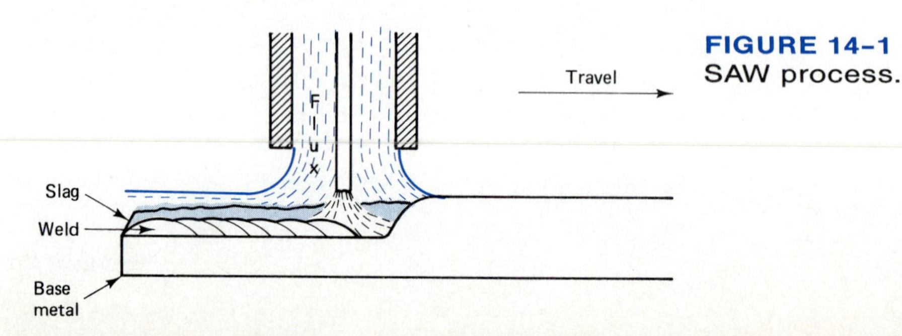

FIGURE 14–1
SAW process.

PHOTO 14–1
Hand-held SAW. Note that the welder has no helmet.

ADVANTAGES OF SAW

Remember, with GMAW the puddle can only get as large and long as the gas shield can successfully shield. Outside this shielding envelope, porosity will result. With SAW, all we have to do is keep the puddle under the flux blanket to keep it shielded. This allows for very large and long molten weld puddles and the resulting large weld deposits all in one pass. The process is generally used on thicker sections ($\frac{1}{2}$ in. and up) of material.

Most SAW today is automatic, with a technician setting up and controlling voltage, amperage, travel speeds, and other essential variables. There are some hand-held applications in use. A unique feature of both the automatic and hand-held applications is that the welder needs no helmet as the arc rays are absorbed by the blanket of flux. The process is limited to the flat and horizontal positions, although experimentation continues on devices that would allow SAW in other positions, such as vertical. Properly set up, SAW produces an exceptionally smooth, uniform bead with little smoke and spatter.

SAW EQUIPMENT

The SAW power source is similar to standard GMAW, but has a higher amperage capacity. It is, of course, constant-voltage and therefore maintains a steady arc length while welding. The wire feeder is heavy duty compared to GMAW types. This is because SAW uses larger-diameter wires ($\frac{3}{32}$ to $\frac{5}{16}$ in.). SAW also requires a flux delivery system. This includes air pressure (from an air compressor) and a flux hopper (Figure 14–2) to contain and supply flux to the weld. Many companies use a flux dryer to remove any moisture and resulting hydrogen that might end up in the weld. A vacuum recovery system is used to pick up any unmelted flux and return it through the dryer and into the flux hopper for reuse. With automatic SAW, the feeder unit and gun are mounted on a side-beam travel carriage. This moves the SAW "head" (feeder unit and gun) along the weld seam.

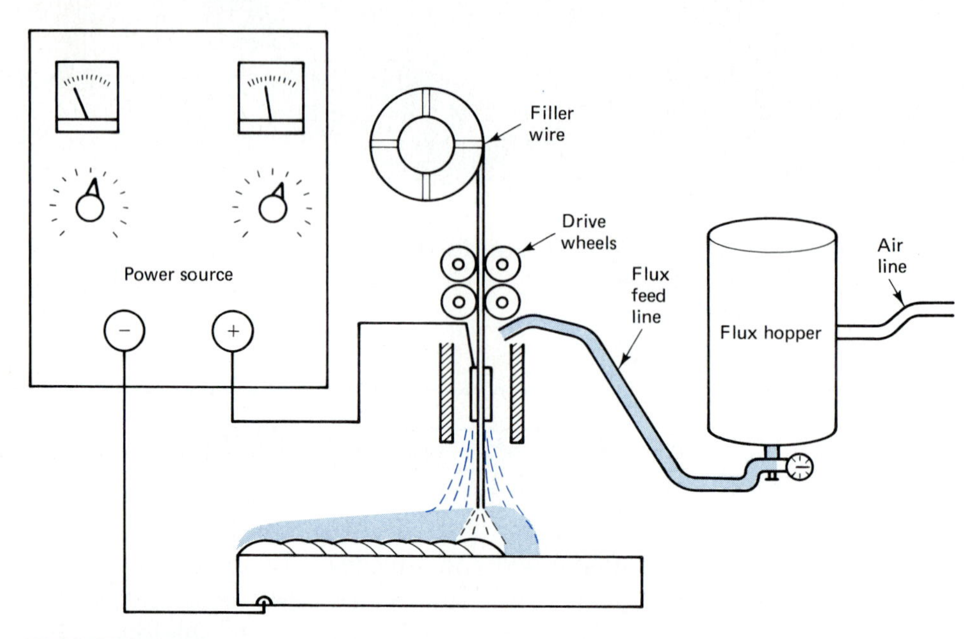

FIGURE 14–2
SAW welding system.

PHOTO 14–2
Automatic side-beam SAW system. (Courtesy of Jetline Engineering, Inc.)

SAW JOINT DESIGN

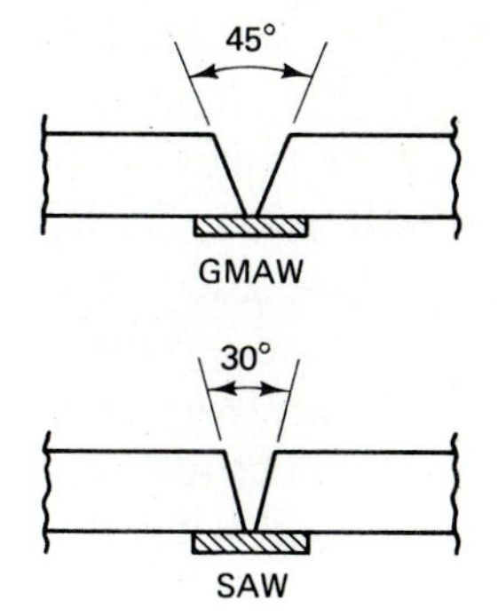

FIGURE 14-3
SAW AND GMAW joint angles.

Joint design for SAW is somewhat different from that for GMAW or other methods. The joints tend to be narrower. For example, on a $\frac{1}{2}$-in. V-groove butt weld, GMAW would use about a 45 to 60° included angle; SAW for the same conditions would use about 30 to 40° (Figure 14-3). This narrow groove angle results in savings in cost and time. Because less filler metal is used, it takes less time to fill the groove.

SAW complete-penetration groove welds usually have a backing strip. This can be a consumable strip, made of the same material as the base metal, or may be a nonconsumable type. For SAW of steel, copper is popularly used. Root support for SAW groove welds can also be accomplished by running a backing bead with a SMAW in the root first, then filling the remaining portion of the groove with SAW (Figure 14-4).

SAW can also use flux backing to contain the molten filler metal in open root passes (Figure 14-5). This is used on thinner sections of material.

Typical joint designs for SAW are shown in Figure 14-6.

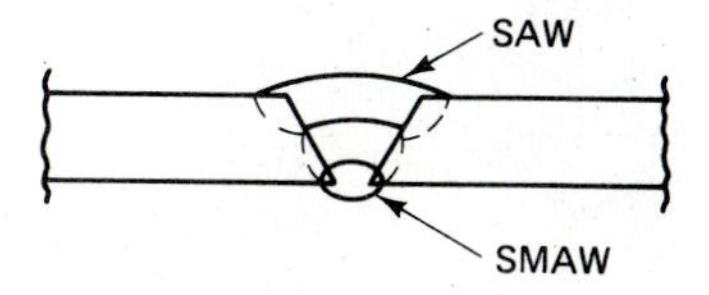

FIGURE 14-4
SMAW root and SAW filler joint.

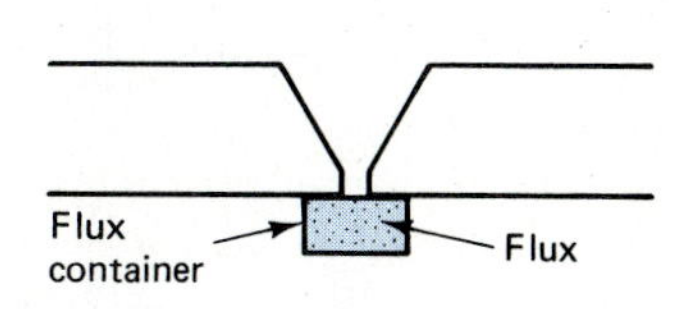

FIGURE 14-5
SAW with flux backing.

FIGURE 14-6
Typical SAW joint designs.

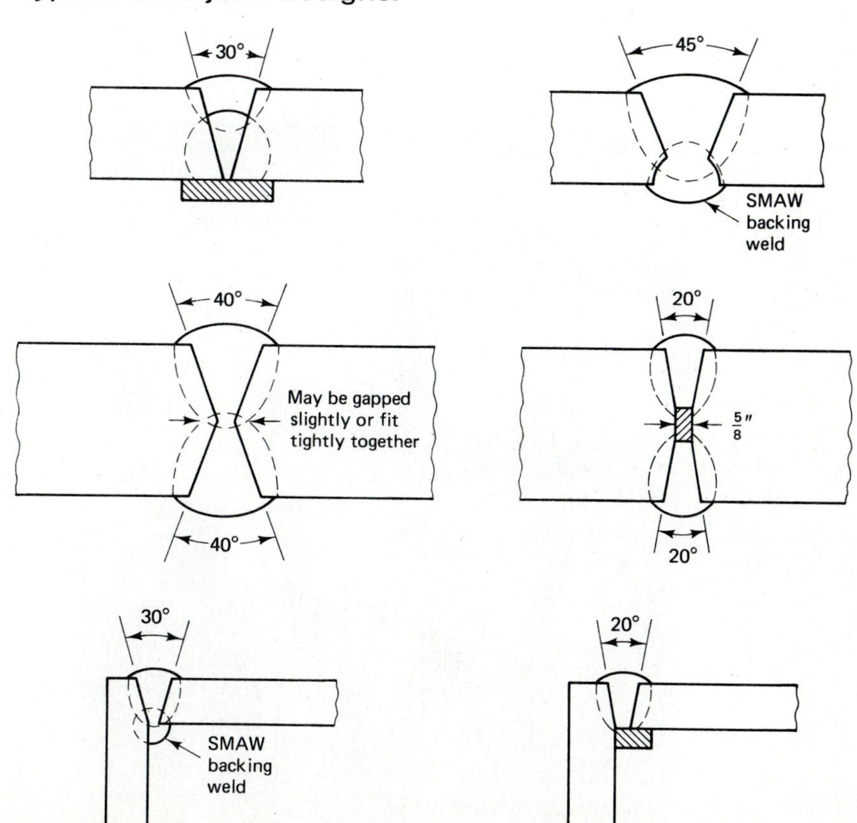

SAW Polarities

Let's first examine polarities that we can use for SAW. DCRP (DCEP) is the most popular for most constant-voltage processes and works well for SAW. It gives a uniform, deeply penetrating bead. If parts to be submerged arc welded are rusty, dirty, or have heavy mill scale, DCRP should be used. DCSP (DCEN) puts more heat into the plate and therefore increases the deposition rate, and should be used where deep penetration is *not* a necessity but where high welding speeds are required. Even higher deposition rates can be achieved with dual-electrode wire systems (Figure 14–7). This uses dc on the first root pass lead wire, and

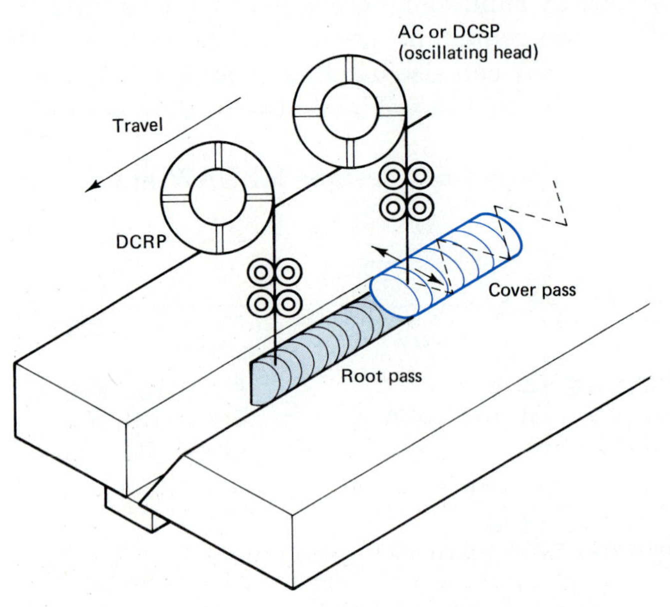

FIGURE 14–7
SAW with dual-head design.

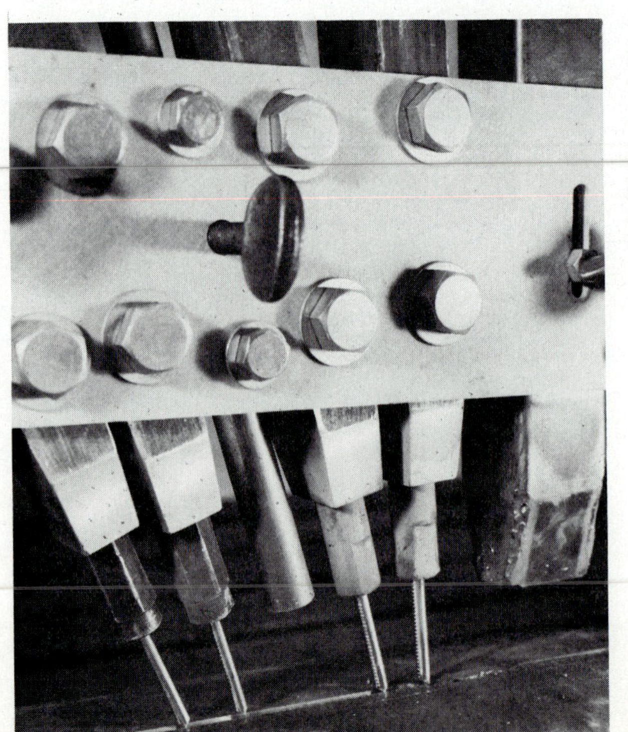

PHOTO 14–3
Double-tandem SAW head. Four $\frac{5}{32}$-in.-diameter wires produce x-ray-quality welds at speeds and deposition rates not obtainable by other single- or multiwire systems. (Courtesy of The Lincoln Electric Co.)

ac on the trailing wire. The ac tends to flatten out on the trailing filler pass. The ac trailing wire can also be oscillated back and forth. This also helps flatten out the trailing pass.

Current Settings

Some suggested current settings are given in Table 14–1. Remember, each manufacturer's machines will vary slightly. Atmospheric humidity, flux dryness, and electrode stickout will all affect exact current settings.

TABLE 14–1
SAW Current Settings

WIRE DIAMETER (IN.)	CURRENT RANGE (A)
$\frac{3}{32}$	150–500
$\frac{1}{8}$	200–900
$\frac{5}{32}$	300–1100
$\frac{3}{16}$	400–1250
$\frac{1}{4}$	550–1400
$\frac{5}{16}$	800–2000

Voltage Settings

Voltage settings are also quite important. Just as in GMAW, the voltage determines the length of the arc. With SAW the arcing is taking place below the blanket of flux, but there is still an arc length. High voltages will lengthen the arc, resulting in a flatter and wider bead; lower volts will shorten the arc length and produce a more convex, narrower bead (Figure 14–8).

Travel Speed

Travel speed is the next essential variable the welding technician can control. If travel speed is increased, heat input decreases and less filler metal is deposited. This results in less bead convexity. Travel speeds that are much too fast will be nonuniform, sparse, and may lack fusion.

FIGURE 14–8
SAW arc length effects.

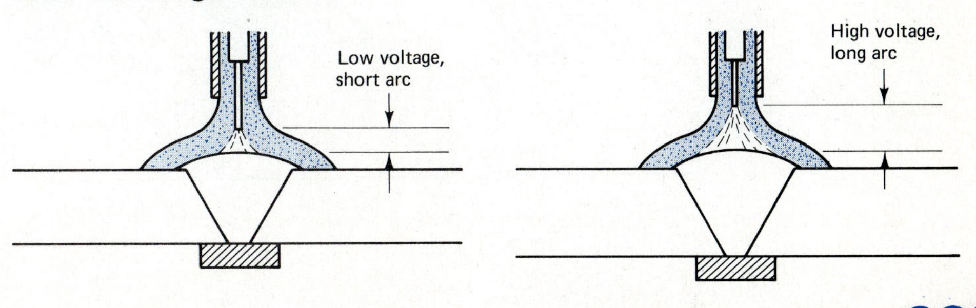

The metal may also contain granules of unmelted flux trapped on the top of the bead. If travel speed is decreased more, more heat, filler, and a more convex bead will result. An extremely slow travel speed will cause cold lap and poor fusion. These relationships are shown in Figure 14-9.

Flux Delivery Rate

If too much flux is deposited, an irregular narrow bead can result. This is due to the inability of the flux gases to escape through the thick layer of flux. If insufficient flux is deposited, you may see the arc flashing through the thin layer of flux. This is usually the first sign that your flux hopper is empty. Not enough flux will also leave a spattery and porous bead, together with numerous other not-so-visible problems. So, how much is correct? We suggest, as with any welding process, that some trial and error be applied before you run too much scrap. But here is a good guideline: About one-third of the flux you deposit should melt into slag, so after running some practice beads, see if the slag is about one-third as thick as the granular flux layer. If you see any white light coming through the flux while welding, increase the flow of flux. You should see only a red glow behind and around the arc.

Cleaning the joint is very critical with SAW. In some processes impurities may be burned out, but forget it with SAW—they won't burn out. They will, however, become permanently trapped in the weld and x-rays won't miss it.

Don't think that just because you do not need to use a hood with SAW that you need not practice. Some of the most highly skilled welders and technicians are SAW welders. If your training facility does not have SAW, practice with GMAW. It is closely related to SAW and will teach you much of the coordination needed for SAW. (*Welding instructors:* Any inexpensive SAW simulation can be carried out by purchasing a bag of SAW flux, pouring it into the groove, and dragging a GMAW gun through it. Use a little higher amperage than normal and turn off the shielding gas.)

FIGURE 14-9
SAW Travel speed effects on beads.

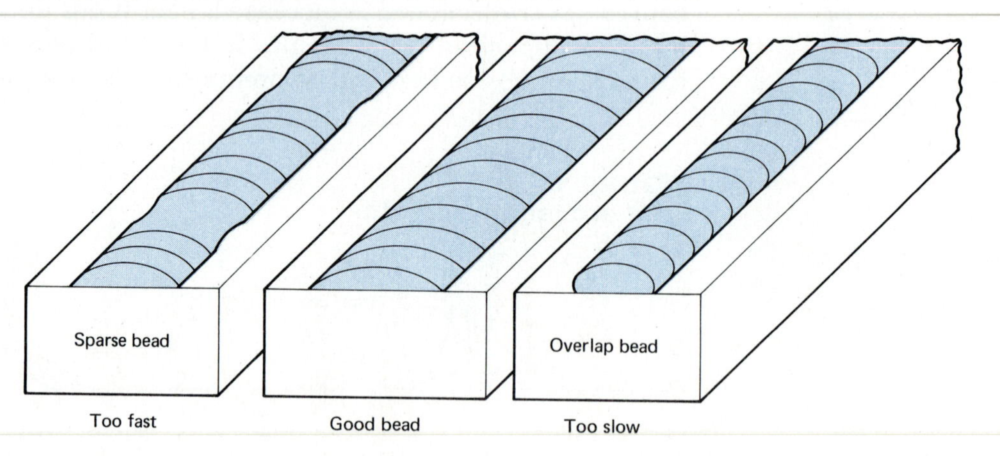

REVIEW QUESTIONS

1. How is atmospheric shielding achieved with SAW?
2. Are there any limitations associated with SAW?
3. What electrode filler wire diameters are possible with SAW?
4. What type of shielding from arc rays is necessary with SAW?
5. What is different about SAW joint designs versus other arc welding joint designs?
6. What polarities can be used for SAW, and for what applications?
7. What effect does travel speed have on the weld bead? Explain.
8. What effect do the amperage and voltage have on the weld bead?

15 SELECTING FILLER WIRE

Whenever you are called on to select a filler metal, there is a checklist of items to review to help you make the best and most economical selection.

1. Match base metal with filler metal composition.
2. Control mechanical and chemical properties.
3. Deoxidize weld metal.
4. Maintain stable arc.
5. Consider economics.

Identification of the metal you are about to weld is your first and primary concern. Consult Chapter 19 for information on metal identification.

Matching the base metal composition with that of the filler metal is the next consideration. This will keep the chemical, physical, and mechanical properties uniform throughout the weldment. If the mechanical properties are not uniform, the weldment will not react uniformly under load. A mismatch of chemical properties may result in corrosion of either base metal or weld. If physical properties do not match, this could result in uneven expansion and contraction in metal exposed to temperature extremes.

Obtaining chemical analysis of filler metals may be necessary.

If the filler metal is an AWS classified filler, its range of elements will be listed in the applicable code section. For example, mild steel GMAW filler wires are listed in Section A5.18. Exact listings can also be provided by the electrode manufacturer.

The amount of weld metal deoxidation is also an important consideration. Rusty, dirty, and oily metals will need fillers with high amounts of deoxidizers, such as ER70S-4, ER70S-5, and E70S-6. These fillers contain aluminum, magnesium, zirconium, or silicon, which absorbs the oxygen and floats it to the surface, keeping it out of the weld.

UNDERSTANDING GMAW AND FCAW FILLER WIRE CLASSIFICATION SYSTEMS

There are three specific numbers on any electrode box (Figure 15–1):

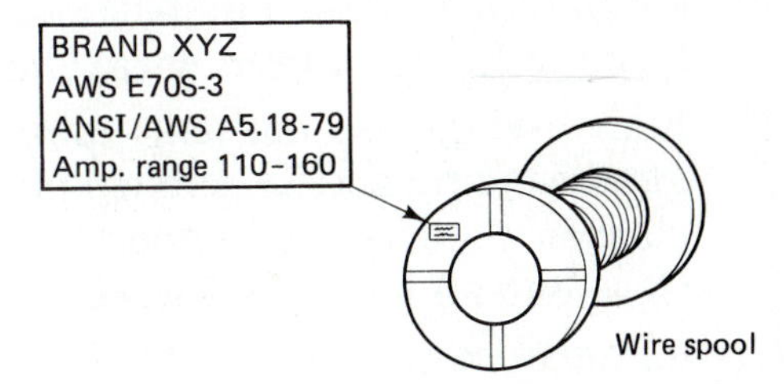

FIGURE 15–1
Electrode box specifications.

1. *AWS classification.* This is the wire classification.
2. *ANSI/AWS classification.* This is the American National Standards Institute/American Welding Society specification to which these electrodes were manufactured.
3. *Amperage range.* This is the suggested range for welding of this type and diameter of electrode. This will get you close, but use the set-by-feel system described earlier.

It is important that you understand what each digit in the AWS classification means. For the example shown in Figure 15–1 (ER70S-3):

E means "electric" or "electrode" and that the wire can conduct electricity when welding.

R means that the wire is also suitable for nonconductive filler and can be dipped into the puddle, as with GTAW, PAW, and so on.

70 indicates the minimum tensile strength of the deposited weld metal in ksi or thousand pounds per square inch (example: 70 = 70,000 psi).

S indicates a solid filler wire.

3 indicates the chemical content range of the wire. To find the range of any wire, check the ANSI/AWS specification.

For example, the specification for ER70S-3 is 0.06 to 0.15 carbon, 0.9 to 1.4 manganese, and 0.45 to 0.70 silicon. Each digit will vary and there may be more elements in the material.

Here is a brief description of the steel AWS-classified GMAW wires.

Mild Steel GMAW Solid Wires

(ER)

ER70S-2: Al, Zr, Ti triple deoxidized—good for welding dirty surfaces. Gases: argon-O_2, argon-CO_2, CO_2.

ER70S-3: high silicon, argon-O_2 or CO_2. Most popular solid wire, similar to E70S-1 but contains more silicon. Gases: argon-O_2, CO_2, argon-CO_2.

ER70S-4: super high silicon, all MIG gases, premium wire. Good for large puddles and large welds. Gases: usually straight CO_2.

ER70S-5: triple deoxidized, all gases similar to E70S-2, but available only in large diameters. Gases: usually straight Co_2.

ER70S-6: similar to E70S-3 but better deoxidized. Good for large welds, premium wire. Gases: straight CO_2.

ER70S-7: high manganese content; gives super deoxidation, good wetting action, and a smooth appearance. Like classes 4 and 6, it is good for large welds. Gases: CO_2, but all gases will work.

ER70S-G: all-purpose wire; open-class wire can have any content, but must meet all requirements of ANSI/AWS A5.18 specification. E70S-G usually contains many deoxidizers, x-ray-quality welds, mild steel, and high-strength low-alloy steel. Some contain molybdenum for pipe welding.

FCAW Tubular Wires

For E70T-3:

E	Electrode
70	Minimum ksi tensile strength
0 or 1	Flat and horizontal only = "0," all position = "1" (usually small diameters)
T	Tubular
3	Alloying elements in flux core (also an indicator of usability and capabilities or wires)

Applications for the Carbon Steel FCAW Wires

T-1: capable of single and multipass welds. Small diameters usually will run in all positions. Runs an excellent bead with spray transfer, and usually low spatter. Gases: CO_2 or argon-CO_2 (75%–25%).

T-3: specifically for the thinner metals in the FCAW

range, single-pass welds only. This wire is self-shielding.

T-4: self-shielding wire that runs globular transfer and will fill poorly fitted joints. It has a high deposition rate but fairly low penetration for a FCAW process.

T-5: also a low-penetration, multipass wire, but it requires CO_2 or argon–CO_2 (75%–25%) shielding. T-5 works best if welded in the 1G, 1F, or 2F position.

T-6: very deep penetration, self-shielding wire. It can run single or multiple passes in flat and horizontal positions.

T-7: runs on DCSP (DCEN), which puts about two-thirds of the heat in the plate. DCSP allows for the use of large electrodes and high deposition rates. High current settings will produce a smooth spray transfer.

T-8: also a DCSP (DCEN), with all the characteristics listed under T-7. This wire has a light slag, which makes it excellent for all-position welding. It can run single- or multiple-pass welds.

T-10: self-shielded DCSP wire that gives high travel speeds due to its slag system. This wire should be limited to single-pass welds in the flat and horizontal positions. It will also run downhill.

T-11: self-shielding and runs on DCSP. It can produce a smooth spray transfer and fairly high travel speeds. It is a general-purpose electrode that will run single or multiple passes in all positions.

T-G or T-GS (no chemical requirements; open classifications): T-G is for new *multiple-pass* electrodes not under any other AWS class; T-GS is for new *single-pass* electrodes not under any other AWS class.

SAW FILLER WIRES

It may seem a little more difficult choosing a SAW wire and flux combination because you are now dealing with two consumables. But you were actually doing the same thing when you selected a certain shielding gas for a particular filler wire. Let's first examine SAW fluxes. They come in two types: (1) active or (2) inactive (neutral). The *active* fluxes actually put elements such as deoxidizers in the weld while welding, thus actively affecting the weld metal strength and properties. The *inactive* or *neutral* fluxes do not add significantly to the strength of the weld metal deposited.

What Do SAW Fluxes Do?

First and most important, fluxes shield molten weld from the surrounding atmosphere, which contains oxygen, nitrogen, and hydrogen, all which will quickly destroy the integrity of a weld. SAW welds are large, long, and hot, so conventional gas shielding systems will not work economically. Flux and slag will remain over the long molten puddle, keeping it shielded until long after it has solidified.

SAW produces a large hot weld, which can result in large metal grains and cracked welds. The slag produced from the flux acts as a slow-cooling blanket, thus allowing the grains to refine and reduce the risk of cracks. Properly made SAW welds will not crack.

Fluxes of the active type also add elements to the weld deposit. These elements are usually deoxidizers such as manganese and silicon. Active fluxes are generally used with lower-grade filler wires to keep welds porosity free. Because of the lower-grade filler wire, the use of active fluxes should be limited to single passes to prevent porosity in the top weld layers.

You may hear the terms "fused" and "bonded" and "mechanically mixed" fluxes. This has to do with the method by which they were manufactured. Fused fluxes combine all elements of the flux together by melting them in a furnace, then grinding the product into a granular flux. These fluxes are quite moisture resistant and can produce a low-hydrogen weld. They can be used with low- and medium-carbon steels to produce a crack-free weld. In bonded-type fluxes, elements are mixed together with a binder and treated so as to produce granulars with well-dispersed elements. This type of flux tends to bite through rust and mill scale quite well, while keeping the weld porosity free.

The mechanical mixed fluxes can be just elements mixed together or with a combination of fused and bonded fluxes. The usability and application of these fluxes will depend on the makeup and combination of elements used.

SAW Wires and Flux Classification System

The SAW wire classification system is somewhat different from the GMAW and FCAW systems in that it uses a separately packaged flux and wire. Flux and wire each has its own system (Figure 15–2). For the flux class F64:

- F Flux
- 6 60,000 psi tensile strength (6 or 7, 60 or 60 ksi) of deposited weld metal, using this flux in combination with a specific welding wire
- 4 Lowest temperature at which weld metal will meet or exceed 20 ft-lb (*impact strength*) on the Charpy V-notch test

FIGURE 15–2 Classification system for SAW wires.

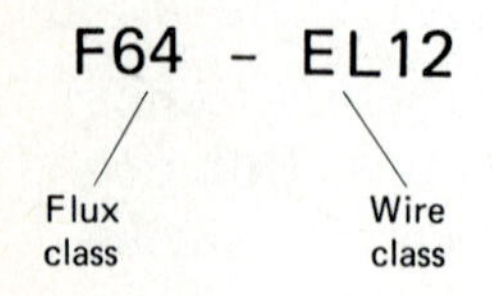

The final digit may be any of the following, depending on the impact strength of the metal deposited.

- Z No impact requirements
- 0 20 ft-lb at 0°F
- 2 20 ft-lb at 20°F
- 4 20 ft-lb at 40°F
- 6 20 ft-lb at 60°F

For the wire class EL12:

E electrode
L low manganese (M = medium and H = high)
12 C, Mn, and Si content (chemical composition) (Deoxidize)

Making the Selection

As stated earlier, begin by examining the job. Are there any rusty surfaces? Do you have medium-carbon or HSLA steels? How many passes have to be made? For oxidized, rusty, or even heavy mill scale, use a bonded flux with many deoxidizers. For medium-carbon, HSLA, or high-sulfur steels, use a fused flux and a wire with a fair amount of deoxidizers. With multipass jobs, remember that active fluxes and low-manganese wires may result in porosity. Keep in mind that manganese is an excellent deoxidizer, but too much will result in a lower-impact-strength weld deposit; in other words, it may be brittle.

The best way to verify that your flux and filler wire combination is suitable for the application is through the procedure qualification tests spelled out by the American Welding Society (see Chapter 27). This is the only way to be assured that the job will have the necessary mechanical properties.

Always consult the flux and wire manufacturers if there is any question regarding the application. They are the people who made the products and know best what they will and will not do. Most manufacturers publish a wire guide that may be helpful.

ESW (ELECTROSLAG WELDING) FILLER WIRES

You can follow most of the same considerations for selecting an electroslag welding (ESW) filler metal as you did for SAW wires, but ESW is very different from SAW, in that it's not an arc welding process, it uses a conductive slag, and uses a square groove joint design (see Fig. 15-3A). Also, it may or may not use a consumable guide tube system which consumes and becomes part of the final weldment. Figure 15-3 shows how ESW filler wires are classified.

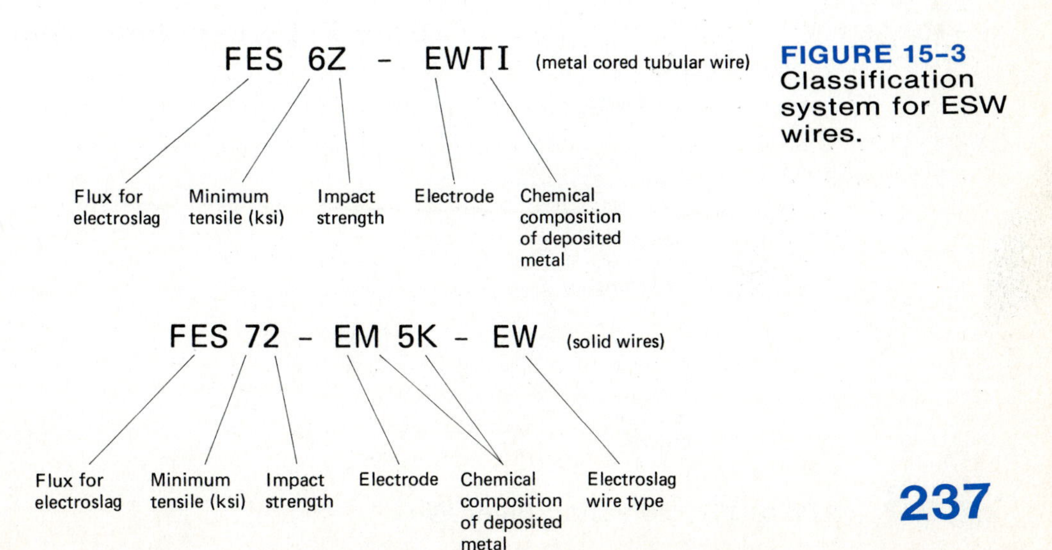

FIGURE 15-3 Classification system for ESW wires.

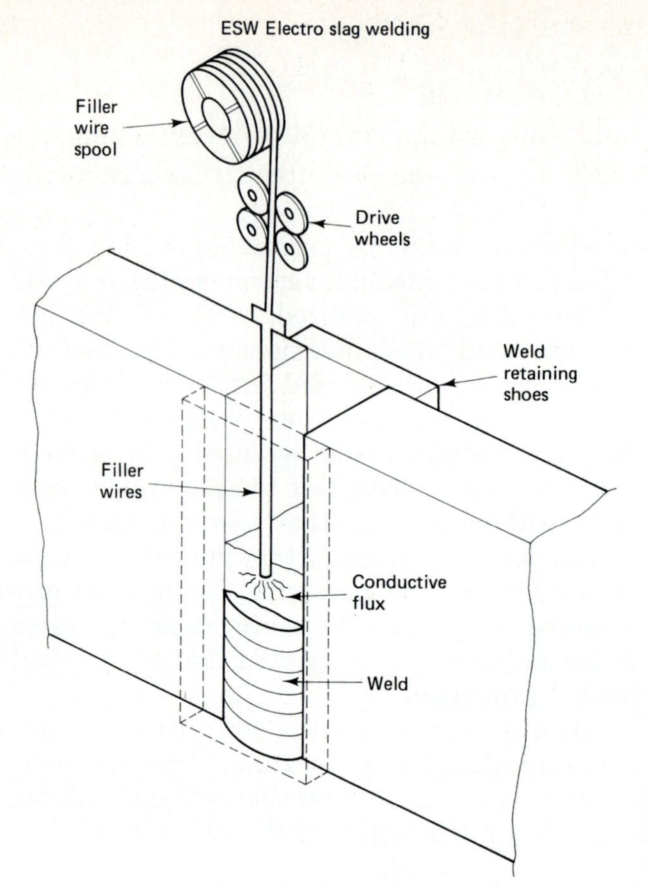

FIGURE 15-4
Classification system for EGW wires.

EGW (ELECTROGAS WELDING) FILLER WIRES

Electrogas welding (EGW) fillers are similar to those for GMAW and FCAW. The process looks similar to ESW, but is an arc welding process, whereas ESW uses the heat of the molten flux to create heat for fusion. Figure 15-5 shows an example of how EGW filler wires are classified.

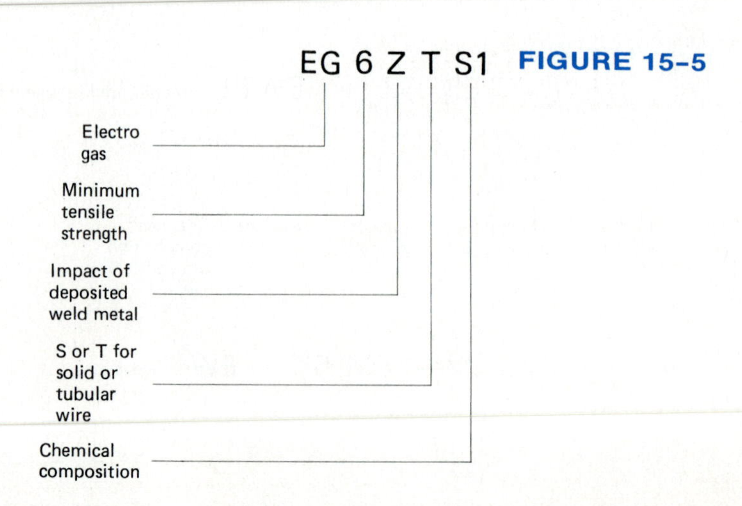

FIGURE 15-5

REVIEW QUESTIONS

1. List the major considerations when selecting a filler wire.
2. In the AWS classification ER70S-3, what does the "ER" stand for?
3. In the AWS classification ER70S-3, what does the "70" stand for?
4. In the AWS classification ER70S-3, what does the "S-3" stand for?
5. How is an all-position FCAW tubular wire indicated in the AWS classification system?

16

PIPE WELDING

Today's pipe welders may be found on jobs involving cross-country petroleum pipelines, gas tungsten arc welding of a coolant line in a nuclear-power plant, or fixing a steam line in a production facility. Pipe welding is a highly skilled trade that only the most persistent welders master. Most pipe welders are certified to the applicable code or standard governing the particular job.

Many novice welders want to become pipe welders immediately, but plate welding skills must be learned before attempting to work on pipe. You do not shoot a few rounds of golf and expect to play in a PGA championship match. The same applies to all forms of pipe welding; many hours of disciplined practice are required if you intend to be a successful pipe welder in today's modern industry. The quality, size, position, and other variables will determine exactly how pipe is welded. Some pipe is simply used for water transmission. Should a failure occur in this pipe, the results would not be as dangerous as would a failure in a high-pressure steam pipe.

These differing quality standards also determine the welding process used. Where extremely high quality is required, the gas tungsten arc welding process is often used. The small-diameter piping in nuclear power plants is often welded with GTAW. Larger diameters use a combination of GTAW on root passes and shielded metal arc welding with low-hydrogen electrodes for the cover passes. This combination usually produces x-ray-quality welds.

PHOTO 16–1
Welders join sections of pipe for the trans-Alaska pipeline in the Chugach Mountains just north of Valdez. [Courtesy of The Standard Oil Co. (Ohio).]

Cross-country pipelines may contain petroleum or natural gas. Welders on these pipelines are usually certified to the American Petroleum Institute standards. Downhill welding with E6010 electrodes at fairly high amperages is common practice among highly skilled pipeline welders. High-pressure pipe welding on steam pipe is often done with shielded metal arc welding. E6010 electrodes, or E6010 in combination with E7018 low-hydrogen electrodes, are commonly used.

Applications of plasma arc welding (PAW) pulsed gas metal arc welding and flux-cored arc welding are being introduced into some areas of pipe welding. These processes improve quality and productivity in some pipe welding applications.

PIPE WELDING POSITIONS

The technique and difficulty of welding pipe will have a great deal to do with what position it is in (Figure 16–1). For example, the 1G or 1GR, which rolls the pipe as it's welded, is not much different from flat welding on plate. In contrast, the 5G and 6G positions are fixed positions and require excellent skill and dexterity.

The 2G, 5G, and 6G positions should also use a special weld pass sequence. The pipe should be beveled and gapped to the specified angle and root opening. The gap can be set to the desired root opening by laying the two sections of pipe in angle iron (Figure 16–2). Now use a piece of filler metal of the specified root opening size to gap the pipe.

For most pipe diameters, four tacks should be used (Figure 16–3). They should be evenly spaced and feathered out. (*Feathering* means to grind the tack down a bit.) Welding should start from on or just to one side of the tack, and proceed through the first quadrant. The sec-

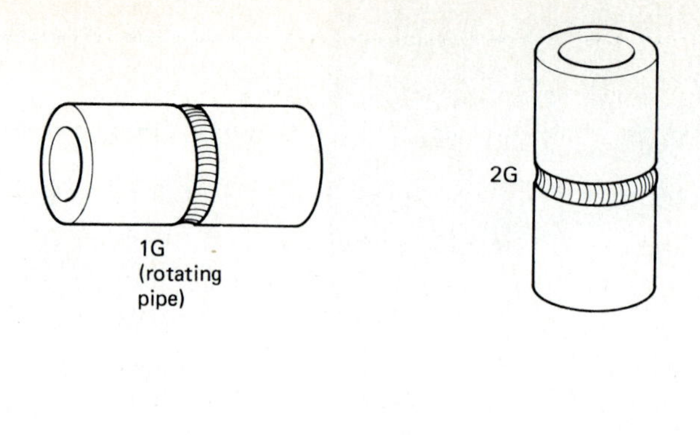

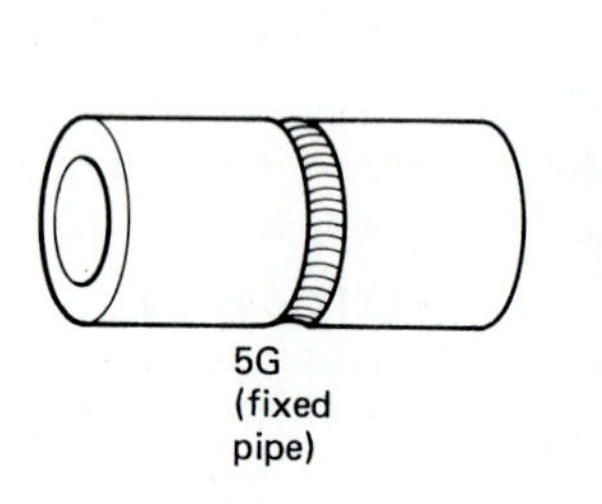

All pipe positions ±15° tol.

FIGURE 16–1
Pipe positions.

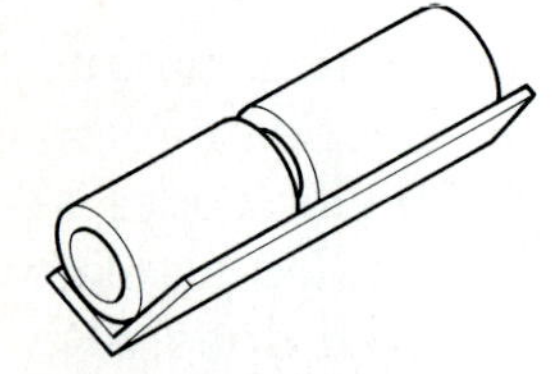

FIGURE 16–2
Angle iron keeps pipe aligned for tacking.

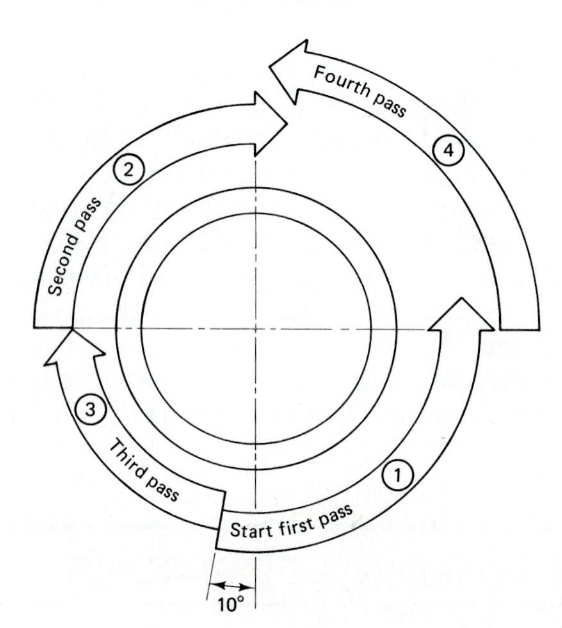

FIGURE 16–3
Pipe sequence for bends: 2G, 5G, and 6G. Overlap at each start and stop.

ond pass should start 180° from the first-pass stopping point. This will allow for even distribution of heat input and lower distortion. Each pass sequence should be welded in an upward direction (except for large diameter, downhill, cross-country pipeline welding). This assures full penetration.

Again, the third pass should start 180° from the stopping point of the second pass, and the fourth pass 180° from the stopping point of the third pass. As each pass wraps around the quadrant, it should be overlapped slightly and tied into the preceding bead.

As you begin your practice joints for pipe welding, remember that you will improve only by practice. Take time to set and align the pipe sections, and prepare the angle, root face, and root opening as specified. Your time spent in preparation will pay off in a sound-quality pipe weld.

TESTING PIPE WELDS

The actual destructive test required to certify a welder depends on the welding code for which the welder is certifying, too. Most codes will allow the test pipe to be radiographed (x-rayed) or destructively tested.

Destructive testing usually involves bend tests; however, the American Petroleum Institute (API) standard 1104 requires tensile sections, nick break, and bend tests. The ASME and AWS code requires two to four tests, depending on the test position. The 1G and 2G positions require one root bend and one face bend (Figure 16-4) or thick sections of pipe require two side bends. The 5G and 6G positions require two root bends and two face bends (Figure 16-5) or four side bends, depending on pipe wall thickness.

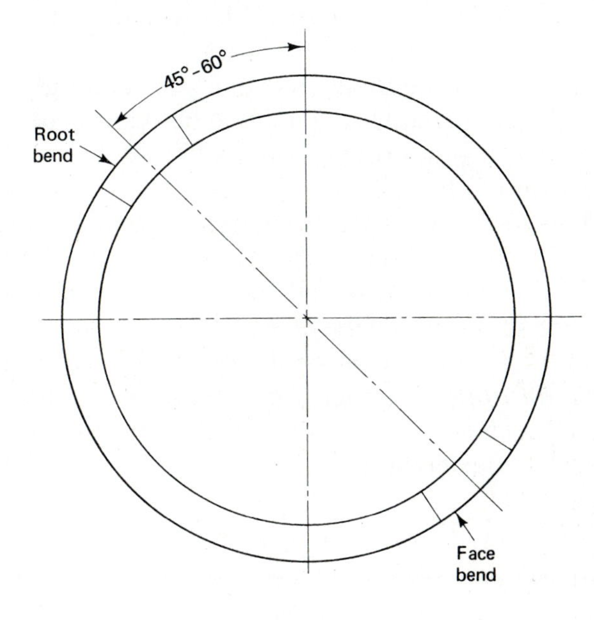

FIGURE 16-4
Pipe bend specimens for 2G.

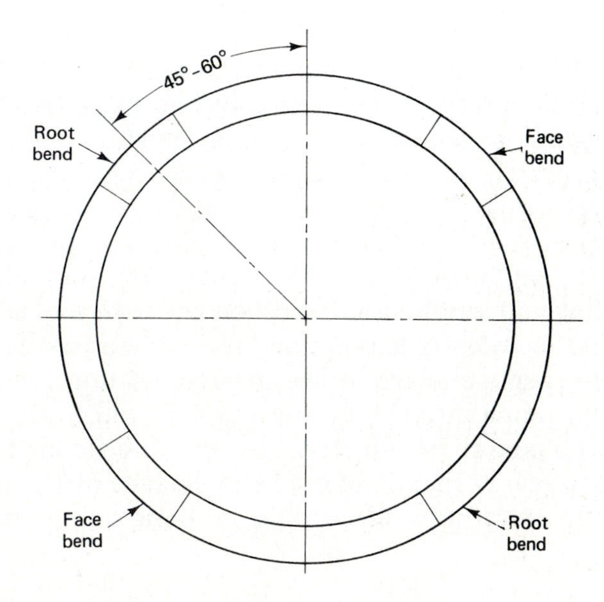

FIGURE 16-5
Pipe bend specimens for 5G and 6G.

Inspecting Bend Test Samples

Guided bend test samples should be bent 180°, or into a U shape (Figure 16-6). The face bend at the weld should have no discontinuities exceeding $\frac{1}{8}$ in. Small cracks on the corners are usually not considered unless they propagate from an internal discontinuity. (See Chapter 25.)

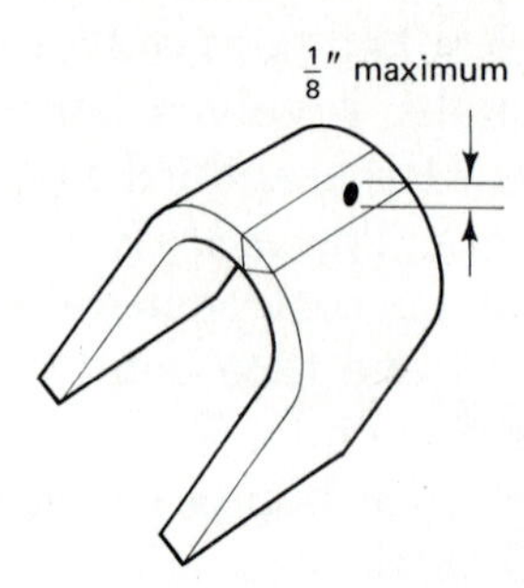

FIGURE 16-6
Bend specimen ($\frac{1}{8}$-in. max. defect.).

Inspecting Pipe Welds

Prior to any destructive testing, the face and root of pipe welds should be visually inspected for any cracks or discontinuities open to the surface. The root can be inspected using a flashlight. These areas should be clear of discontinuities. The face convexity of the weld should have no more than a $\frac{1}{8}$-in. crown (convexity) and should be filled to the full cross section of the joint.

The root penetration should protrude no more than $\frac{1}{16}$ in. and should be at least flush with the surface of the pipe (no suck-back). There should be no unfused edges at the root. Mild spatter is acceptable provided that it can easily be removed.

It is recommended that you consult the specific code or standard you are working under, for specific code requirements. Remember, too, that the codes often update their requirements, so stay up to date and consult your code book.

SMAW PIPE WELDING

Uphill SMAW

Today, SMAW of pipe is one of the oldest and most reliable methods. It gives high quality that usually has little trouble passing an x-ray test. Again, it cannot be overemphasized that pipe welding demands a lot of practice, and shielded metal arc welding of pipe is no exception.

FIGURE 16-7
Electrode angles on pipe. Keep electrode perpendicular (90°) to the pipe.

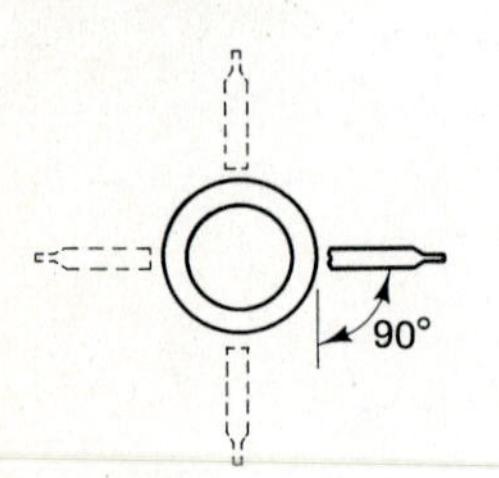

Set up pipe sections using angle iron or pipe alignment pictures, to keep aligned for tacking. Tack pipe and prepare to sequence weld beads as explained earlier. Get into a comfortable position and make sure that you will be able to reach the entire quadrant from that position. By "comfortable position" we mean a position that will not cause fatigue. Strike the arc in the groove; do not strike on the side of the pipe and move into the groove, as this will leave arc strikes that may cause the pipe to be rejected. As you proceed along the quadrant, keep the electrode perpendicular to the surface of the pipe (Figure 16-7).

Restarting welds should be done using the scratch start technique. The bead should actually start just behind the stopping point of the previous weld. This will assure proper tie-in.

For successful root passes use the three-variable system: $\frac{1}{8}$-in. electrode, $\frac{1}{8}$-in. root opening, $\frac{1}{8}$-in. root face. Any practical size, such as $\frac{3}{32}$ in., can be substituted as long as the electrode diameter, root opening, and root face are equal. Other root-pass setups are used, however. The three-variable system works well for the new pipe welders.

The completed pipe groove weld should be filled to at least its full cross section but not have a convexity that exceeds $\frac{1}{8}$ in. The root should be completely penetrated, with no unfused edge. The root penetration should not protrude more than $\frac{1}{16}$ in.

If your pipe welds are not running as good as they should, adjust one or more of the five variables you learned about in Chapter 7. Let's review them. The five variables are five basic changes or adjustments that we can easily make in the application of SMAW to correct the bead:

1. Electrode size
2. Amperage or heat
3. Electrode angle
4. Speed of travel
5. Arc length

Electrode sizes that are too large cause electrodes to stick or may be hard to start. The final bead will usually be too convex and overlapped. Electrodes that are too small may spatter and may even catch fire. It will also be difficult to hold a steady arc length because the electrode is being consumed too fast.

If the amperage (heat) is too high, spatter, undercut, and overheating of the plate will occur. If the heat is too low, the arc will be difficult to strike and the bead will be overlapped.

There are two types of electrode angles: (1) travel angle and (2) angle of attack. If the travel angle is too great from perpendicular to the plate, a bead with poor penetration that is too convex will result. The correct travel angle is 5 to 15° in the direction of travel. The angle of attack should always be about one-half the joint angle. For example, for a 90° tee joint, the electrode angle should be 45°. An incorrect angle will result in uneven leg length, poor fusion, and possibly, undercut on the top leg.

If the travel speed is too slow, the bead will pile up and overlap will result. If it is too fast, the bead will be sparse and have poor fusion. Gaining skill in adjusting to just the right travel speed will come only from practice and watching the puddle flow out.

The arc length will be changing continually as the electrode is consumed. If it gets too long, an erratic arc and spatter will develop. It is good to hold a close arc, especially with low-hydrogen electrodes, but if it gets much too close, the rod will not arc properly. Together with improper shielding, sticking and an erratic arc may result.

Any time beads are not coming out correctly, analyze these five variables first. Chances are that an adjustment of one of them will solve the problem (see Table 16-1).

TABLE 16-1
Quick Reference Machine Setting Table for SMAW

Problem	Remedy
Spatter	Decrease arc length (shorten) Decrease amperage
Overlap	Increase amperage Increase travel speed
Bead too convex	Increase amperage Increase travel speed
Undercut	Adjust electrode angle Decrease amperage Decrease travel speed Decrease arc length (shorten)
Lack of penetration	Increase amperage Increase root opening Decrease root face
Lack of fusion	Increase amperage Use slight side to side technique Adjust electrode angle
Porosity	Clean joint of any oil, grease, or moisture Keep close arc length Keep electrodes dry
Other variables to check	1. Correct polarity for electrodes used 2. Condition of electrodes 3. Size and type of electrodes 4. Angle and amperage of electrodes

Downhill SMAW

Downhill pipe welding is commonly used on cross-country pipelines. These pipes are usually 12 in. or larger in diameter (36 in. in diameter is not unusual). The groove design is similar to the uphill design; however, downhill work uses a smaller root face. For $\frac{1}{8}$-in. electrodes a $\frac{1}{16}$-in. root face should be used. Align pipe sections using angle iron or pipe alignment fixtures, and tack the pipe in four places, as explained previously. Tacks should be feathered out. Downhill pipe uses a two-stroke vertical-down sequence, with a 10 to 15° electrode lean angle (Figure 16-8).

E6010 electrodes and high amperages are commonly used. This is to provide good penetration and offset the lack of penetration sometimes common on downhill welds. On the root pass a slight keyhole is established just ahead of the electrode. Cover passes usually use E6010 $\frac{5}{32}$- or $\frac{3}{16}$-in.-diameter electrodes at relatively high amperage settings.

The arc should be started at the top of the pipe in about the 12 o'clock position. The weld should progress downhill until the 6 o'clock (bottom of the pipe) is reached. The second sequence should start at the top of the pipe, overlapping the first bead slightly, and progress downhill until it reaches the 6 o'clock position again.

With large-diameter pipes you will need to stop and restart

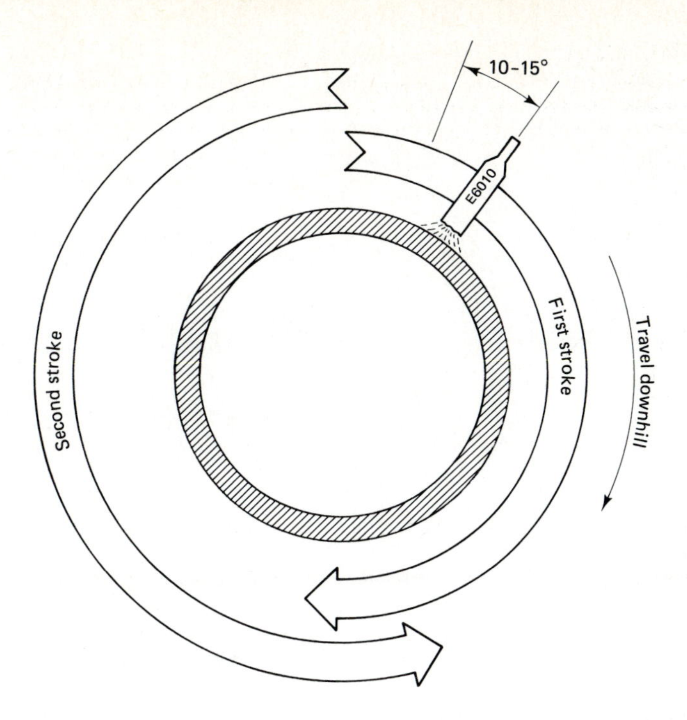

FIGURE 16-8
Downhill welding sequences.

beads before the sequence is completed. To stop the bead and break off the arc, quickly move the electrode down and off to one side.

When restarting the bead, you should strike the arc just above ($\frac{3}{8}$ to $\frac{1}{2}$ in.) the stopping point crater. A long arc should be held momentarily to reestablish a puddle. The groove should be filled to its full cross section and should not protrude more than $\frac{1}{16}$ in. The root pass should exhibit a completely fused, uniform bead.

Welders are qualified and certified for downhill pipeline welds a bit differently than for uphill pipe welds. Tensile sections, nick breaks, and bend tests are used (Figure 16-9). The test straps must meet the following requirements to pass. Bend tests should be bent into a U-shape (180°). There must be no crack or open discontinuity larger than $\frac{1}{8}$ in., or one half of the pipe wall thickness visible on the face of the bend. Bend test should be face and root bends on pipe with wall thicknesses of $\frac{1}{2}$ in. and under, and side bends for wall thicknesses over $\frac{1}{2}$ in.

Tensile sections must be machined or ground, then pulled with the tensile machining until failure. To pass, the ultimate tensile strength of the tensile section must meet or exceed that of the base metal.

Nick break specimens should have a $\frac{1}{8}$-in. notch cut into both sides of the specimens at the center of the weld (Figure 16-10). This can be cut with a hacksaw. One end of the specimen is then put into a vice and struck with a hammer. This should break off and expose the exterior of the weld.

The exposed interior of the weld should have no more than $\frac{1}{16}$ in. porosity or total more than 2% of the exposed surface area. Slag inclusions should not exceed $\frac{1}{32}$ in. in length or $\frac{1}{8}$ in. in depth, or one-half of the wall thickness (whichever is less). There should be at least $\frac{1}{2}$ in. between any two slag inclusions that might be present.

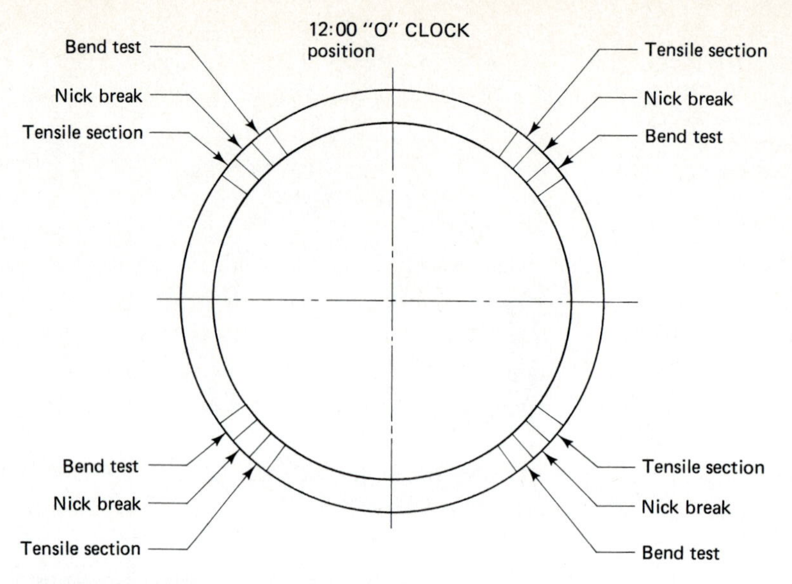

FIGURE 16-9
Bend test specimens for API pipe.

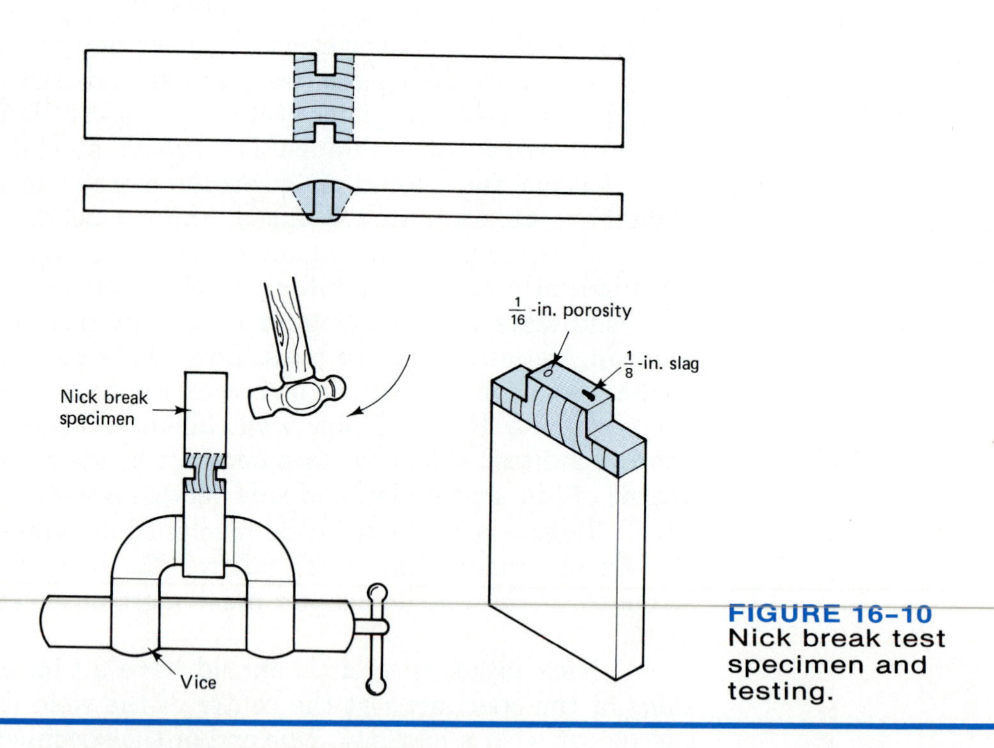

FIGURE 16-10
Nick break test specimen and testing.

GTAW PIPE WELDING

One of the highest-quality pipe welds is produced with the gas tungsten arc welding process. It gives clean, high-quality welds with a minimum amount of distortion.

Consult Chapter 8 for your basic setup procedures for GTAW, and make sure that you have completed all the GTAW plate projects before attempting any GTAW pipe projects.

When GTAW is used on pipe, it often requires a back purge, which is a way of getting shielding gas coverage on the inside root of the weld. There are a number of ways to provide shielding at the root. One of the simplest is simply to tape off the ends, leaving just enough space to insert an inert-gas hose (Figure 16–11). Special pipe caps can also be purchased or fabricated.

GTAW of pipe also requires a special tungsten taper (Figure 16–12). After sharpening a 2% thorium tungsten, put a small flat spot (about 0.020 in.) on the end of the tungsten. The tungsten helps distribute the arc evenly at the joint edges.

Terminating beads should be done by walking the arc off to the side of the bead and slowly reducing the amperage. This technique reduces the chance of crater cracks and fisheyes at the end of the bead.

One of the most popular techniques for GTAW pipe welds is the walk-the-cup technique. This technique will allow you to produce good welds consistently, with a minimum of welder fatigue. The procedure is fairly simple. Set up the pipe sections using angle iron to keep the pipes aligned, as explained earlier in the chapter. Gap the sections slightly less

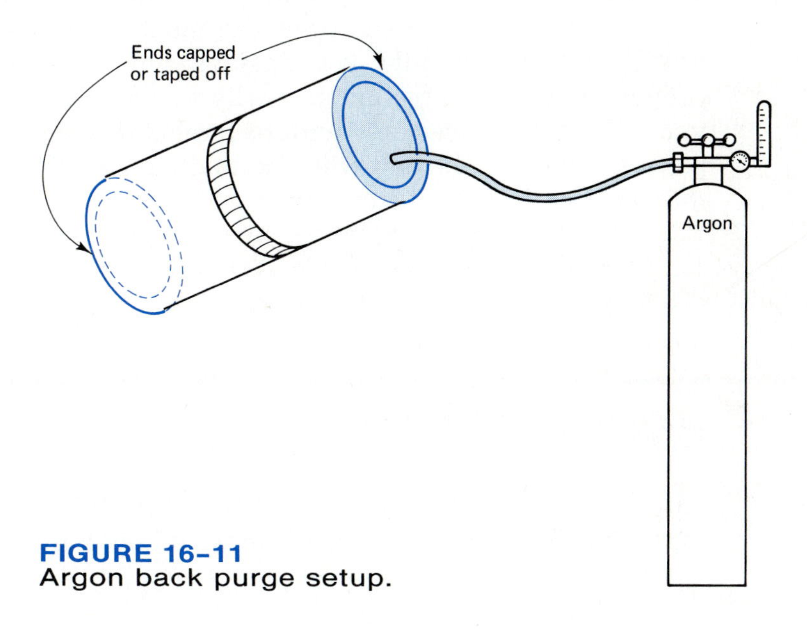

FIGURE 16–11
Argon back purge setup.

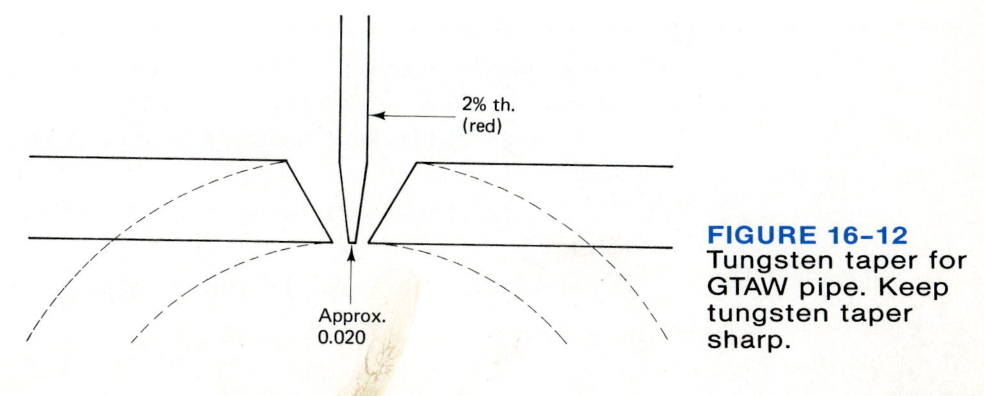

FIGURE 16–12
Tungsten taper for GTAW pipe. Keep tungsten taper sharp.

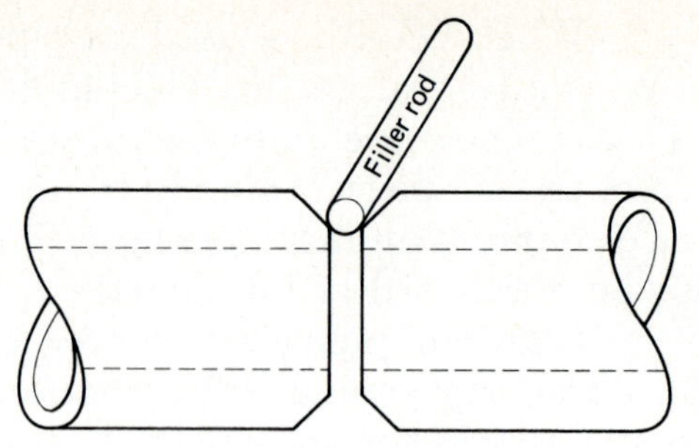

FIGURE 16–13 Filler rod position for walking the cup.

FIGURE 16–14 Walk-the-cup and freehand passes.

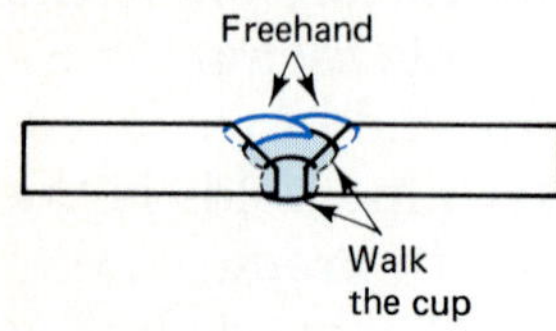

than the diameter of the filler rod you will be using. The filler rod should rest in the groove without slipping through (Figure 16–13).

Now rest the cup at a slight angle in the groove. Using a circular motion, rock the cup around the groove, transferring slight pressure from side to side. The puddle should flow from side to side, consuming the filler rod into the puddle. Add slight forward pressure and the torch should progress forward—you are now walking the cup (Figure 16–14).

You will notice the tungsten must be adjusted so that it does not dip into the puddle, yet must be close enough to the puddle for proper control. This technique works well on root and lower passes, but you will have to use the standard freehand technique on the outer fill passes. Walking the cup will also make difficult positions such as 6G much easier.

REVIEW QUESTIONS

1. Where might you find pipe welding in industry today?
2. What does quality have to do with how a pipe might be welded?
3. What type of pipe might be welded uphill, and what type might be welded downhill?
4. Sketch the 6G pipe position.
5. How might pipe welds be tested?
6. What type, and how many, test coupons must be removed for 5G or 6G pipe position welder qualification?
7. With SMAW, how should the electrode be angled in relation to the pipe?
8. What type of starting techniques should be used with SMAW of pipe?
9. What is the three-variable system of setting up for the root pass of pipe?
10. What are the five variables for correcting a SMAW bead?

11. What adjustment would you use to correct undercut on a SMAW pipe weld?
12. What adjustment would you use to correct lack of penetration on a SMAW pipe weld?
13. Explain the walk-the-cup technique of pipe welding.
14. What is the maximum convexity at the face and the maximum penetration at the root of pipe welds?

PIPE PROJECT 1

Uphill SMAW: 1GR Position

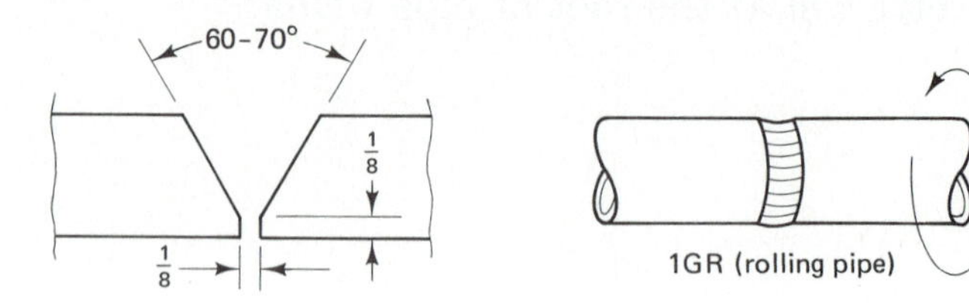

Material: A-53 pipe

Size: 6-in. diameter, schedule 40

Joint Design: Groove weld, 60–70° included angle

Position: Flat groove, rotating pipe (1GR)

Technique: Stringers are recommended throughout (weaves may be used on outer fill passes)

Polarity: DCRP

Electrode: Root pass, E6010, $\frac{1}{8}$-in. diameter
Fill passes, E7018, $\frac{1}{8}$-in. diameter

Amperage: Root pass, 65–80 A
Fill and cover passes, 95–120 A

Cleaning: Chipping hammer, wire brush

Procedure: Using angle iron to align the pipe, gap $\frac{1}{8}$ in. and deposit four equally spaced tacks. Deposit the root passes, rotating the pipe to keep the bead at a flat-to-slightly-uphill angle. Sequence-weld the pipe to prevent distortion.

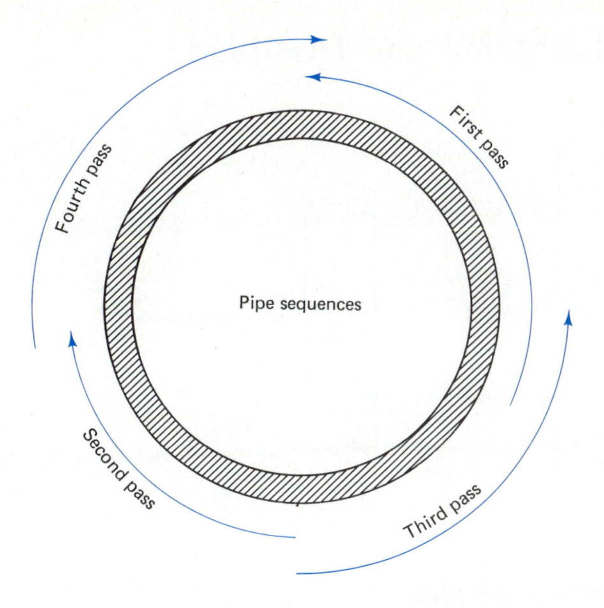

The second pass, known as the "hot pass," is run 10 to 15% hotter to burn out any slag anchored into the top of the root pass. Fill the remainder of the groove to its full cross section, sequencing the passes. Thoroughly clean each layer before depositing the next.

Visual Inspection: Using a flashlight, examine the root side of the pipe for unfused edges, excessive burn-through, or internal concavity. No visible unfused edges are acceptable. A maximum of $\frac{1}{16}$ in. of burn-through and $\frac{1}{16}$-in.-deep internal concavity is acceptable as long as the edges are fused. The face of the weld should show complete fusion, with no visible cracks, porosity, slag inclusions, underfill, or undercut.

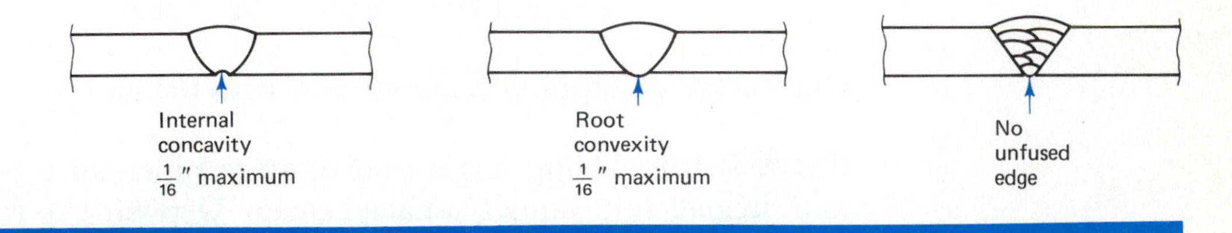

PIPE PROJECT 2

Uphill SMAW: 2G Position

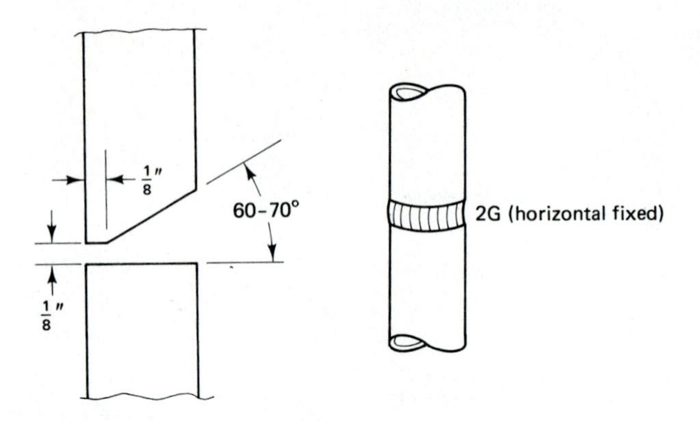

Material: A-53 pipe

Size: 6-in. diameter, schedule 40

Joint Design: Groove weld, 60–70° included angle

Position: Horizontal fixed position (2G)

Technique: Stringer beads only!

Polarity: DCRP

Electrode: Root pass, E6010, $\frac{1}{8}$-in. diameter
Fill and cover passes, E7018, $\frac{1}{8}$-in. diameter

Amperage: Root pass, 65–80A
Fill and cover passes, 95–120A

Cleaning: Chipping hammer and wire brush

Procedure: Using angle iron to align pipe, set a $\frac{1}{8}$-in. root opening, and deposit four equally spaced tacks. Deposit the root pass, keeping the electrode 90° perpendicular to the pipe surface as you proceed around the pipe. Sequence-weld the pipe to reduce distortion.

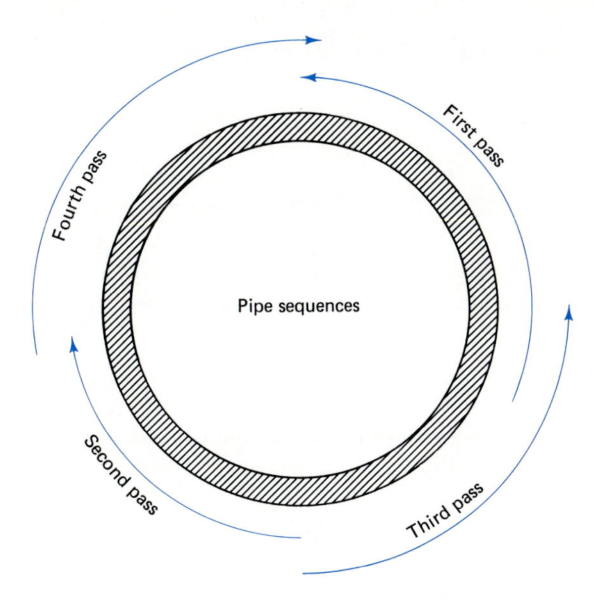

The second pass, known as the "hot pass," is run 10 to 15% hotter to burn out any slag anchored into the top of the root pass. Fill the remainder of the groove to its full cross section sequencing the passes. Thoroughly clean each layer before depositing the next.

Visual Inspection: Using a flashlight, examine the root side of the pipe for unfused edges, excessive burnthrough, or internal concavity. Visible unfused edges are not acceptable. A maximum of $\frac{1}{16}$ in. of burnthrough, and $\frac{1}{16}$-in.-deep internal concavity is acceptable as long as the edges are fused. The face of the weld should show complete fusion, with no visible cracks, porosity, slag inclusions, underfill, or undercut.

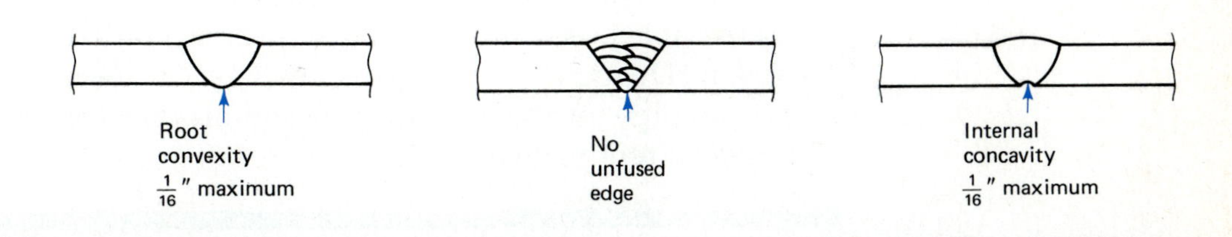

PIPE PROJECT 3

Uphill SMAW: 5G Position

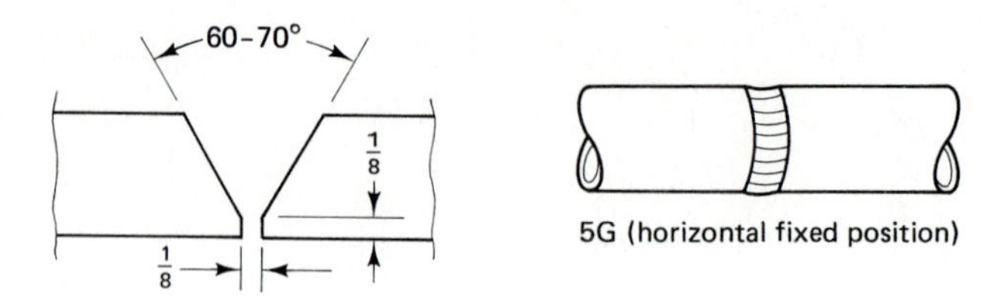

Material: A-53 pipe

Size: 6-in. diameter, schedule 40

Joint Design: Groove weld, 60–70° included angle

Position: Vertical fixed position (5G)

Technique: Stringers (weaves can be used on the outer fill passes and cover pass if bead profiles are acceptable)

Polarity: DCRP

Electrode: Root pass, E6010, $\frac{1}{8}$-in. diameter
Fill and cover passes, E7018, $\frac{1}{8}$-in. diameter

Amperage: Root pass, 65–80A
Fill and cover passes, 95–120A

Cleaning: Chipping hammer and wire brush

Procedure: Using angle iron to align the pipe, set a $\frac{1}{8}$-in. root opening and deposit four equally spaced tacks. Now position yourself into a comfortable position and make sure that you are able to reach the full quadrant. Deposit the root pass keeping the electrode 90° perpendicular to the pipe surface as you proceed around the pipe. Sequence-weld the pipe to reduce distortion.

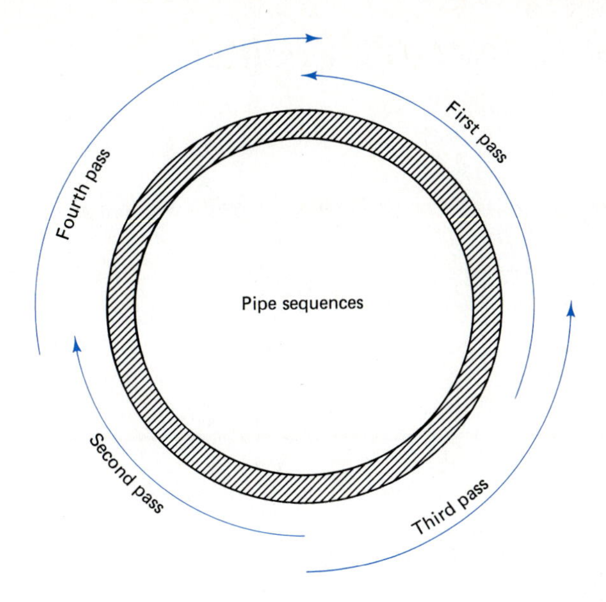

The second pass, known as the "hot pass," is run 10 to 15% hotter to burn out any slag anchored into the top of the root pass. Fill the remainder of the groove to its full cross section, sequencing the passes. Throughly clean each layer before depositing the next.

Visual Inspection: Using a flashlight, examine the root side of the pipe for unfused edge, excessive burn-through, or internal concavity. Visible unfused edges are not acceptable. A maximum of $\frac{1}{16}$ in. burn-through and $\frac{1}{16}$-in.-deep internal concavity are acceptable as long as the edges are fused. The face of the weld should show complete fusion, with no visible cracks, porosity, slag inclusions, underfill, or undercut.

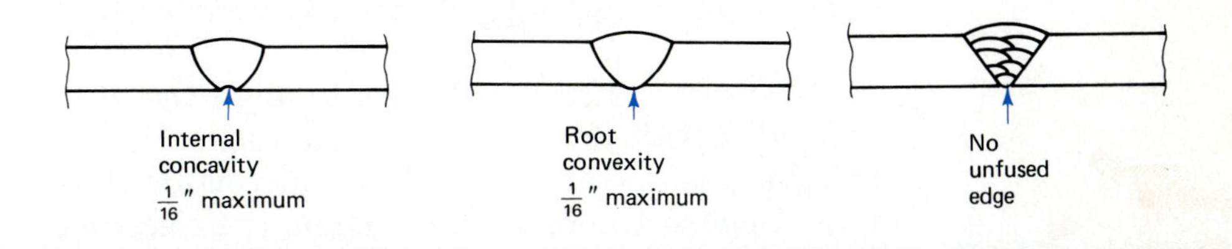

PIPE PROJECT 4

Uphill SMAW: 6G Position

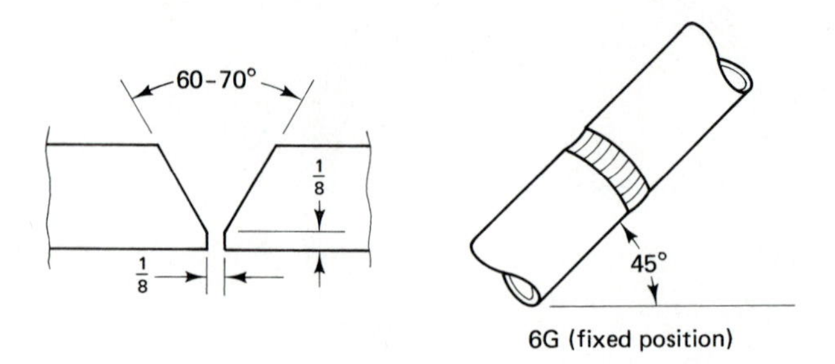

Material: A-53 pipe

Size: 6-in. diameter, schedule 40

Joint Design: Groove weld, 60–70° included angle

Position: 45° inclined between 2G and 5G (6G)

Technique: Stringers (weaves can be used on the outer fill passes or cover pass if bead profiles are acceptable)

Polarity: DCRP

Electrode: Root pass, E6010, $\frac{1}{8}$-in. diameter
Filler passes, E7018, $\frac{1}{8}$-in. diameter

Amperage: Root pass, 65–80A
Fill and cover passes, 95–120A

Cleaning: Chipping hammer and wire brush

Procedure: Using angle iron to align the pipe, set a $\frac{1}{8}$-in. root opening and deposit four equally spaced tacks. Position yourself into a comfortable position and make sure that you are able to reach the full quadrant. Deposit the root pass keeping the electrode 90° perpendicular to the pipe surface as you proceed around the pipe. Sequence-weld the pipe to reduce distortion.

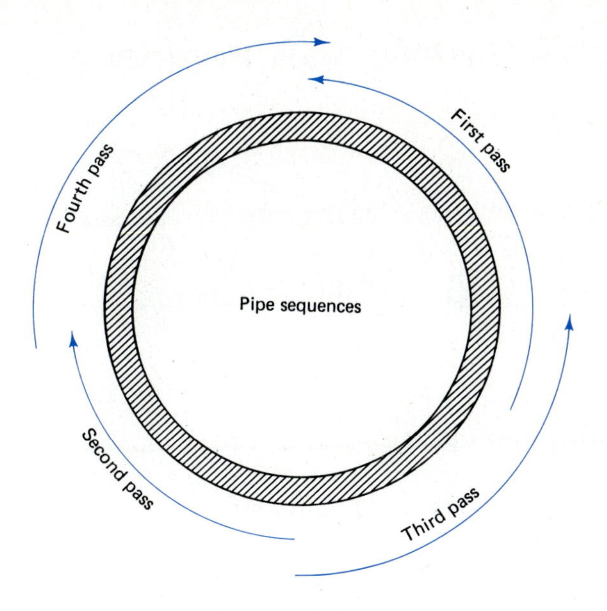

The second pass, known as the "hot pass," is run 10 to 15% hotter to burn out any slag inclusions anchored in the top of the root pass. Fill the remainder of the groove to its full cross section sequencing the passes. Thoroughly clean each layer before depositing the next.

Visual Inspection: Using a flashlight, examine the root side of the pipe for unfused edges, excessive burn-through, or internal concavity. Visible unfused edges are not acceptable. A maximum of $\frac{1}{16}$-in.-deep internal concavity is acceptable as long as edges are fused. The face of the weld should show complete fusion, with no visible cracks, porosity, slag inclusions, underfill, or undercut.

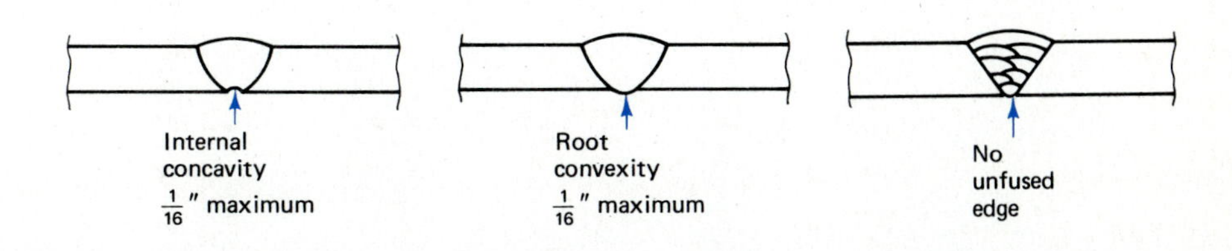

PIPE PROJECT 5

Downhill SMAW: 1GR Position

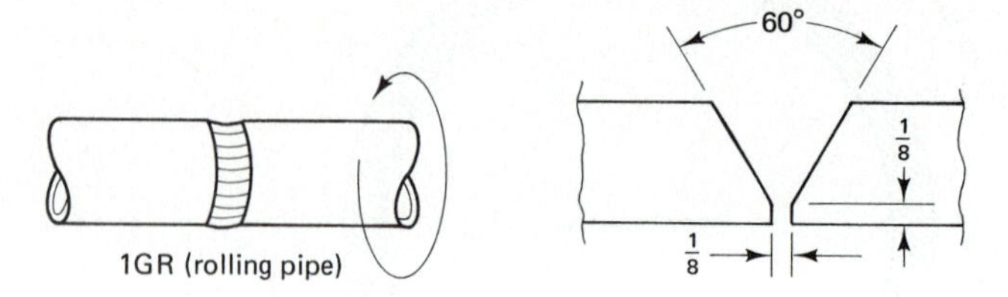

Material: A-53 pipe

Size: 8-in. diameter, schedule 40

Joint Design: V-groove, 60° included angle

Position: Roll pipe (1GR)

Technique: Stringers and weave beads

Polarity: DCRP

Electrode: E6010, $\frac{5}{32}$-in. diameter

Amperage: Root pass, 80–100A
Fill and cover passes, 95–140A

Cleaning: Chipping hammer and wire brush

Procedure: Use angle iron to keep the pipes aligned. Deposit four small tacks equally spaced around the pipe. Using a small grinder or pencil grinder, feather out each tack. Run the root coming downhill. Roll pipe into position for each downhill pass.

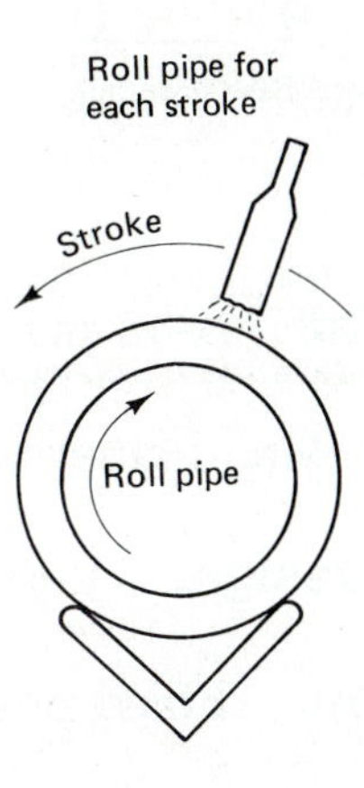

Fill the groove to its full cross section. Use the side-to-side weave technique for the filler and cover passes.

Visual Inspection: Examine the root of the pipe using a flashlight. There should be no more than 1 in. of unfused edges, and no more than $\frac{1}{4}$ in. of burn-through. Internal concavity is acceptable as long as the edges are completely fused. The cover pass should have more than a $\frac{1}{16}$-in. crown, and no more than a $\frac{1}{32}$-in. deep undercut and a maximum of 2 in. in length.

PIPE PROJECT 6

Downhill SMAW: 5G Position

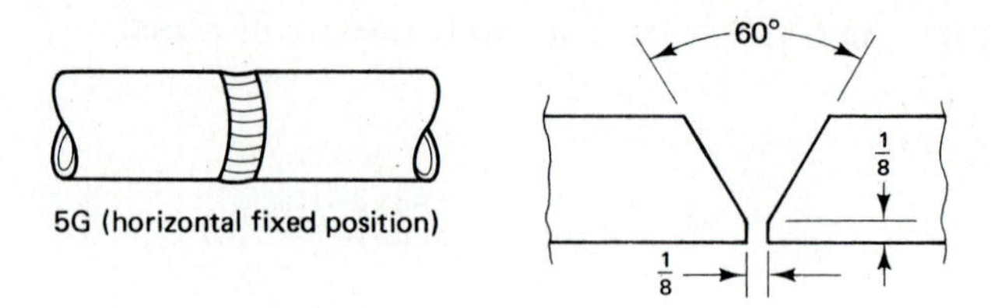

Material: A-53 pipe

Size: 8-in. diameter, schedule 40

Joint Design: V-groove, 60° included angle

Position: Fixed position (5G)

Technique: Stringer and weave beads

Polarity: DCRP

Electrode: E6010, $\frac{5}{32}$-in. diameter

Amperage: Root pass, 80–100A
Fill and cover passes, 95–140A

Cleaning: Chipping hammer and wire brush

Procedure: Use angle iron to keep the pipes aligned. Deposit four small tacks equally spaced around the pipe. Using a small grinder or pencil grinder, feather out each tack. Run a two-stroke downhill root pass: one stroke down each side.

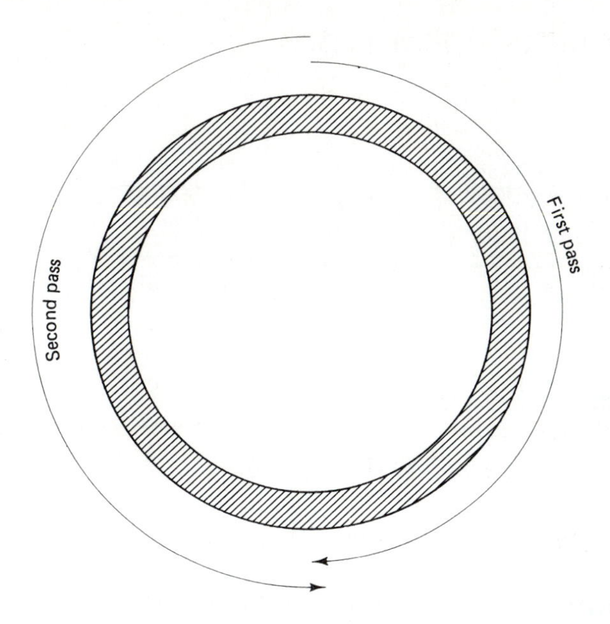

Use the two-stroke downhill system on the remaining filler passes. Stagger the starting points at the top of the pipe so as not to start a new layer at the same place twice. Fill the groove to its full cross section using the side-to-side weave technique.

Visual Inspection: Examine the root of the pipe using a flashlight. There should be no more than a 1-in. unfused edge, and no more than $\frac{1}{4}$ in. of burn-through. Internal concavity is acceptable as long as the edges are completely fused. The cover pass should have no more than a $\frac{1}{16}$-in. crown and no more than a $\frac{1}{32}$-in.-deep undercut and be a maximum of 2 in. in length.

PIPE PROJECT 7

GTAW: 1GR Position

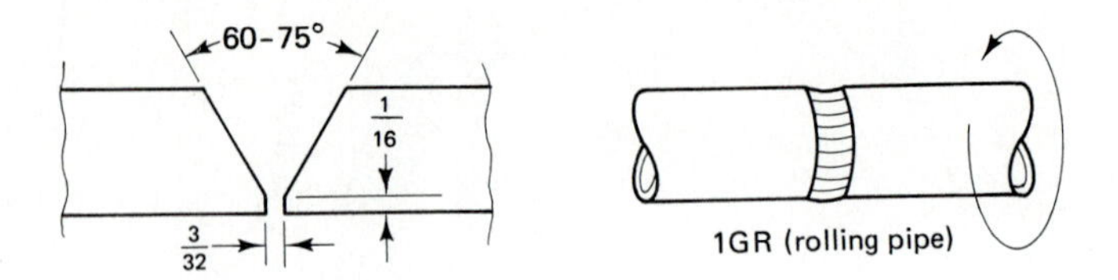

Material: A-53 pipe

Size: 4-in. diameter, schedule 40

Joint Design: V-groove, 60–75° included angle

Position: Roll pipe into position (1GR)

Technique: Walk the cup stringers and freehand stringers

Polarity: DCSP

Tungsten: 2% Thorium (EWTH2)

Filler Rod: E70S-2

Cleaning: Wire brush

Procedure: Grind the tungsten to a sharp point, then grind a 0.020-in. flat spot on the tungsten tip. Use a $\frac{3}{32}$-in. root face, slightly less than a $\frac{3}{32}$-in. root opening, and $\frac{3}{32}$-in. diameter filler rod. Deposit four equally spaced tacks. Using the walk-the-cup technique, deposit the root pass. Use the four-stroke sequence procedure to reduce distortion.

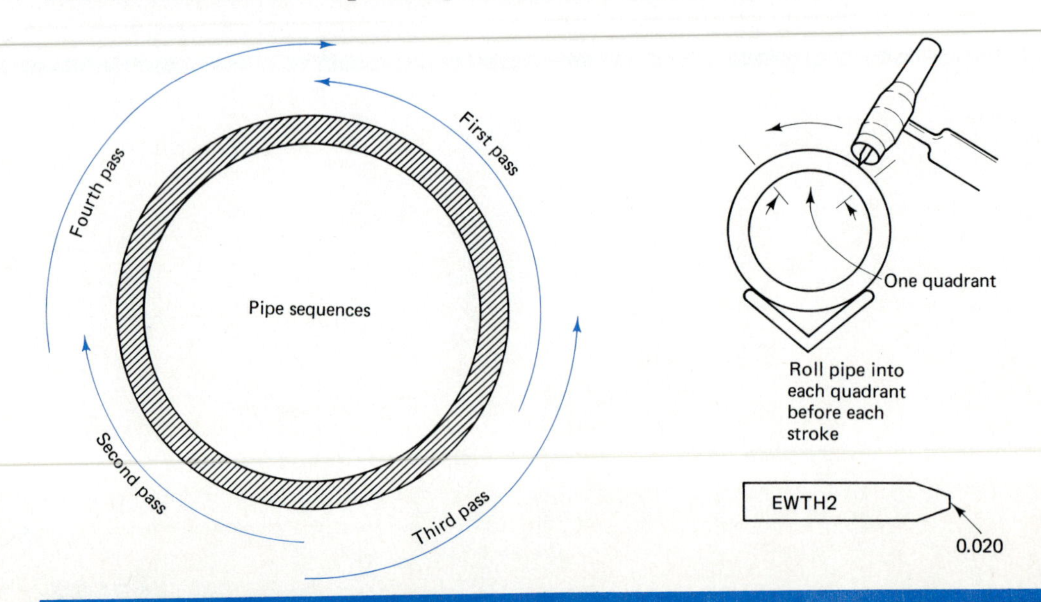

Continue to deposit filler passes walking the cup. The outer fill passes and cover passes will have to be done freehand.

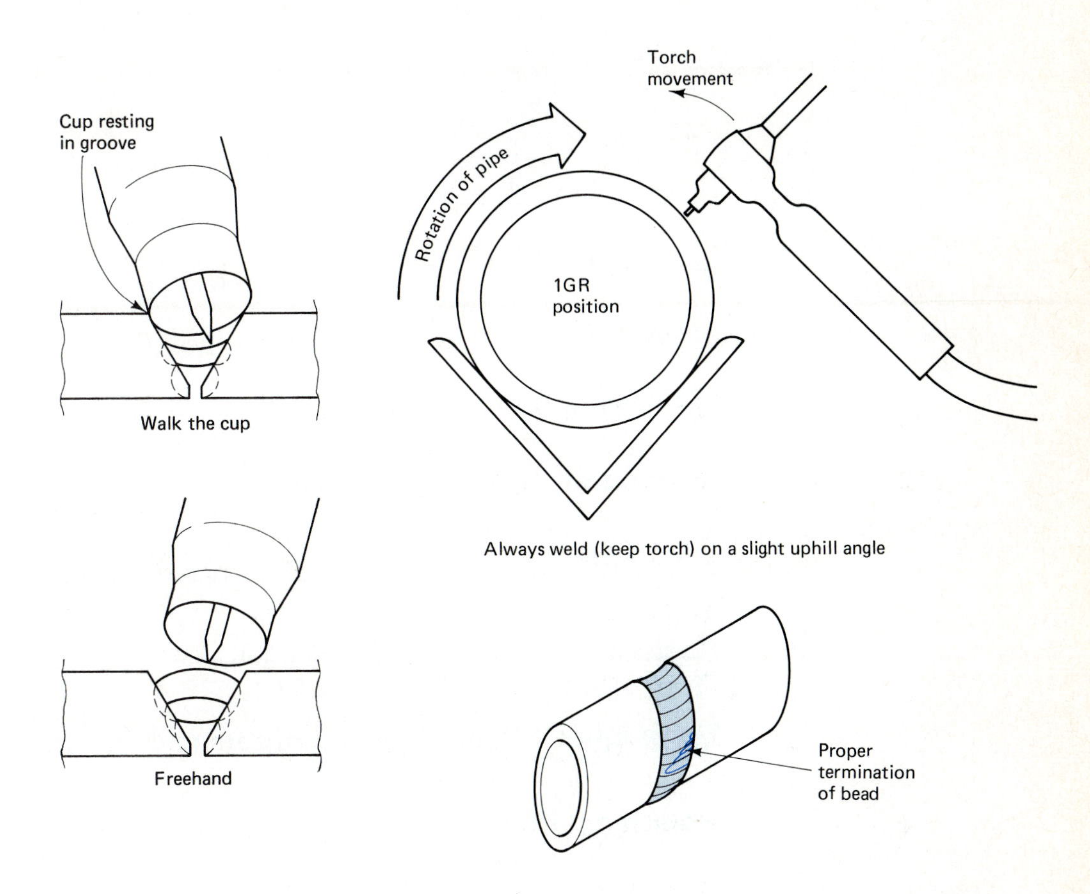

Visual Inspection: The root can be inspected using a flashlight or flashlight and mirror. The root should be very smooth and almost flush, with no more than a $\frac{1}{16}$-in. convexity. The cover should appear very smooth, with smoothly tied-in edges. There should be no cracks, porosity, undercut, or underfill on the root or cover pass.

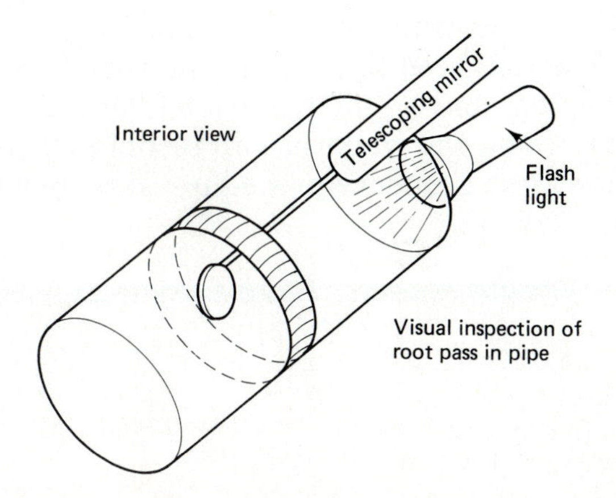

PIPE PROJECT 8

GTAW: 2G Position

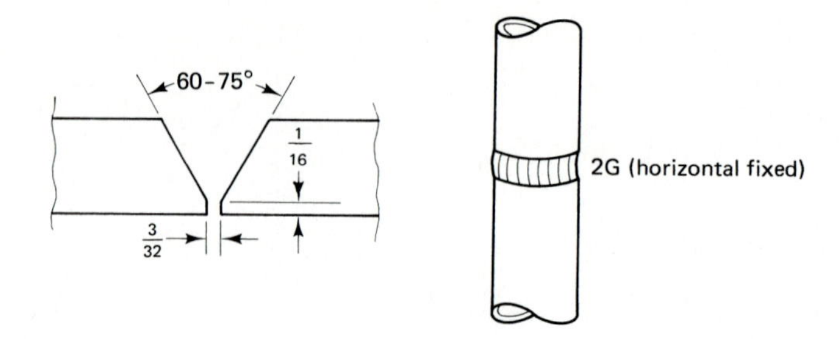

Material: A-53

Size: 4-in. diameter, schedule 40

Joint Design: V-groove, 60–75° included angle

Position: Horizontal groove (2G)

Technique: Walk-the-cup stringers and freehand stringers

Polarity: DCSP

Tungsten: 2% Thorium (EWTH2)

Filler Rod: E70S-2

Cleaning: Wire brush

Procedure: Grind tungsten to a sharp point, then grind a 0.20-in. flat spot on the tungsten tip. Use a $\frac{3}{32}$-in. root face, slightly less than a $\frac{3}{32}$-in. root opening, and a $\frac{3}{32}$-in.-diameter filler rod. Deposit four equally spaced tacks. Using the walk-the-cup technique, deposit the root pass. Use the four-stroke sequence procedure to reduce distortion.

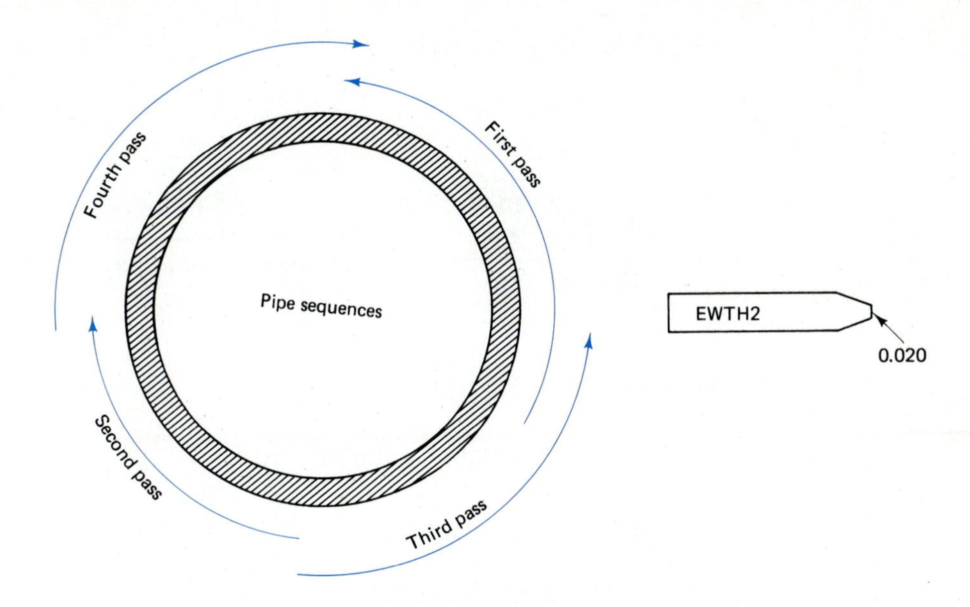

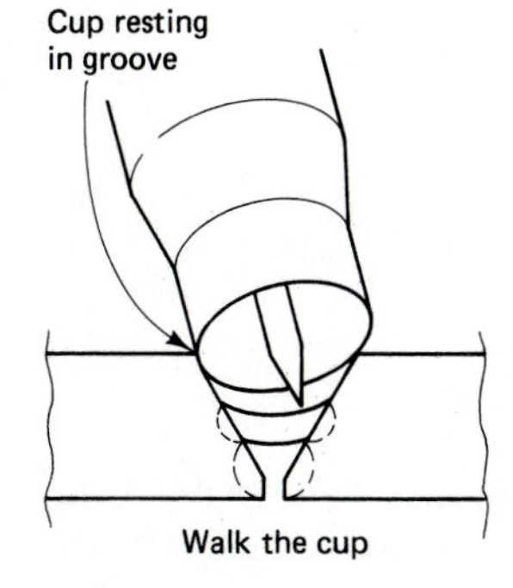

Continue to deposit filler passes, starting from the bottom up, walking the cup as long as possible. The outer fill passes and cover passes will have to be done freehand.

Visual Inspection: The root can be inspected using a flashlight or flashlight and mirror. The root should be very smooth and almost flush, with no more than a $\frac{1}{16}$-in. convexity. The cover should appear very smooth, with no overlap on the bottom edges and smoothly tied-in beads. There should be no cracks, porosity, undercut, or underfill on the root or cover pass.

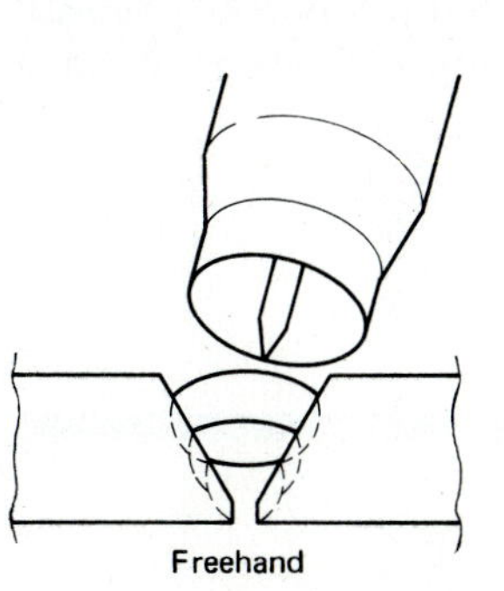

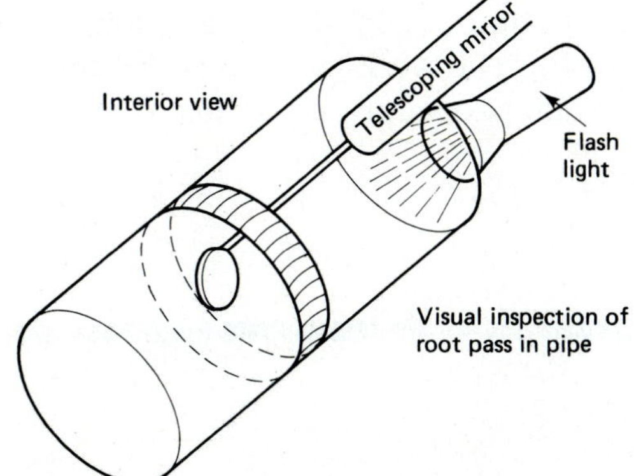

PIPE PROJECT 9

GTAW: 5G Position

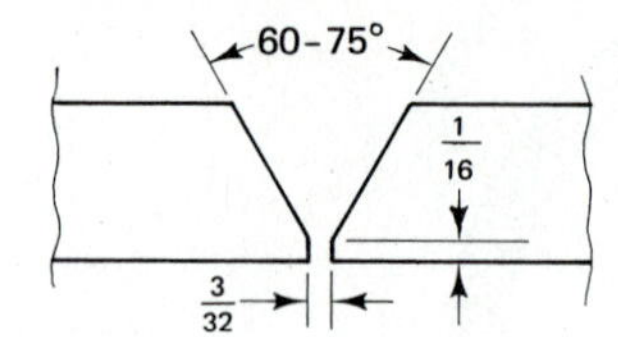

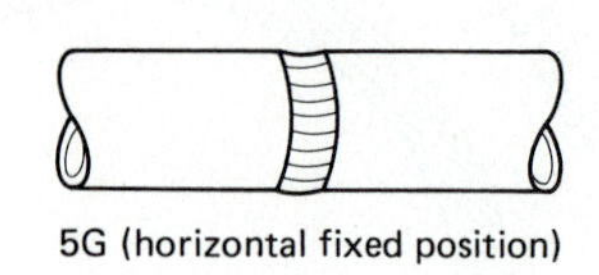

5G (horizontal fixed position)

Material: A-53 pipe

Size: 4-in. diameter, schedule 40

Joint Design: V-groove, 60–76° included angle

Position: Fixed pipe (5G)

Technique: Walk-the-cup stringers and freehand stringers

Polarity: OCSP

Tungsten: 2% Thorium (EWTH2)

Filler Rod: E70S-2

Cleaning: Wire brush

Procedure: Grind tungsten to a sharp point, then grind a 0.20-in. flat spot on the tungsten tip. Use a $\frac{3}{32}$-in. root face, slightly less than a $\frac{3}{32}$-in. root opening, and a $\frac{3}{32}$-in.-diameter filler rod. Deposit four equally spaced tacks. Using the walk-the-cup technique, deposit the root pass. Use the four-stroke procedure to reduce distortion.

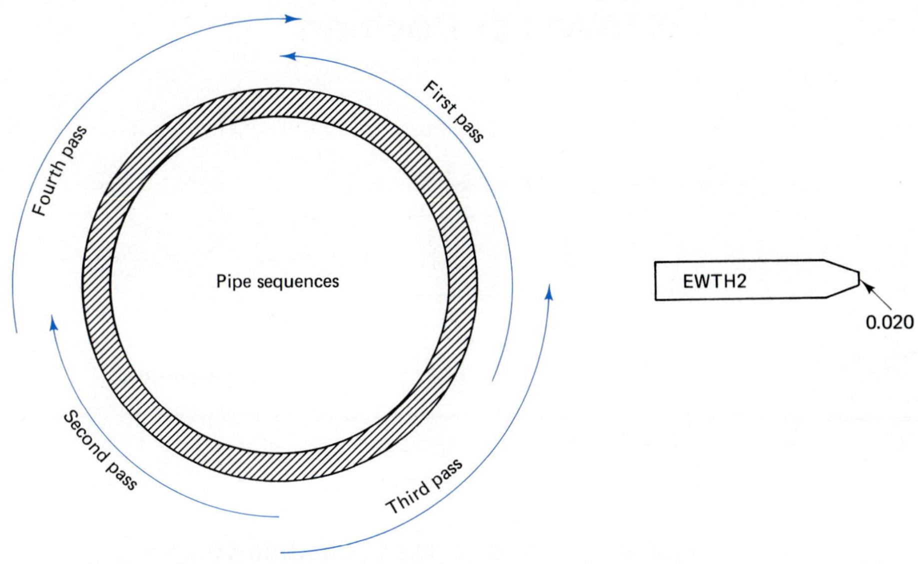

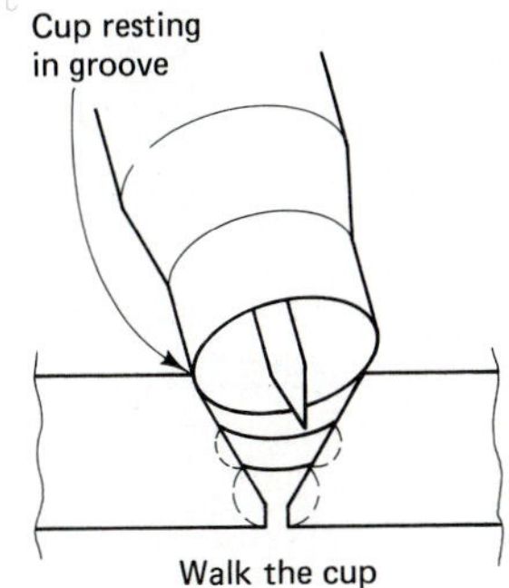

Continue to deposit filler passes walking the cup. The outer and cover passes will have to be done freehand.

Visual Inspection: The root can be inspected using a flashlight or flashlight and mirror. The root should be very smooth and almost flush, with no more than a $\frac{1}{16}$-in. convexity. The cover should appear very smooth, with smoothly tied-in edges. There should be no cracks, porosity, undercut, or underfill on the root or cover pass.

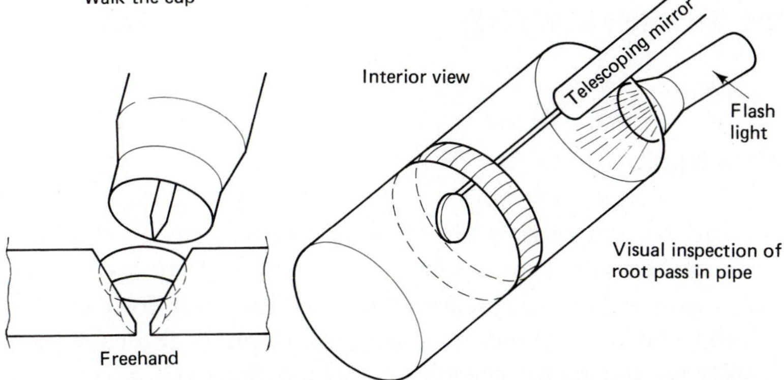

PIPE PROJECT 10

GTAW: 6G Position

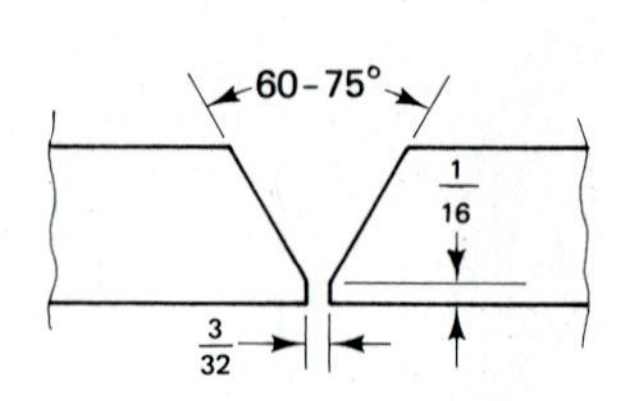

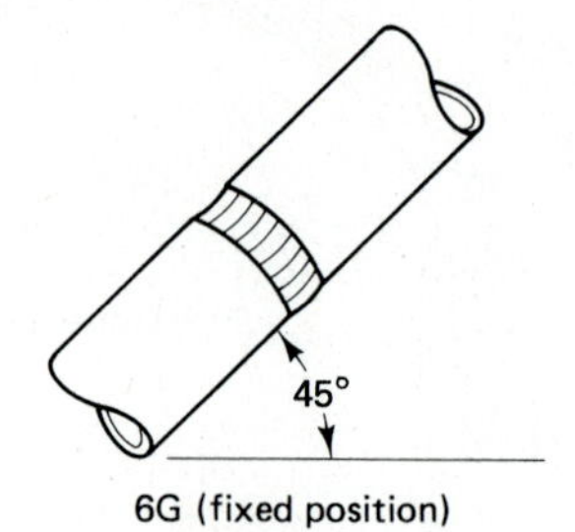

6G (fixed position)

Material: A-53 pipe

Size: 4-in. diameter, schedule 40

Joint Design: V-groove, 60–76° included angle

Position: 45° inclined fixed pipe (6G)

Technique: Walk-the-cup stringers and freehand stringers

Polarity: DCSP

Tungsten: 2% Thorium (EWTH2)

Filler Rod: E70S-2

Cleaning: Wire brush

Procedure: Grind tungsten to a sharp point, then grind a 0.020-in. flat spot on the tungsten tip. Use a $\frac{3}{32}$-in. root face, slightly less than a $\frac{3}{32}$-in. root opening, and a $\frac{3}{32}$-in.-diameter filler rod. Deposit four equally spaced tacks. Using the walk-the-cup technique, deposit the root pass. Use the four-stroke sequence procedure to reduce distortion.

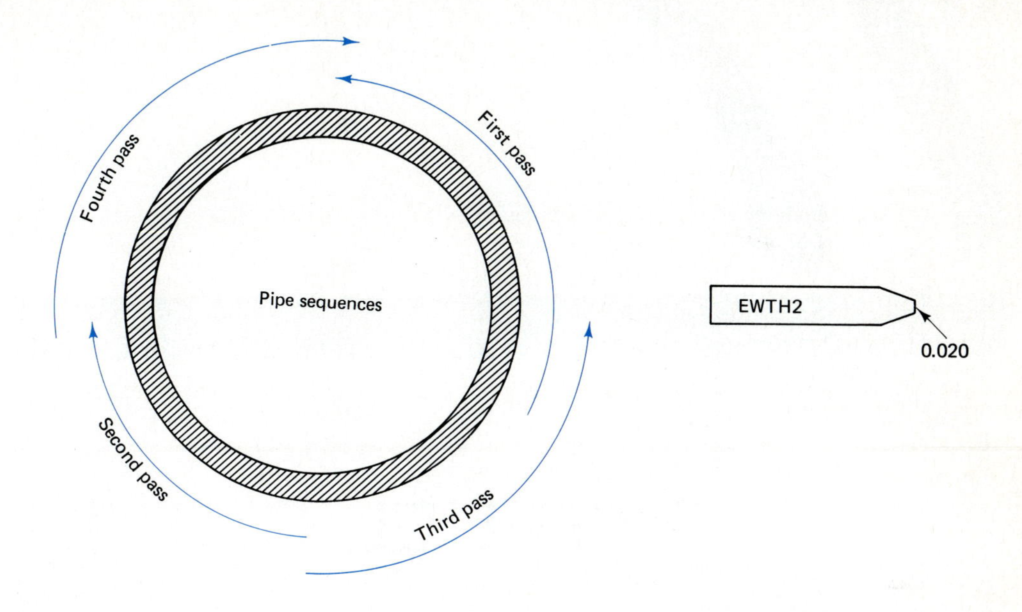

Continue to deposit filler passes walking the cup. The outer and cover passes will have to be done freehand.

Visual Inspection: The root can be inspected using a flashlight or flashlight and mirror. The root should be very smooth and almost flush, with no more than a $\frac{1}{16}$-in. convexity. The cover should appear very smooth, with smoothly tied-in edges. There should be no cracks, porosity, undercut, or underfill on the root or cover passes.

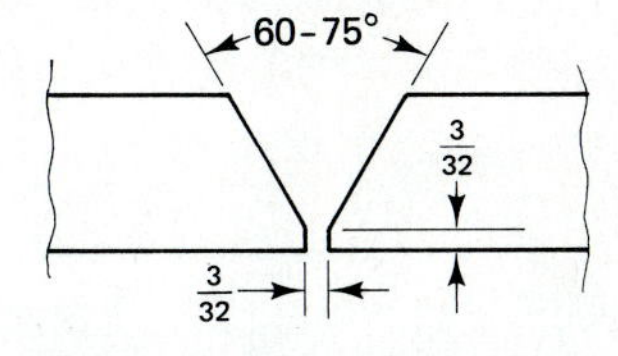

OXYFUEL CUTTING

Oxyfuel cutting (OFC) has for many years been the primary means of cutting medium and heavy plate. Even with the introduction of plasma, laser, and mechanical cutting processes, OFC remains the fastest and most economical means of cutting medium and heavy thicknesses.

PHOTO 17–1
Six-torch automatic oxyfuel cutting. (Courtesy of Airco Welding Products, a division of The BOC Group, Inc.)

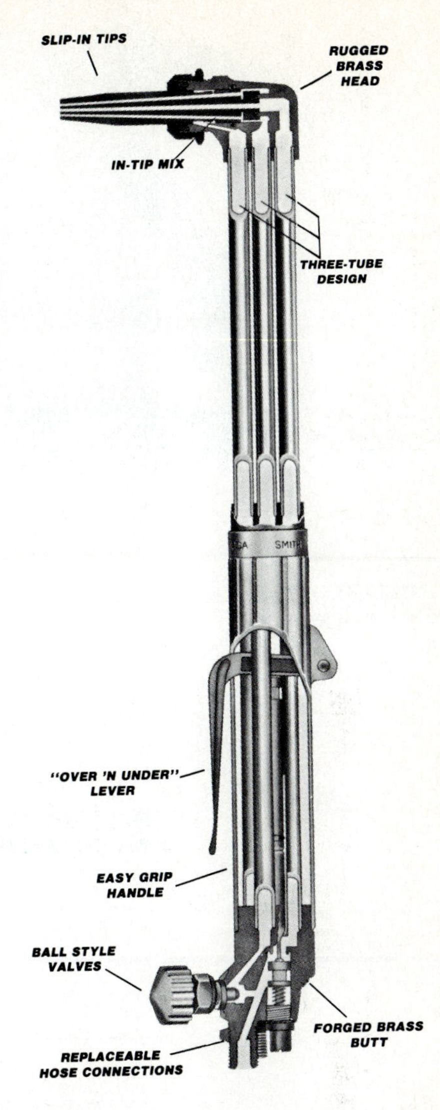

PHOTO 17–2
Oxyfuel cutting torch. (Courtesy of Smith Welding Equipment.)

OFC is used primarily on carbon steels, but some limited success is possible on stainless steels with special surface-applied powders. OFC maintains tolerances of $\pm\frac{1}{32}$ in. and better on fully automatic OFC applications. With the right equipment OFC can cut thickness ranges from $\frac{1}{32}$ to 10 in. and more. Speeds can range as high as 35 in. per minute, and 20 to 25 in. per minute on 1″ plate.

OFC THEORY

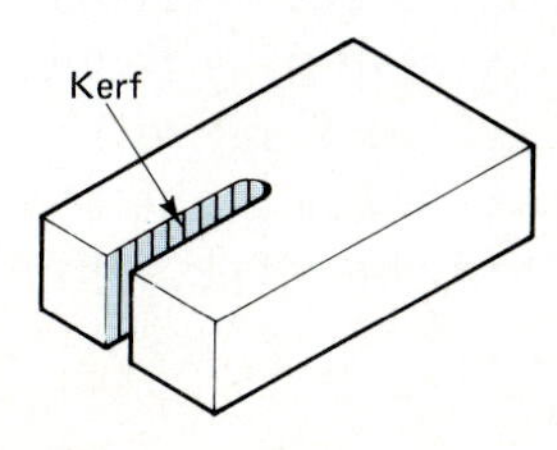

FIGURE 17–1
Kerf.

Oxyfuel cutting utilizes a principle known as oxidation. Oxidation occurs in metals when its temperature reaches its "burning" temperature. When burning occurs, iron oxides and metal impurities are formed in the kerf. The kerf (cut) is the oxidized slot made from the oxidizing action and oxygen pressure of the oxyfuel cutting torch (Figure 17–1). Pure oxygen increases this oxidation action. The oxygen pressure from the torch tip blows the iron oxide–slag mixture from the kerf, leaving a clean narrow slot (cut) through the metal.

PHOTO 17-3
Mapp gas kerf. (Courtesy of Airco Welding Products, a division of The BOC Group, Inc.)

FIGURE 17-2
Cut edge.

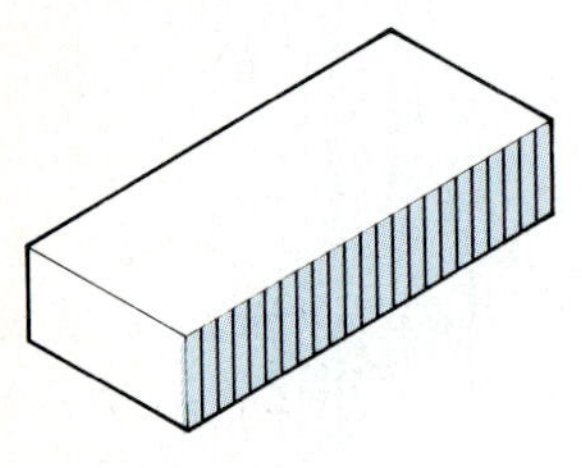

As long as there is sufficient heat in the plate, the pure O_2 stream will allow the cut to continue. This is why OFW is so fast and efficient when cutting steels. There is usually a thin slag layer left on the surface of a freshly made cut. This should fall off easily if the cut is properly made (Figure 17-2). Always remove this layer before attempting to weld the plate.

OFC TERMS

Drag lines are the small ripples, if done correctly, left on the plate cut edge (Figure 17-3). They may be sweeping or straight down, depending on the speed of the cut. The cutting jet is the high-pressure oxygen stream coming from the center orifice of the tip. *Preheating orifices* on cutting tips provide and maintain the heat to achieve oxidation. They also provide a heat shield to protect the molten steel edges of the kerf from the atmosphere (Figure 17-4).

Flatness is the depth difference between the highest point on the cut edge and the deepest notches or gouges (Figure 17-5). *Roughness* describes the total series of high-low points on the cut edge surface (Figure 17-6).

Notches are occasionally deep gouges that exceed the average depth of the normal surface roughness (Figure 17-7). *Angularity* is any deviation from the specified angle. It is usually measured in bevel-angle degrees (Figure 17-8).

Slag is oxidized metal that forms on the bottom edges of the cut edge. A light slag is normal on a good cut, but heavy slag may result from improper technique, torch settings, or travel speeds (Figure 17-9). *Edge rounding* (Figure 17-10) is the result of overheating the top corners of the cut edge plates. It is usually the result of the torch being too close or too hot.

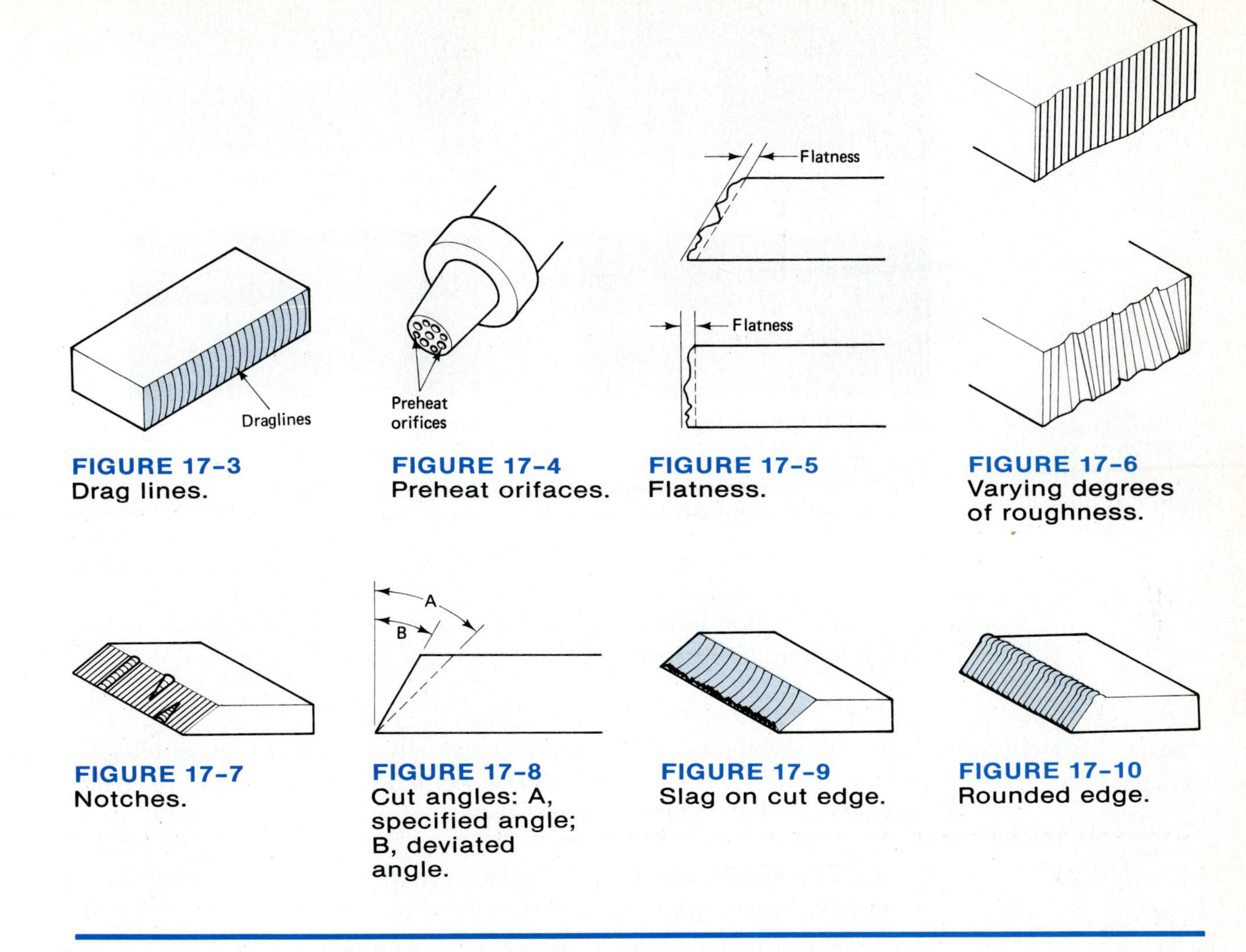

FIGURE 17–3
Drag lines.

FIGURE 17–4
Preheat orifaces.

FIGURE 17–5
Flatness.

FIGURE 17–6
Varying degrees of roughness.

FIGURE 17–7
Notches.

FIGURE 17–8
Cut angles: A, specified angle; B, deviated angle.

FIGURE 17–9
Slag on cut edge.

FIGURE 17–10
Rounded edge.

OFC GASES

Various gases and mixtures have been experimented with to provide the fastest, highest-quality cuts, yet be safe for use in industry. The gas mixes described below are the most popular in industry today.

O_2-acetylene produces the hottest flame known on earth, 5900°F. This makes this mixture excellent for cutting, but its Btu (British thermal unit) rating, which is the measure of the volume of heat produced, is slightly lower than Mapp's. It is still an excellent selection, especially where the tanks must be used for cutting and welding.

Oxy-Mapp cutting has gained rapid popularity. It is fast, hot, and has the highest Btu volume rating of all industrial gases used for cutting. It is good for stack cutting, and 70-lb Mapp gas cylinders last about five times longer than the standard 300-cubic foot acetylene cylinder. Handling Mapp gas cylinders is also lighter and safer than handling acetylene.

Oxypropane requires high volumes of oxygen to reach the temperature required for cutting, about three times as much as oxyacetylene, but the gas itself is relatively cheap.

Oxynatural gas can also be used for cutting, but its flame tem-

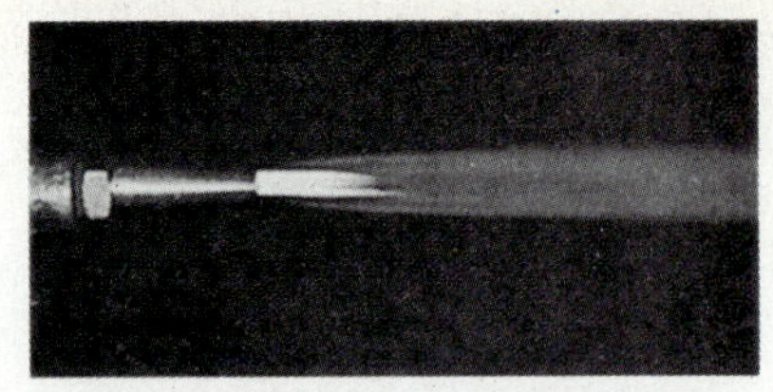

Carburizing: For Stack Cutting Only

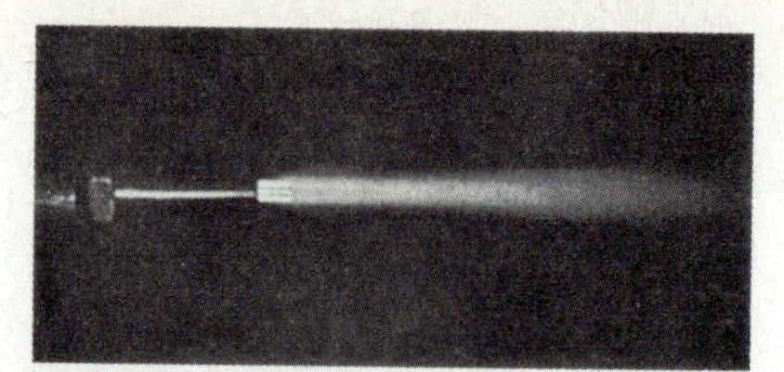

Slightly Oxidizing: For Hand Cutting Or Fast Starts and Beveling

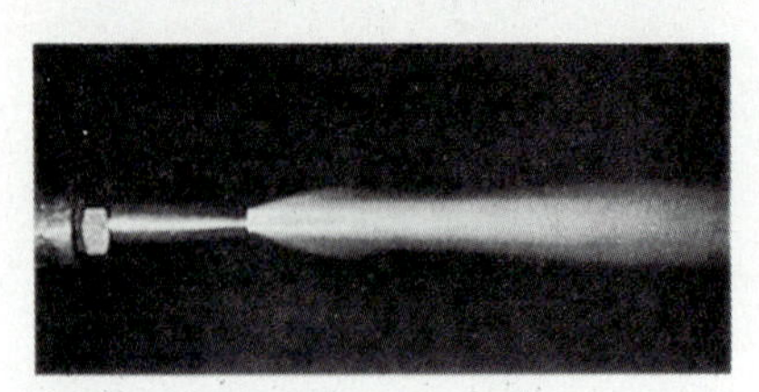

Neutral: For Machine Cutting

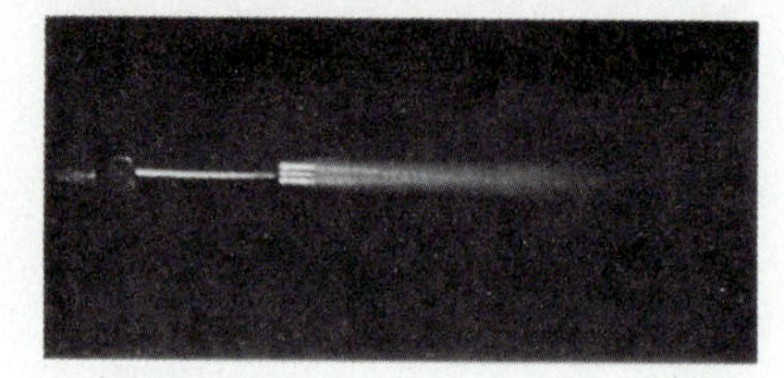

Oxidizing: Not Recommended For Cutting

PHOTO 17-4
Mapp gas flame settings. (Courtesy of Airco Welding Products, a division of The BOC Group, Inc.)

perature and Btu rating are considerably lower than that of Mapp or oxyacetylene. Natural gas is also a bit of a safety problem because it generates carbon monoxide. In some cases it may be cheaper to use, as many manufacturing facilities use it for heating or in heat-treating ovens.

Safe Handling of Cylinders

Most industrial gas cylinders are constructed of rigid high-grade steel plate and must meet ICC and DOT specifications, but there are special precautions that must be taken when handling these cylinders:

1. Transport only in cylinder carts with cylinders chained.
2. Transport only with cylinder caps on.
3. Never leave cylinders freestanding; chain them to a wall or stationary fixture.
4. Open slowly and always crack the valve first. This removes any dirt that might blow into and damage the regulator diaphragm before it is connected to a regulator.
5. Never weld, strike, or use the cylinder for any purpose other than that for which it was intended.

Should you ever attempt to move a cylinder without the cap on, and the cylinder tips over and the neck breaks off, the cylinder will turn into an unguided missile and do plenty of damage before it finally comes to rest. So do not take chances; use the cap and chain them up.

OFC STARTUP PROCEDURE

When preparing to make an oxyfuel cut, make sure that all equipment, bottle pressure regulators, and safety gear are within reach and ready to go. Your safety glasses and goggles should be on. Flip your goggles up so that you can see to light the torch.

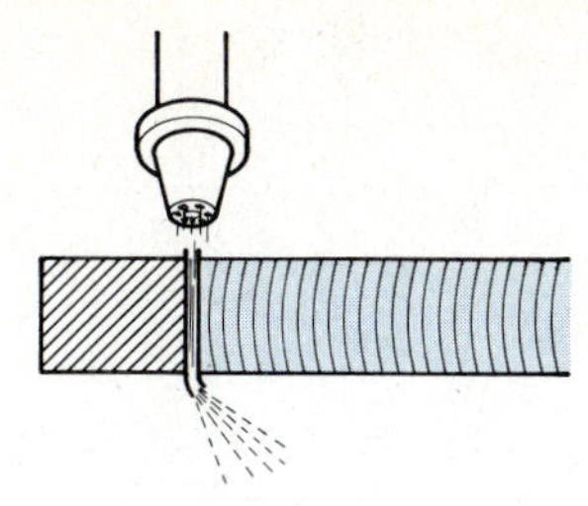

FIGURE 17–11
Cross section of cut. Sparks should flow straight down and slightly to the back.

FIGURE 17–12
Stop and restarts.

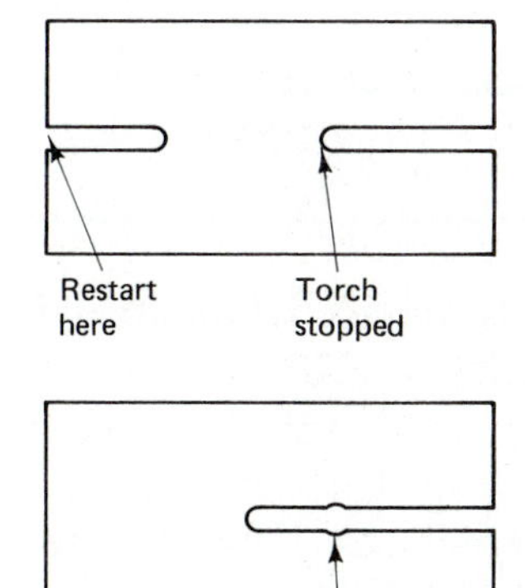

1. Position the torches on the track for automatic cutting, and position yourself in a comfortable position where you can reach a full stroke cut for manual cutting.
2. Light the fuel gas first. Check the photos and tables for the exact fuel gas setting and adjustment.
3. Add oxygen until a neutral flame is achieved. Use a slightly oxidizing flame for Mapp gas.
4. Adjust the preheat control to about $\frac{1}{8}$ to $\frac{1}{4}$ in. above the surface of the plate.
5. Turn on the oxygen jet (cutting jet) to check for any feather in the flame; if there is a feather, add pure oxygen until it is gone. Turn off the oxygen set.
6. Make one pass over the plate, right over your cutting line. This will preheat the plate slightly and make sure that you are right on your cutting line.
7. Move the torch to the edge of the plate—about half on and half off.
8. When the plate turns dull red, turn on the oxygen jet and start moving down your cut line.
9. Adjust your travel speed, looking straight down through your cut, watching the oxidation stream slice through the plate (Figure 17–11).
10. If you loose the cut or the torch goes out, restart at the opposite end of the cut and work in toward where the cut ended. Attempting to restart in the same place may leave a keyhole gouge (Figure 17–12).
11. When you are finished, be sure to shut off the torch, close the bottles, and bleed the hose lines down.
12. Use soapstone to mark all cut metal "hot."

OFC OF SHEET METAL

Oxyfuel cutting of sheet metal presents some new problems. First, it is so thin that there is rapid heat input, therefore a lot of thermal expansion and the resulting distortion. Another problem posed by the high heat input into sheet metal is the rapid rate of oxidation. You will recall that oxidation is the principle behind the oxyfuel cutting process. In cutting, if the oxidation path is too wide, the oxygen jet will not blow it all clear of the kerf, leaving slag (oxidized metal) on the sides of the cut metal joint (Figure 17–13).

There are some solutions to these problems! First of all we need to lower the heat input by using a smaller oxyfuel cutting tip. There are many designs, some with only one or two preheat orifices. The next step is to use a greater torch angle when making the cut. It makes no difference whether your hand cutting or using an automatic burning unit. The steeper torch angle allows the oxidation process to proceed along the joint faster and makes it actually cut a thicker oxidation

FIGURE 17–13
Warped plate and slagged edge.

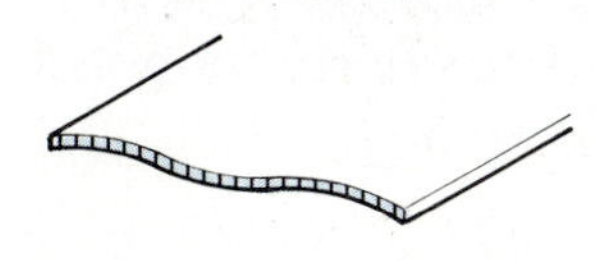

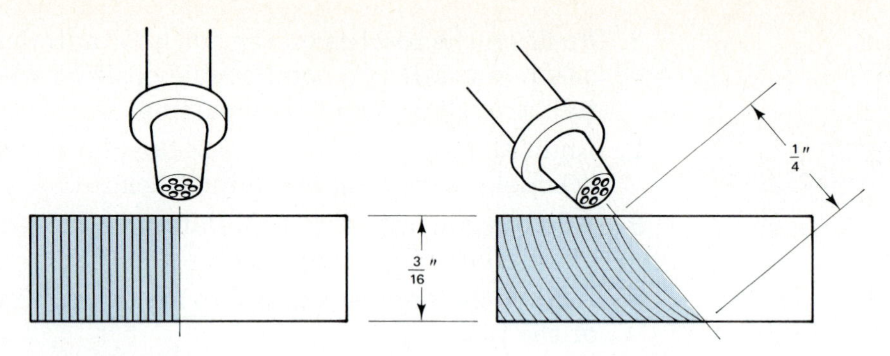

FIGURE 17-14
Torch angles.

stream (Figure 17-14). Increasing the travel speed will result in a cleaner cut and faster cut with less distortion.

FINE TUNING THE CUT

Examining the cut edge of plate is a good way to check what adjustments are necessary to fine tune the cut. Practice various torch settings; this will help you to recognize adjustments necessary for correction. Analyze the cuts shown in Figure 17-15.

Figure 17-15a shows an example of a correct cut. The dragline is just about vertical, with just a slight sweep at the bottom of the cut edge. This cut would require no additional grinding or machining before welding.

In Figure 17-15b, the sweeping draglines and layer of slag at the bottom of the plate indicate too high a travel speed or possibly not enough oxygen pressure. The slag layer can usually be chipped off easily.

In Figure 17-15c, the gouges at the bottom half of the cut edge resulted from too much oxidation action. Traveling too slow, or too much oxygen, will usually cause this problem.

If the torch tip is too close, combustion and oxidation may result from the preheat cones, leaving small gouges at the top of the cut edge plate (Figure 17-15d). If the torch tip is too far from the plate, the flame heat will round the top edge of the cut surfaces (Figure 17-15e). The preheat cones should be $\frac{1}{8}$ to $\frac{1}{4}$ in. off the plate, depending on the tip size.

Too much oxygen pressure causes a turbulent oxygen jet and as a result, erratic kerf and draglines (Figure 17-15f). Oxygen jet pressure ranges from 30 to 120 psi, depending on the tip size.

Too much preheat flame will tend to overheat and round the top of the cut edge plate (Figure 17-15g). Preheat gas pressure ranges from 5 to 50 psi depending on the tip size. Always use setup charts for exact

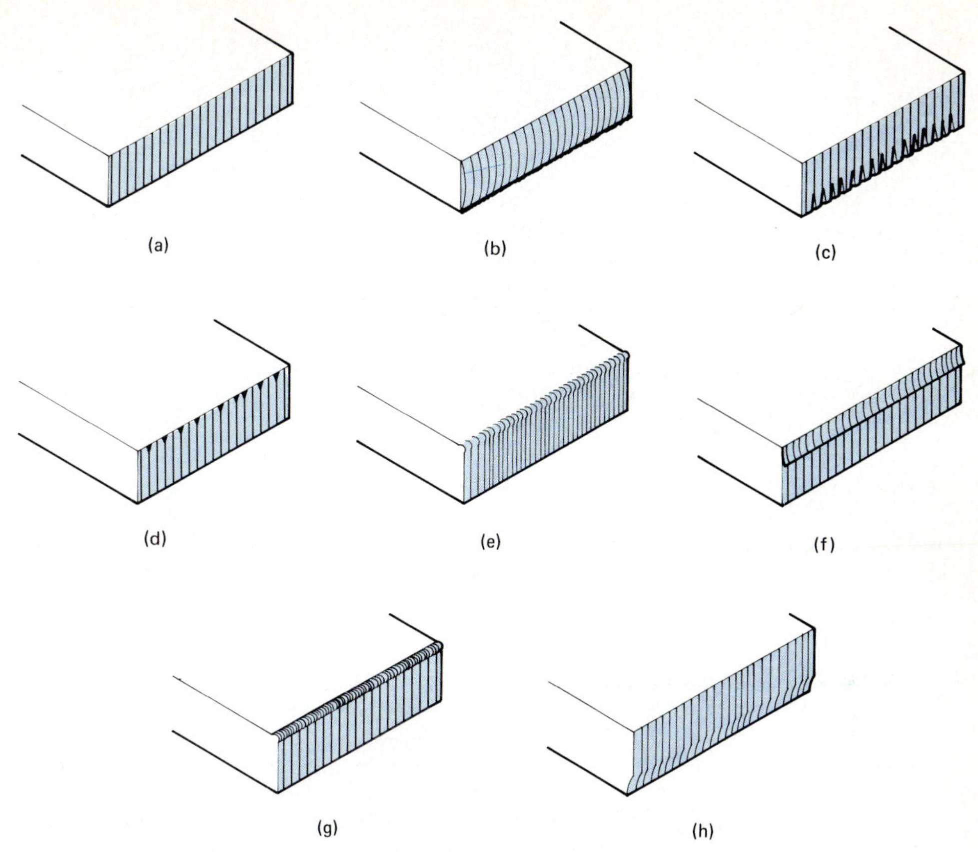

FIGURE 17–15
Cut edges.

oxygen, fuel gas, and tip size for the thickness and variables used for cutting.

The cutting tip condition is an important consideration for making high-quality cuts. A slag-covered, dirty tip will restrict the oxygen jet and cause turbulences (Figure 17–15h). Careful use of standard tip cleaners will work fine for cleaning cutting tips.

Bevel cuts are accomplished in much the same way as standard straight cuts, but a little more preheat and oxygen cutting jet pressure are required. A straight cut on 1-in. plate requires that the torch cut 1 in. of material. But to cut a 45° bevel through the same 1-in. plate, we must cut through a 1.4-in. section of material; this is because the angle increases the area of the cut (Figure 17–16). So we must adjust the torch as if we were cutting through a thicker section of material.

FIGURE 17–16
Torch angles and cutting distances.

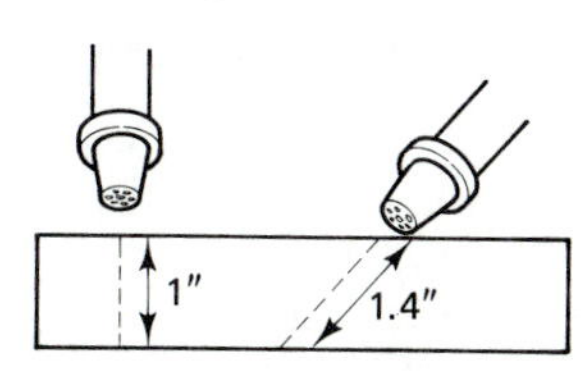

A good bevel cut has smooth uniform draglines on both cut edges (Figure 17–17a). If travel speed is too fast, gouges may result (Figure 17–17b). Insufficient preheat will produce the same result. Too much preheat will roll both top cut edges (Figure 17–17c). Travel speeds may also have to be reduced for beveled cuts.

Typical oxygen and Mapp gas pressures are listed in Table 17–1.

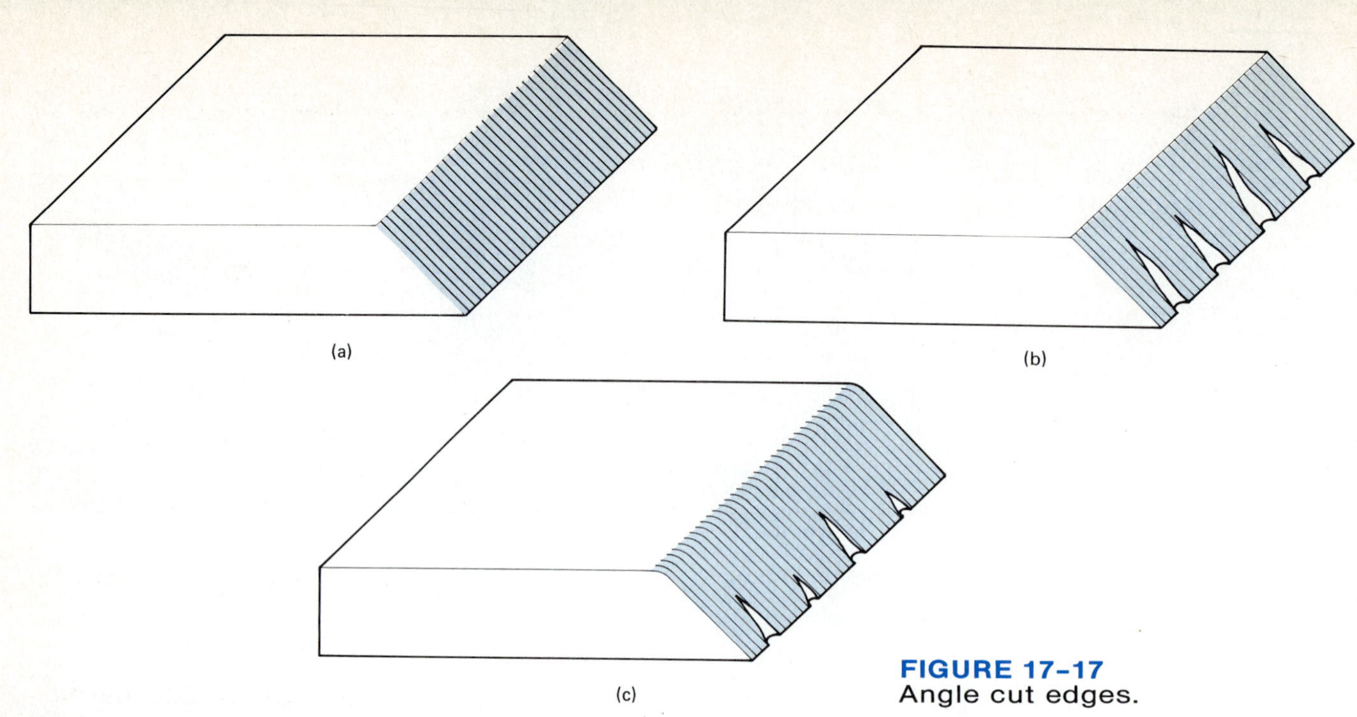

FIGURE 17–17
Angle cut edges.

TABLE 17–1
Typical Oxygen and Mapp Gas Pressures

PLATE THK. INCHES	TIP SIZE NO.*	CUTTING SPEED IN./MIN.	OXYGEN CUTTING PRESS P.S.I.G.	OXYGEN CUTTING FLOW C.F.H.	OXYGEN PREHEAT PRESS P.S.I.G.	OXYGEN PREHEAT FLOW C.F.H.	MAPP GAS PRESS P.S.I.G.	MAPP GAS FLOW C.F.H.	KERF WIDTH INCHES
$\frac{3}{16}$	00/72	22–26	20	45	2–10	8–20	2–6	4–8	.05
$\frac{1}{4}$	0/68	22–26	35	60	2–10	6–20	2–6	4–8	.06
$\frac{3}{8}$	0/68	20–26	40	60	5–10	8–20	2–8	5–10	.06
$\frac{1}{2}$	1/62	20–26	45	85	5–10	8–20	2–8	5–10	.07
$\frac{5}{8}$	1/62	18–24	50	90	5–10	8–20	2–8	5–10	.07
$\frac{3}{4}$	2/56	16–21	45	115	5–10	10–20	2–10	6–10	.09
1	2/56	14–19	50	120	5–10	10–20	2–10	10–15	.09
$1\frac{1}{4}$	3/54	13–18	45	160	10–20	10–20	2–10	10–15	.11
$1\frac{1}{2}$	3/54	12–16	50	170	10–20	20–40	2–10	10–15	.11
2	4/52	10–14	50	220	10–20	20–40	2–10	10–15	.12
$2\frac{1}{2}$	4/52	9–13	55	235	10–20	20–40	6–10	12–20	.12
3	5/49	8–11	50	290	10–30	20–40	6–10	12–20	.14
4	5/49	7–10	55	310	10–30	30–40	6–10	12–20	.14
6	6/39	5–8	60	675	10–30	30–40	8–15	12–20	.20
8	6/39	4–6	65	720	20–40	30–50	10–15	12–20	.20
10	7/30	3–5	60	940	20–40	36–60	10–15	15–25	.26
12	7/30	3–5	65	990	20–40	36–60	10–15	15–25	.26
14	8/18	2–4	65	1080	20–40	36–600	10–15	15–25	.34

*Tip No. → 2/56 ← No. Drill Size
Source: Arco Welding Products.

REVIEW QUESTIONS

1. What type of material will oxyfuel cut?
2. How hot must material be for flame cutting to take place?
3. What are draglines?
4. What is the actual oxyfuel cut called?
5. What is edge rounding?
6. What is different about oxyfuel cutting sheet metal than cutting plate?
7. Describe how you would correct a cut with heave slag and sweeping draglines.
8. Describe how you would correct a cut that has gouges at the bottom of the cut edge.
9. What problems might a dirty cutting tip cause?
10. What additional problems are encountered when making beveled cuts?

18 PLASMA ARC CUTTING

An incredible advancement in the welding and cutting field has been the introduction of plasma arc cutting (PAC). In the past, accurate thermal cutting of nonferrous metals such as aluminum, stainless steels, and coppers were, at best, fair. With the plasma arc cutting we can now cut these metals accurately and very rapidly. Plasma also cuts thin sections of carbon steel very quickly. It is not uncommon to cut $\frac{1}{4}$-in. carbon steel as fast as 150 ipm (inches per minute).

There are two types of plasma arc cutting systems: (1) transferred and (2) nontransferred. The transferred system uses the work to complete the circuit, whereas in the nontransferred method the circuit is completed through the tip and the plasma gas is projected out of the tip by high primary gas pressures. Most cutting systems use the transferred system. The nontransferred system is used for plasma spray surfacing and cutting on nonconductor substances such as glass, ceramics, and plastics.

Plasma arc cutting can be automatic or manual. Automatic systems use high amperages with a heavy-duty plasma torch head. The system is usually mounted on an automatic *X-Y* axis travel system. Due to the high level of noise and smoke associated with these large systems, the tip of the torch head is usually submerged under water. This not only muffles noise and smoke, but prevents sparks from traveling into the work area and keeps the workpiece temperature down.

The hand-held plasma arc cutting systems are fairly lightweight

PHOTO 18-1
PAC of stainless steel. (Courtesy of Thermal Dynamics Corp.)

PHOTO 18-2
This heavy-duty system can be used for hand or automatic plasma arc cutting. (Courtesy of Thermal Dynamics Corp.)

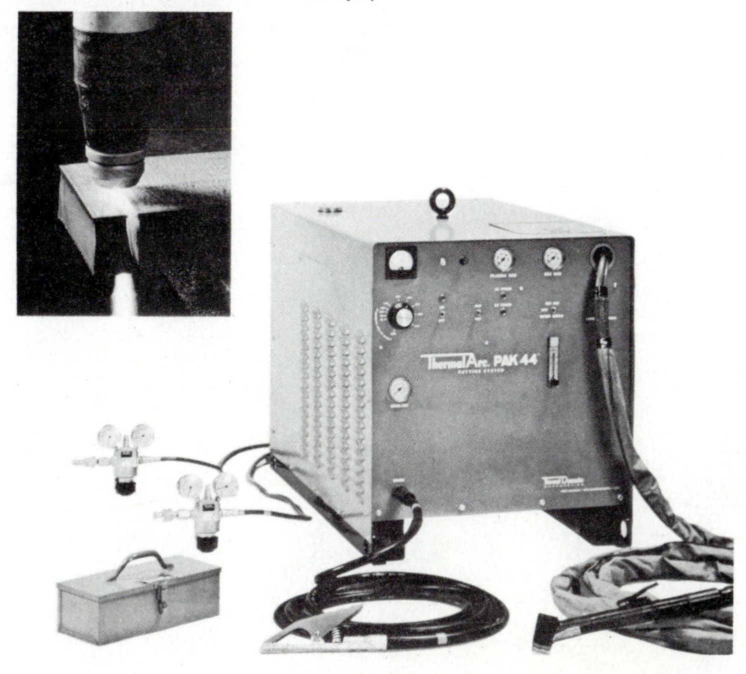

and portable. These systems have simplified maintenance and repair work as well as production jobs, especially when it comes to cutting difficult-to-cut metals such as stainless steels, coppers, aluminums, cast irons, and other metals.

HOW PLASMA ARC CUTTING WORKS

As its name implies, plasma arc cutting uses a high-velocity plasma gas to melt and blow away metal from the kerf. The arc is first established with a pilot arc. This high-frequency pilot arc eliminates any need for striking to start the arc. Once the arc is started, the primary current conducts across the primary inert gas flow. This is the actual plasma, the flow of current (ionization) in this gas. When this ionized gas is constricted and forced out a small orifice, it creates a high-velocity, high-energy "plasma" that easily melts and cuts many metals that cannot be thermally cut with other processes.

PAC EQUIPMENT

A special power source (Figure 18–1), torch (Figure 18–2), and gases are used in plasma arc cutting. The power source is a special constant-current power source, equipped with a high-frequency pilot arc. The primary controls on the power source will be the amperage control, the run/set switch, an on/off switch, and the gas flow rate (the gas flow rate may be on the gas bottle regulators). The torch is a specially designed torch for PAC. It may be water cooled, or may rely on gas flow and convection to cool the torch. The torch is made up of a number of complex parts, but the basic components include the cup, tip, gas distributor, liner or insulator, torch head, collet, tungsten, and cap.

There are also a number of O-rings for sealing out the atmosphere. Most torches also come with a special torch wrench for changing tips. The tungsten is a specially machined 2% thoriated tungsten.

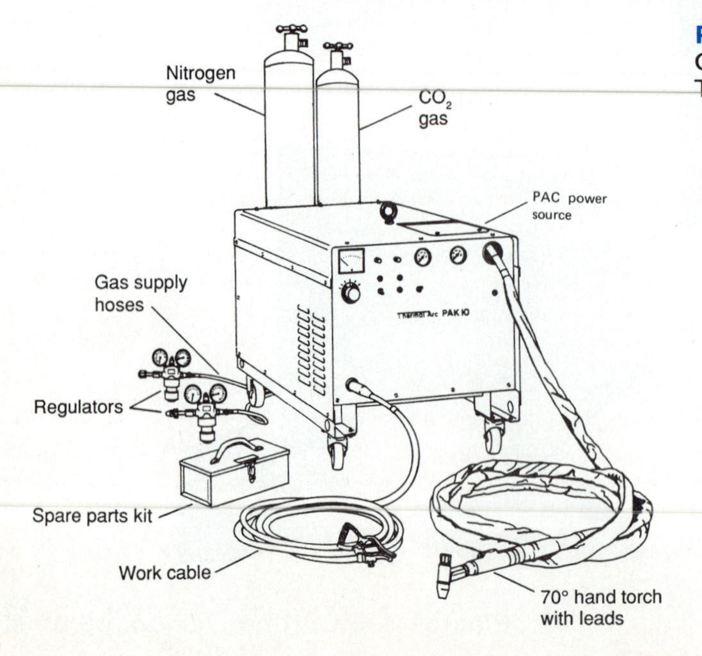

PHOTO 18–3
Complete PAC system. (Courtesy of Thermal Dynamics Corp.)

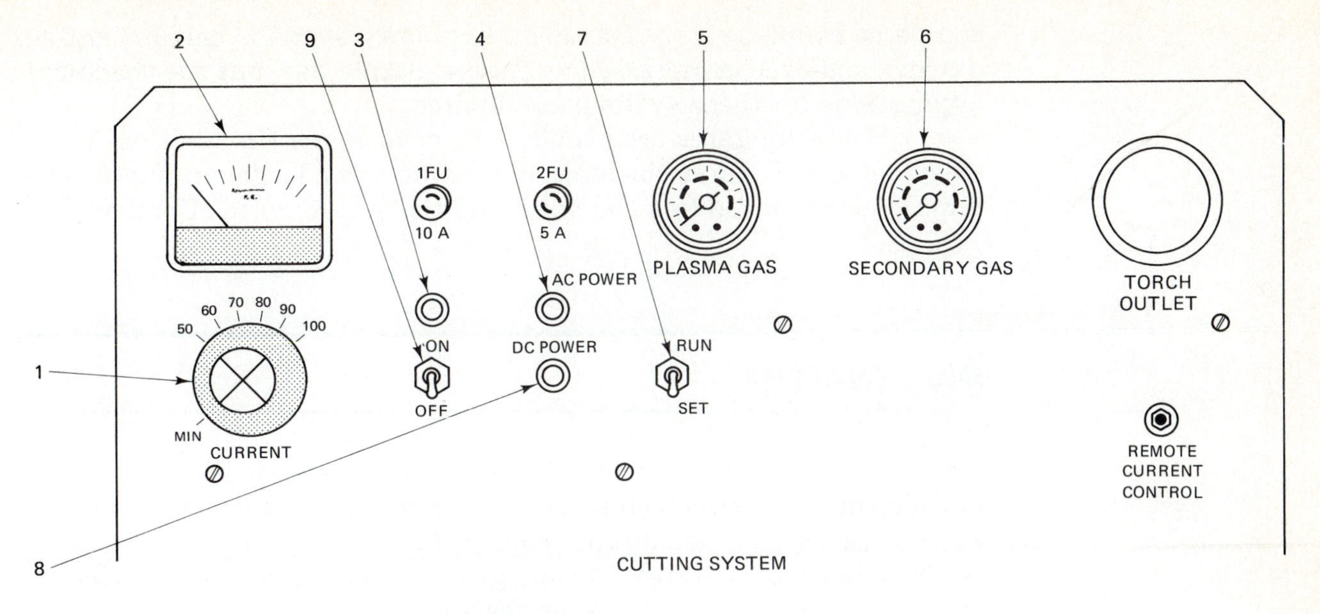

1. Current control
2. Ammeter
3. ON light
4. Ac power ON light (indicates ac power to the machine)
5. Plasma gas working pressure
6. Secondary gas working pressure
7. RUN/SET – SET for setting gas pressure, RUN for coming
8. Dc power ON light (indicates dc power for coming)

FIGURE 18–1
PAC torch control panel. (Courtesy of Thermal Dynamics Corp.)

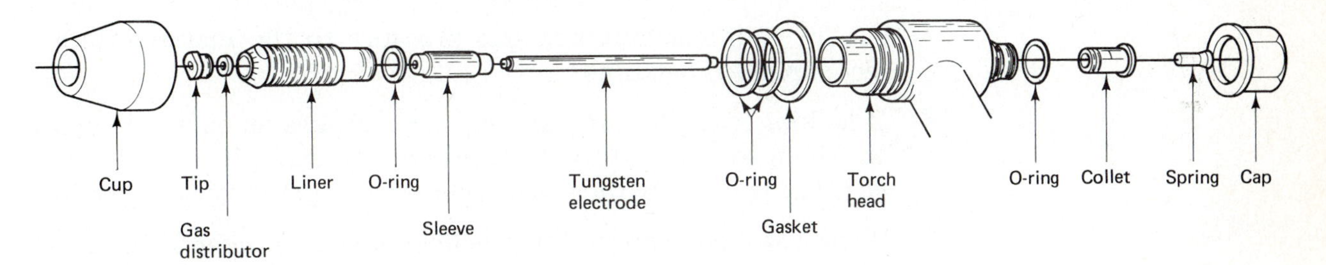

FIGURE 18–2
PAC torch control panel. (Courtesy of Thermal Dynamics Corp.)

PAC GASES

There are two gases needed for PAC. The first is the primary gas; this gas is the actual plasma gas that does the cutting. The other is called the secondary or shielding gas; it is designed to shield both the kerf area and the primary plasma gas. It also serves to cool the torch.

Primary plasma gases are usually nitrogen or a 70% argon–30% hydrogen mixture. Nitrogen is very popular because it is inexpensive and versatile.

Argon–hydrogen is a hotter plasma gas but is quite expensive compared to nitrogen. Some small, light-duty systems can use argon, oxygen, and even compressed air as the plasma gas, but the thickness applications for these systems are limited.

Shielding gases are usually CO_2 or nitrogen. In addition to the shielding gases, water shielding is used on some large automatic systems. The water reduces the noise and the smoke quite effectively.

PAC SETUP

Gas flow rates and machine settings are recommended by most PAC machine manufacturers, so we suggest that you consult your manufacturer's operational manual that came with the machine. Here are some common machine settings; Plasma gas flow usually ranges between 30 and 60 psi, but will vary based on the size of the material thickness and the type of material. Secondary gas flow ranges from about 50 to 80 psi, and may also vary a bit. Both the travel speed and amperage are important variables in the correct setup and in the quality of the cut. Recommendations are given in Table 18–1.

PAC PROCEDURE

1. Set up your machine and gas pressure to the manufacturer's recommended settings.
2. If making straight cuts, an angle iron makes an excellent guide on which to rest the torch. If other than straight cuts, lay out your cut with soapstone or metal-marking paint stick or marker.
3. Wear a standard arc welding helmet with a number 9, 10, 11, or 12 shaded lens (the shade you use will depend on your personal experience). You should also wear welding gloves and a welding jacket, together with standard arc welding clothing.
4. Take a practice cut without depressing the start button. As you do this, make sure that you can reach the entire length of the cut. If you cannot, reposition yourself until you can.
5. To start the cut, depress and hold the start button on the torch. At first you will hear and see the pilot arc. This will help you find your starting point and will start the cut.
6. As you're traveling across your cut, watch the kerf for a clean-cut edge and make sure that you are getting completely through the part. With special "stand-off" cups the PAC torch can actually be dragged across the workpiece. With standard cups a $\frac{1}{8}$- to $\frac{3}{8}$-in. stand-off should be used (stand-off is the cup-to-work distance).

TABLE 18–1
Travel Speeds (Inches per Minute)

THICK-NESS (IN.)	100 A		200 A		300 A		400 A	
	HIGHEST QUALITY	MAXIMUM SPEED	HIGHEST QUALITY	MAXIMUM SPEED	HIGHEST QUALITY	MAXIMUM SPEED	HIGHEST QUALITY	MAXIMUM SPEED
				Aluminum				
$\frac{1}{8}$	150	200						
$\frac{1}{4}$	85	115	120	180	150	200		
$\frac{1}{2}$	35	50	85	100	105	130	120	180
$\frac{3}{4}$			50	0	60	85	100	120
1					40	60	60	80
2							40	45
3							15	20
				Stainless Steel				
$\frac{1}{8}$	105	140						
$\frac{1}{4}$	60	80	90	110	100	130		
$\frac{1}{2}$	20	30	60	80	65	85	95	105
$\frac{3}{4}$			40	55	50	65	60	80
1					35	45	40	55
2							15	18
3							6	8
				Carbon Steel				
$\frac{1}{8}$	85	120						
$\frac{1}{4}$	40	50	75	95	80	105		
$\frac{1}{2}$	15	20	40	55	40	65	70	90
$\frac{3}{4}$			20	30	30	40	40	50
1					20	30	25	30
2							14	17
3							5	5

7. If you would like to calculate your speed in inches per minute (ipm), simply mark off a 12-in. section of your material. Now have someone time you as you make the 12-in. cut. The time (in seconds) × 5 will be your speed in ipm.

$$T \text{ (for 12-in. cut)} \times 5 = \text{speed (ipm)}$$

8. Examine the cut edge surface for heavy slag, rough areas, and uncut areas. Adjust your amperage or speed accordingly.

Problems and Solutions

Rough areas on cut edges. Increase the speed of the cut; change the tip; use a piece of angle iron or guide to keep the cut straight.

Heavy slag. Increase the travel speed; change the tip; check the gas pressures.

Uncut areas. Slow the travel speed or increase the amperage; change the tip; check the gas pressures.

REVIEW QUESTIONS

1. What type of material can be cut with plasma arc cutting?
2. Can plasma cutting be automated? Explain.
3. What are the two main types of plasma cutting?
4. What does the primary gas do? What does the secondary gas do?
5. What does water do for plasma arc cutting?
6. What type of clothing is required for making plasma arc cuts?
7. What might you check if you are getting heavy slag on the cut edge surface?

METALLURGY

In this chapter we give you a basic understanding of metals you will be working with throughout your career. We have no intention of making you a metallurgical engineer, but you will begin to realize that having knowledge of these "metallurgy basics" will help you through your career and enable you to troubleshoot welding problems. Remember, knowing why and how something works always helps to fix it when it does not work.

As an aid in reading the chapter, a list of the common symbols for various elements is given in Table 19-1.

TABLE 19-1
Symbols for Various Elements

Element	Symbol	Element	Symbol
Aluminum	Al	Molybdenum	Mo
Beryllium	Be	Nickel	Ni
Carbon	C	Silicon	Si
Chromium	Cr	Silver	Ag
Cobalt	Co	Tantalum	Ta
Copper	Cu	Thorium	Th
Gold	Au	Titanium	Ti
Hydrogen	H	Tungsten	W
Iron	Fe	Vanadium	V
Lead	Pb	Zinc	Zn
Magnesium	Mg	Zirconium	Zr
Manganese	Mn		

Carbon steels are still the most common metals used in our civilization today. Chances are that you will learn to weld carbon steel before attempting to tackle any other metal—aluminum, stainless steels, coppers, and so on.

The majority of welding is done on carbon steels. That is not to say that the new metals, such as aluminum, are not growing rapidly in popularity and applications, but pound for pound carbon steels are still the most heavily used metals.

COMPOSITION OF STEEL

Before we go into detail on how steel is made, let's examine what it is:

iron (Fe) + carbon (C) = steel

If we take iron and mix in some carbon for strength, we have steel. But it is not quite that simple; if we do not add something to remove the oxygen (a deoxidizer) it would be so porous that we would not be able to call it steel. So we add silicon (Si) (and sometimes other deoxidizers) to absorb the oxygen. So the basic ingredients in steel are:

iron (Fe) + carbon + silicon (Si) = steel

Pure iron without carbon is not very strong. Its uses can be traced back to around 3500 to 4000 B.C. Iron itself cannot be used as structural shapes (H-beams and wide-flanged beams, standard I-beams, etc.) because it would be too soft and weak for support.

It was found that when small amounts of carbon were added, the strength and hardness increased rapidly. Structural grades of carbon steels now have tensile strengths ranging from 58,000 to 100,000 psi and high-strength low-alloy, and boiler and pressure vessel plates from 65,000 to 135,000 psi. (We discuss high-strength low-alloy steel further later in the chapter.) It can now be seen that carbon is the key element in the strengthening and hardening of steel. We also classify steels as to their carbon content (Table 19-2).

TABLE 19-2
Carbon Content for Steel Classifications

CLASSIFICATION	PERCENT CARBON
Very-low-carbon steel	Below 0.05
Low-carbon steel	0.05–0.14
Mild steel	0.15–0.29
Medium-carbon steel	0.30–0.50
High-carbon steel	0.51–1.50
Very-high-carbon steel	1.51–1.75
Cast iron	Over 1.75

Alloying Elements

Additional alloying elements may be added to steel to strengthen, to improve mechanical properties, and for corrosion resistance. Following is a summary of alloying elements added to steel and what they do for the steel.

Carbon (C) (even in small amounts) has the greatest strengthening and hardening effect on steel. As carbon is added to steel, strength and hardness will increase until 0.83% is reached. At this point strength (tensile strength) will start to decrease. The metal's ductility and melting temperature will always decrease as carbon is added (Figure 19-1). Because of its potent effect on steel, carbon is considered the most important element added to steel.

Chromium (Cr) does much the same thing as carbon does. It strengthens and hardens with the additional property of giving corrosion resistance. To get full corrosion resistance from chromium, in steel, it requires at least 12% chromium be added to steel. Of course, chromium is one of the basic elements found in stainless steels. When enough chromium is added to steel, it forms a chromium oxide shield over the steel, which prevents further oxidation (such as iron-oxide rust) (Figure 19-2).

Nickel increases the strength of steel while maintaining its ductility. This property is called toughness. It also maintains its toughness in steel at low temperatures. Hardenability and corrosion resistance are increased with the addition of nickel.

Manganese in steel combines and breaks down sulfur and phosphorus (which for the most part are impurities), to prevent brittle weldments and cracking. It will also increase hardenability. It is also a deoxidizer, but too much manganese can cause cracking.

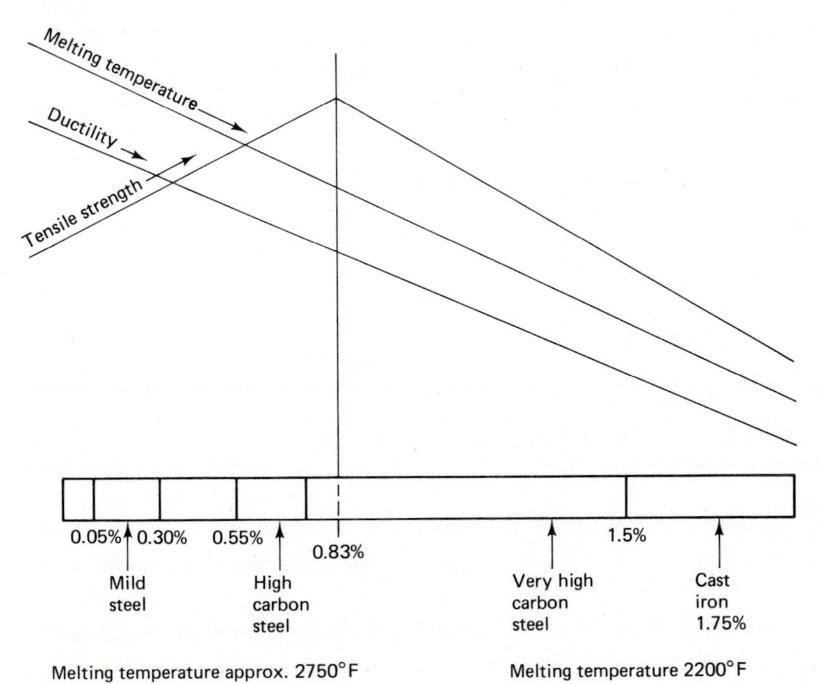

FIGURE 19-1
Effects of carbon on steel.

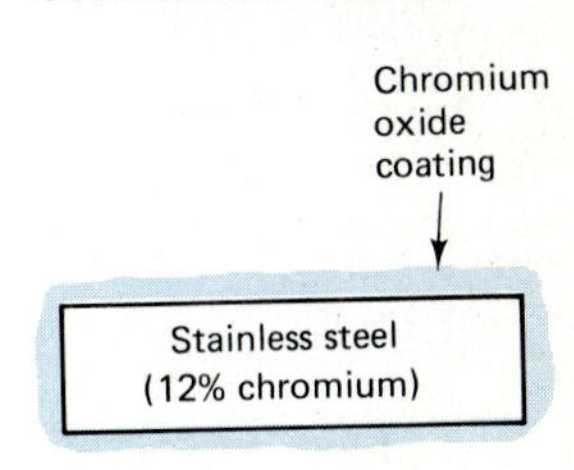

FIGURE 19-2
Oxide coating on stainless steel.

Molybdenum is a very strategic alloying element used in material such as aircraft tubing, boilers, and boiler piping to make them stronger and lighter. Molybdenum (also known as "moly") strengthens and hardens steel, at both normal and elevated temperatures. It also tends to reduce creep in steels under constant stress and to increase corrosion resistance slightly.

Cobalt gives steel a property known as "red hardness," that is, good strength at high temperatures.

Vanadium is added to steel to control grain size during heat treatment. It is commonly mixed with chromium alloys. When the alloy is heat treated, the chromium hardens the alloy and the vanadium refines the grains, thus preventing brittleness and cracking.

Tungsten is added to steel both as a grain refiner and to give strength at high temperatures ("red hardness"). Most tungsten steel alloys are considered heat-resistant steels.

Sulfur aids slightly to machinability but is primarily an impurity, because it causes brittleness and poor weldability (more than 0.05% is considered an impurity).

Phosphorus in proportions over 0.04% is considered an impurity because it causes cracking. Lower amounts help make steel easier to machine and provide a little corrosion resistance to steel.

Copper is added to steel for added corrosion resistance. Such steels are called weathering steels because some structural grades can be exposed to the atmosphere without heavy oxidation (rusting).

Deoxidizers

The next group of elements found in basic carbon steel are the deoxidizers. Silicon is one of the most commonly used deoxidizers, but others, such as aluminum, magnesium, and titanium, are also used. The deoxidizers remove oxygen absorbed into the steel. Unshielded molten steel will absorb oxygen rapidly and make the steel porous and weak. Some deoxidizers, such as aluminum, form a crust over the steel and blanket it from the atmosphere during steelmaking.

Silicon is the primary deoxidizer used in making steel. It improves the fluidity of steel and increases hardenability.

Aluminum is also an excellent deoxidizer and controls the grain size in steel.

Zirconium functions much as aluminum. It is a superdeoxidizer and refines grain structure.

Titanium is a deoxidizer, a grain refiner, and will disperse sulfur. It is often used in high-strength low-alloy (HSLAs) steels.

HOW STEEL IS MADE

Now that we understand what is in steel, let's examine how we make it. The first step in making steel is purifying the base element iron. Where does iron come from? Unfortunately, we do not just go out and dig up a chunk of pure iron. It is out there, but it is mixed up with other rock compounds that contain iron.

PHOTO 19–1
The Number Eight blast furnace at U.S. Steel's Fairfield works is the largest and most modern in the South. The computer-controlled, conveyor-charged furnace consumes approximately 8000 net tons of iron ore, 3000 net tons of coke, and 500 net tons of limestone to produce 5000 net tons of molten iron each day. (Courtesy of U.S. Steel Corporation.)

PHOTO 19–2
Molten iron is cast (drained) from a blast furnace at a U.S. Steel corporation plant near Pittsburgh, Pennsylvania. (Courtesy of U.S. Steel Corporation.)

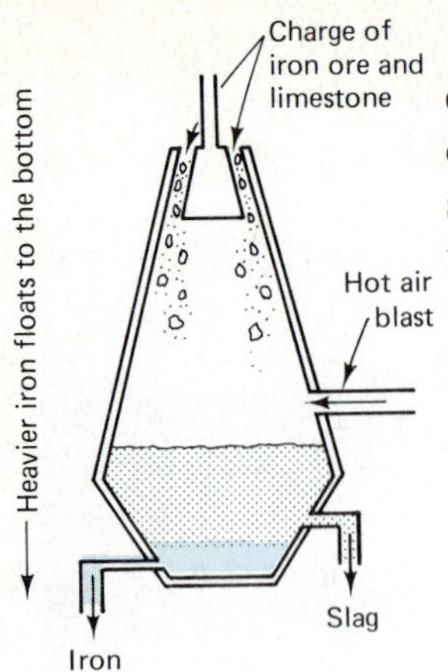

FIGURE 19–3
Blast furnace.

How do we get the iron out? Separating the iron from its ore is done in a blast furnace (Figure 19–3). The furnace simply heats the iron ore to the point at which the iron is liquefied. Being heavier, the iron sinks to the bottom, where it is "tapped off." Lighter rock compounds float to the top.

The following elements are used to separate iron from its ore:

1. *Coke:* acts as the fuel for heat in the furnace; made from dried coal
2. Iron ore: the rock compound from which the iron is drawn
3. *Limestone:* slagging compound that removes impurities from molten iron
4. *Air or oxygen:* added to accelerate combustion in the furnace and get the ore up to the liquefication temperature

Most blast furnaces run continuously. A new charge is fed into the furnace after the iron is tapped off from the bottom. This cycle may

PHOTO 19–3
Steelmaking flowchart. (Courtesy of the American Iron and Steel Institute.)

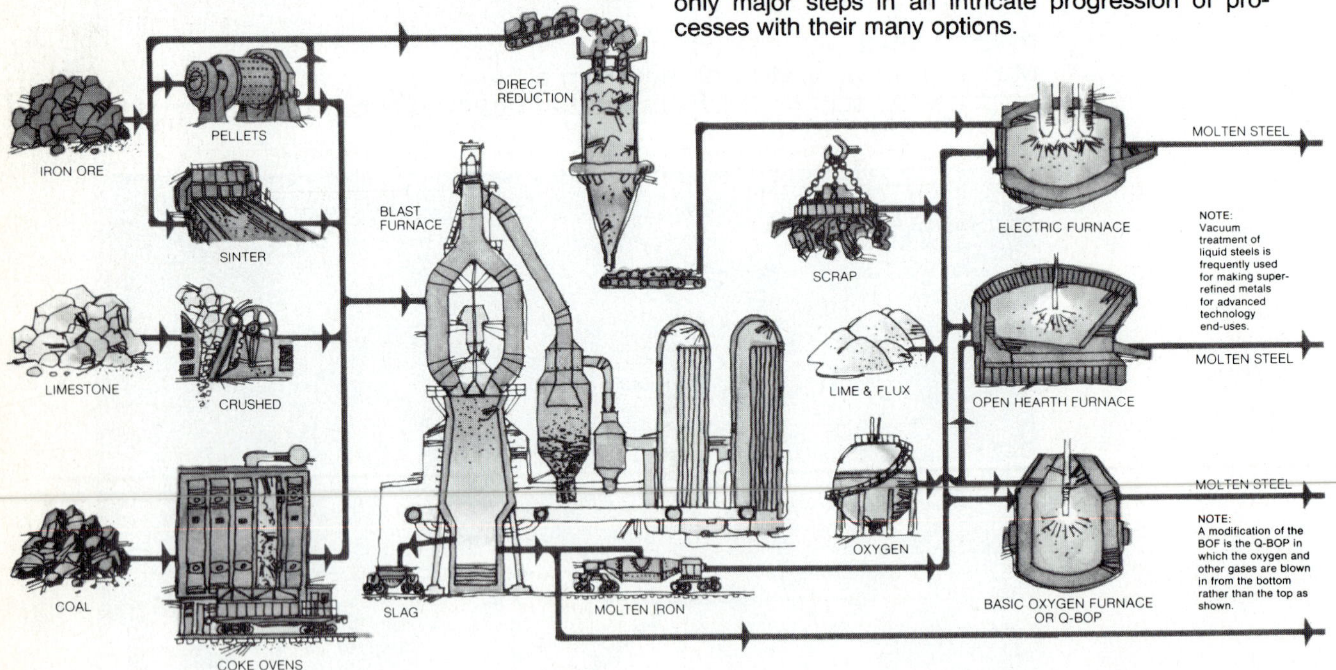

some environmental systems parallel to steelmaking

Land reclamation restores mines and quarries to natural state. Tree-planting is one method.

Stack cleaners capture dust from numerous steelmaking processes, keeping it out of the atmosphere.

Venturi scrubbers spray water into dust-laden gases. Recovered solid particles may often be recycled.

continue for months at a time. As combustion is taking place in the furnace, carbon monoxide is being produced. The carbon monoxide combines with oxygen, thus removing it from the iron as it floats toward the bottom of the furnace. The limestone acts as a fluxing agent, absorbing impurities and floating them to the top and away from the liquid iron. Coke is the actual fuel for the whole operation. It burns at about 3500°F, which is enough to melt and separate the iron from its ore. The oxygen in an air blast that enters from the side of the furnace accelerates the combustion of the coke and increases the temperature.

In the furnace, a typical charge consists of about 4000 lb of iron ore, 2000 lb of coke, 1000 lb of limestone, and 2000 lb of air to net 2000 lb of raw iron. The raw iron is called *pig iron.* In this state iron is not good for much of anything.

The next step is to turn the pig iron into steel. This step is usually accomplished by one of four basic steelmaking processes: open hearth, Bessemer, oxygen furnace, or electric furnace. This step accomplishes two important functions:

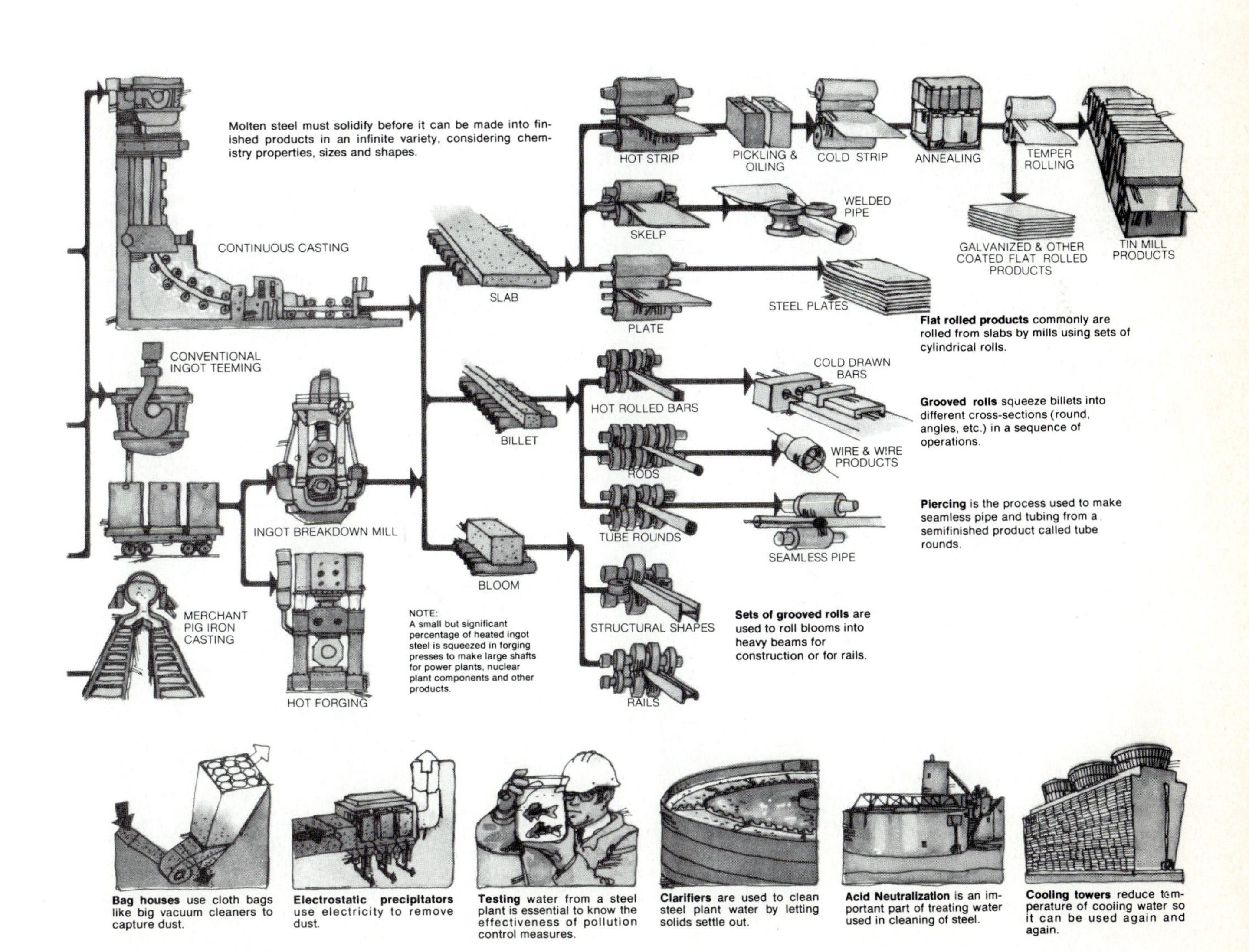

1. Deoxidizes and removes impurities
2. Adds carbon and alloys to the steel.

Deoxidizing is done by either the vacuum degreasing method or the chemical method using deoxidizers such as aluminum, titanium, zirconium, or silicon. Vacuum degasing requires the extra step of the degasing unit.

PROPERTIES OF METALS

It is very important that you understand the various properties that metals may possess. Knowing the properties of metals will help you to select and identify various types of metals.

Metals possess three basic properties:

1. physical properties
2. chemical properties
3. mechanical properties

Physical Properties

Physical properties include melting point, allotropic forms, magnetic properties, weight, and color. The melting point of steel will vary depending on what is in the steel. More carbon will lower the melting point. Some elements, such as cobalt or tungsten, will raise the melting temperature of steels. The allotropic forms of steel refer to the space lattice structure of the atoms as the steel is heated. These forms are important when heat treating. Figure 19-4 shows what happens to steel when it is heated.

FIGURE 19-4
What happens to steel when it is heated.

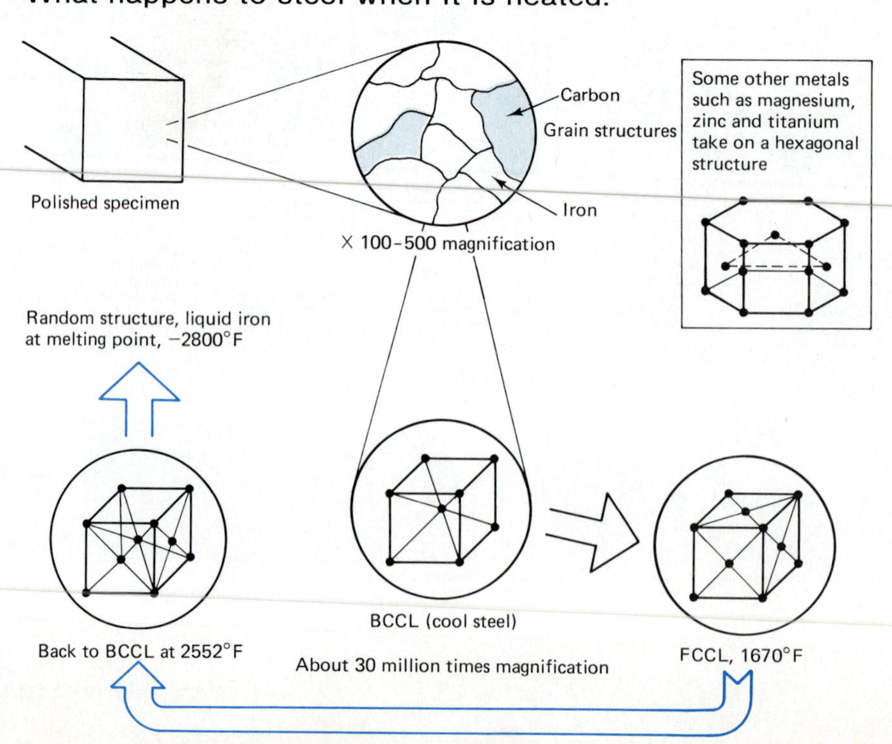

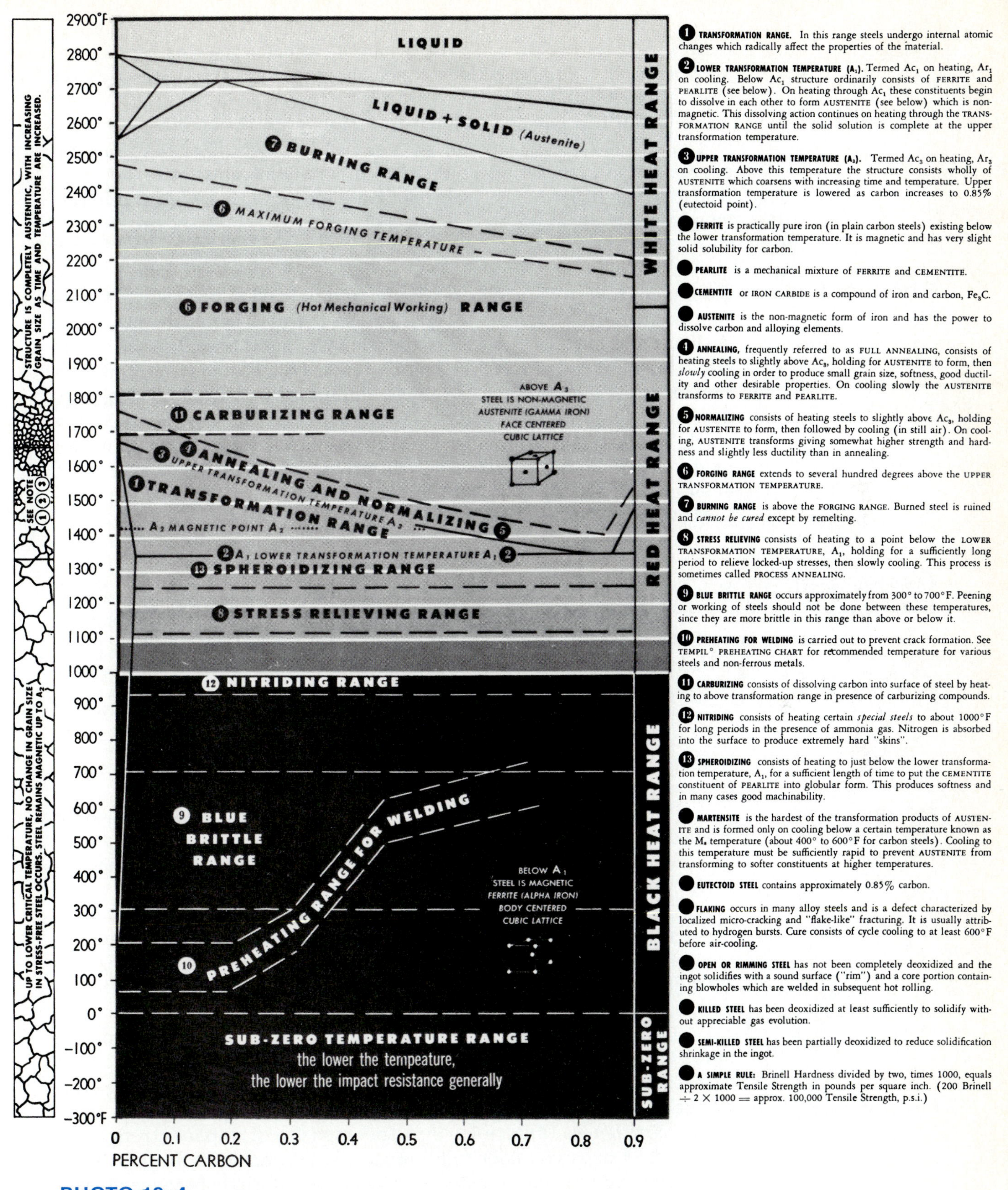

PHOTO 19–4
Color and heat treatment chart. (Courtesy of Tempil Division, Big Three Industries, Inc.)

Examining the *weight* of a metal will help you identify various metals. Always use a *known* metal weight when comparing to your unknown metal. Examine different metal's densities to determine weight (see Table 19-13).

Color can be used to identify metals both at room temperature and when heated. Some metals such as carbon steels, change to varying degrees of red, whereas aluminum and magnesiums have no color change when heated. The surface of fractured sections will exhibit various colors, depending on the metal or alloy.

Chemical Properties

Chemical properties include corrosion and oxidation properties. *Corrosion* is actually a chemical attack on the metal. Chemicals may react with various metals. In a corrosive environment stainless or nickel steels may be used to resist chemical attack. *Oxidation* is a normal reaction for many metals, such as copper, aluminum and stainless steel (copper oxide, aluminum oxide, and chromium oxide). These oxides actually shield the metals from further harmful oxidation, by forming their own oxide shield on the surfaces of the metal. *Iron oxide,* however, does eventually consume itself from continuous oxidation if exposed to moisture and air. We call it rust.

Mechanical Properties

It is important that you understand the mechanical properties of metals. It will help you in designing for various *loads* that weldments must withstand. Welding inspectors must understand both mechanical properties and the direction of various stresses affecting these properties.

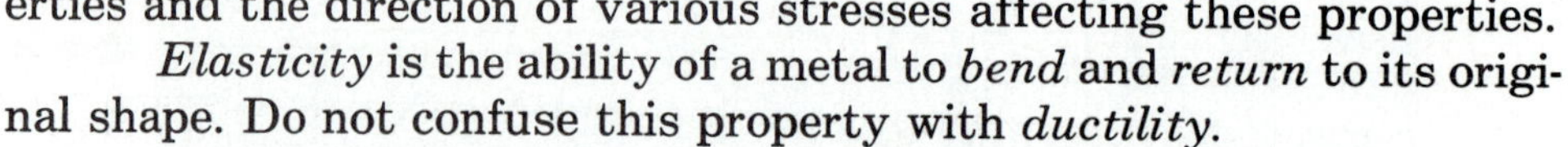

Elasticity is the ability of a metal to *bend* and *return* to its original shape. Do not confuse this property with *ductility.*

Ductility is the ability of a metal to bend without breaking, and without returning to its original shape. Aluminum and copper are ductile metals (Figure 19-5).

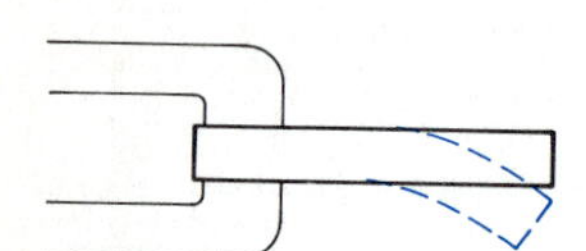

FIGURE 19-5
Ductile property.

Malleability is the ability of metal to be easily rolled, formed, or shaped.

Compressive strength refers to a metal's ability to be forced together or compressed (Figure 19-6).

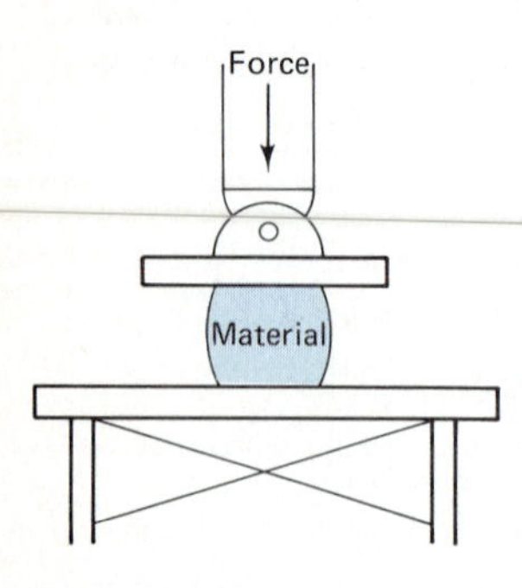

FIGURE 19-6
Compression.

Tensile strength (or tension) is the property that indicates a material's resistance to being pulled apart (Figure 19-7). It is used to compare a material's ultimate strength because it actually gives the strength of a material's molecular bond. We usually rate metals based on the tensile strengths.

Impact strength is a material's ability to withstand sudden blows (Figure 19-8). Machine dies are good examples of metal that withstand impact as parts are pressed into and shaped by the die. Tremendous impact forces are exerted on the die.

Brittleness (Figure 19-9) is the opposite of impact strength. If a material fractures under mild impact, it is considered brittle. Some cast irons and high-carbon steels are considered brittle because of their high carbon content.

Toughness has been described as a combination of impact strength and ductility. A material that is tough has the ability to re-

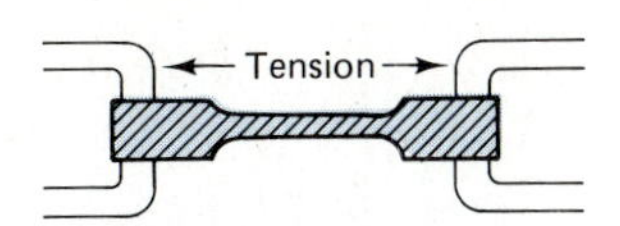

FIGURE 19–7
Tension property.

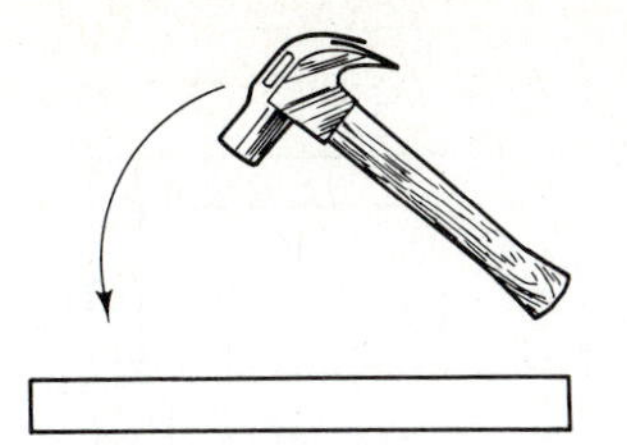

FIGURE 19–8
Impact property.

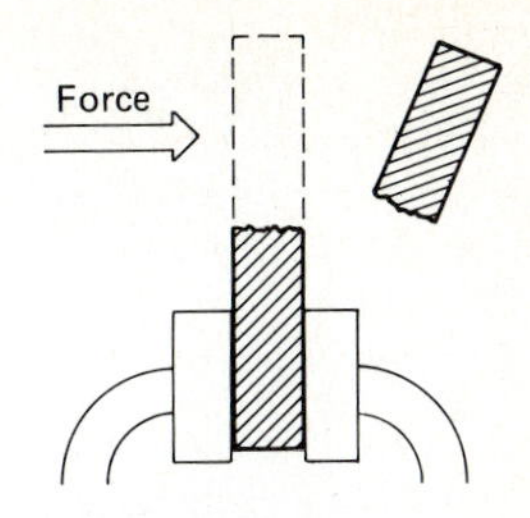

FIGURE 19–9
Brittleness property.

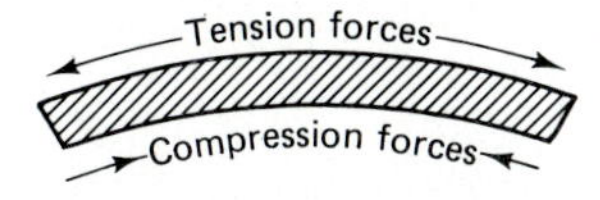

FIGURE 19–10
Bending property.

ceive a blow without fracturing, but yet is ductile enough to bend a bit under loads. The hull of a ship is a good example of toughness. It must absorb tremendous forces from wind and waves during rough seas, but remain ductile enough to prevent fatigue of the hull structure.

Bending strength is a combination of tension and compression. When bending material the outer fibers are under tension and the inner fibers are under compression (Figure 19–10). Beams in buildings undergo bending loads.

Fatigue failure (Figure 19–11) is the result of loads cycling on and off or in opposite directions. This property is very important to the designer, the inspector, and the welder. The reason for this concern is because it is the one type of failure that can result *without* overloading the material. Fatigue failure may result even if the tensile strength limits of the material have not been exceeded. Plant maintenance welders deal with fatigue continuously, because machines and parts are under continuous changing loads. Bridges are classic examples of fatigue loading, during rush-hour traffic, heavy moving loads are exerted on the bridge, whereas during light traffic hours the loads are much less. It is this cycling that leads to fatigue failure in metals. It is said that fatigue accounts for most mechanical breakdowns.

Stress is simply the load or force applied to a part; *strain* is the resultant deformation due to the stress (Figure 19–12). A term known as the *modulus of elasticity* (E) uses a stress–strain curve (Figure 19–13) to show a material's stiffness based on the type of material:

$$\frac{\text{stress}}{\text{strain}} = \text{modulus of elasticity}$$

FIGURE 19–11
Fatigue property; usually a progressive failure as shown here.

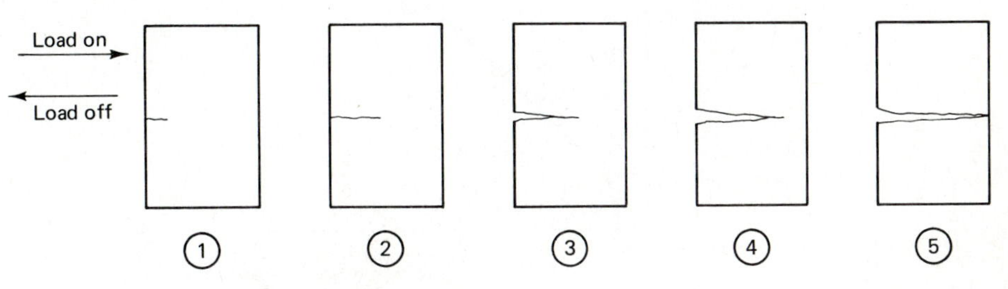

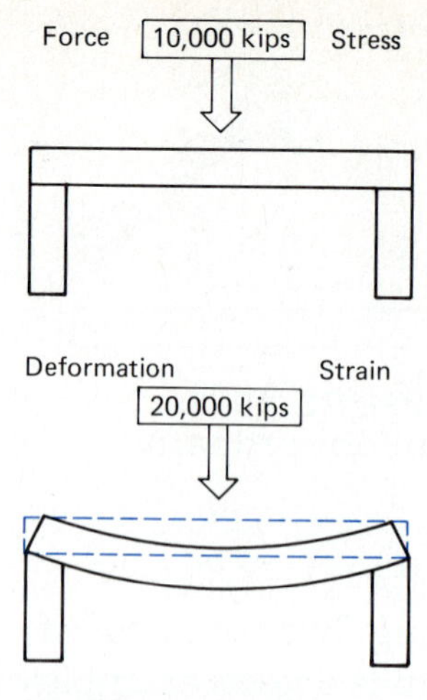

FIGURE 19–12
Stress versus strain.

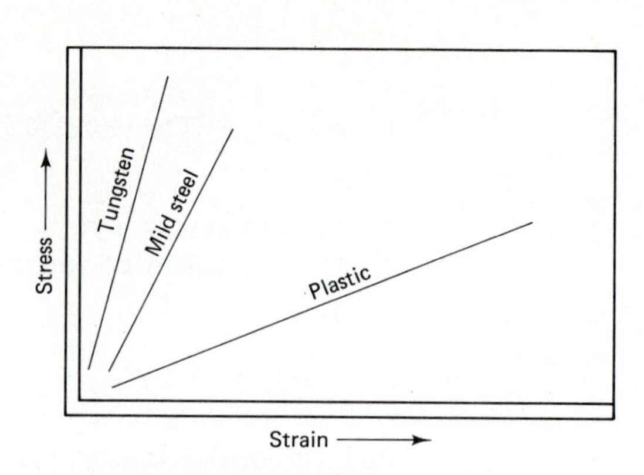

FIGURE 19–13
Modulus of elasticity.

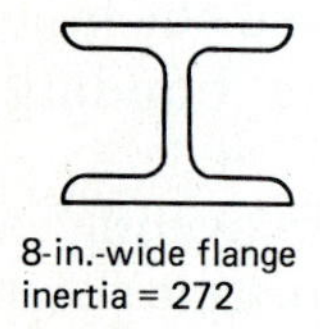

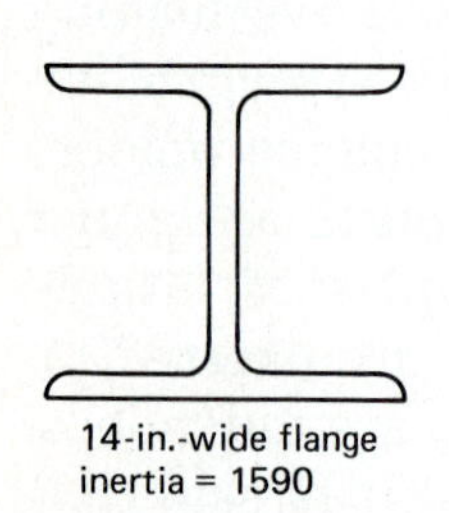

FIGURE 19–14
Inertia of beams.

Moment of inertia (I) is a beam's stiffness based on its geometric *shape* (Figure 19–14).

Elastic limit and yield point are confusing terms for most welders and technicians, so let us sort them out here. Let's take a piece of metal and slowly and steadily add a tensile force to it. This force is usually provided by a tensile puller or tensile testing machine. If we plot stress versus strain, we will see the amount of movement of the metal compared to the load on it (Figure 19–15). What is happening? As the load increases, the material evenly stretches until it reaches its elastic limit. Once it exceeds the elastic limit it will never return to its original shape.

Theoretically, if the load is released before it reaches the elastic limit, it will return to its original shape. As we continue adding more stress, we almost immediately reach the yield point. Yield means to give or stretch a bit, and that is exactly what the metal does.

Yield is a phenomenon unique to carbon steels. Once a material yields, it is usually considered to be unusable because it is deformed permanently. As the load continues to increase, it will finally reach its ultimate tensile strength and the failure.

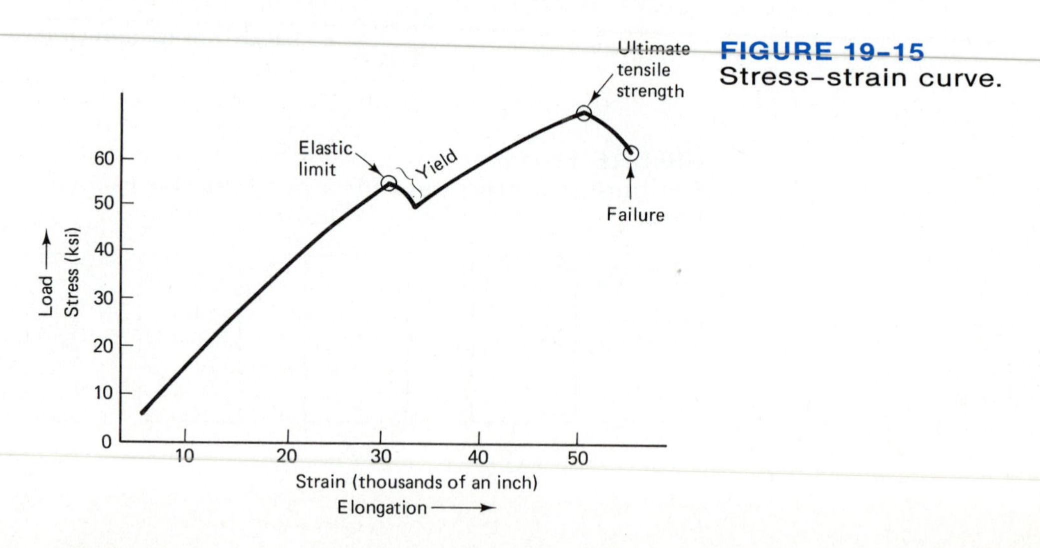

FIGURE 19–15
Stress–strain curve.

CLASSIFYING STEELS

Extracting iron from its ore and turning it into steel is only the beginning for some steels. Many steels go on to become various steel alloys. As we mentioned earlier, various alloying elements can affect the physical, chemical, and mechanical properties of steels. Let's look at some steels and alloys and their uses.

Mild Steel

Mild steel has a 0.15 to 0.30% carbon content. It is one of the most commonly used steels in industry. It is rolled into most structural shapes, such as beams, plates, bars, and pipes. Mild steel can be heated and cooled without much hardening. Its weldability is extremely good because of the lower carbon content. Remember that the lower the carbon content, the better the weldability.

Medium-Carbon Steel

Medium-carbon steel has 0.30 to 0.50% carbon content. It is used for machinery parts, leaf springs, and components that require a little extra strength and hardness. Weldability is still fair in the carbon range 0.30 to 0.49%, but from 0.40% carbon and up you are liable to get weld

PHOTO 19–5
Accurate preheat temperature can be checked with temperature crayons. When the crayons melt, the temperature is at or above the crayon test temperature. (Courtesy of Tempil Division, Big Three Industries, Inc.)

crack if you do not preheat or use low hydrogen electrodes (see Chapter 7). Preheating is done to preexpand and lower the thermal shock and resulting stress placed on the metal during welding (Figure 19–16). It will also allow the metal to cool much more *slowly*. This slow cooling allows the brittle grain structure to shrink back to its normal size. Medium-carbon steels are also quite hardenable by heat treatment and may need postweld heat treatment, such as normalizing or annealing.

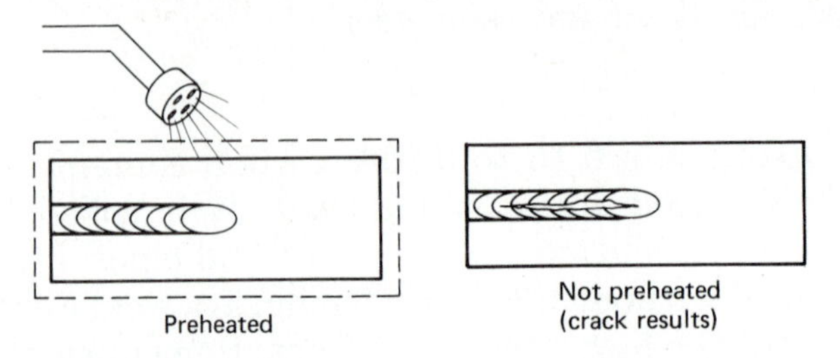

FIGURE 19–16
Preheating versus no preheating.

High-Carbon Steel

High-carbon steels contain 0.51 to 1.50% carbon content. They are used as tool steel, for shaping and forming steel parts. High-carbon steel is used for these jobs because the shaping and forming material must be harder and stronger than the material it is shaping and forming. Springs, metal files, and railroad rails are samples of the other uses of high-carbon steel. Successful welding of high-carbon steel takes extreme care. Some high-carbon steel may even be considered unweldable, but most can be welded if all the precautions are taken.

Welding High-Carbon Steel The main problem with welding high-carbon steel is cracking. Cracking is due to high carbons and high hardenability. If the weld is not cooled slowly, it will harden and crack. Another problem is stress. When welds are deposited on high-carbon steel, residual stress builds in the weld area from the heating and cooling of the weld deposit. Lower-carbon steels can absorb these weld-induced stresses, but high-carbon steel is harder and much less ductile, and unfortunately, any stress is often relieved in the form of a crack.

The *solution?* Unfortunately, we do not have a guaranteed solution, but here are some steps to take that will tilt the odds in your favor:

1. Use *low-hydrogen* electrodes. Hydrogen not only makes electrodes run hot, it absorbs into the weld and heat-affected zone at certain temperatures. This hydrogen sets up stresses in the weld area which may lead to weld and underbead (heat-affected zone) cracks (see Chapter 7).
2. *Preheat* an area of at least 3 in. around the area to be welded. If the entire part can be preheated that will even be better. A preheat temperature of 125° to 750°F is recommended.
3. Some of the most difficult-to-weld high-carbon steels may be

welded with special electrodes that are designed specifically for difficult-to-weld steels (see Chapter 7).

4. *Peening* is a technique used to prevent cracks by relieving stress built up in the material due to welding. It sends shock waves into metal, thus relieving the stress areas. Using the rounded end of a ball peen hammer, strike the weld with moderate force. Start peening immediately after welding, and stop when the hammer starts to spring back a little. Overpeening will only work-harden and rebuild stress in the weld area.

High-Strength Low-Alloy Steel

High-strength low-alloy steels (HSLAs) are a classification of steels designed for higher strength, better mechanical properties, and more corrosion resistance than standard mild steels. Although used in many areas of industry, they are very popular for structural steels, because they provide more strength in thinner sections. This allows us to build higher buildings at lower cost because of the reduced weight. HSLAs are used in shipbuilding, boiler, and other applications. HSLAs are classified by their chemical composition. Their tensile strength range is 60,000 to 90,000 psi and their yield is 42,000 to 80,000 psi.

HSLAs get their added strength from various combinations of chromium, nickel, copper, manganese, molybdenum, nitrogen, titanium, zirconium, and vanadium. The amount of these alloying elements added to HSLA is quite low.

HSLAs are quite weldable; however, it is suggested that you follow the precautions outlined under the welding of medium-carbon steels. The reason for following the medium-carbon steel chart is that these steels are susceptible to hydrogen embrittlement and cracking.

Alloy Steels and High-Alloy Steels

There are many varieties of alloy steels. Alloys are added to give the mechanical, chemical, or physical properties needed for various service requirements. Stainless, high-strength low-alloy, and some tool steels are typical alloy steels. A metal is considered an alloy when it is mixed with two or more additional elements. For example:

iron (Fe) + carbon (C) = steel + additional elements = alloy

Stainless Steel

Stainless steels are special corrosion-resistant steels that are used in environments where oxidation and chemical attack are problems. Stainless steel uses include: chemical transmission pipes; chemical tanks; lab equipment; food industry processing equipment; tanker trucks; and yes, even the kitchen sink may be made of stainless!

Stainless steel comes in three general classifications:

1. Austenitic (300 series)
2. Ferritic (400 series)
3. Martensitic (400 and 500 series)

The terms *austenitic, ferritic,* and *martensitic* come from the mi-

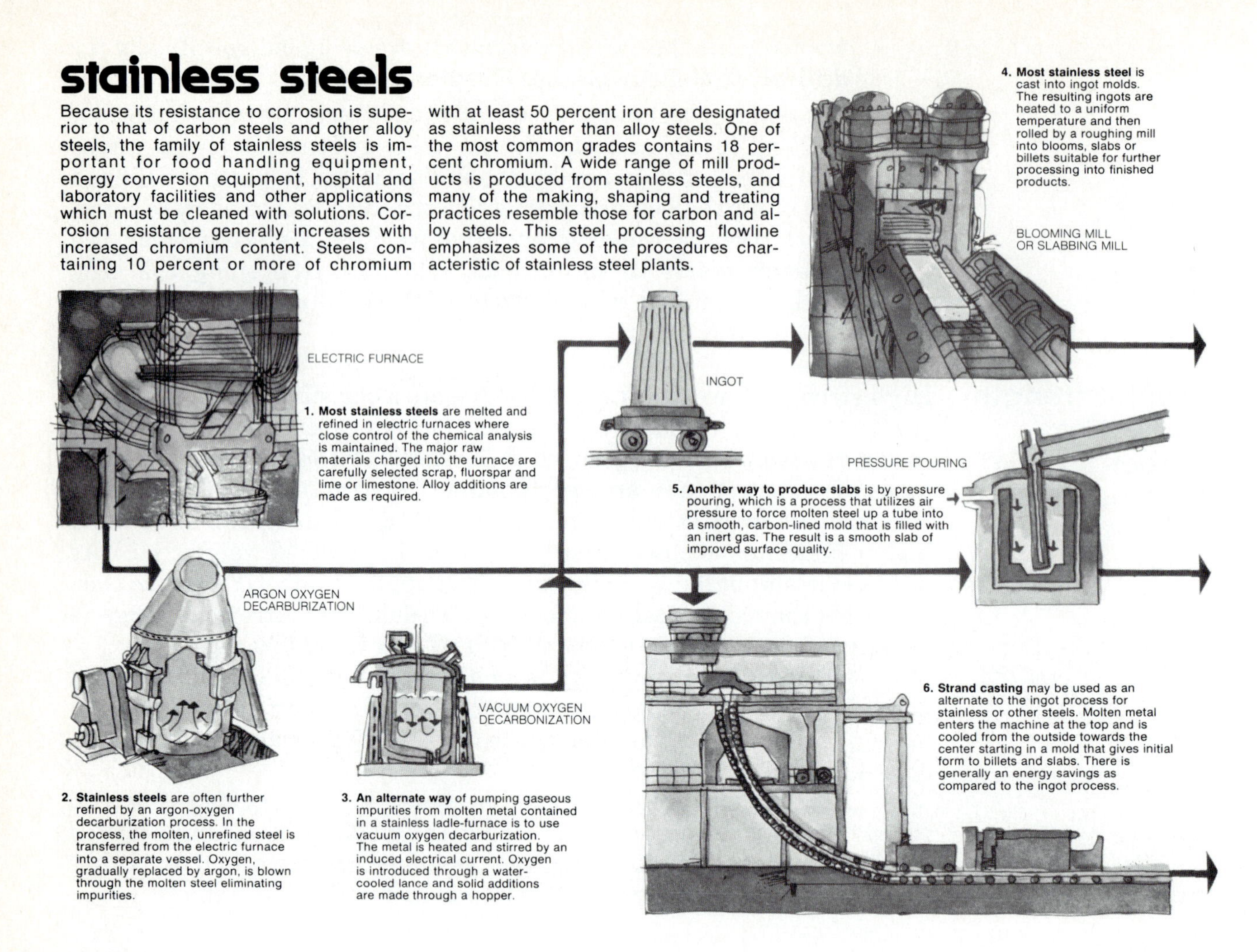

PHOTO 19-6
Stainless steelmaking flowchart. (Courtesy of the American Iron and Steel Institute.)

crostructure these stainless steels form at room temperature. Austenitic grades are the nonmagnetic grades, which have *nickel* and *chromium.* They are *not* hardenable by heat treatment, and because they contain both nickel and chrome, they are very corrosion resistant.

Austenitic stainless steels have low thermal (heat) conductivity, so they tend to insulate. There may also be distortion when welding or cutting these stainless steels (Figure 19-17). This distortion is due to the nonuniform heat distribution and the resulting uneven expansion and contraction. The low thermal conductivity also makes austenitic stainless steels good for piping of cryogenic fluids such as liquid oxygen. The pipe insulates preventing the cryogenic fluid from warming and boiling off.

Austenitic stainless also has good toughness properties because of the nickel and chromium content. They also offer the most corrosion resistance of all the grades of stainless steel and are very weldable.

The ferritic or 400 series stainless steels use straight chromium for their corrosion resistance; there is no nickel added. Like the austenitic grades, the ferritic stainlesses cannot be hardened by heat treat-

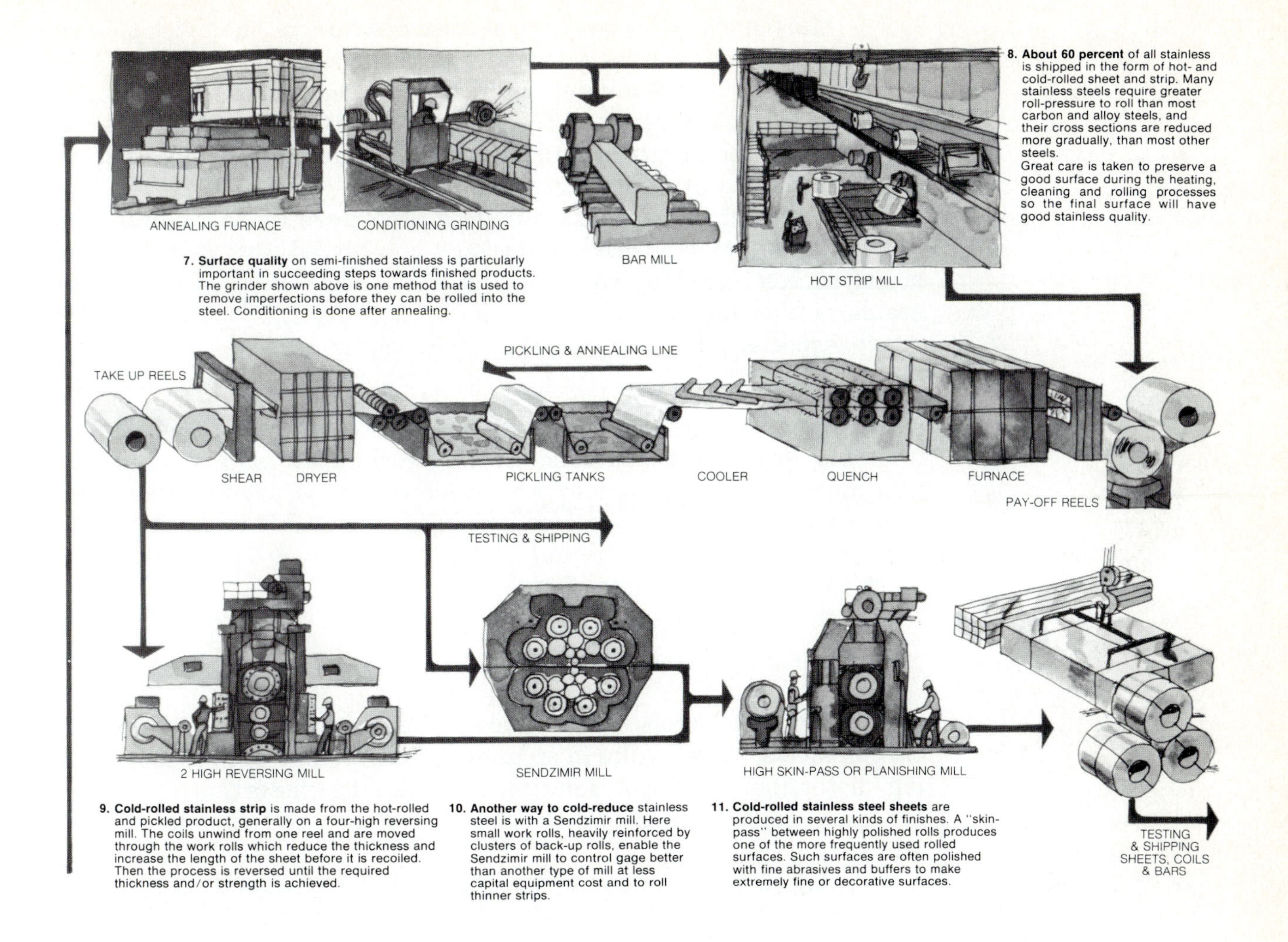

ment. Ferritic stainless steels also have fairly low thermal conductivity and good corrosion resistance. The degree of corrosion resistance will depend on the percentage of chromium content. Chromium is the principal corrosion-resisting element. Chromium oxide forms over the metal, shielding it from further harmful oxidation such as iron oxide (rust).

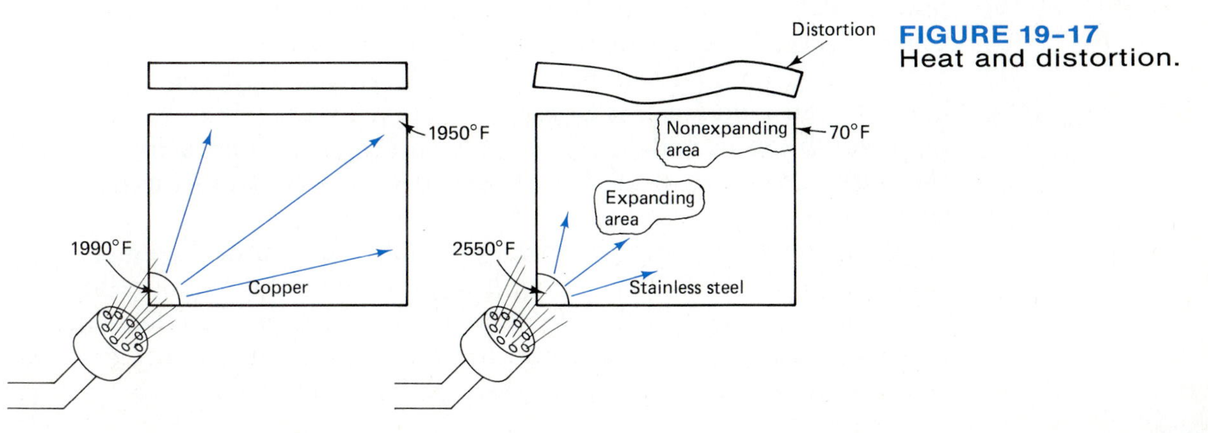

FIGURE 19–17
Heat and distortion.

Martensitic (400 and 500) stainless steels for the most part use straight chromium for their corrosion resistance. However, the 414 and 431 series have small amounts of nickel. These stainless steels will retain their corrosion-resistant properties above 1000°F, so they are used in many high-temperature applications. They *are* slightly magnetic and can be hardened by heat treatment.

Although martensitic stainless steels are quite weldable, the higher-carbon martensitic grades are considered the most difficult stainless steel grades. You will find that using the austenitic grade stainless filler metals may give you the most success on any stainless grade, especially the higher-carbon martensitic grades.

Welding Stainless Steel Stainless steels are considered quite weldable; however, the welder, welding technician, and welding engineer must all be concerned with the following aspects of welding stainless steels:

1. Warpage and distortion
2. Carbide precipitation
3. Chemical properties

Stainless steels tend to have lower thermal conductivity than carbon steels. This results in an uneven distribution of heat and resulting distortion. The best solution for reducing distortion is to lower the heat input. This lowers the heat imbalance and expansion differences.

Another area of concern is carbide precipitation. Have you ever observed a stainless steel weld that has rusted or has oxidation (rust) along the sides of the weld (probably in the heat-affected zone)? This is where carbon precipitates into the weld zone and combines with the chromium atom to form chromium carbide. This new substance will not shield with a protective oxide as pure chromium does. The result is eventual oxidation (rust) in the chromium carbide areas.

There are a few ways of eliminating this problem. One of the easiest and cheapest ways is to use an "L" class electrode, for example, E308-16-L (the "L" stands for "low carbon"). Less carbon in the weld area means less chance of chromium carbide forming. Carbon precipitates into the weld zone due to heat. Carbon loves heat and precipitates toward it. Therefore, keeping the heat input low helps reduce carbide precipitation. Another method involves using "stabilized" electrodes. These electrodes have elements such as columbium and titanium that act as stabilizers. Here is how they work: As the heat precipitates carbon into the weld zone, it combines with the stabilizing element (forming columbium carbide) instead of with the chromium atoms, leaving the pure chromium oxide free to shield the metal from further oxidation.

Another concern involves matching up chemical elements. For example, if the weld has more chromium than the surrounding base metal, the weld would be harder (and may be more brittle). Since we usually want the weld and base metal to act in unison under stress, this would be an undesirable property. The reverse could also happen, producing a "too soft" weld, or a weld that would be more susceptible to oxidation and corrosion. The solution is quite simple; use the same filler metal grade as the base metal to be welded. However, be aware that some engineers do require a higher-grade filler to be used where it may be beneficial. For example, a stainless container, where it is known that corrosion starts in welded corners and welded seams, may use a 308 base metal but require E309 filler metal to overcome the problem.

Steel Classifying Systems

Steels are classified under a number of different systems, depending on the uses. The American Society of Testing Materials classifies structural steels by the specification number and grade, each of which has a specific yield and tensile strength (Table 19-3).

TABLE 19-3
Strengths for ASTM Steel Classifications

ASTM SPECIFICATION	YIELD STENGTH (PSI)	TENSILE STRENGTH (PSI)
A-36	36,000	58,000
A-53. grade B	35,000	60,000
A-131 grades A and B	32,000	58,000
A-588 grade 50	50,000	70,000
A-709 grade 50	50,000	65,000
A-572 grade 60	60,000	75,000

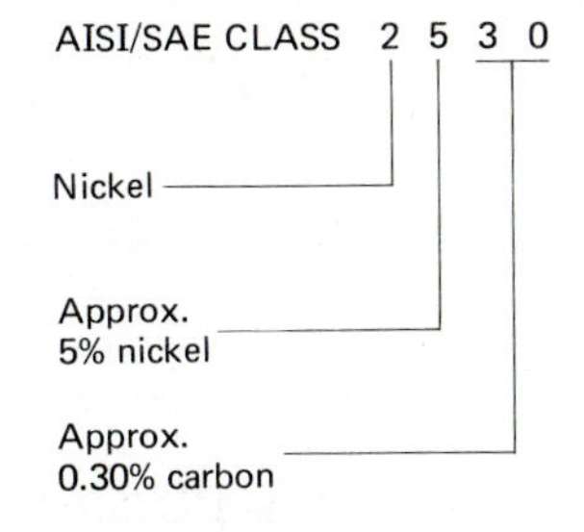

FIGURE 19-18 AISI/SAE classification system for steel.

The AISI/SAE (American Iron and Steel Institute/Society of Automotive Engineers) uses a four-digit system which indicates the primary alloying element followed by the approximate percentage of that element, then the carbon content. An example is shown in Figure 19-18. The first digit designations are as follows:

1 Carbon steel
2 Nickel steel
3 Nickel-chromium steel
4 Molybdenum-chromium (or nickel) steel
5 Straight chromium steel
6 Chromium-vanadium steel
7 Tungsten steel
8 Silicon-manganese steel

The first four digits may also be followed by a letter to indicate the process used to make that steel.

A Alloy steel, open-hearth process
B Carbon steel, acid Bessemer process
C Carbon steel, open-hearth process
D Carbon steel, acid open-hearth process
E Carbon and alloy steel, electric furnace

STEEL WELDABILITY

As stated previously, the more carbon in steel, the more difficult it will be to weld without cracking. When other alloying elements are added to some steels, they will harden and make welding difficult. Elements such as chromium, manganese, nickel, and others will do much the same thing that carbon does to reduce weldability. It does, however, take a *lot more* of these elements to reduce weldability than it does carbon.

We can calculate approximately how difficult it will be to weld a given steel alloy if we know what alloying elements it contains. A carbon-equivalent formula will give you an approximate equivalency of these other elements.

$$CE = C\% + \frac{Mn\%}{6} + \frac{Ni\%}{20} + \frac{Cr\%}{10} + \frac{Cu\%}{40} - \frac{Mo\%}{50} - \frac{V\%}{10}$$

Note: This CE formula is used for alloy steel. Notice the negative values for molybdenum and vanadium; this is because in high alloys they may prevent hardenability. However, in carbon and low-alloy steels, molybdenum would add to hardenability and should be added, not subtracted; use $+ \frac{MO\%}{4}$.

If the carbon equivalent (CE) is 0.40 or over, difficulty in making crack-free welds can be expected. Steps to prevent cracking, such as preheating and the use of low-hydrogen electrodes, are suggested, as explained on page 302.

WHAT CAN BE DONE TO STEEL

Rolling is one of the most common forming methods for structural steels. Two or more hardened steel rollers press the hot metal gradually into the shape desired (Figure 19–19).

Extrusion or *drawing* pulls the metal through a die, which shapes the metal to the shape of the die (Figure 19–20). Some diameters of filler wires and electrodes are extruded.

FIGURE 19–19
Rolling.

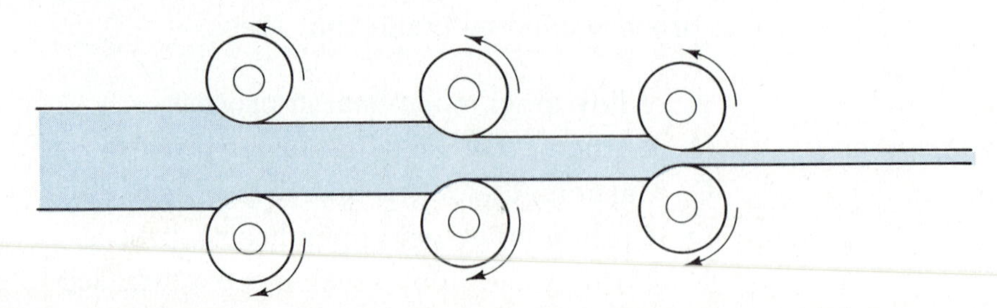

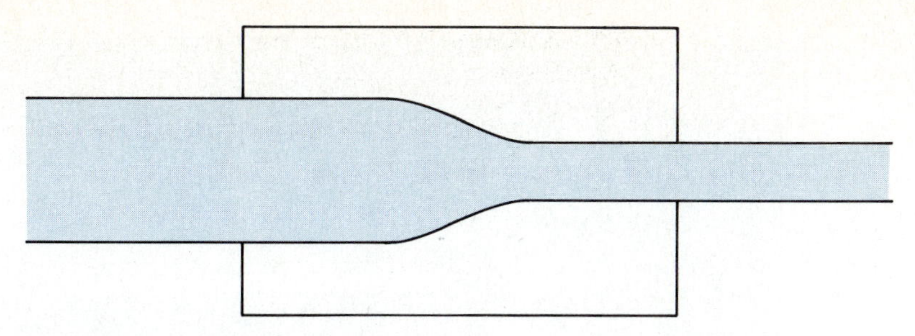

FIGURE 19–20
Drawing or extrusion.

Casting is a method of shaping steel by pouring molten steel into a mold. Upon cooling, the steel will take on the shape of the mold. Sand castings and die castings are methods used today.

Cold working includes bending or shearing steel into the desired shape. Bending can also be done a little easier if the metal is annealed.

Quenching is a method of rapid cooling of steel from above its critical temperature (about 1330–1670°F) to produce various degrees of hardness and strength.

There are three standard quenching methods:

1. A *water* quench gives the fastest cooling rate and therefore a very hard but, in many cases, a brittle structure.
2. An *oil* quench is a bit slower but will give most heat-treatable steels the required hardness.
3. An *air* quench, or *air blast* quench, is the slowest, but still hardens more than cooling in still air.

Each of these quenches forms hard, strong microstructures. The fastest cooling forms the hardest structure, called martensite. A little slower cooling will form bainite, and slow cooling forms ferrite and pearlite in steel.

Now that we have gone through the hardening of steel to form the desired microstructure, we must now adjust the hardness so that it has the exact mechanical properties we want. This process is called *tempering.* Tempering is done just below a metal's critical temperature (below 1333°F). The steel is heated, then cooled at a precise speed to produce varying degrees of softness in the steel.

Annealing is done to make steel very soft for cold working. It is accomplished by heating the steel above the critical point and allowing the steel to cool very slowly over a long period of time.

Normalizing brings grain size back to "normal" size. Welding, cold working, and some heating processes distort the grain size and give weldments undesirable mechanical properties. After a weldment has been normalized, grain size will be fairly uniform throughout. Normalizing cools steel in still air.

Stress-relief heat treatment is done for the sole purpose of relieving stresses that may have built up during welding or forming. It is always done below the critical temperature. (For building and bridge construction the maximum temperature for stress-relief heat treatment is 1200°F.)

PHOTO 19–7
A red-hot ribbon of steel emerges from U.S. Steel's continuous slab caster at Gary, Indiana, works. This is the cutoff point, where slabs are torch-cut to the desired length by remote control from the operator's booth at the upper right. Continuous casters are an energy-efficient way to bypass the teeming, ingot stripping, and soaking pit operations. (Courtesy of U.S. Steel Corporation.)

PHOTO 19–8
Steel plate emerges from the four-story-high finishing stand on the 160-in. plate mill at U.S. Steel's Texas works, near Baytown. Shown here, the arm of the x-ray thickness gage extends over the plate as one of the plate mill crew rechecks the thickness of the plate. In the background, a slab from the roughing stand moves toward the finishing stand. (Courtesy of U.S. Steel Corporation.)

PHOTO 19–9
Six strands of hot steel exit the continuous caster at U.S. Steel's Lorain, Ohio, works. The round shapes are destined to be made into pipe at the plant's seamless pipe mills. (Courtesy of U.S. Steel Corporation.)

METAL IDENTIFICATION

As we stated earlier, understanding basic metallurgical principles will help you identify unknown metals and alloys. Maintenance and repair welders may have to identify many different types of metals they will come across. Some of these materials are dirty, greasy, and difficult to identify. We cannot select a welding process, let alone the correct filler metal, if we do not know what metal we have.

Color and Appearance

One of the quickest and simplest methods of identifying a metal is by its color. The color may vary slightly depending on how smooth its surface is; for example, newly machined parts will reflect light off their surfaces better than will a rough sand-casted surface, so the rough surface will appear darker. Color will not always lead you to the exact content of a metal, but it may lead you to the category of the material, or at least eliminate what material it is not. Tables 19–4 to 19–7 will help you identify various metals and alloys by their color and appearance.

TABLE 19-4
Metal Identification by Color

METAL OR ALLOY	COLOR
Aluminum bronze	Yellow gold
Aluminum magnesium	Silver
Brass	Yellow Reddish
Bronze	Reddish Yellow
Cast Iron	Dark Silvery
Malleable iron	Dark silvery
Mild steel	Dark silvery
Monel (nickel alloy)	Silvery stainless
Nickel silver	Silver
Pewter	Silver
Stainless steel	Silver stainless
Zinc alloy (white, metal)	Silver

TABLE 19-5
Identification by Surface Appearance

METAL	FRACTURE	UNFINISHED SURFACE	NEWLY MACHINED
Alloy steel	Medium gray	Rough, dark gray	
Monel	Light gray	Smooth, dark gray	
Nickel	White	Smooth, dark gray	
Copper	Red	Red-brown, green (oxidation)	Bright red
Brass	Red-Yellow	Yellow-brown	Bright-yellow-red
Aluminum	White	Light gray	Light white
Lead	Crystal white	Smooth, white-gray	White
Wrought iron	Bright gray	Smooth, light gray	Very smooth, light gray
Low-carbon steel	Bright gray	Dark gray	Bright gray
High-carbon steel	Very light gray	Dark gray	Bright gray
Malleable iron	Dark gray	Dull gray	Light gray
Gray cast iron	Dark gray	Very dull gray (sand)	Light gray
White cast iron	Silvery crystalline	Dull gray (sand)	Difficult to machine

TABLE 19-6
Casting Identification by Appearance

SMOOTH-DIE-CASTED SURFACE	ROUGH SAND-CASTED SURFACE
Aluminum	Cast iron
Magnesium	Cast steel
Zinc and white metal	Stainless steel
	Copper, brass, and bronze

TABLE 19–7
Castings Identification by Color

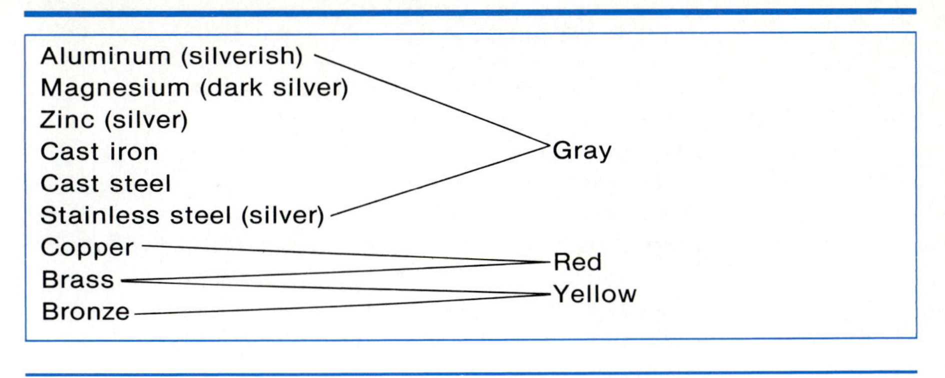

Flame Test

Another handy method for identification is the flame test. Various metals react differently when melted under an oxyacetylene flame. Table 19–8 will give you an idea of what some metals will do.

TABLE 19–8
Metal Identification by Flame Test

METAL	SPEED OF MELTING	COLOR CHANGE	SLAG ACTION	PUDDLE ACTION
Monel	Slower than steel	Red	Quiet, hard	Quiet
Nickel	Slower than steel	Red	Quiet, hard	Quiet
Copper[a]	Slow	Black-red, if any	Quiet	Bubble, solidifies slow
Brass	Medium-fast	Slightly red	White fumes	Beading, bubble with high heat
Aluminum[a]	Faster than steel	None	Quiet	Quiet
Lead	Very fast	None	Quiet	Quiet, can boil
Wrought iron	Fast	Bright red	Quiet	
Low-carbon steel	Fast	Bright red	Quiet	
High-carbon steel	Fast	Bright red	Quiet	
Malleable iron	Moderate	Dull red	Quiet, tough	
Gray, cast iron	Moderate	Dull red	Quiet, tough	
White cast iron	Moderate	Dull red	Quiet, tough	

[a] Hot-shortness characteristics.

Melting Temperature

Knowing the melting temperature of certain metals may help you isolate the unknown metal (Table 19–9). Hold the known metal(s) and the unknown metal under a flame at the same time—which melted first? If the known metal did, then the unknown metal may have a higher melting point. If your unknown melted first, the reverse would be true. Thermal conductivity may throw this test off slightly, because metals with high thermal conductivity may melt more slowly, even if they have a low melting temperature. However, in most cases this test can be helpful.

TABLE 19-9
Melting Temperatures for Common Metals

METAL OR ALLOY	° F
Aluminum and some aluminum alloys	1220
Brass	1660
Bronze	1625
Cast iron	2300
Copper	1981
Gold	1945
Iron (pure)	2780
Lead	621
Magnesium and some magnesium alloys	1240
Molybdenum	4532
Nickel	2650
Nickel copper (monel)	2400
Nickel silver	1706
Silver (pure)	1762
Steel	
Low-carbon	2700
Medium-carbon	2600
High-carbon	2500
Stainless steel	
Ferritic	2650
Martensitic	2600
Austenitic	2550
Tantalum	5162
Tin	450
Titanium	3270
Tungsten	6170
Vanadium	3182
Zinc	787
Zirconium	3090

Magnetic Properties

Magnetic properties of metals can help us distinguish ferromagnetic metals from nonferromagnetic metals (Table 19–10). The simple magnet test is used to check if a metal is magnetic. If the magnet readily sticks to the part, you probably have carbon steel. If it does not stick at all, you may have a nonferrous metal such as aluminum or copper.

Some metals may have a slight magnetic pull, such as some stainless steels. Many plated steels look like stainless steel when placed beside each other. By using the magnet test you can quickly distinguish them. The plated steel will draw the magnet, whereas the stainless will have little or no pull on the magnet.

Hardness

The hardness of a metal can be used to help identify the metal. For example, the various grades of carbon steel (mild steel, medium- and high-carbon steel) look very much alike, but their hardness when tested will tell the difference. The hardness is tested by using a hardness tester such as the Rockwell or the Brinell. If you do not have a hardness tester at your immediate disposal, a file will do just fine. Find an edge on the part that you can file without damaging it and file the edge.

TABLE 19-10
Magnetic Properties of Metals

Magnetic
- Carbon steels
- Cast iron
- Malleable iron

Slightly magnetic
- Stainless steel (ferritic)
- Stainless steel (martenistic)

Nonmagnetic
- Stainless steel (austenitic)
- Nickel and nickel alloys
- Brass, bronze, and copper alloys
- Aluminum and magnesium

TABLE 19-11
File Hardness Test

TYPE OF METAL	REACTION TO FILE	BRINELL (KG)	ROCKWELL
M/S Low carbon steel	Readily cut	143	79 (B) scale
Medium-carbon steel	Cut with moderate pressure	187	91 (B) scale
High-alloy steel	Difficult but will cut	295	31 (C) scale
High-carbon steel	Cut only with greatest effort	402	42 (C) scale
Tool steel	Nearly impossible to cut	510	51 (C) scale
Hardened tool steel	Cannot be cut with a file	603	58 (C) scale

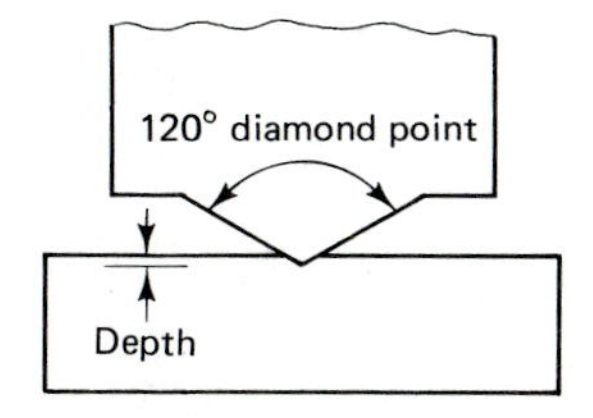

FIGURE 19-21 Rockwell hardness test.

FIGURE 19-22 Brinell hardness test.

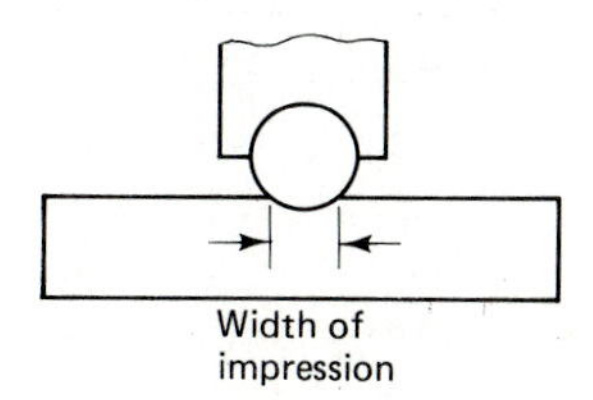

When filing the edge, use a firm and steady stroke, observing the file reaction as you cut into the piece (Table 19-11).

For very accurate hardness testing a Rockwell, Brinell, or scleroscope hardness tester is used. Metals can be fairly accurately identified if we know exactly the hardness of the metal.

In the Rockwell system a diamond-tipped cone is driven into the part; the depth of the indentation is measured (Figure 19-21). The harder the material, the less the cone will penetrate. The hardness is usually read out directly on the dial. The Rockwell system uses two systems, the C scale for *hard* metals and the B scale for the *softer* metals. A metal ball instead of a cone is used with the B scale.

The *Brinell* hardness testing process uses a hardened steel ball. The ball falls into the test area at a given load (Figure 19-22). The diameter of the impression left by the ball is then measured (usually with the aid of a magnified diameter-measuring gage). The larger the diameter, the softer the test material.

The scleroscope measures hardness by dropping a weighted hammer down on the test material and measures the height to which it rebounds (Figure 19-23). Although this may be a slightly less accurate test, it is quick, easy, and usually portable.

The Rockwell, Brinell, and scleroscope hardness scales are listed in Table 19-12.

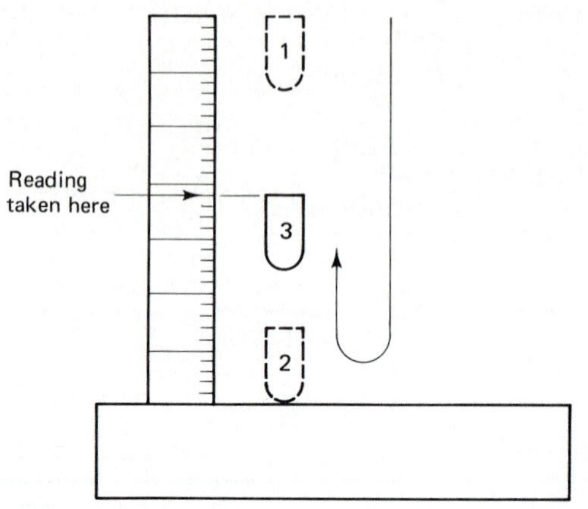

FIGURE 19-23
Scleroscope hardness test.

TABLE 19-12
Hardness Scales

ROCKWELL C	ROCKWELL B	BRINELL SCALE	SCLEROSCOPE SCALE		TYPICAL MATERIAL HARDNESS
84					Diamond
77				Harder ↑	Topaz
75					Tungsten
70					Quartz
66		12	93		
62		657	86		
58		603	81		Hardened tool steel
55		562	75		
51		510	71		Tool steel
50		497	68		Crankshaft
44		426	60		
43		415	58		White cast iron
42		402	56		High-carbon steel
38		359	51		Chromium
35		330	46		Manganese
31		295	42		High-alloy steel
25		252	35		302 stainless steel
24	100	245	34		
22	99	233	32		
12	91	187	27		Medium-carbon steel
10	89	179	25		Gray cast iron
8	87	170	24		
6	85	163	23	↓ Softer	
2	81	149	22		
1	80	146			
0	79	143	21		Mild steel

Weight

Observing the weight of a metal will also help you to determine what the material is. Metals such as aluminum and stainless steel look alike, but stainless steel will be observed to be much heavier than aluminum. When using this system, pick up various known metals that are about the same size as your unknown. Compare these metals to the ones listed on Table 19–13. Use the process of elimination to determine what metal or alloy you might have.

TABLE 19–13
Metal Identification by Weight

Alloy	Specific Gravity[a]	
Aluminum magnesium	2.0–2.5	Lighter ↑
Aluminum alloy	2.7–3.0	
Zinc alloy (white metal)	6.7–6.8	
Pewter	7.0–7.6	
Cast iron	7.0–7.6	
Malleable iron	7.6	
Mild steel	7.6– .8	
Stainless steel	7.6–7.9	
Aluminum bronze	7.5–8.2	
Brass	8.7–8.8	↓
Nickel silver	8.7–8.8	Heavier
Bronze	8.8–8.9	
Monel (nickel alloy)	8.9	

[a]To find the specific gravity, weigh the object in air, then weight submerged in water. Find the difference between the two, then use this formula:

$$\frac{\text{air weight}}{\text{air weight minus water weight}} = \text{specific gravity}$$

Spark Test

One of the most popular systems of metal identification is the *spark test* (Figure 19–24). The material is ground on a pedestal grinder and the spark stream is observed. This system takes a carefully trained eye, and until you become well seasoned at spark testing, you may find it difficult. The key is experimenting and practicing with known metals first. It is also a good idea to have a known metal there such as mild steel, to compare the spark stream of your unknown metal. *Caution:* Do not take your eyes off what you are doing while watching the spark stream, as grinders can grind skin off very fast. Also, never grind or use the spark test on metals that are nonferrous, as they will build up in the wheel. This buildup of nonferrous metal in the grinding wheel will cause an imbalance and can make the wheel explode.

Spectrograph Test

Other, more sophisticated systems are used to identify metals if extreme accuracy is required. A *spectrograph* burns a portion of the material and analyzes the light spectrum it produces. Every material produces its own unique light spectrum, and by analyzing these spectrums we can tell exactly what a metal consists of. Most spectrographs today

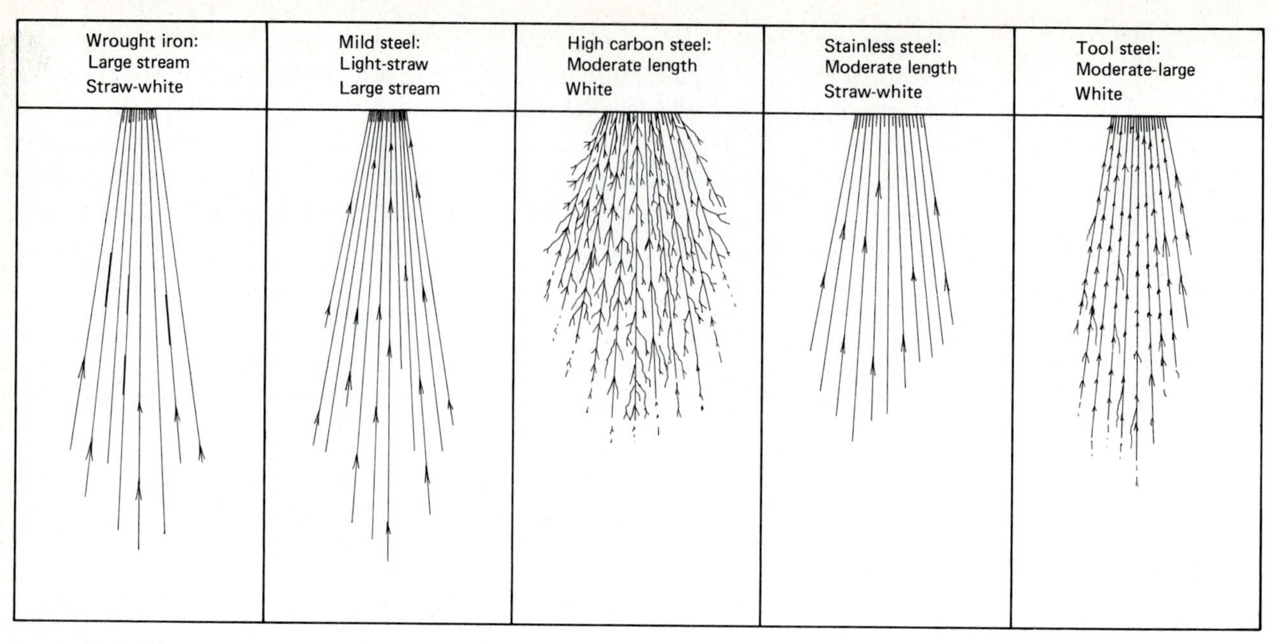

FIGURE 19-24
Spark test.

are very highly sophisticated. Most spectrographs have a computer printout or a computer screen indicating the exact elements that a metal or alloy contains.

ALUMINUM

One of the most amazing metals used in industry today is aluminum. Aluminum products include everything from aircraft and space vehicles to food wraps and pop cans. Industries such as the automotive industry have turned to aluminum to decrease the weight and increase the fuel efficiency of cars. Air travel and space travel would be much different than we know it today if it were not for aluminum.

Aluminum is easily shaped by rolling, extrusion, and casting. It is welded by the GTAW, PAW, GMAW, and some limited applications of SMAW and OFW. The highest-quality aluminum welds are achieved with GTAW and PAW, while the fastest method of welding aluminum is the GMAW process. (See the individual welding processes for the exact aluminum welding procedures.)

The production of aluminum is somewhat complex, but the key operation in extracting aluminum from its ore (bauxite) is the electrolysis process. Recycling aluminum is much easier than extracting it from its ore. The used aluminum must simply be remelted, alloyed, and deoxidized before rolling, casting, or extruding shapes.

Your ability to weld this metal of the future will certainly be an advantage for you.

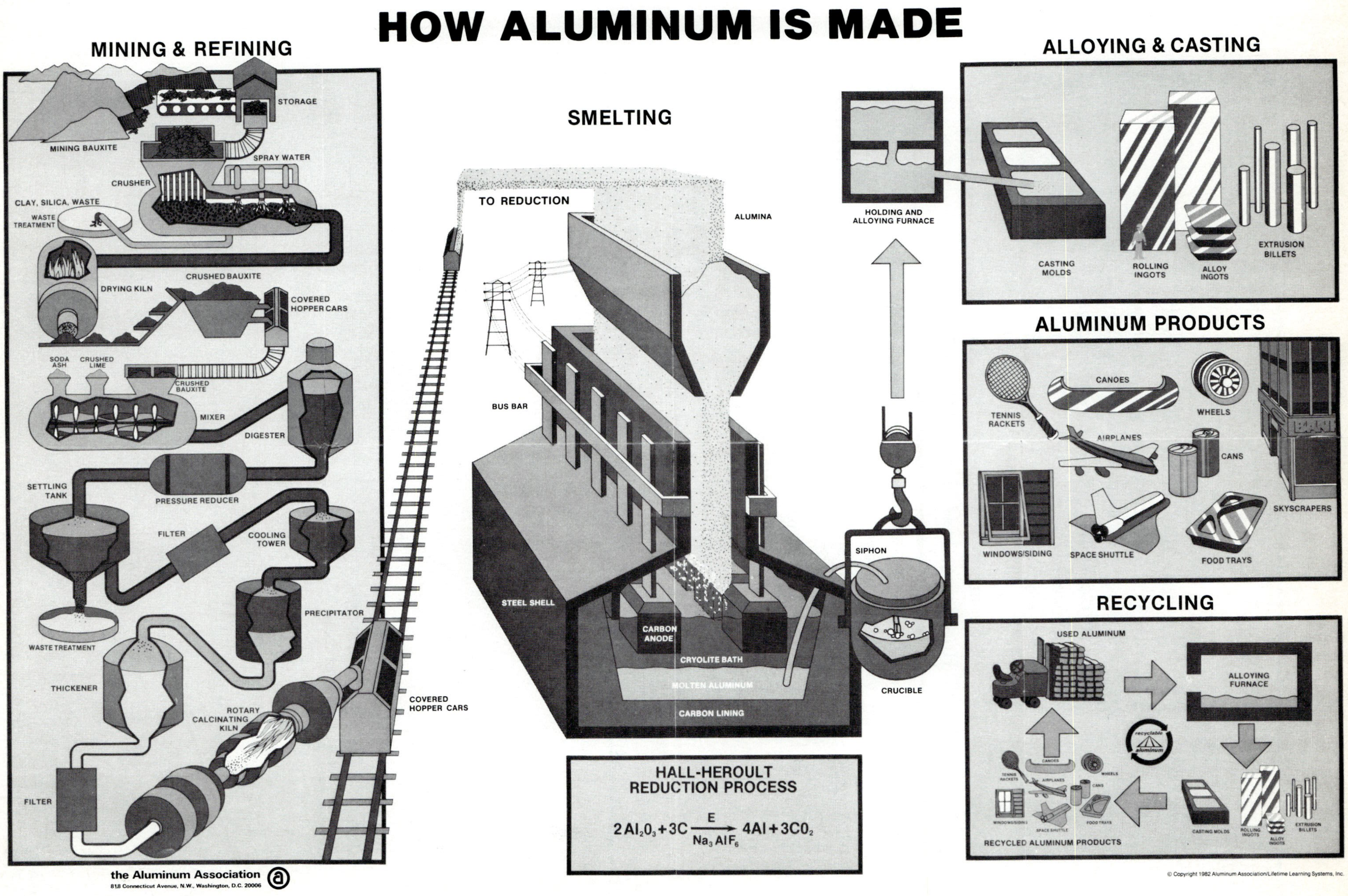

HOW ALUMINUM IS MADE

MINING & REFINING

MINING BAUXITE
STORAGE
SPRAY WATER
CRUSHER
CLAY, SILICA, WASTE
WASTE TREATMENT
DRYING KILN
CRUSHED BAUXITE
COVERED HOPPER CARS
SODA ASH
CRUSHED LIME
CRUSHED BAUXITE
MIXER
DIGESTER
SETTLING TANK
PRESSURE REDUCER
FILTER
COOLING TOWER
PRECIPITATOR
WASTE TREATMENT
THICKENER
ROTARY CALCINATING KILN
FILTER

the Aluminum Association
818 Connecticut Avenue, N.W., Washington, D.C. 20006

COVERED HOPPER CARS

SMELTING

TO REDUCTION
ALUMINA
BUS BAR
STEEL SHELL
CARBON ANODE
CRYOLITE BATH
MOLTEN ALUMINUM
CARBON LINING
SIPHON
CRUCIBLE
HOLDING AND ALLOYING FURNACE

HALL-HEROULT REDUCTION PROCESS

$$2Al_2O_3 + 3C \xrightarrow[Na_3AlF_6]{E} 4Al + 3CO_2$$

ALLOYING & CASTING

CASTING MOLDS
ROLLING INGOTS
ALLOY INGOTS
EXTRUSION BILLETS

ALUMINUM PRODUCTS

TENNIS RACKETS
CANOES
WHEELS
AIRPLANES
CANS
SKYSCRAPERS
WINDOWS/SIDING
SPACE SHUTTLE
FOOD TRAYS

RECYCLING

USED ALUMINUM
ALLOYING FURNACE
recyclable aluminum
CANOES
WHEELS
TENNIS RACKETS
AIRPLANES
CANS
WINDOWS/SIDING
SPACE SHUTTLE
FOOD TRAYS
CASTING MOLDS
ROLLING INGOTS
ALLOY INGOTS
EXTRUSION BILLETS
RECYCLED ALUMINUM PRODUCTS

PHOTO 19–10
Flowchart of aluminum making. (Courtesy of the Aluminum Association.)

REVIEW QUESTIONS

1. Of what is steel made up?
2. What effect does carbon have on steel?
3. What is the effect of adding chromium to steel?
4. What might be considered an impurity in steel?
5. What elements are considered deoxidizers?
6. What would be considered a physical property of steel?
7. What mechanical property accounts for most failure of equipment?
8. What happens at a material's elastic limit?
9. List some applications of high-carbon steel.
10. How do you weld high-carbon steels?
11. Where are stainless steels used?
12. What problems are encountered when welding stainless steels? How are these problems overcome?
13. What is the approximate chemical content of AISI/SAE class 4130 alloy?
14. What is an alloy's carbon equivalent?
15. What is the difference between annealing and normalizing?
16. List some ways of identifying metals.
17. What are some applications of aluminum?
18. By what processes can aluminum be welded and cut?

WELDING CAST IRON

One of the handiest skills that you can learn is how to weld cast iron. Considering that we produce about 18 million tons of cast iron products in the United States each year, your chances of having to repair cast iron are quite good.

COMPOSITION OF CAST IRON

There are many different types of cast irons. Changing the elements put into the cast iron will change the type, but changing the cooling rate can also produce a completely new cast iron. Cast iron is made up mostly of iron, with carbon and silicon added. You have probably noticed that it is just about the same mixture that makes up steel, but the key is the amount of carbon added. To be classified as cast iron, it must have at least 1.75% carbon or greater. Many cast irons have greater amounts of carbon and additional elements. Here is a list of what you will find in most cast irons:

- *Iron:* basic ingredient
- *Carbon:* (1.75%+) lowers the melting temperature and forms graphite, which improves machineability

- *Silicon:* deoxidizes (removes air) and increases fluidity
- *Manganese:* also deoxidizes the "cast"
- *Chromium:* adds hardness and helps fight corrosion
- *Nickel:* improves toughness and also helps fight corrosion
- *Cerium:* helps turn carbon into graphite flakes
- *Calcium silicide:* also helps turn carbon into graphite flakes
- *Aluminum:* helps deoxidize
- *Sulfur:* actually an impurity, but it does help machining
- *Phosphorus:* mainly an impurity

FIGURE 20-1 Possible cast combinations.

FIGURE 20-2 Classifications for possible types of cast irons.

White
Gray
Malleable
Ductile
High alloy

Remember, there are thousands of different types of cast iron, because we can take the same mixture, alter the cooling rate slightly, and we have a brand-new cast iron. This fact makes thousands of combinations possible.

Each dot in Figure 20-1 represents a possible "cast" combination. Incredible! But fear not, we have broken them down into general categories to make it a little easier (Figure 20-2). This looks much better!

But we still must realize that when it comes to welding one type of gray cast iron, for example, it may not act the same as another gray cast iron. This is because there are so many combinations in each category. Let's look specifically at each general class of cast iron.

CLASSES OF CAST IRON

White cast iron is hard, brittle, and one of the most difficult to weld. It is used for things like rolling mills, rollers, and break shoes. When white cast iron is produced, there is no slow cooling after it has been poured into its mold, so it is fairly cheap but not very strong. It will actually appear white-silver and crystal-like. By all means, try to braze this metal first. Your chances for success are much greater than they would be trying to arc-weld white cast iron.

Gray cast iron is a good strong wear-resistant "iron." It is used in engine blocks, gears, and pulleys. When it is poured, it is slow cooled about 18 to 36 hours. If the precautions we are going to mention in a few minutes are followed, you will probably have good success welding "gray."

Malleable cast iron, as the name implies, is actually quite malleable for a cast iron. It will absorb medium torsional forces placed on it. It is used for auto rear ends (differentials) and conveyor components. When it it poured, it is held above 1600°F for about 6 days and then slow cooled. It has a good tensile strength of about 60,000 to 100,000 psi. It is also quite weldable if precautions are observed.

Ductile cast iron, also known as nodular and meehanite, has fairly good ductility and is shock resistant. It is used on heavy equipment, farm equipment, and machine tools. It is similar to gray iron but has magnesium and cerium in it, and a special pouring technique is used that controls cooling. Its tensile strength can reach as high as 150,000 psi.

High-alloy cast irons can be one of many types. They are usually full of nickel, chromium, molybdenum, aluminum, silicon, and other alloying elements. Depending on what the cast was designed for, they usually exhibit very good mechanical properties and high tensile strength. Of course, they are usually quite expensive, too. Fortunately, they are usually quite weldable. You may find them used as pump impellers and housings or any high-corrosion application.

PREPARING TO WELD CAST IRON

Carbon loves heat, and cast irons are full of carbon. You will remember from Chapter 19 that carbon also makes metals brittle and welds crack. So the more heat you apply to an isolated area, the more carbon it will absorb. The key here is going to be keeping the heat down or at least well dispersed—we will show you how in just a few minutes.

Second, cast irons are hard and brittle (although many have good compressive strength and abrasion resistance). So internal stresses will build rapidly from any heat input; the result is that small cracks can become large ones fast once heat is applied. The key here is that precautions must be taken to stop further cracking if you are repairing a crack. With these items in mind, here is the procedure for fixing your "cast."

Identification

Is it cast iron or cast steel? Surprisingly, some freshly broken cast steel sections look just like cast iron. Rub your finger over the freshly broken area. If it comes away clean, it may be cast steel. If you feel a gritty graphite texture and your finger now looks dirty, it is cast iron. Once we have decided that it is cast iron, try to identify the type. If you know what the casting was used for, that should help you identify it. Check some of the applications we mentioned earlier. Knowing what type you have and how weldable it is may let you know what you are in for and how careful you are going to have to be.

Process Selection

Brazing can be the answer to many cast iron jobs. Using this low-temperature process can reduce your worries of too much heat and high stresses and the resulting cracks. Be aware, though, that castings that are exposed to high-temperature or high-stress conditions should not be brazed. (Engine blocks, exhaust manifolds, and so on, should not be brazed.) It may also require large volumes of oxyfuel heat to get a large casting up to brazing temperature. If you do decide that your project can be brazed, you should follow the same joint preparation procedure and preheat, described for arc welding of the cast.

Joint Preparation

Joint preparation is extremely important whether arc welding or brazing cast iron. For cracks or joining separate parts, you should "vee" out the entire joint. Grinding or air carbon arc cutting will work fine. Be sure to vee it completely. If any portion of the joint is not completely fused when you are done, it will simply be an area of high stress concentration and re-break. So take the time to prepare the joint correctly (Figure 20–3).

If the area to be repaired is a crack, in addition to vee preparation, you should drill a hole a both extremes of the crack (Figure 20–4). This will disperse the stress from welding heat over a larger surface area, and prevent the crack from spreading further. Make sure that you have found the end of the crack. If a small hairline crack extends past your drilled hole, your efforts will have been wasted. You can ascertain the end of a crack by using an acid etch or dye-penetrant test (see Chapters 26 and 27). The holes should be about the same diameter as the thickness of the part, but not exceed $\frac{1}{2}$-in. diameter. Make sure that you drill the holes first before you apply any heat to the casting (such as with the air carbon arc). If the hole is large enough that filling the hole with weld may be difficult, plug it with a bolt and cut the bolt off flush with the casting surface.

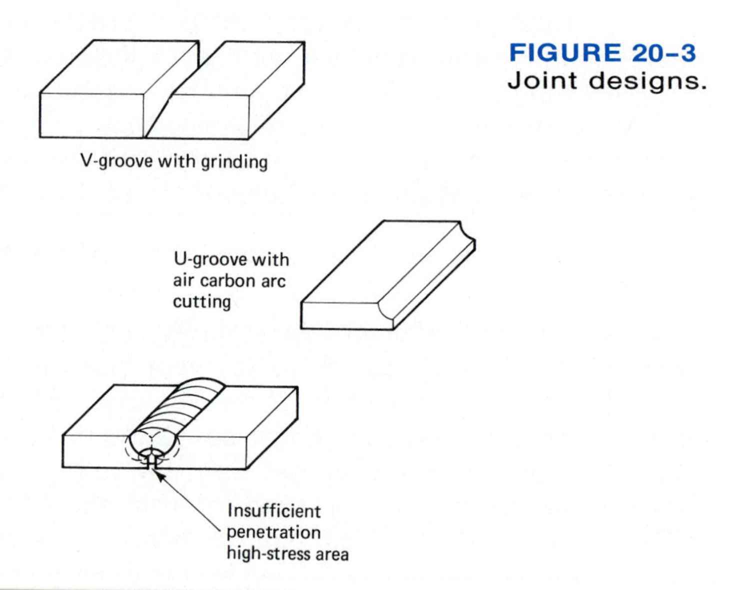

FIGURE 20–3
Joint designs.

FIGURE 20–4
Crack repair preparation.

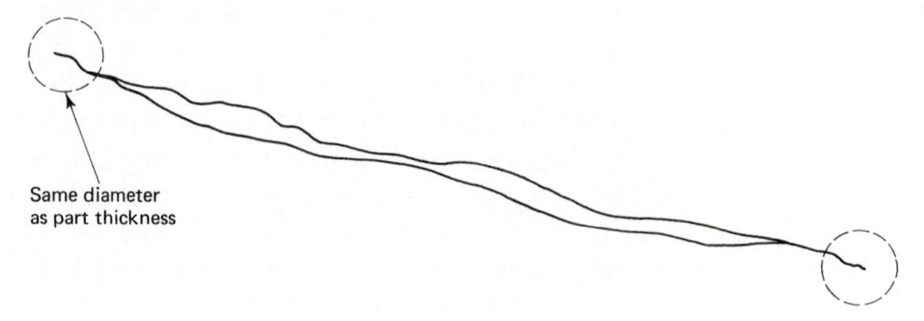

Layout Procedure

You are now ready for the layout procedure. The key here is to keep the heat from welding, well dispersed, and not let it build up in any one area. Using soapstone or other metal-marking device, draw short lines perpendicular to the joint or crack; keep them about 2 in. apart (Figure 20-5a). Now number every other space and ends last (Figure 20-5b). This is your welding sequence—but do not start welding just yet.

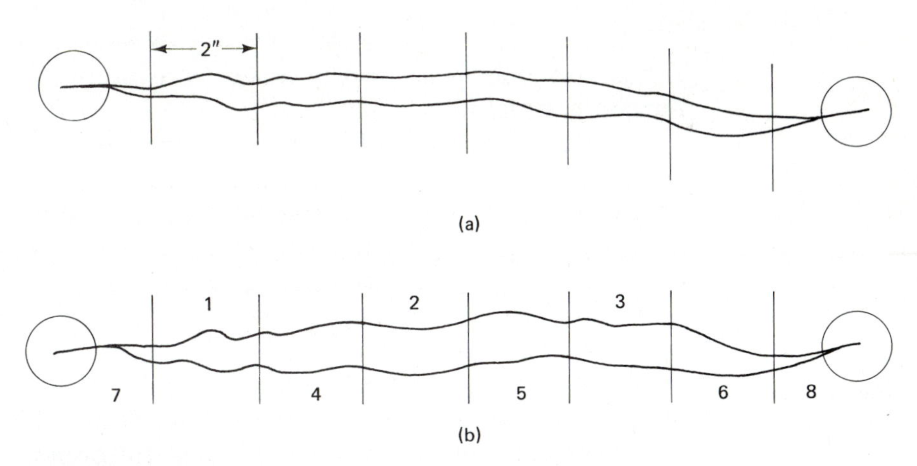

FIGURE 20-5
Crack repair layout.

Preheating

The oxyfuel torch works fine, but if you happen to have a heat-treating oven available, that is great! Use it to preheat. Most welders do not have one readily available for their use, so we will discuss the oxyfuel torch method. Using a large tip, heat a wide area around the repair area (at least 3 to 5 in. radius around the broken area). Do not get too carried away; we only want the cast to reach about 175 to 225°F. If you have a 200°F temperature crayon, that will be fine. If you have no way to measure the casting temperature, try the "spit and roll" technique. Just drop a little moisture on the surface; if it balls up and rolls off, you are just about at the temperature you want. If you do have a heat-treat oven at your disposal, soak the casting at about 200°F. Make sure that your helmet, machine, and electrodes are right nearby and ready to go when you start preheating. If you waste time gathering equipment after preheating, your cast will be cooling.

Welding

Once preheated, start welding immediately. Weld 2 in. at a time in the stagger-step sequence you laid out. Run the electrodes within the recommended amperage range on the electrode container. Upon completing each 2-in. deposit, immediately start peening the weld with a ball peen hammer. Peening sends shock waves into the casting and helps relieve stress (Figure 20-6). When you start peening, the weld is still hot and soft, so the hammer will fall dead as it strikes the weld. When the weld starts to cool, it will cause the hammer to "spring back."

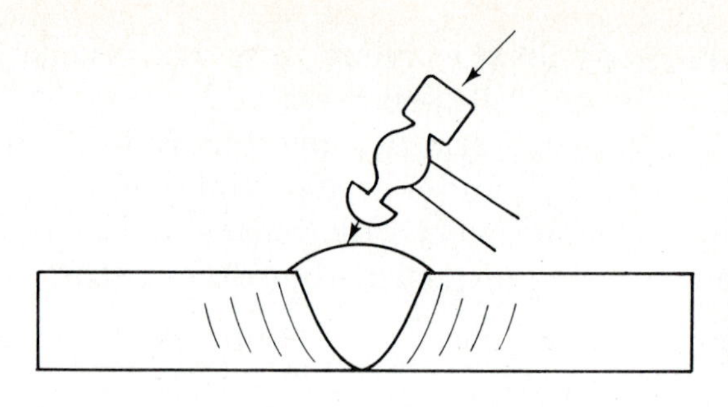

FIGURE 20-6
Peening produces shock waves.

When this happens, *stop peening;* continuing to peen will only work-harden the weld.

Continue to weld in the numerical sequence that you have laid out, peening each 2-in. section. After you have completed three to four 2-in. sections, or if the casting seems to be radiating a lot of heat, stop and let the casting cool a bit. Only let the temperature fall back to about 200°F. Use your temperature crayon to check. When the original 200°F preheat temperature is reached, restart welding and peening.

When welding is complete, we must slow-cool the weld. Slow cooling can be accomplished by immersing the cast in sand or lime, or simply by covering the cast with a nonflammable blanket. Do not remove the cast until it has cooled back to room temperature.

ELECTRODES FOR CAST IRON

Your first choice in the selection of an electrode should be the nickel-type cast iron electrodes. There are special AWS classes of SMAW electrodes exclusively for welding cast irons (AWS Section A5.15). These should be your first choice for arc welding of cast iron. These electrodes contain nickel because of its soft and ductile deposit. Some also contain nickel and copper and iron, which have an improved fluidity.

E NIFE CI

Electrode

Alloy: nickel and iron in this case

Cast iron electrode application (an "A" or "B" indicates that the electrode can also be used for other applications)

FIGURE 20-7
Classification system for cast iron electrodes.

Figure 20-7 shows how the cast iron electrode classification works. The high-nickel electrodes (ENI-CI) work well and are easily machined after welding, whereas the nickel-iron electrodes (ENIFE-CI) are the best choice where high strength is required. The smoothest-flowing cast electrode is the nickel-copper class (ENICU-A and ENICU-B).

When cast iron SMAW electrodes are not available, the next best choice is high-nickel stainless steel electrodes (E310-16) or low-hydrogen electrodes. Your chances for success may be a bit lower with these electrodes, but for many weldable cast iron jobs, they have been known to work fine.

Things to remember when welding cast iron:

1. Never weave the electrode, stringer beads only.
2. Do not let the cast iron get too hot while welding.
3. Do not let the cast iron fall below the preheat temperature while welding.

FIGURE 20-8 Classification system for OFW filler metals for cast iron.

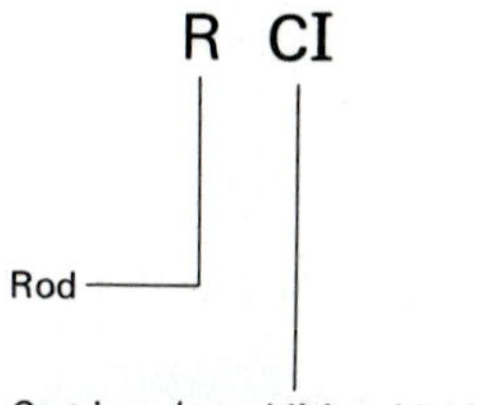

4. Slow-cool the cast when you are finished. Never quench a cast iron.
5. Do not peen too long.
6. Do not get too frustrated if you fail; many cast irons just do not weld.

Up until now we have examined only SMAW of cast iron, but cast iron can also be oxyfuel *fusion* welded. There are special AWS-classified filler rods specifically for this purpose, and they are themselves "cast iron." Although many people prefer the SMAW system because of strength requirements, good results can be obtained with the OFW method. The puddle can be somewhat difficult to see, but the filler is dipped in the same manner as in OFW of mild stee. Figure 20-8 shows the classification of OFW filler metals for cast iron.

REVIEW QUESTIONS

1. How many types of cast iron are there?
2. How much carbon must cast iron contain to be classified as cast iron?
3. How difficult is white cast iron to weld compared to other cast irons?
4. What happens to cast iron when heated?
5. How should you repair a cast that is not under high stress or high heat?
6. When repairing a cast iron crack, what should be done before welding?
7. How can you find a small crack that may not be visible to the eye?
8. What is the recommended preheat for cast iron?
9. What is peening?
10. How long should peening be performed after the weld is completed?

DRAWINGS AND BLUEPRINTS

Detail drawings and blueprints are the language of the industry; they are the methods of conveying the ideas and calculations of the designer or engineer to welders and fitters, who must in turn fabricate or assemble the product. Welding technicians and inspectors must also be blueprint and welding symbol experts if they are to be effective and respected in their positions.

We cannot overstress the need for understanding this extremely important area of the welding industry. In some circumstances the welder may be required to carry out every operation involved in fabricating the entire part from start to finish. However, in some company systems and policies, the welder may be concerned only with the actual welding operations involved with construction of the part. In other words, another department might be concerned with machining, grinding, or drilling or other operations in manufacturing the part.

Blueprints are often called whiteprints because white background paper is now used. The drawing starts on paper known as drawing vellum. This paper is thin and allows light to pass through it. It looks something like tracing paper. The final draft is then run through a machine known as a diazo that will produce copies of the original in fine detail.

TYPES OF PROJECTIONS

Prints are usually displayed in one of three types of basic projections:

1. isometric
2. pictorial
3. orthographic

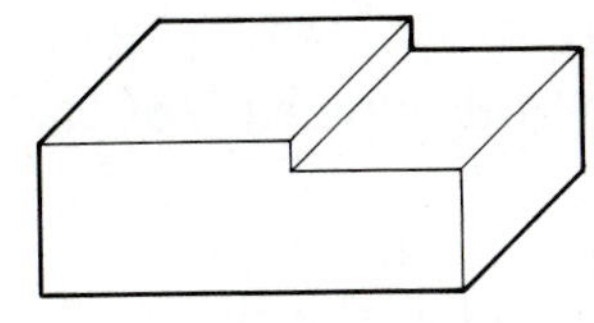
(a) Pictorial view

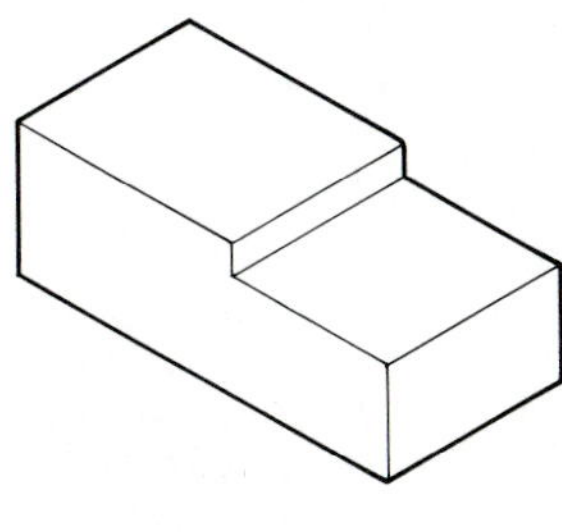
(b) Isometric

FIGURE 21-1 Isometric and pictorial views.

Pictorial or oblique projections (Figure 21-1a) are drawn on a 45° angle. They give the reader a three-sided, three-dimensional view of the object. The front of the object is viewed straight in at no angle. The other two views are projected at a 45° angle from the front view. These types of projections are not as popular because they do not give an equal presentation of all three sides.

Isometric projections (Figure 21-1b) are drawn on a 30° angle, using the 30°-60°-90° triangle. Isometrics give the reader a three-sided, three-dimensional view of the object, but unlike pictorial projections, isometric give an equal view of each of the three sides. It is very commonly used where three-dimensional views are needed to show the part better. Variations of isometric projections are the exploded views. They are very commonly used on products with many parts to illustrate the sequence of their assembly.

Orthographic projections (Figure 21-2) are by far the most commonly used in the welding fabrication industry. They give the reader a straight-in view of the object. No angular views are used with orthographic. Orthographic presentations can use as many views as necessary to facilitate fabrication of the part. Each view is rotated 90° from the view next to it. Views must also be correctly lined up and parallel with each other. The front, top, and right-side views are the most commonly used for orthographic projections. But many views, such as section, auxiliary, and detail views, may be used. The key is to include as many views as necessary to allow production of the part. In some cases

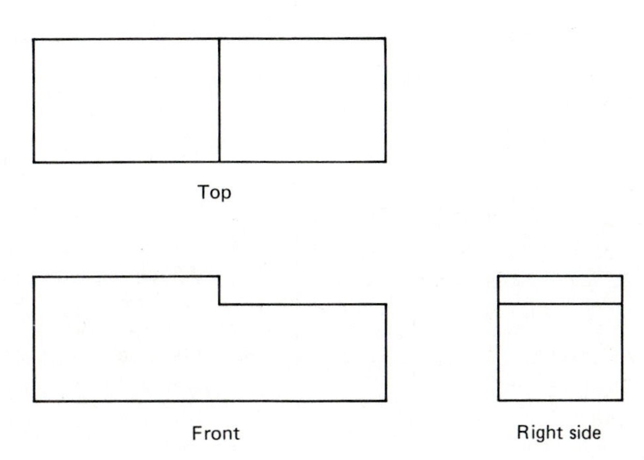

FIGURE 21-2 Orthographic projection.

this can require thousands of drawings. Imagine the drawings that are necessary in designing all of the components on an automobile or modern aircraft. Remember, many of these components are welded, and you must understand these drawings if you plan to work in this high-tech industry.

When first trying to make sense of a complex print, it is important to note that the front view should give you the best view of the actual shape of the object. Each neighboring view will be at a 90° rotation exposing that corresponding side (Figure 21-3). The reader almost always pictures in his or her mind what the object looks like, or imagines the object as if it were setting right before him or her. This is a bit difficult, being that orthographic projections show only one side of the object at a time. It just takes practice.

The next step is to read all notes and specifications (Figure 21-4). These may contain important information about how the part is to

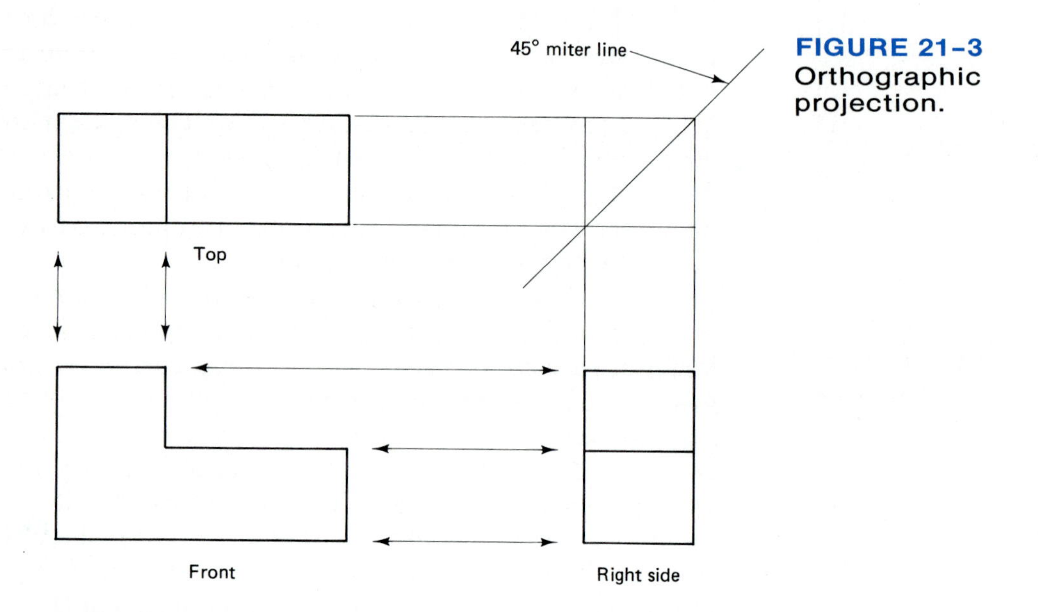

FIGURE 21-3 Orthographic projection.

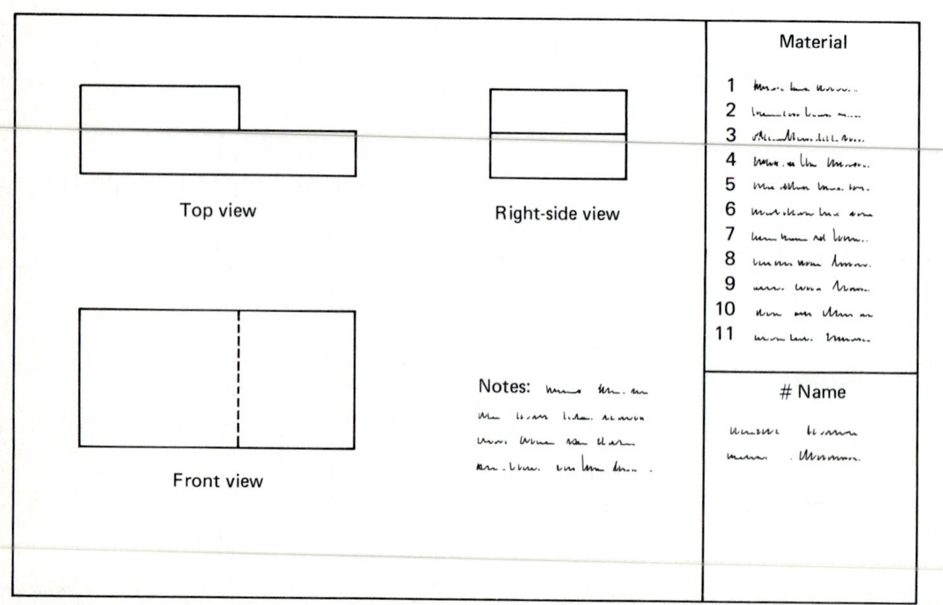

FIGURE 21-4 Orthographic projection layout.

be fabricated. The name of the part may also give you a clue as to the shape and the ultimate application of the part. For example, if a print is labeled "conveyor frame" or "engine mount," you may have seen similar items before. Immediately, you start picturing these parts in your mind. This will help you to identify your print.

BLUEPRINT LINES AND SYMBOLS

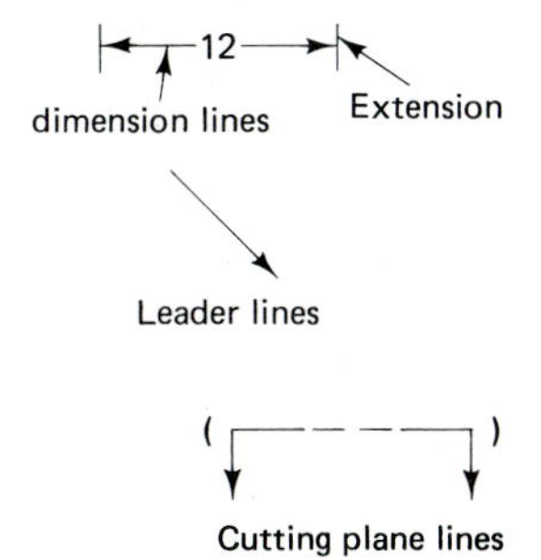

These lines outline all visible edge surfaces, showing the shape of the part. They are the heavyweight lines on the print.

These lines represent edge surfaces that are not visible, but are behind the surface in the view. They are medium-weight lines.

These lines cross through the center of drilled holes or show the center of anything put on the part (Figure 21-5). They are intended for dimensioning the exact center location of holes and prints. These are medium-weight lines. Notice the "long-short-long" sequence.

Dimension and extension lines are used for showing the exact extent and location of a dimension.

These lines are used to point out dimensions, radiuses, diameters, or any notes that refer to a specific area on the part (Figure 21-6).

These lines show where a section view will be exposed. The use of these will be clearer later in the chapter when we discuss section views in detail. They are heavyweight lines with a long-short-long sequence, with arrows pointing in the direction of the section-view exposure (Figure 21-7).

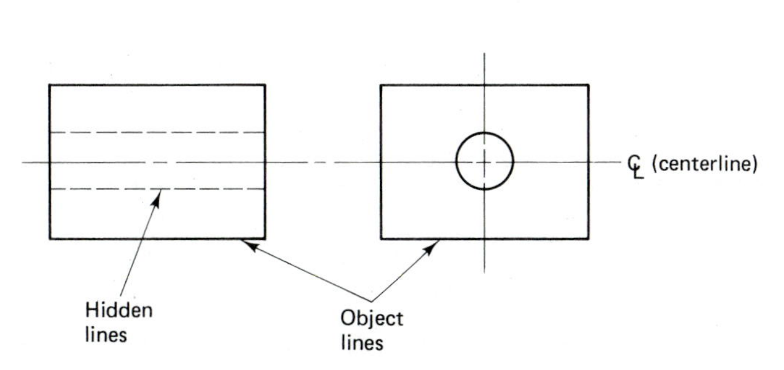

FIGURE 21-5
Hidden, object, and center lines.

FIGURE 21-6
Leader, dimension, and extension lines.

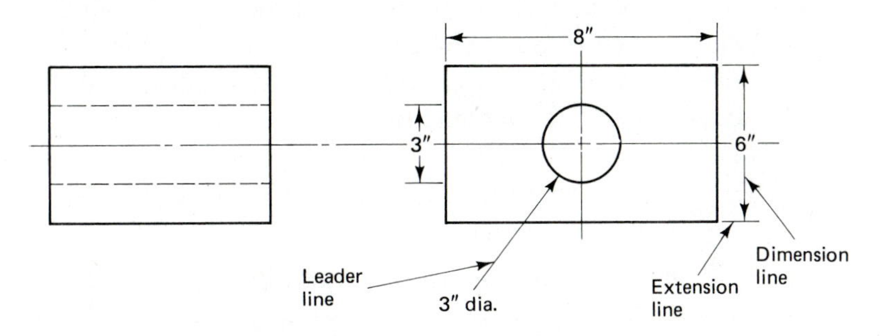

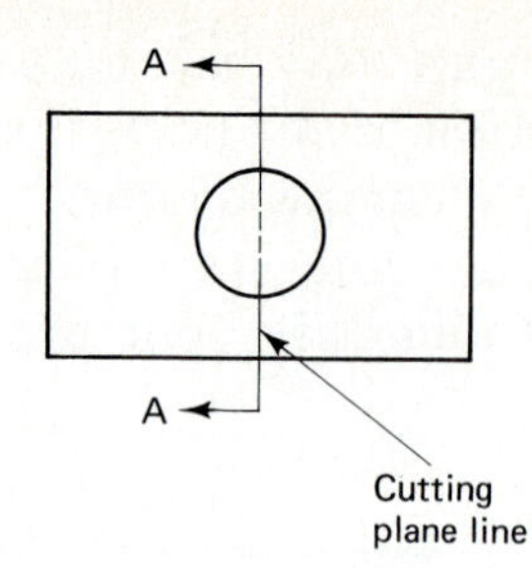

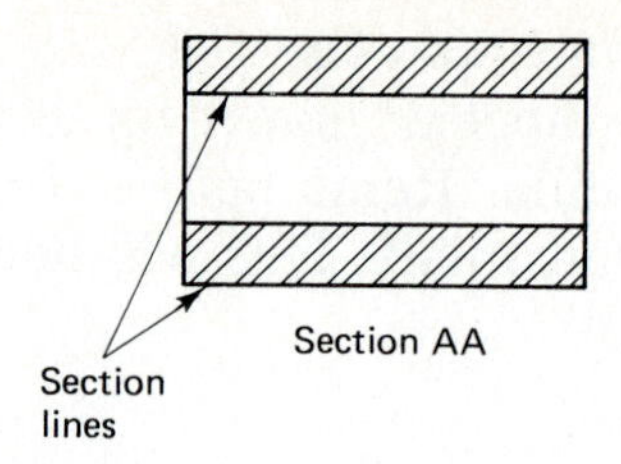

FIGURE 21-7
Cutting plane and section lines.

Short break lines

These lines are used to show an imaginary break, so a revolved section view or other data can be inserted between the break lines. They are drawn as medium-weight lines.

Long break lines

These lines are used to conserve paper when showing a long part that is the same shape over its length. They are drawn as medium-weight lines.

Section lines are used in section views to show the content of the material.

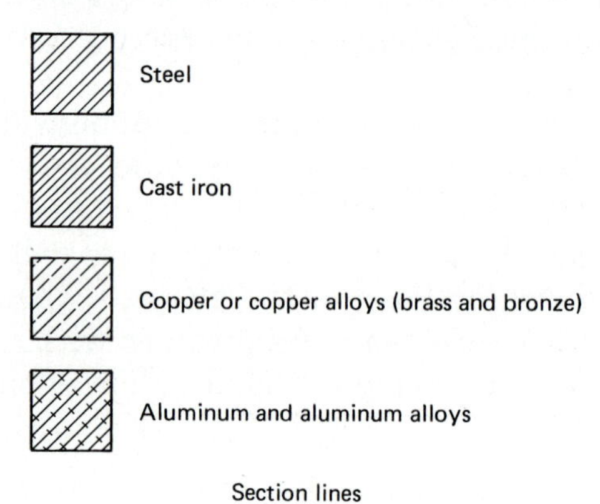

MAKING DETAIL DRAWINGS

A very effective way to learn to read prints is to learn to make them yourself. We do not intend to make you a design engineer or even a draftsperson, but some basic drafting skills will help you learn to solve complicated prints, and you will find the skill handy in many areas of your work.

Notice that the block shown in Figure 21-8 is drawn in isometric. Try to redraw the block in orthographic. Your drawing should look like Figure 21-2. The following equipment will be helpful to make your drawings look professional.

- 45° triangle and 30°-60°-90° triangles
- Graph paper or typing paper
- Lead holder (mechanical pencil) or pencil
- Erasers and erasing shield

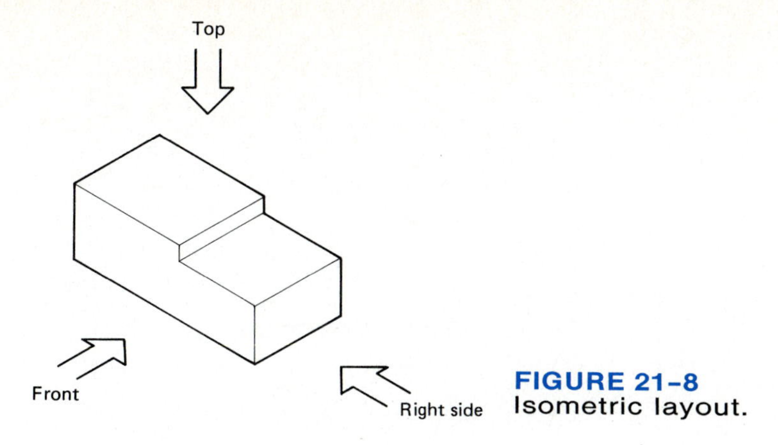

FIGURE 21-8
Isometric layout.

- Elliptical templates (optional)
- Circle templates (optional)
- Other templates you might have
- Drafting board and T-square (optional)

If you do not have all of this drawing equipment at your immediate disposal, that is okay; a sheet of paper and a triangle are enough to start with.

Starting Your Drawing

Decide which side of the object gives the best overall shape. Start with this view, and call it the front view. Do not be afraid to get started. Start putting lines on your paper; you can always erase. Work from the front view, extending lines to the top and right-side views. Keep all surfaces and edges in line from view to view (Figure 21-9).

FIGURE 21-9
Orthographic layout procedure.

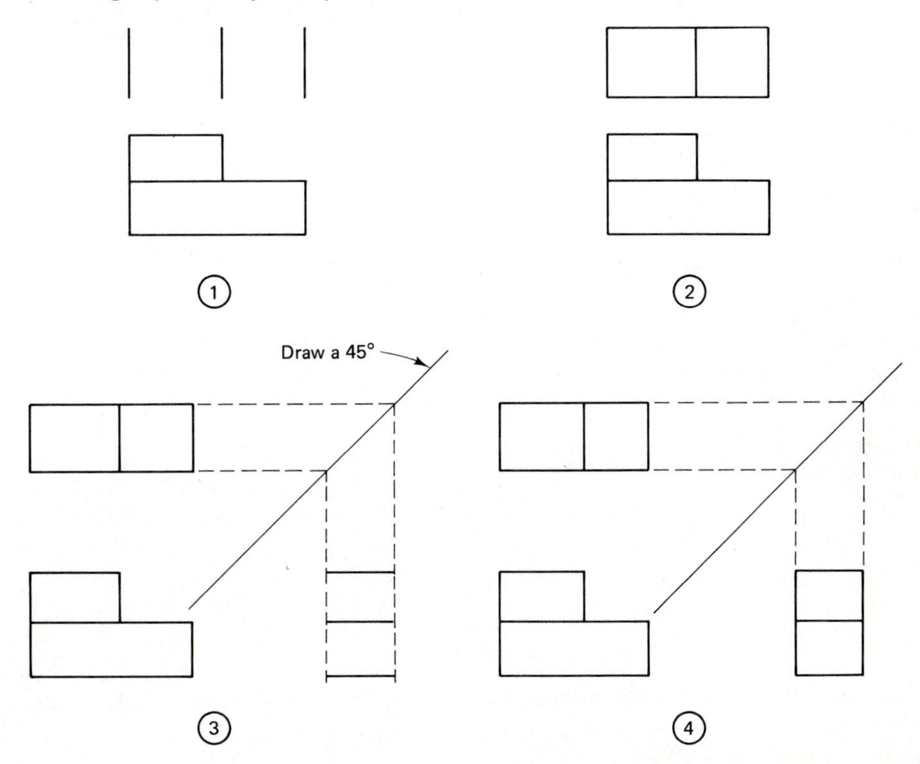

FIGURE 21-10a Isometric to orthographic drawing practice.

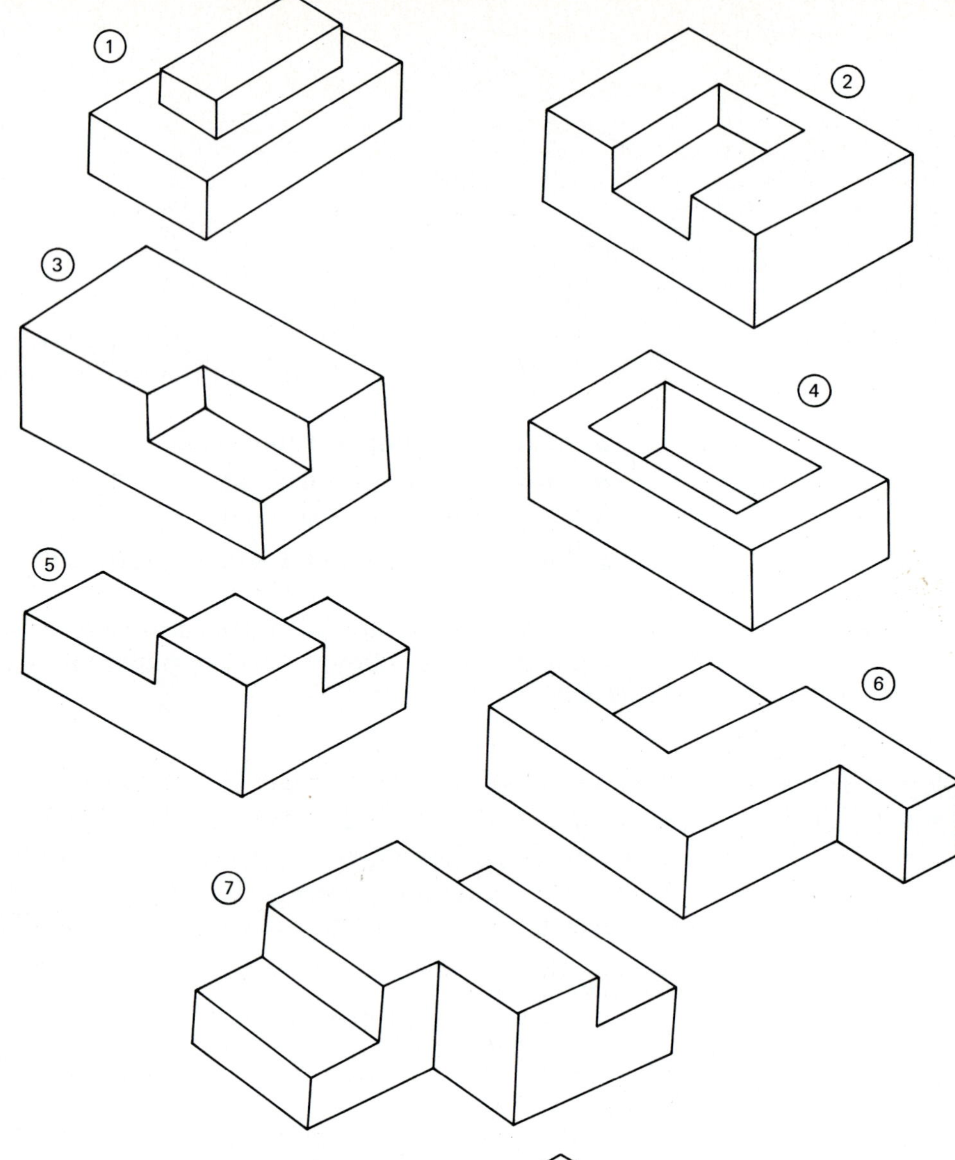

FIGURE 21-10b Drawing practice. Draw an orthographic: add any and all information needed to fabricate the part, including notes, specs, bill of mat., leader lines with drilled hole dimensions, and spec box.

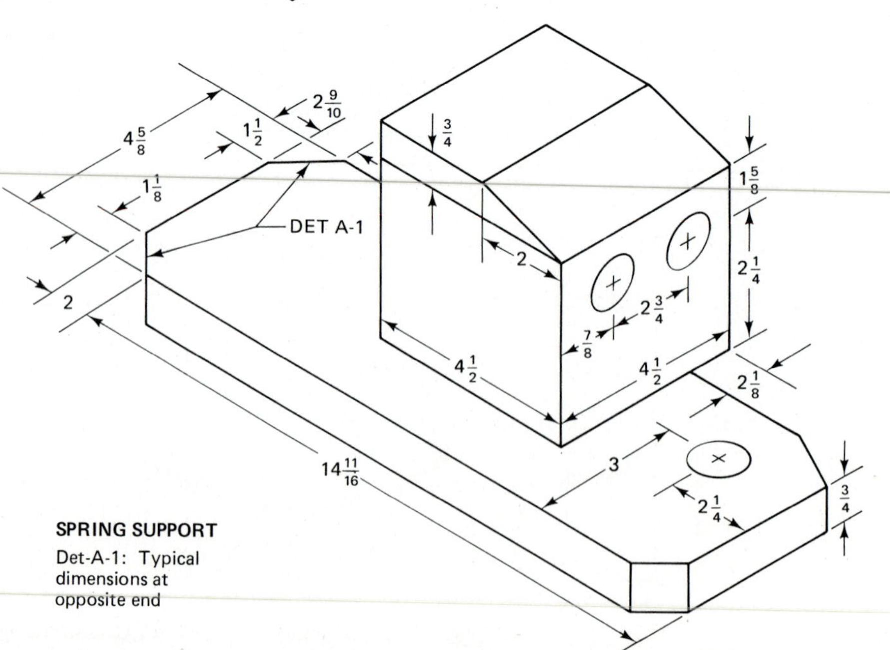

FIGURE 21-10c Drawing practice. Draw an orthographic: add any and all information needed to fabricate the part, including notes, specs, bill of mat., leader lines with drilled hole dimensions, and spec box.

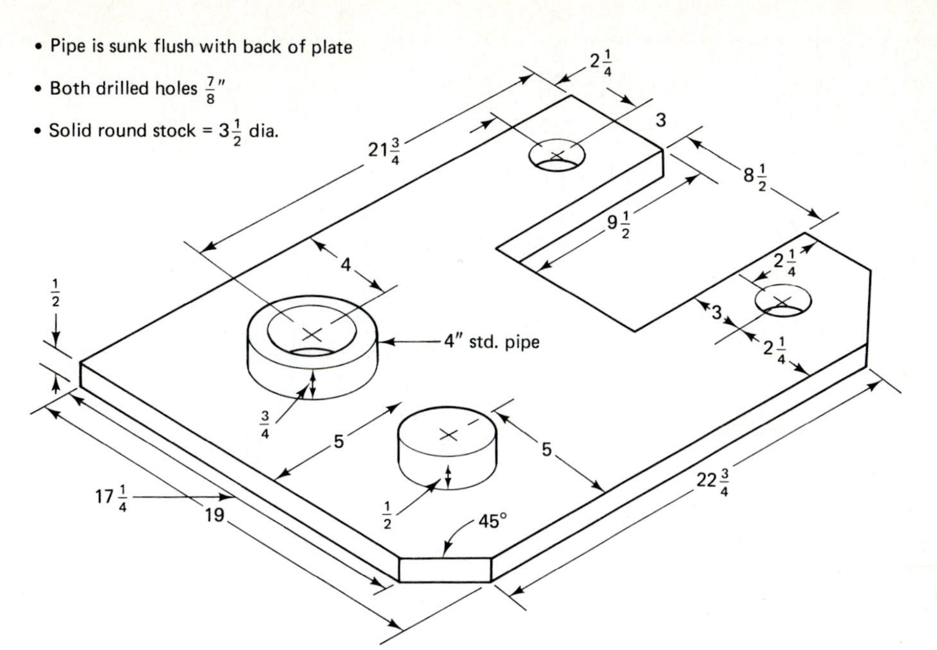

FIGURE 21-10d Drawing practice. Draw an orthographic: add any and all information needed to fabricate the part, including notes, specs, bill of mat., leader lines with drilled hole dimensions, and spec box.

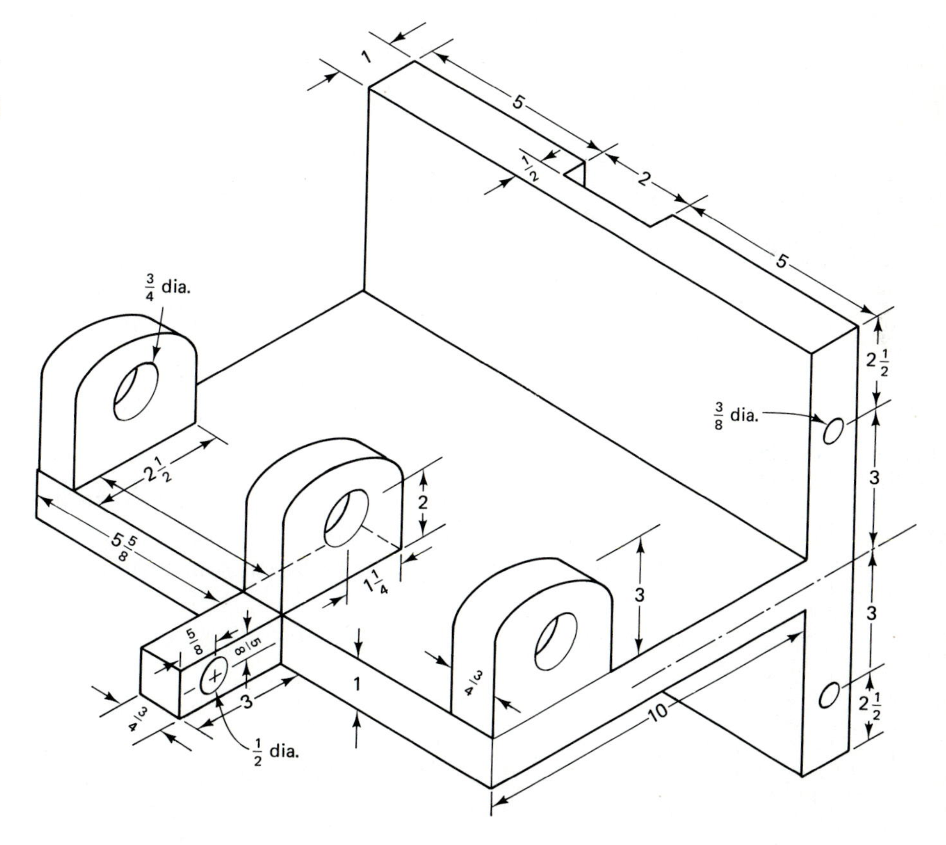

Redraw these isometric shapes in orthographic projections. Start in numerical order from the easiest to the hardest. Do not try to tackle them all at once; take a few each day and you will see steady progress in your drafting and blueprint reading skills.

DIMENSIONS

When working with blueprints you will find that dimensions or sizes will be given in numbers, fractions, decimals, or degrees. Examples of these are shown in Figures 21-11 and 21-12.

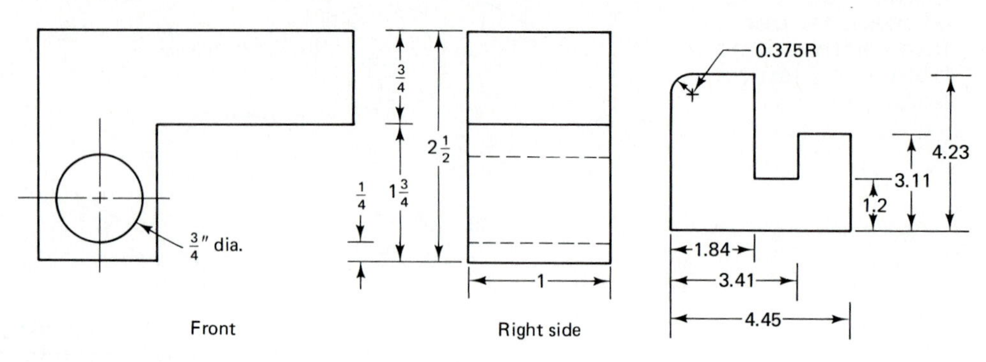

FIGURE 21-11
Dimensioning systems.

Angular Dimensions

Angular dimensioning (degrees or dimensions) is shown in Figure 21-12. Note the two systems used to show angles, the dimensioning system, and the angle-in-degrees system.

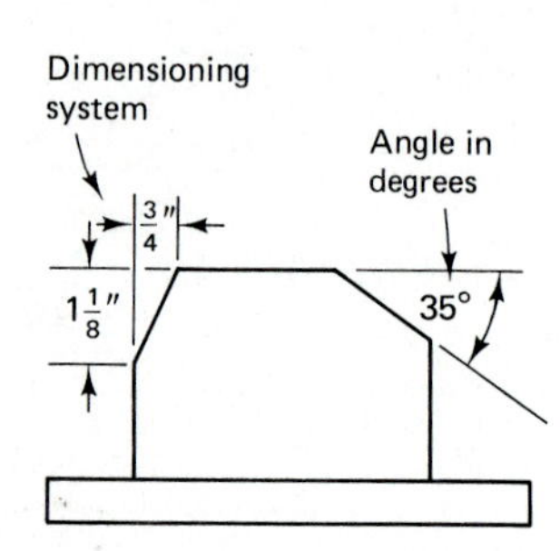

FIGURE 21-12
Dimensioning system.

Circular Dimensions

Circles and portions of a circle are used on radiuses, arcs, and holes on detail drawings. Let's understand what can be done with the circle and parts of the circle. Parts of the circle are labeled in Figure 21-13.

Formulas

1. To find the diameter of a circle:

$$D = 2R \qquad R = \text{radius}$$

2. To find the radius of a circle:

$$R = D/2$$

3. To find the circumference of a circle:

$$C = \pi D$$

Pi (symbol π) is a constant that has the value 3.1416.

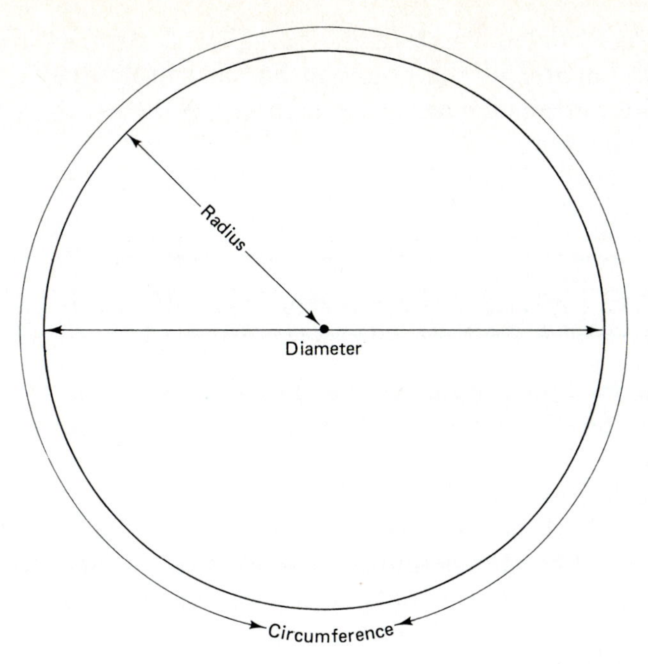

FIGURE 21–13
Parts of a circle.

Using Parts of the Circle for Layout and Fabrication Notice portions of a full circle in the shapes shown in Figure 21–14. Drilled holes are dimensioned by their diameter (Figure 21–15). A leader line is used to indicate which hole the data are referencing. Usually, the fractional and decimal diameter is given for the hole. When countersink (CSK) or counterbore (C Bore) holes are indicated, a depth of hole must

FIGURE 21–14
Circles in layouts.

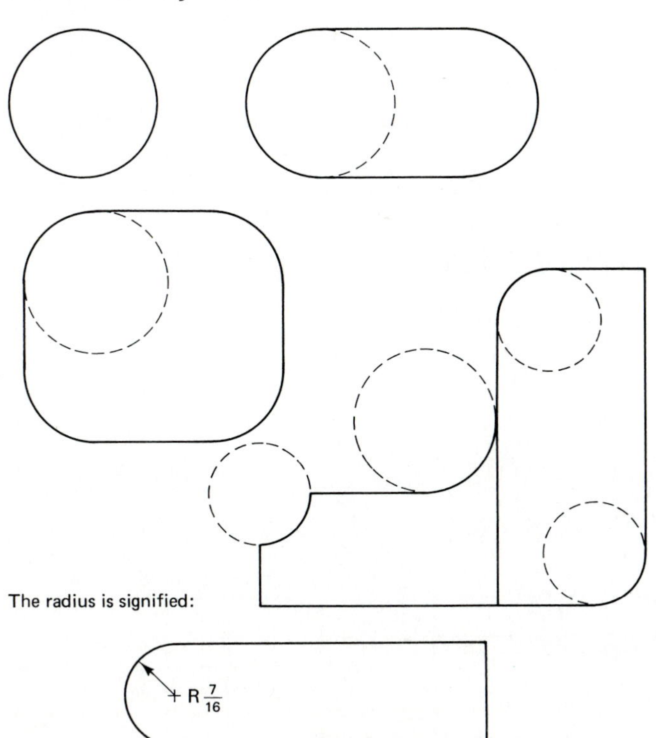

FIGURE 21–15
Drilled holes and countersinks.

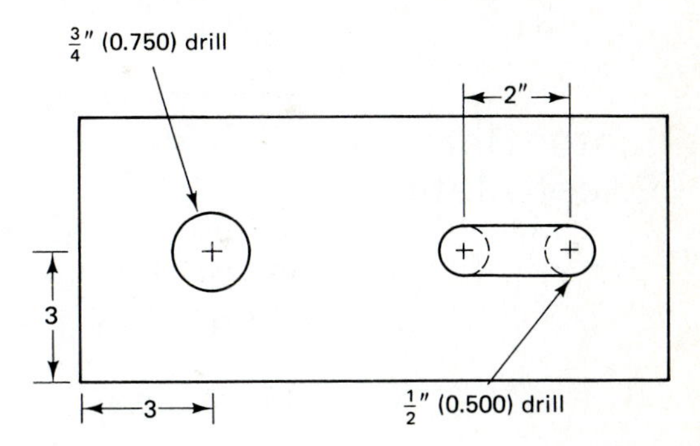

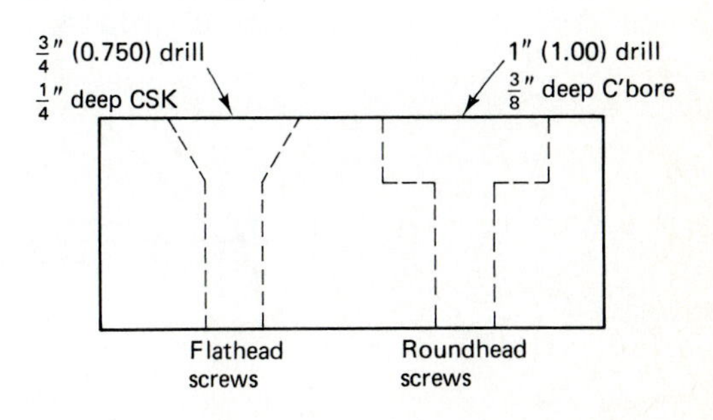

also be shown. Centerlines (CL) should also be dimensioned so that the exact center of the hole will be accurately drilled on the part. Slotted holes are made simply by drilling two holes, then cutting the space between them.

SYSTEMS FOR DIMENSIONING PARTS

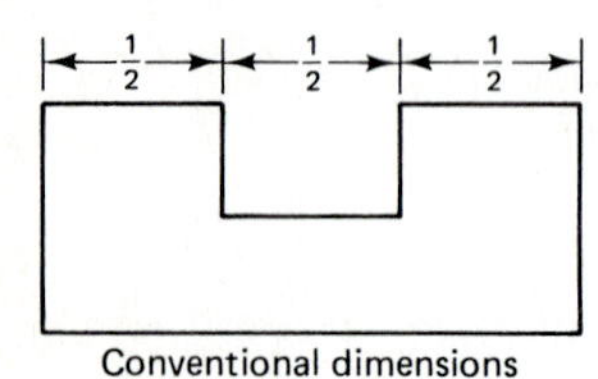

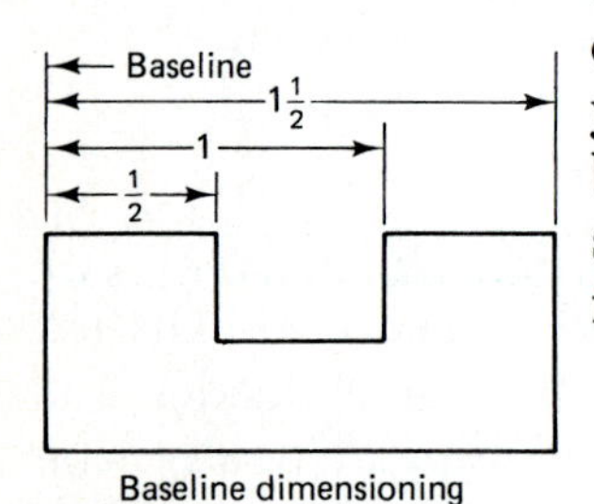

FIGURE 21-16 Conventional and base line dimensioning.

The two basic systems used to dimension parts on drawings are conventional and baseline (Figure 21-16). The convention system is most widely used for general fabrication and construction drawings. The baseline system is used on machined parts and surfaces for better accuracy. With baseline there is less chance for accumulative error.

The dimensioning of plate stock or structural shapes (standard beams, wide-flange beams, etc.) is done in a special sequence depending on the shape. It is important that you understand these systems so that you can read and understand the material list on prints. You may also be called upon to order steel stock. Knowing the size sequence will assure you of getting the stock you intended. The dimensions used are as follows:

T	Thickness
W	Width
L	Length
lb/ft	Weight per linear foot
F	Flange width
D	Depth or diameter on pipe
W	Leg length

Plate, Bars, or Sheet Metal

The difference between plate, bars, and sheet metal (Figure 21-17) is:

1. $\frac{3}{16}$ in. thick and under is considered sheet metal.
2. 6 in. wide and under is considered bar stock.
3. Over 6 in. wide and $\frac{3}{16}$ in. thick is considered plate.

FIGURE 21-17 Sheet, plate, and bar stock.

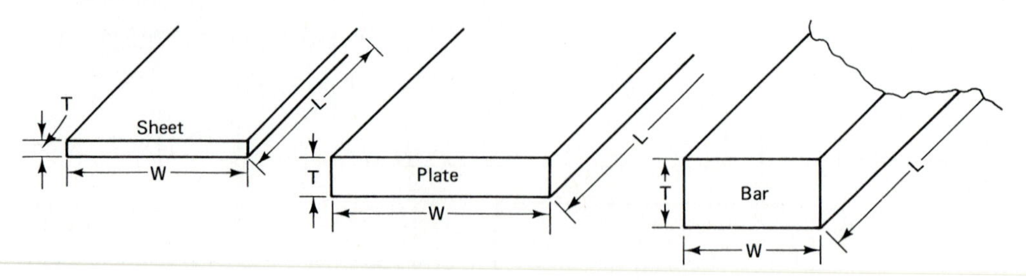

Regardless of whether it's plate, bar, or sheet metal, it is all dimensioned

$$T \times W \times L \text{ (thickness} \times \text{width} \times \text{length)}$$

Pipe and Solid Round Stock

Pipe is always referred to by its inside diameter (ID) of it's 12 in. and under, or outside diameter (OD) if it's over 12 in. The wall thickness of pipe must also be included on any pipe dimensions. Pipe is classified by a schedule system. Standard schedules and their corresponding thicknesses are given in Table 21-1. Pipe or solid round stock is dimensioned as $D \times L$ (diameter $\times$ length; Figure 21-18).

Structural Shapes

Structural shapes are usually dimensioned by lb/ft (weight per linear foot) because that is what determines their strength. For example, the 27 × 177 lb/ft, 27 × 160 lb/ft, and 27 × 145 lb/ft WF beams all have

TABLE 21-1
Approximate Wall Thickness

PIPE DIAMETER (IN.)[a]	SCH. 20	SCH. 40	SCH. 80	SCH. 120	SCH. 160
$\frac{3}{8}$		0.091	0.120		
$\frac{1}{2}$		1.09	0.147		0.188
1		0.133	0.179		0.250
$1\frac{1}{2}$		0.145	0.200		0.281
2		0.154	0.218		0.344
3		0.216	0.300		0.438
4		0.237	0.337	0.438	0.531
5		0.258	0.375	0.500	0.625
6		0.280	0.432	0.562	0.719
8	0.250	0.322	0.500	0.719	0.906
10	0.250	0.365	0.594	0.844	1.125
12	0.250	0.406	0.688	1.000	1.312
14	0.312	0.438	0.750	1.094	1.406
16	0.312	0.500	0.844	1.219	1.594
18	0.312	0.562	0.938	1.375	1.781
20	0.375	0.594	1.031	1.500	1.969
22	0.375		1.125	1.625	2.125
24	0.375	0.688	1.219	1.812	2.344

[a] ID. up to and including 12 in.; OD for pipe over 12 in.

FIGURE 21-18
Pipe and round stock.

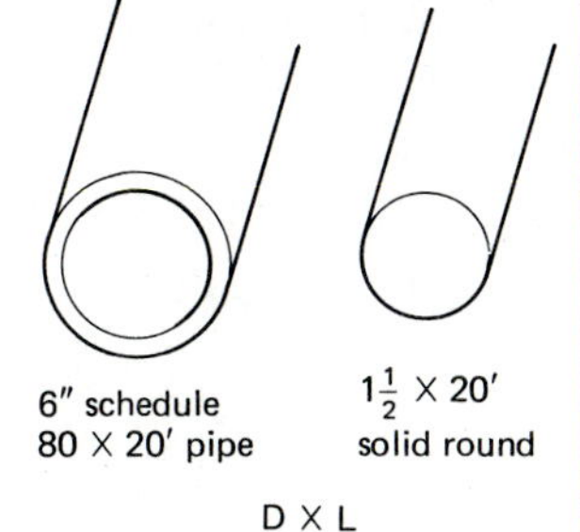

different strength levels, but they are all 27-in.-deep beams. This is why the weight per foot is given together with the beam depth. The American Institute of Steel Construction (AISC) publishes information regarding beam strengths and other data about structural shapes used in construction.

Wide-flange beams are sized by $D \times$ lb/ft $\times L$ (depth × weight per linear foot × length; Figure 21–19a). Other size specifications can be found in the AISC tables. *S-beams* (previously called I-beams) are sized by $D \times$ lb/ft $\times L$ (depth × weight per linear foot × length; Figure 21–19b). *M-beams* (previously called H-beams) are sized by $D \times$ lb/ft $\times L$ (depth × weight per linear foot × length; Figure 21–19c).

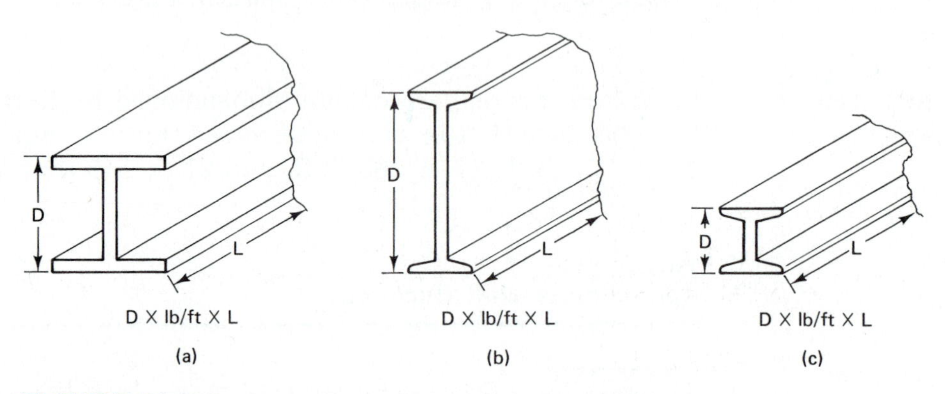

FIGURE 21–19
Wide-flange, S, and M beams.

Miscellaneous structural shapes include angle iron and channel iron or shapes commonly used in connecting the larger structural shapes, such as WF, M, and S beams that are used in construction. *Angle iron* is sized by $W \times W \times T \times L$ (width of one leg × width of the other leg × thickness × length; Figure 21–20). This system is used because some angle iron has unequal leg lengths. *Channel iron* is sized two different ways. If used as a support member, where its exact strength must be known, it will be sized $D \times$ lb/ft $\times L$ (depth × weight per linear foot × length). If it is going to be used for its shape and its exact strength is not of critical importance, it will be sized $D \times F \times T$ × L (depth × flange × thickness × length; Figure 21–21).

FIGURE 21–20
Angle iron.

FIGURE 21–21
Channel iron.

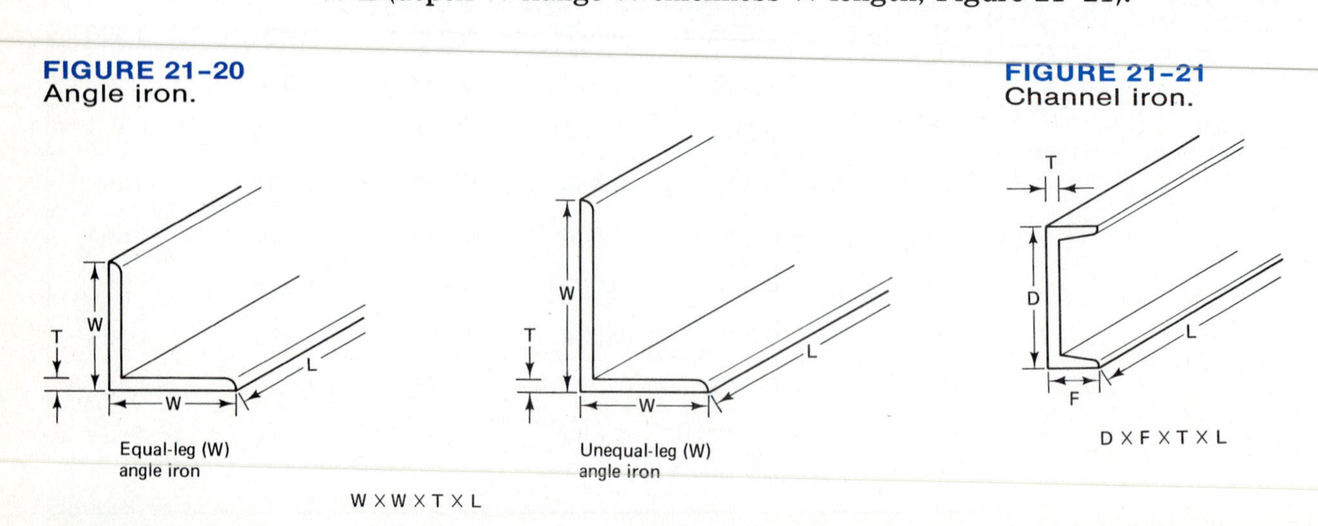

TOLERANCES

Tolerances represent the dimensional accuracy required, given as a plus (+) or minus (−) value. A value of 0.002 would be considered a close tolerance and might be found on a machine surface or machining print. One-eighth (0.125) inch is usually considered a fairly loose tolerance, depending on what is being measured and how it was fabricated. If we are fabricating from a print that gives a tolerance limit of $\pm \frac{1}{16}$ (0.0625) in. and a dimension is shown to be 16 in., this means that when fabrication is completed, that dimension can be between $16\frac{1}{16}$ and $15\frac{15}{16}$ and still be within the acceptable tolerance specifications. Several tolerance examples are given in Table 21-2. Tolerances are usually found in the specification box on the lower right of the print.

TABLE 21-2
Tolerance Examples

DIMENSION	TOLERANCE	MAX. SIZE	MIN. SIZE
24 in.	$\pm \frac{1}{8}$ (0.125)	$24\frac{1}{8}$ in.	$23\frac{7}{8}$ in.
5 ft	$\pm \frac{1}{16}$ (0.0625)	5 ft $\frac{1}{16}$ in.	4 ft $11\frac{15}{16}$ in.
30°	± 5°	35°	25°
0.750	± 0.003	0.753	0.747
12 in.	$+\frac{1}{4}$(0.250) in.	$12\frac{1}{4}$ in.	12 in.

SECTION VIEWS

One of the most common types of views on prints, other than the standard top, front, and right-side views, is the section view (Figure 21-22). Section views are used to show interior details and expose parts that

FIGURE 21-22
Section views.

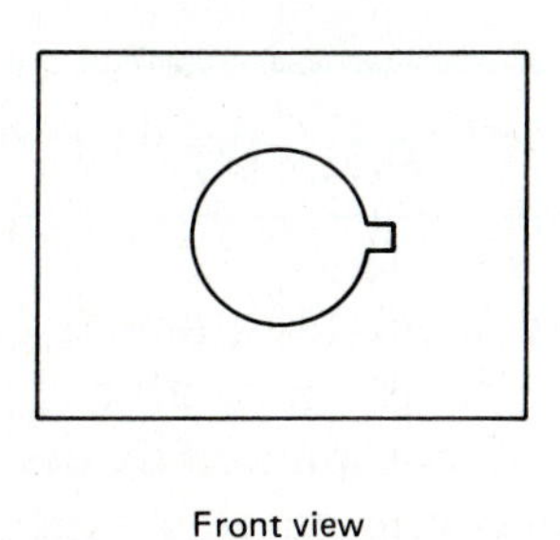

Front view

A
A
Right-side view

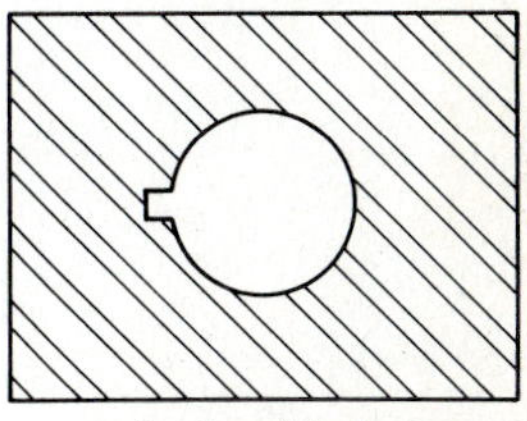

Section AA view

would otherwise not be seen clearly with standard views. There are two types of lines used specifically for section views: (1) the cutting plane line (↓‾ ‾ ‾↓) and (2) section lines (▨). The cutting plane line denotes the surface that is imagined to be cut; the letters at the ends of the arrows indicate the name of the section view. The arrow on the end points in the direction of the surface that will be exposed in the section view. The section lines indicate the type of material of which the part is made.

Revolved sections are usually used specifically for showing the shape and content of a beam or long member. The revolved section view clarifies what type of member is being viewed. Notice the drawings in Figure 21–23; it would be difficult to tell whether the part was a channel or an S-beam without the revolved section view.

An *enlarged view* is drawn to give the reader a better, more detailed view of a complex area or assembly. It is usually a separate drawing or view of the detailed area. A leader line is used to indicate and identify the area that will be enlarged and detailed (Figure 21–24).

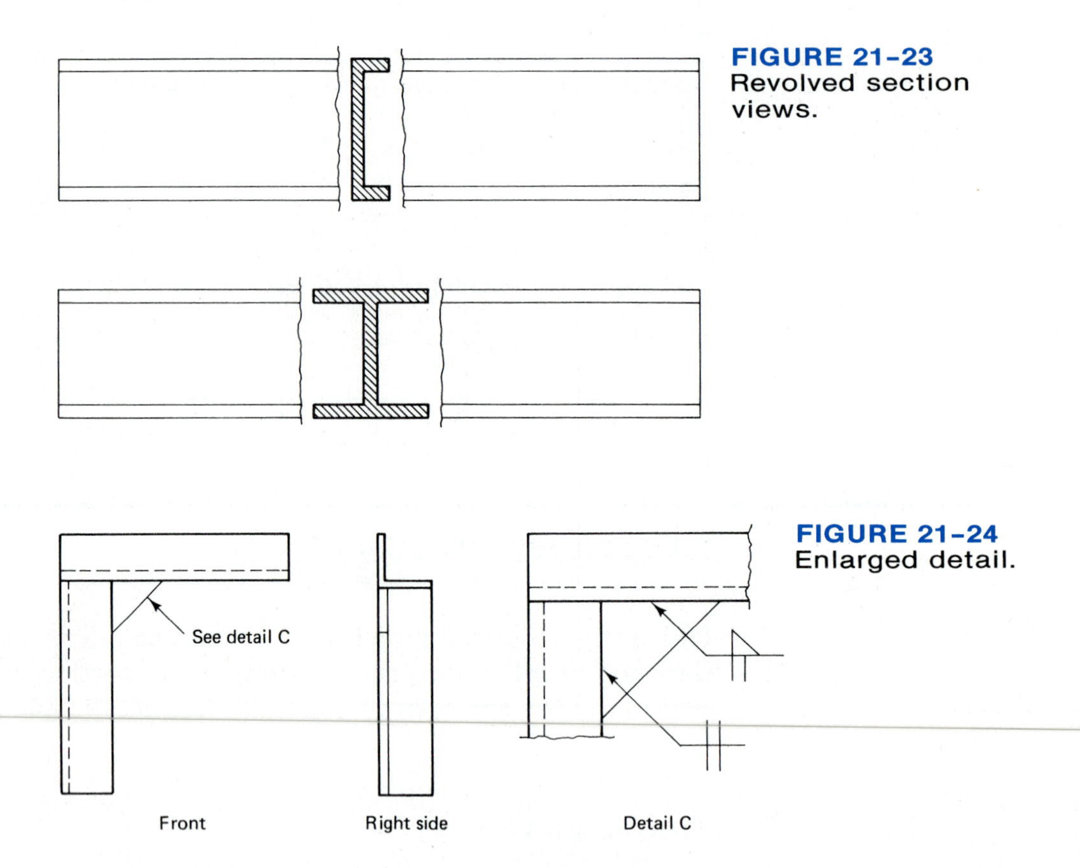

FIGURE 21–23 Revolved section views.

FIGURE 21–24 Enlarged detail.

ASSEMBLY AND DETAIL PRINTS

In many cases parts require a special sequence of assembly (Figure 21–25). This print will show specifically how all the parts in the assembly are to be fabricated. It will be accompanied by detailed prints that show the fabrication of all the individual smaller parts (Figure 21–26).

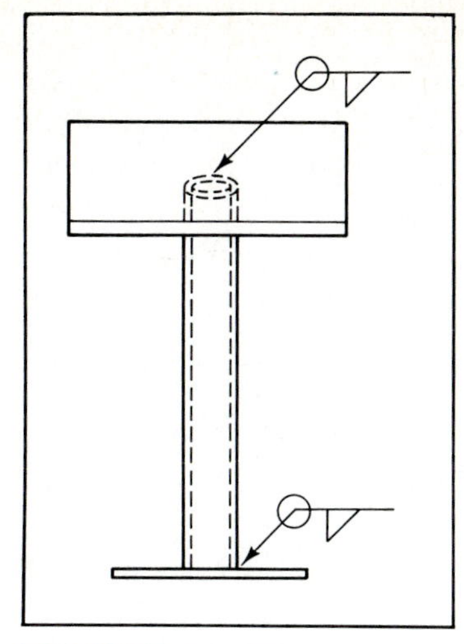

FIGURE 21-25
Assembly view.

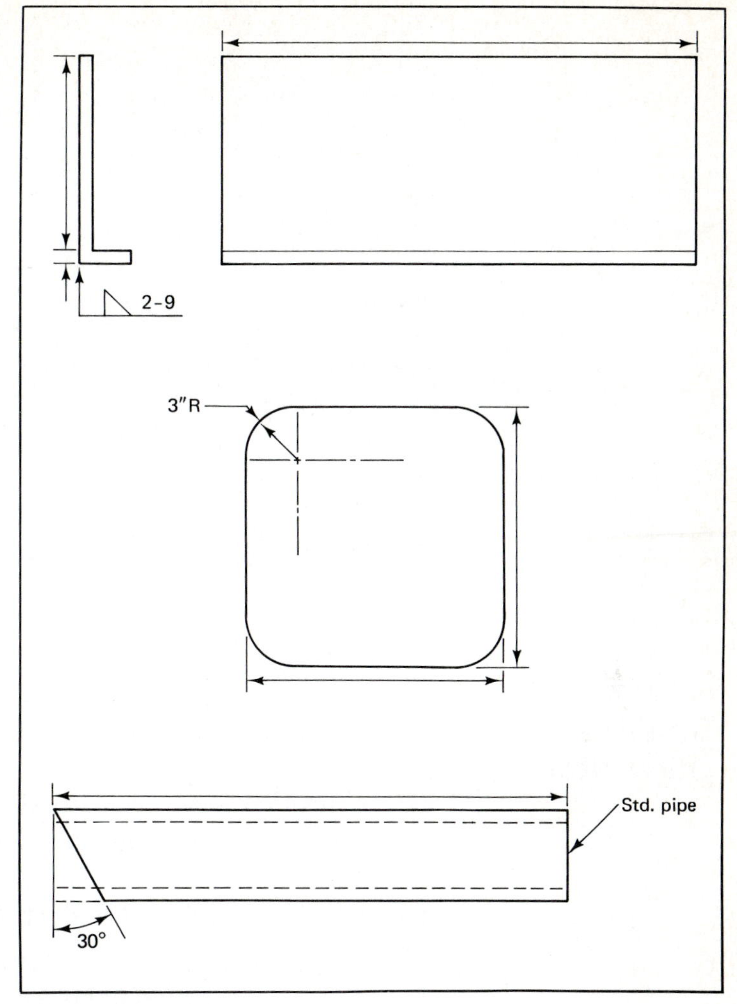

FIGURE 21-26
Detail view.

REVIEW QUESTIONS

The drawings shown in Figures 21-27 to 21-29 are typical of what you can expect in the welding industry. Study and answer the questions, which will help improve your blueprint reading skills.

Welding Machine Stand

Use Figure 21-27 to answer the following questions.

1. How many machine stands are required?
2. What types of views, and how many, are shown?
3. How many parts are required to assemble one welding machine stand?
4. How can you check the squareness of the machine stand?
5. From what type of stock is the machine stand made?
6. What is the total length of stock needed to fabricate one machine stand?
7. What is the total length needed to fill the order?

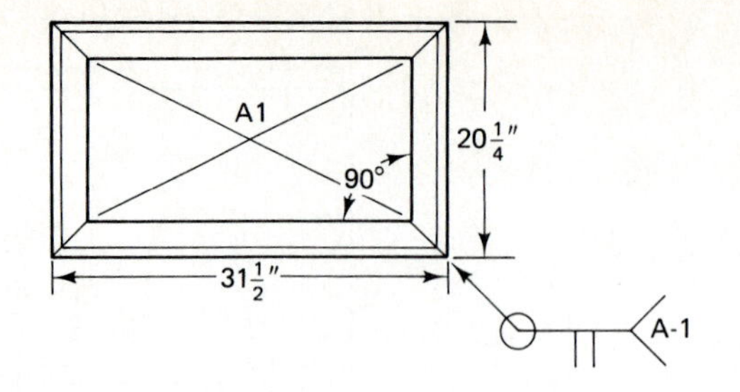

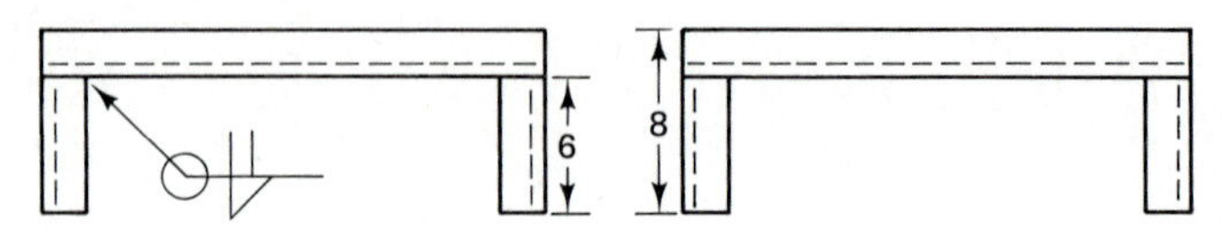

All angle = 2 X 2 X $\frac{3}{16}$

Note A-1: Tack first, square, then weld

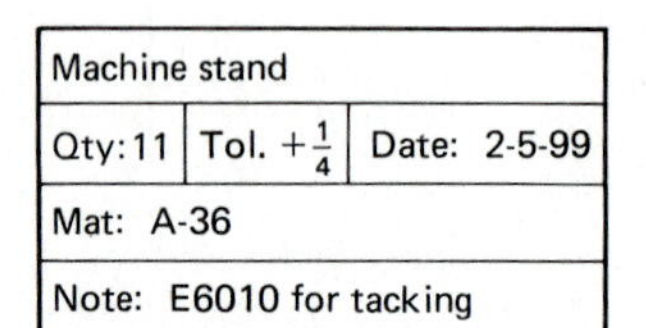

Machine stand		
Qty: 11	Tol. $+\frac{1}{4}$	Date: 2-5-99
Mat: A-36		
Note: E6010 for tacking		

FIGURE 21-27 Typical detail drawing: machine stand.

Steel Dumpster

Use Figure 21-28 to answer the following questions.

8. What views are used to show the dumpster?
9. How many sections of channel iron are used to fabricate one dumpster frame?
10. What is the overall length of channel used?
11. How many sections of plate are used to fabricate one dumpster?
12. What is the capacity of the dumpster?
13. How many sections of pipe are used?
14. Give the dimensions for the following:
    A:__________
    B:__________
    C:__________
    D:__________

Air Handler Platform

Use Figure 21-29 (page 346) to answer the following questions.

15. What types of views, and how many, are used to show the air handler platform?
16. What types of structural shapes are used to fabricate the air handler platform?
17. What type of welds are used for fabrication?
18. What is the overall size (length, width, height) of the air handler platform?
19. Give the dimensions for the following:
    A:__________
    B:__________
    C:__________
    D:__________

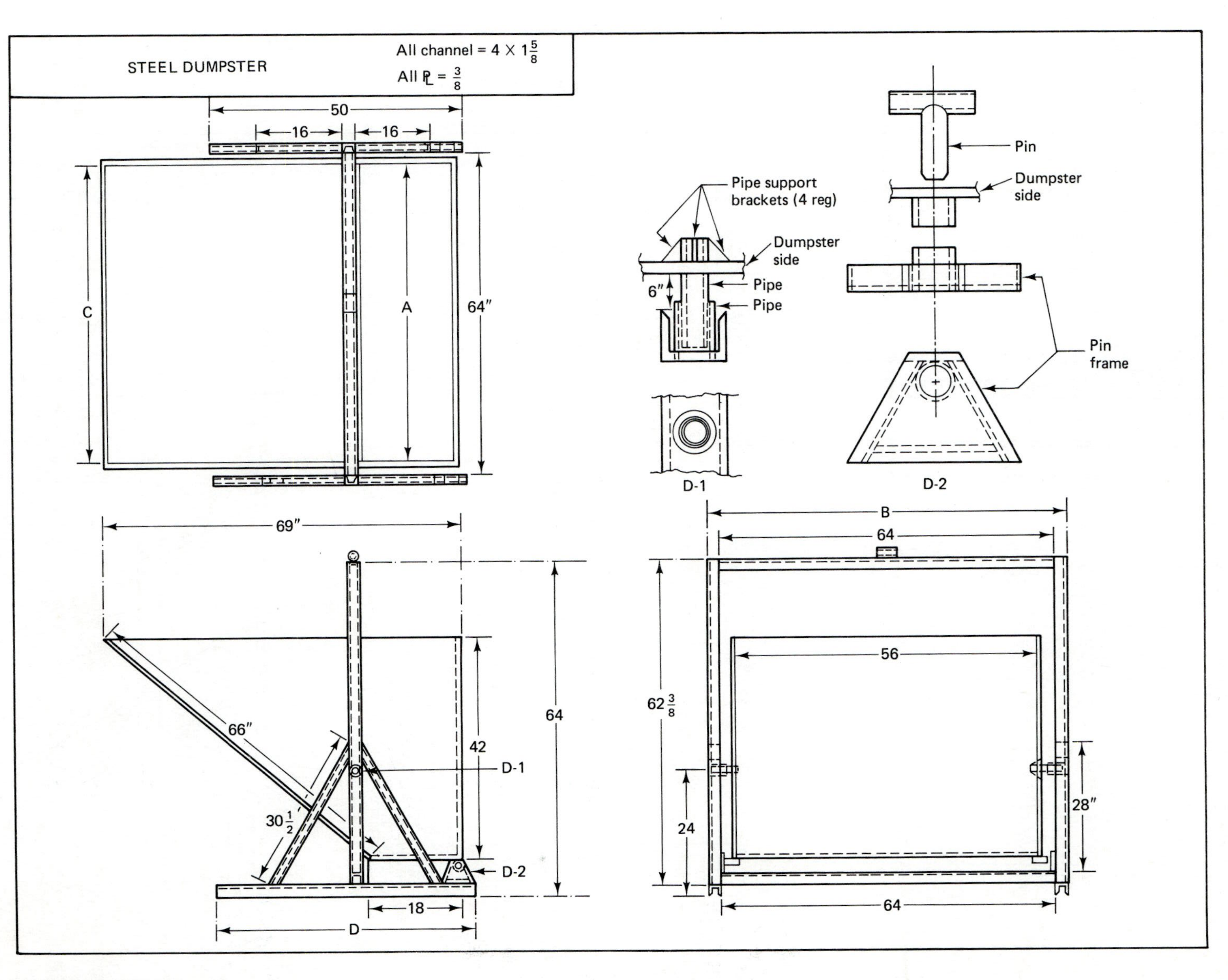

FIGURE 21–28
Typical detail drawing: dumpster.

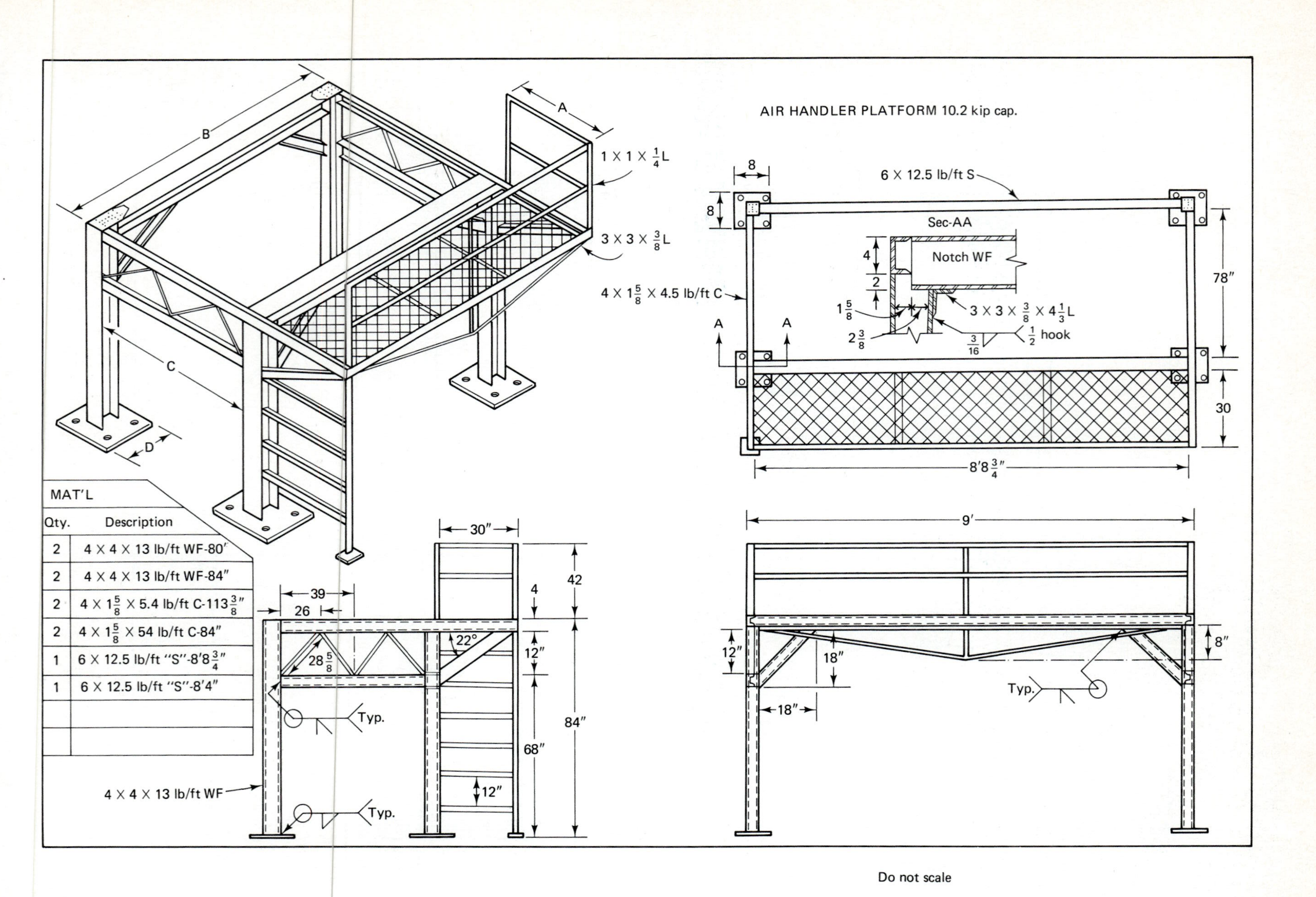

MAT'L

Qty.	Description
2	4 X 4 X 13 lb/ft WF-80″
2	4 X 4 X 13 lb/ft WF-84″
2	4 X 1$\frac{5}{8}$ X 5.4 lb/ft C-113$\frac{3}{8}$″
2	4 X 1$\frac{5}{8}$ X 54 lb/ft C-84″
1	6 X 12.5 lb/ft "S"-8′8$\frac{3}{4}$″
1	6 X 12.5 lb/ft "S"-8′4″

FIGURE 21-29
Typical detail drawing: air handler platform.

WELDING SYMBOLS

One of the most dangerous mistakes that designers and engineers can make is to assume that a joint will be welded properly without specifying the welding process, joint design, weld size, or type of filler metal. Many weld failures can be traced to such engineering assumptions. Using the standardized AWS/ANSI welding symbols will at least verify to the welder the engineer's intent.

The next step is to make sure that welders, technicians, and inspectors also fully understand welding symbols. The American Welding Society has spent many years standardizing and refining these symbols. Your knowledge of the symbols will greatly enhance your understanding of blueprints and the ultimate intent of the designers whose prints you will be interpreting.

Dimensions put on welding symbols to indicate length, pitch, size, diameter, and so on, will always be in inches, unless it is stated that metric units (SI units) are to be used. In the case of metric units, millimeters will be used as units of measure on welding symbols. The American Welding Society states that dual units should not be used on symbols; this is to prevent confusion and error.

WELDING SYMBOL TERMINOLOGY

First recognize the difference between a *weld symbol* and a *welding symbol.* Figure 22–1 shows examples of weld symbols: they indicate the type of joint preparation and weld. Figure 22–2 shows examples of welding symbols. They contain all the information needed for the weld, including the weld symbol and any additional information necessary. The three basic parts of the welding symbol are the arrow, the tail, and the reference line (Figure 22–3). The arrow will point to one of the sides of the joint. The reference line will contain the weld symbol and any numerical data, such as size, effective throat, length, etc. The tail will be added only if additional notes, specifications, or welding process information needs to be added to the symbol.

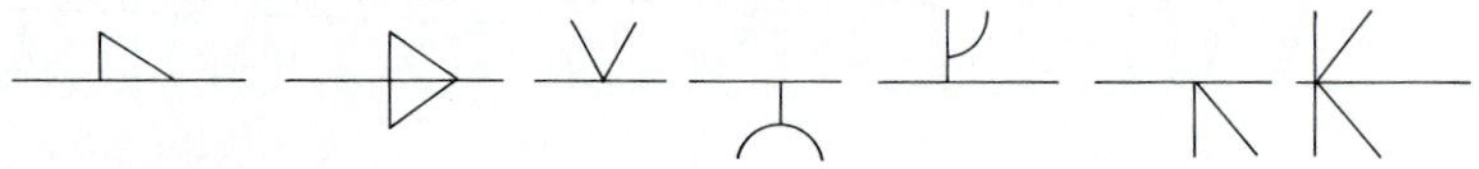

FIGURE 22–1
Weld symbols.

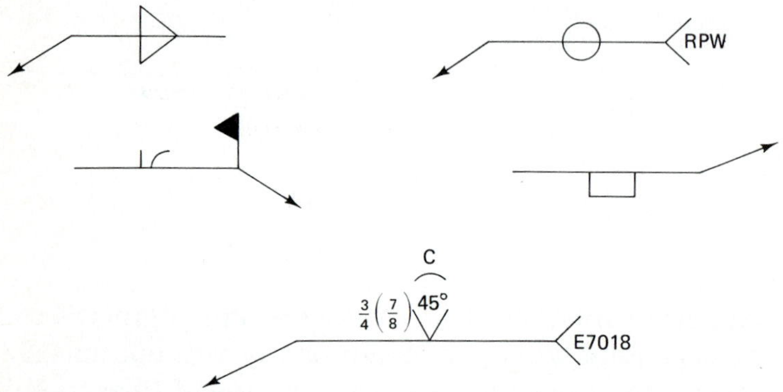

RPW

C

$\frac{3}{4}\left(\frac{7}{8}\right)$ 45°

E7018

FIGURE 22–2
Welding symbols.

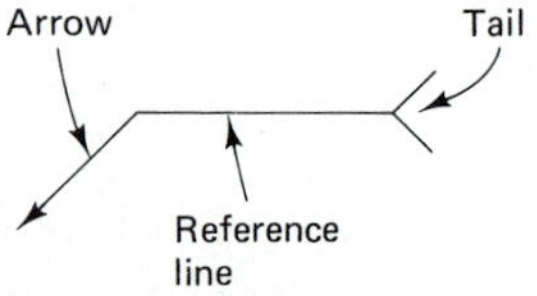

Arrow

Tail

Reference line

FIGURE 22–3
Welding symbol parts.

WELDING SYMBOL PARTS

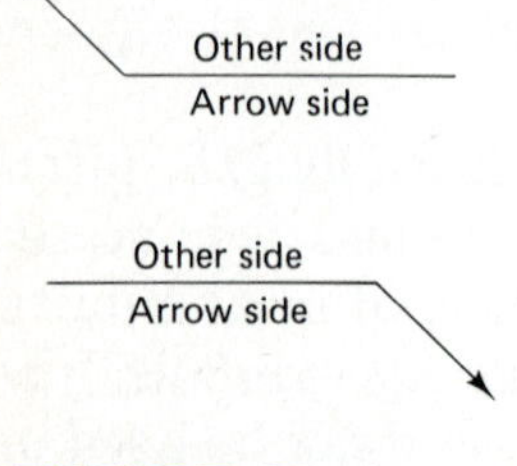

FIGURE 22–4
Arrow and other side.

Arrow and *other side* are terms applied to indicate on which side the weld is to be applied. On the symbol, the best way to remember which side is the arrow side: it is always the side toward the bottom of the print; the "other side" will always be toward the top of the print (Figure 22–4). Notice that the direction of the arrow has no significance on the reference line or arrow and "other side." Examples of arrow and other-side welds are shown in Figure 22–5.

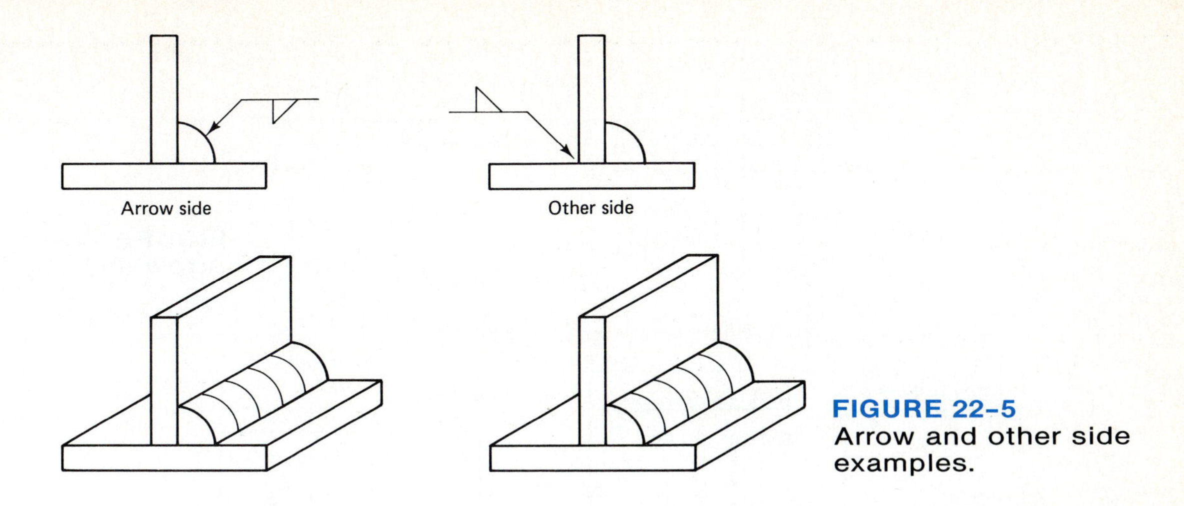

FIGURE 22-5
Arrow and other side examples.

If welds are wanted on both sides of a joint, the symbol is applied to the arrow and the other side (Figure 22-6). If welds are wanted on all four sides of a joint, one of the techniques shown in Figure 22-7 might be used.

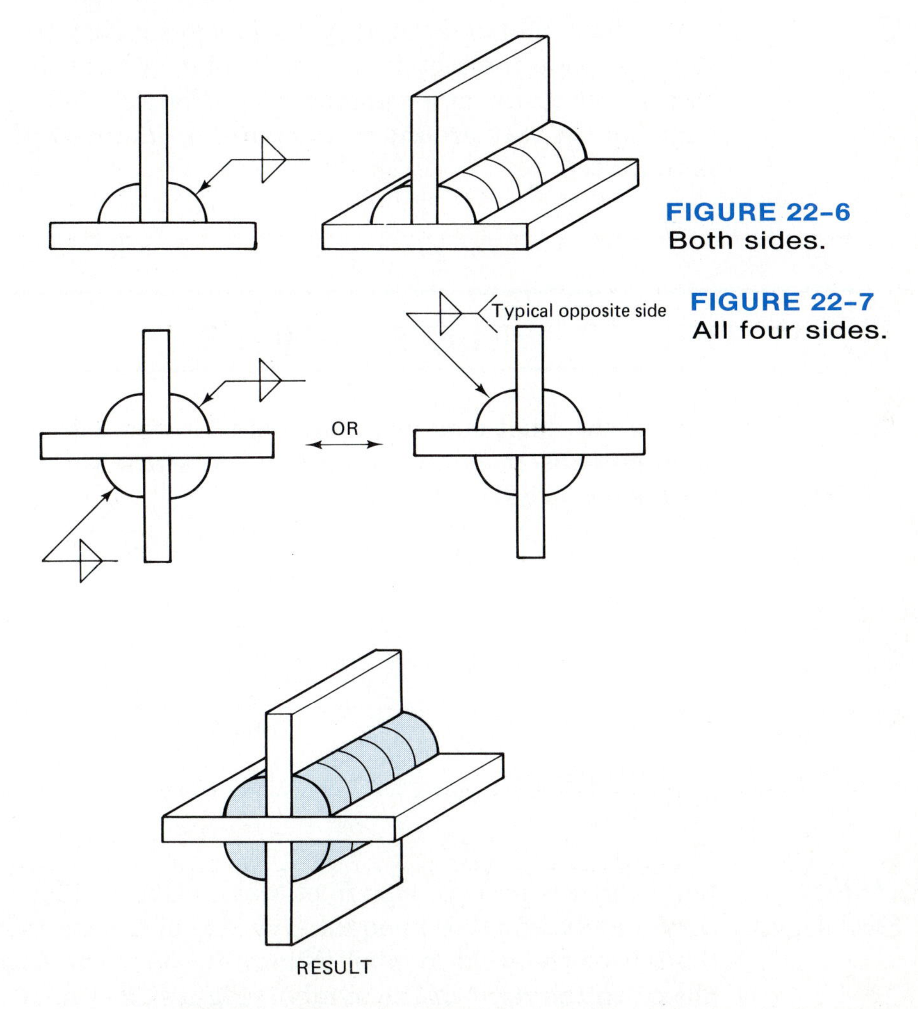

FIGURE 22-6
Both sides.

FIGURE 22-7
All four sides.

Study the examples in Figure 22-8, which may be a bit confusing on a print: Notice that the "other side" is not always at the actual horizontal opposing side of the joint. Note the position of the longer plate in each example.

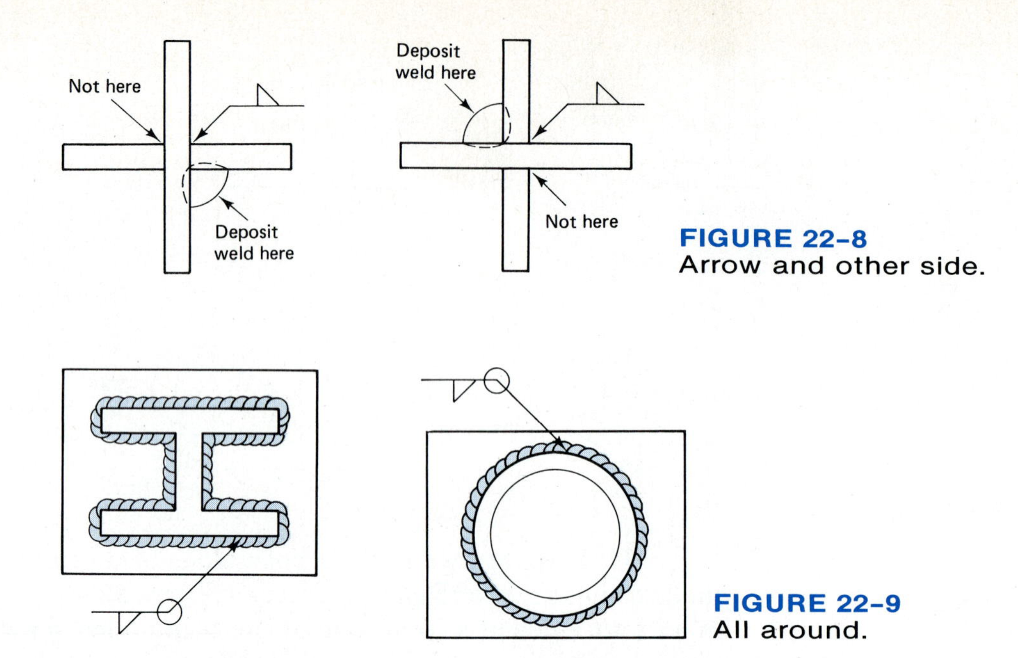

FIGURE 22-8
Arrow and other side.

FIGURE 22-9
All around.

The weld-all-around symbol makes it easy to indicate a weld that is to continue completely around a joint. A circle is put at the intersection of the arrow and reference line (Figure 22-9). The weld is to continue all the way around *to the point of its origin* if "weld all around" is indicated.

FILLET WELD SYMBOLS

One of the most common symbols that you will encounter is the fillet weld symbol. It can be applied to the arrow side, the other side, or to both sides (Figure 22-10).

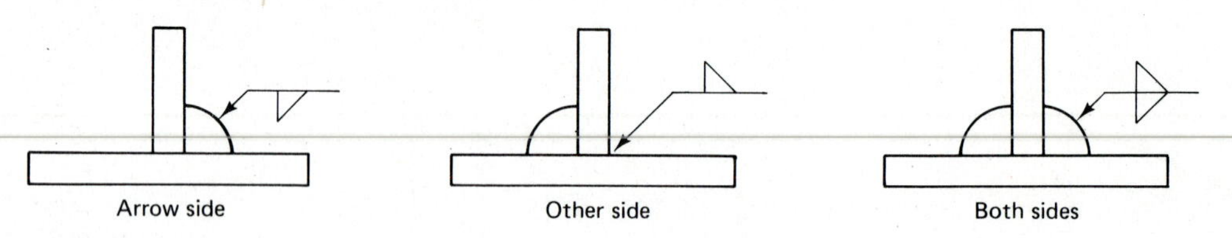

FIGURE 22-10
Fillets.

Sizing Fillet Weld Symbols

Leg length is used to size fillet welds. Every fillet weld has two legs; they may be equal or unequal. The size of the leg in inches is placed to the *left* of the weld symbol (Figure 22-11). The *length* of the weld is placed to the *right* of the symbol (Figure 22-12). If there is no dimension to the right of the symbol, the weld is to be made the full length of the joint.

Length and *pitch* are both indicated where intermittent welds

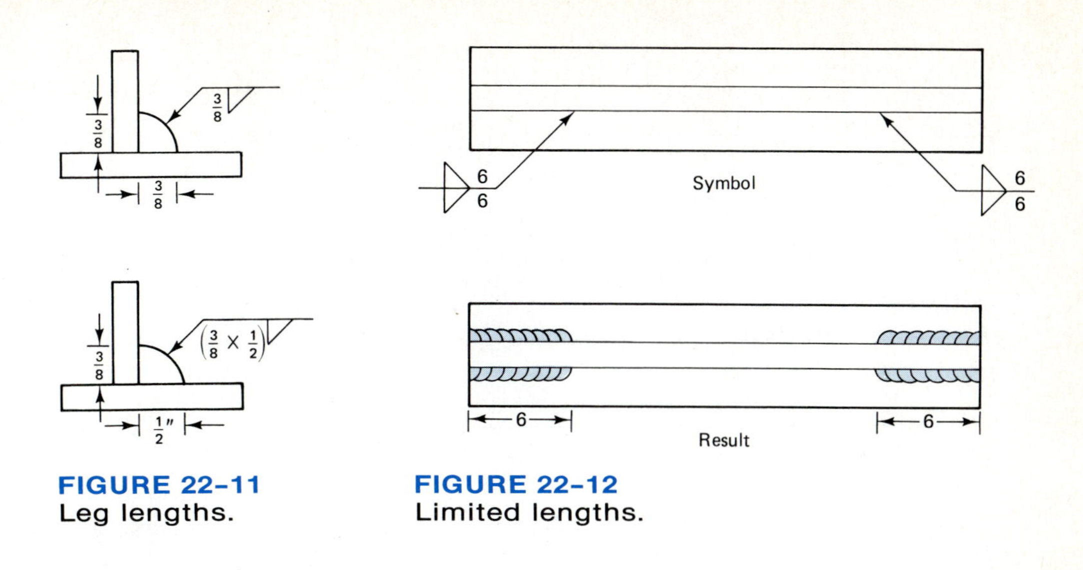

FIGURE 22–11
Leg lengths.

FIGURE 22–12
Limited lengths.

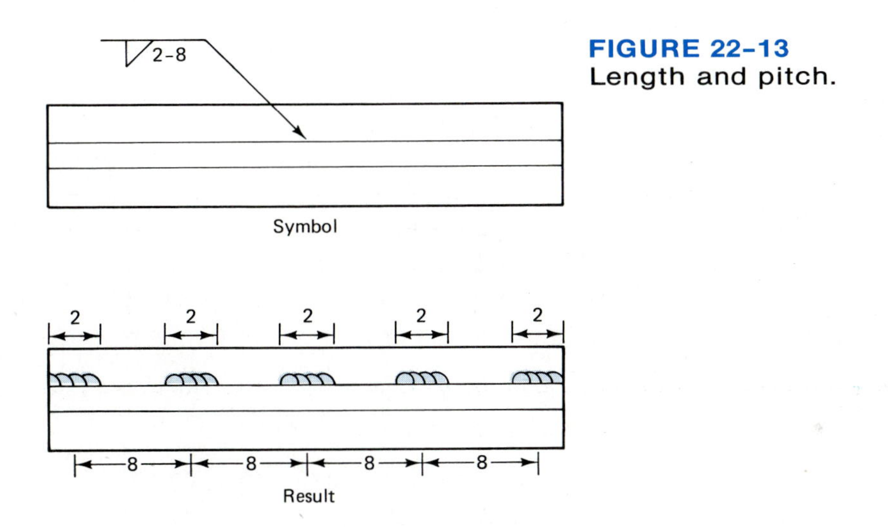

FIGURE 22–13
Length and pitch.

are specified (Figure 22–13). On the right side of the symbol there will be two numbers separated by a dash; the first is the weld length and the second is the pitch. The pitch is simply the distance from one weld center to the next weld center.

If an intermittent weld is used on two sides opposite each other, it is called a chain intermittent weld (Figure 22–14). If the welds are offset on the opposite side, it is known as a staggered intermittent weld (Figure 22–15). To produce an evenly spaced stagger, use one-half the pitch distance for the starting point of the opposite-side weld. Always start on the side whose weld symbol is farthest to the left (Figure 22–16). In the drawing the arrow side is farthest to the left, so it is started first.

Intermittent welds are often combined with continuous welds. This requires some type of spacing between the continuous and intermittent welds. Simply use the pitch minus the weld length to reach this spacing distance (Figure 22–17).

Some typical fillet weld symbols and joint applications are shown in Figure 22–18.

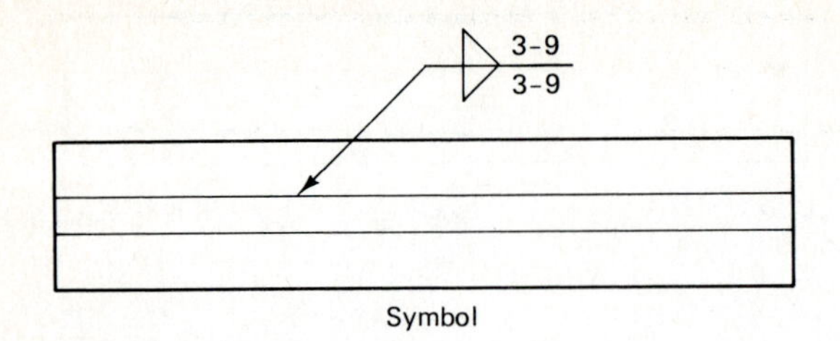

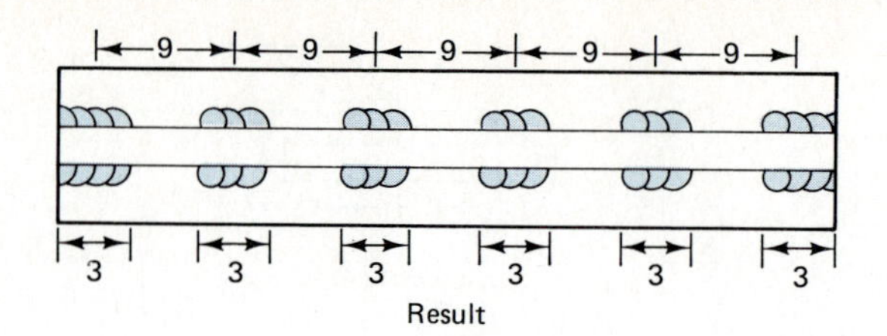

FIGURE 22-14
Chain intermittent.

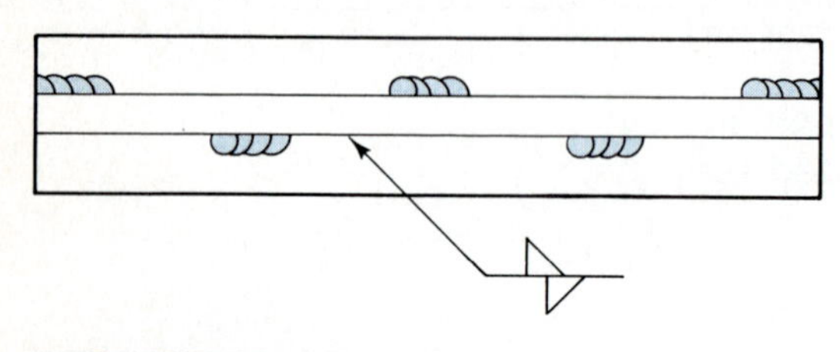

FIGURE 22-15
Staggered intermittent.

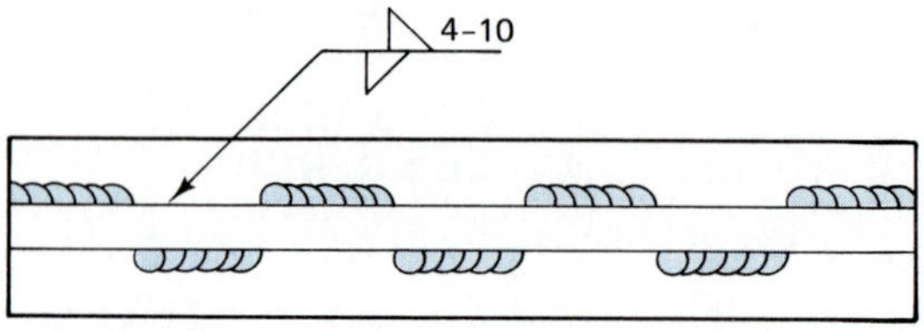

FIGURE 22-16
Staggered intermittent.

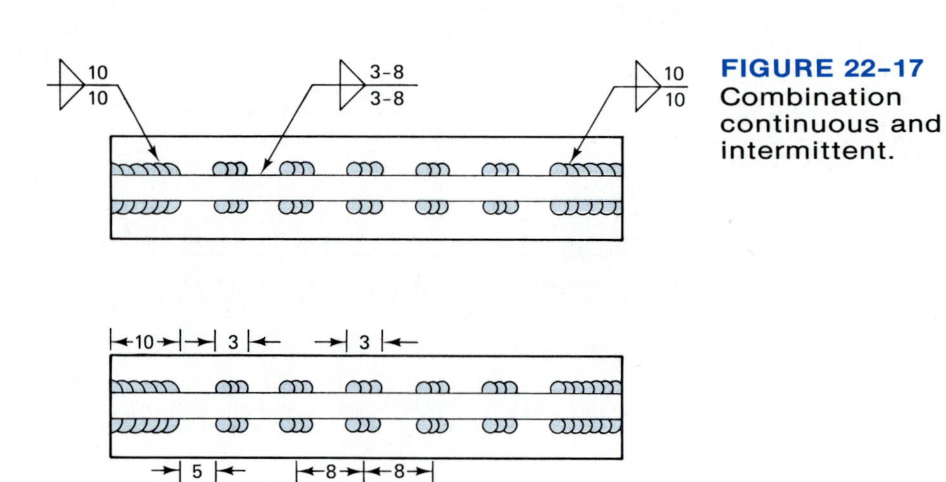

Key: 8-3 = 5

FIGURE 22-17
Combination continuous and intermittent.

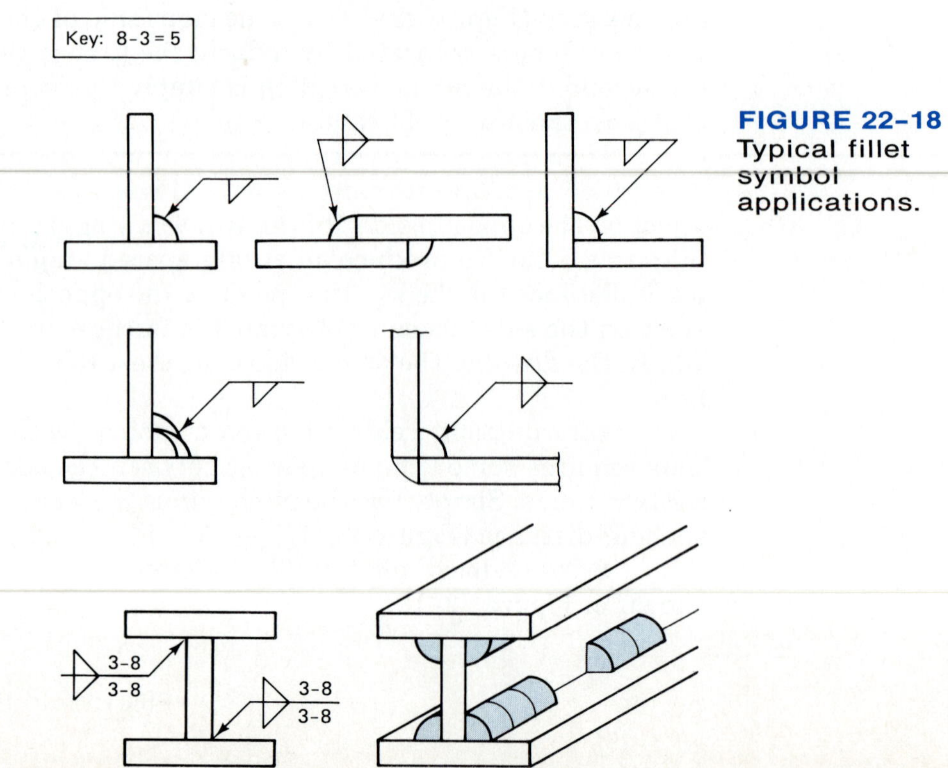

FIGURE 22-18
Typical fillet symbol applications.

GROOVE WELD SYMBOLS

There are eight basic groove weld joint designs used in today's welding industry. Just as in fillet welds, they can be used on the arrow side, the other side, or both sides. They may also be used in combination with each other, or in combination with fillet welds. Examples of groove weld joint design possibilities are shown in Figures 22-19 to 22-27.

Square Grooves (Figure 22-19) Most codes limit the maximum thickness of square groove joints to $\frac{1}{4}$ (0.250) in. This is due to the lack of edge preparation and the resulting poor penetration on thicker metals.

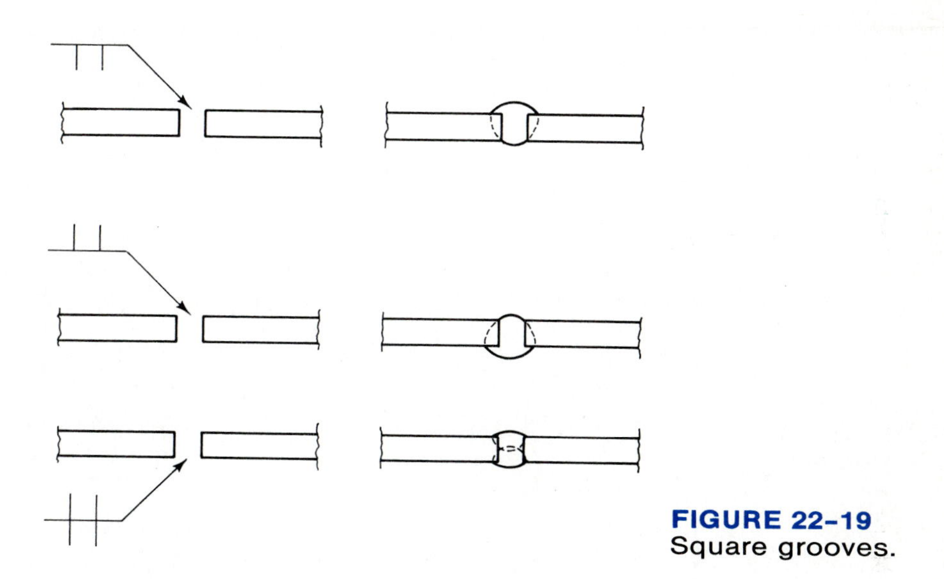

FIGURE 22-19
Square grooves.

V-Grooves (Figure 22-20) V-grooves are among the most common groove designs used when 100% fusion is required. The v-groove design gives the welder good accessibility into the joint and the root. V-grooves are usually used on material over $\frac{1}{4}$ in. (0.250 in.).

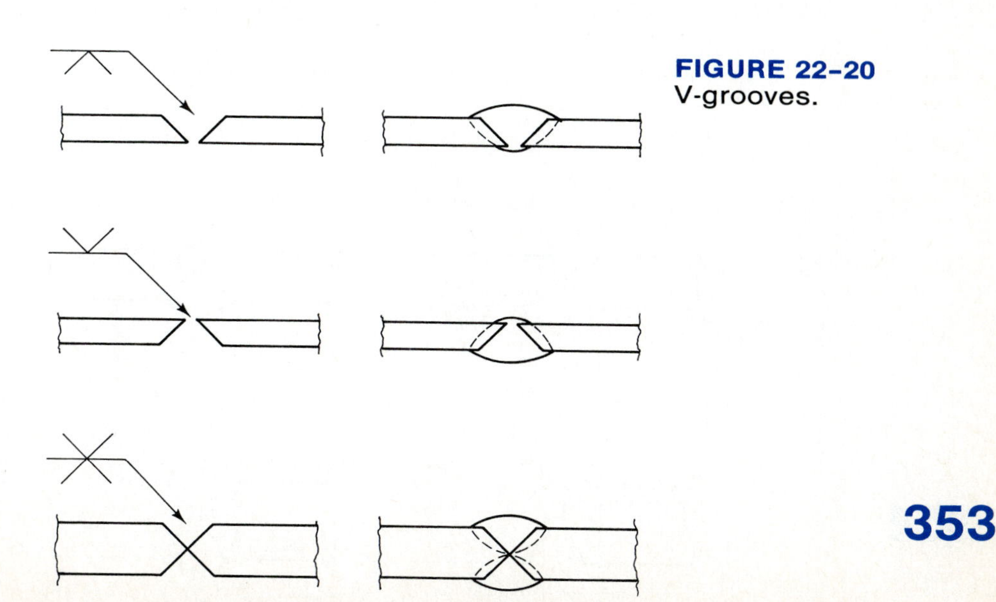

FIGURE 22-20
V-grooves.

Bevel Grooves (Figure 22–21) Note the extra break in the arrow with bevel groove symbols. This is required to indicate which plate is to be prepared by beveling. The arrow will *break toward* the plate (or pipe) that is to be beveled (Figure 22–22).

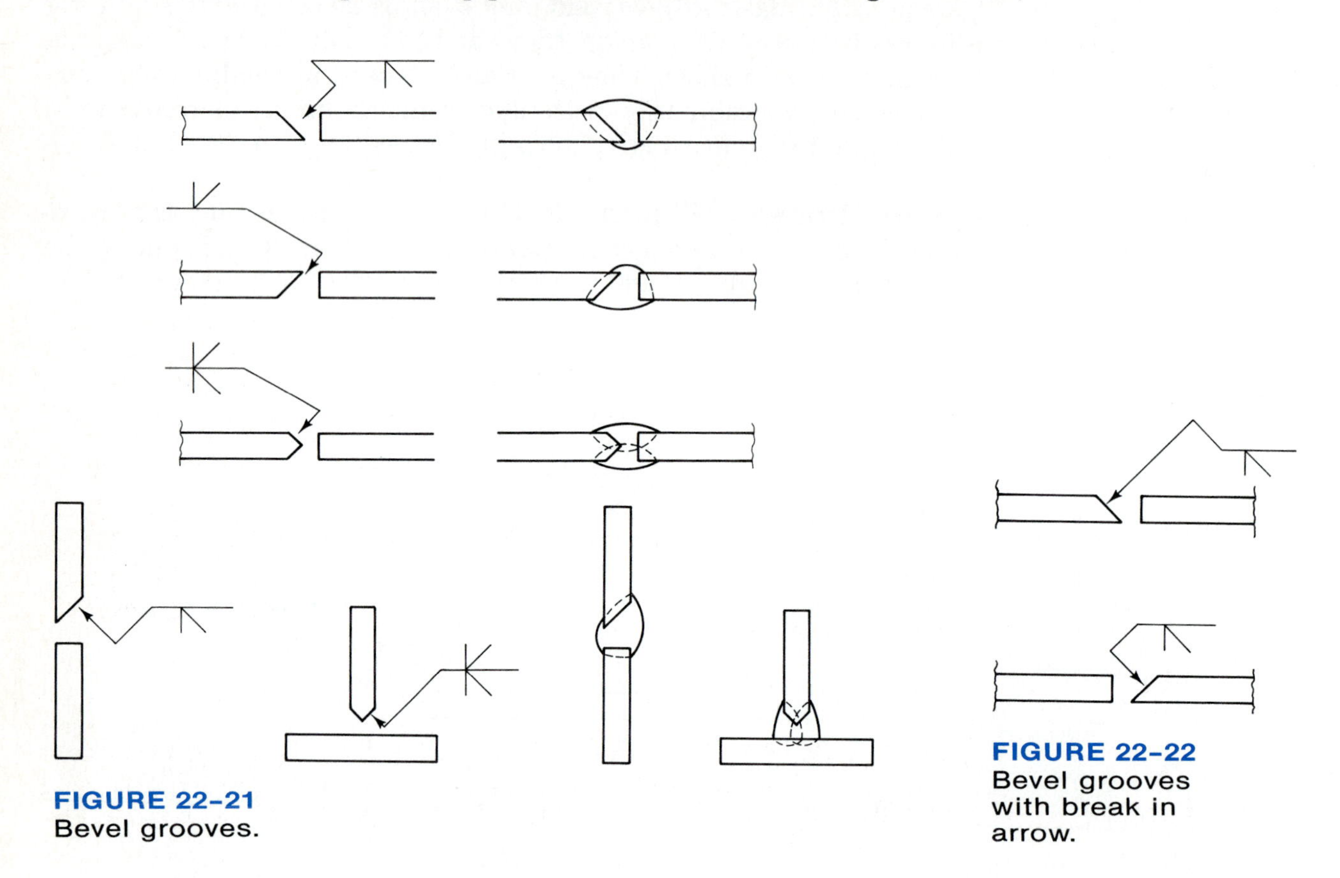

FIGURE 22–21
Bevel grooves.

FIGURE 22–22
Bevel grooves with break in arrow.

U-Grooves (Figure 22–23) U-grooves are similar to Vee-grooves but are produced by using the air carbon arc process. Air carbon arc tends to leave the cut surface with a concave surface. When the two plates are put together, they appear as a U-shape. This system of joint preparation is popular in field welding and construction because it is quite fast and portable.

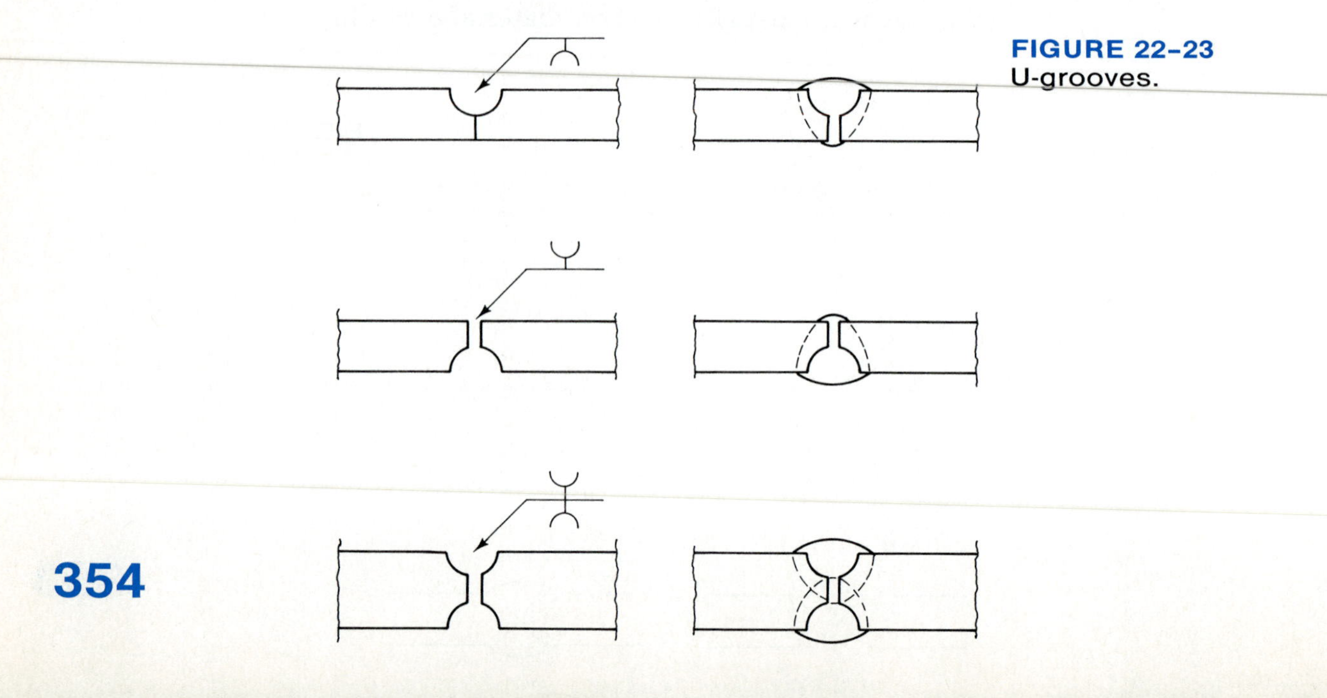

FIGURE 22–23
U-grooves.

J-Grooves (Figure 22–24) J-Grooves are variations of the U-grooves. With the J-groove, only one side is prepared with the air carbon arc. You will notice that as in the bevel groove, the J-groove requires a break in the arrow.

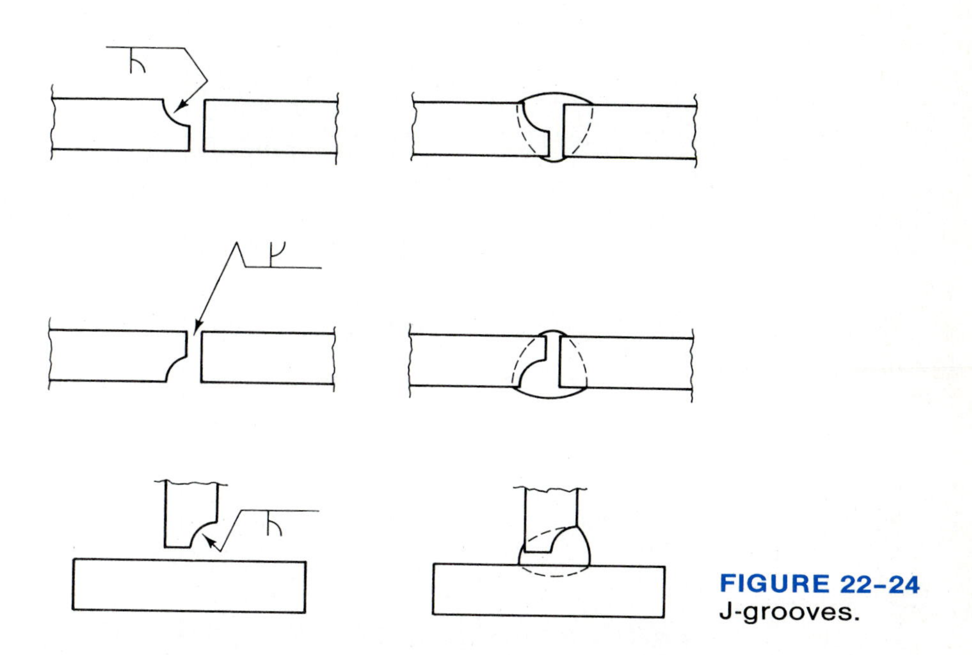

FIGURE 22–24
J-grooves.

Flair V-Grooves (Figure 22–25) The flair V-groove is used to indicate welds on pipe-to-pipe or round stock-to-round stock. The weld joint looks similar to a fillet configuration except for the curved surface.

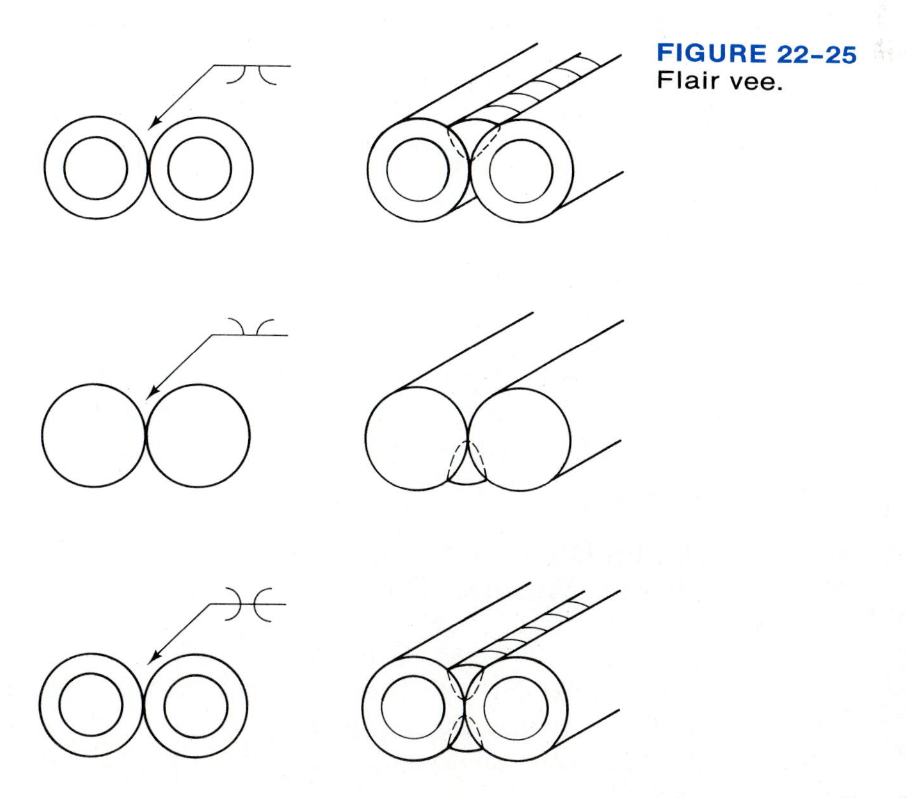

FIGURE 22–25
Flair vee.

Flair Bevel Grooves (Figure 22–26) The flair bevel is similar to the flair Vee-groove except that it is a pipe-to-plate or round stock-to-plate weld and only one side is flaired.

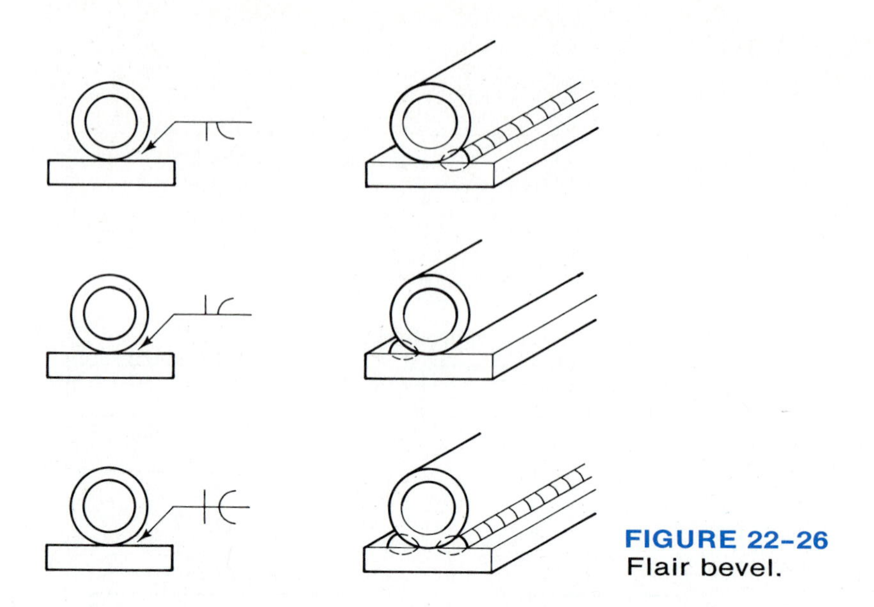

FIGURE 22–26
Flair bevel.

Scarf Joints (Figure 22–27) The scarf joint is one of the newest symbols. It is used exclusively on brazed or bronzed welded joints. It differs from the square groove in that it is cut at an angle. This allows more surface area for the brass to grip, yet we have the same thickness. Remember, brazing uses capillary action to bond plates together, not fusion. With capillary action, the more surface area, the stronger the bond.

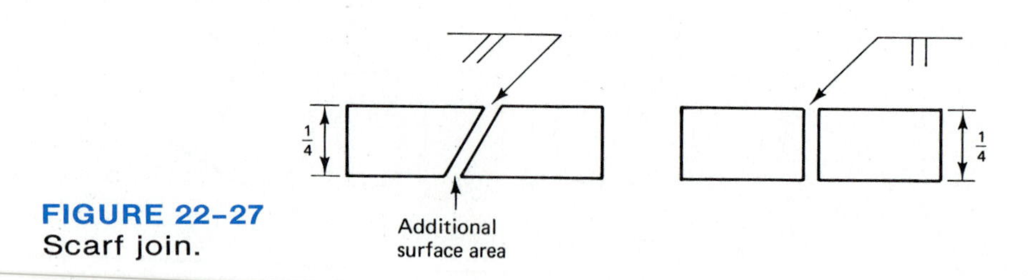

FIGURE 22–27
Scarf join.

Sizing Groove Welds

The size of groove welds is placed on the left side of the symbol and refers to the depth of the groove. It is important to note that depth is measured from the surface of the plate, not from the edge (Figure 22–28).

FIGURE 22–28
Surface, depth, and plate edge.

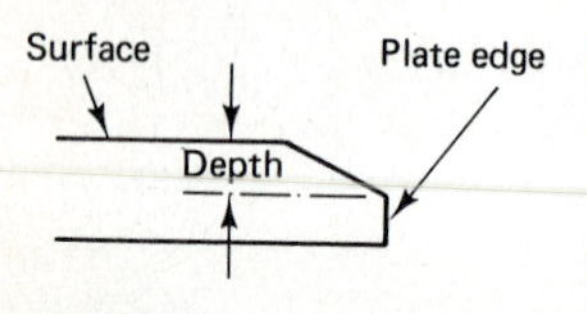

Depth Dimensioning Depth dimensioning of groove welds is shown in Figure 22–29.

Effective Throat Dimensioning The effective throat of a groove weld is the distance straight through the weld, starting in line with the surface and going to the deepest point of penetration (Figure 22–30). Notice that you do not include any weld reinforcement, but you

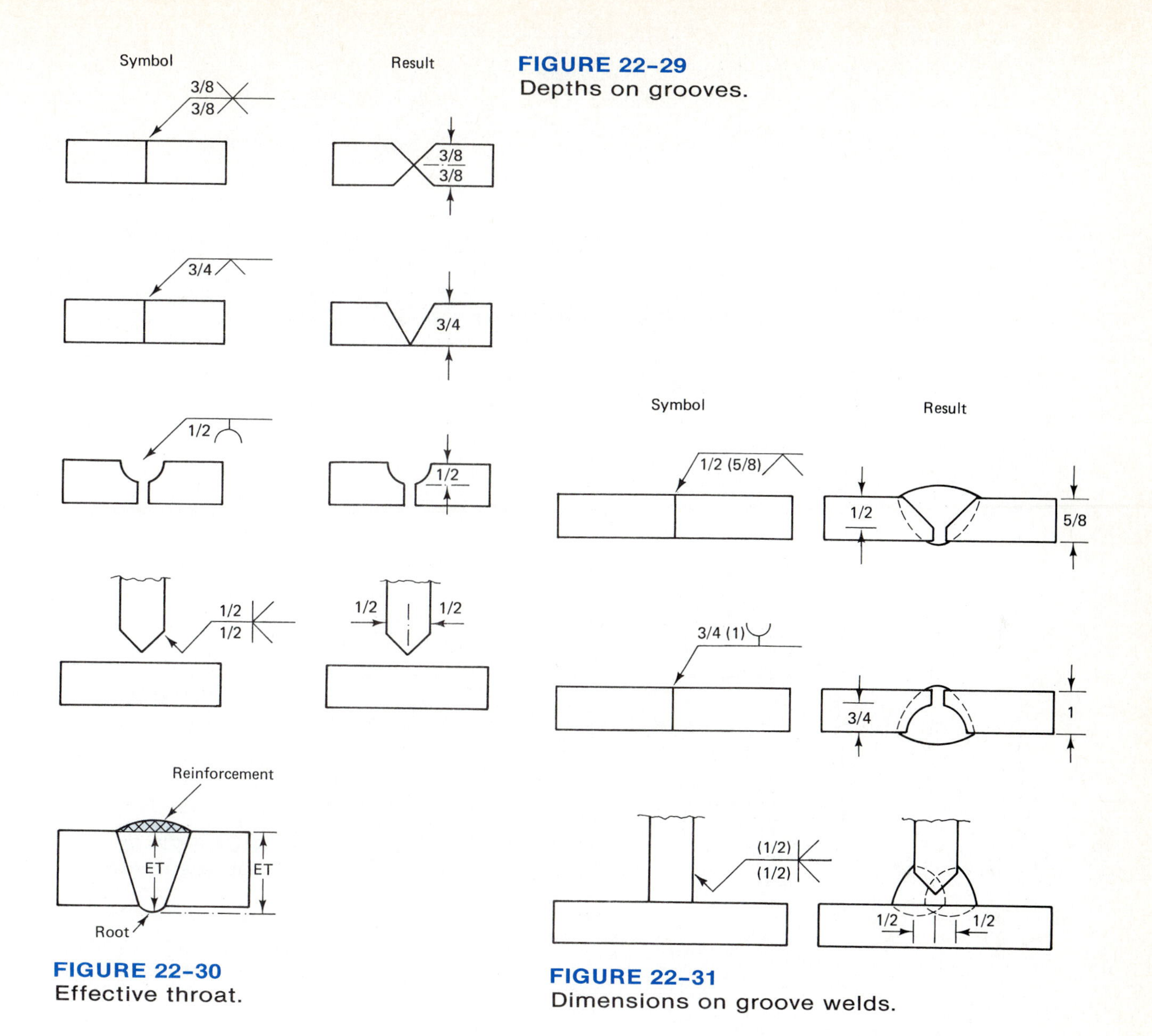

FIGURE 22-29
Depths on grooves.

FIGURE 22-30
Effective throat.

FIGURE 22-31
Dimensions on groove welds.

do include the root penetration. The effective throat is shown on the welding symbol to the left of the symbol and in parentheses (Figure 22-31).

Root Opening

There are many situations where groove welds require a root opening (GAP) between the abutting plates, to allow for adequate penetration. The root-opening dimension is placed in the center of the symbol (Figure 22-32).

Included Angle

An included angle (given in degrees) is the total angle produced by adding the bevel angle of both plates. The included angle is added to a groove welding symbol when it is necessary to accurately indicate the angle in which the welder will have to weld in (Figure 22-33). Engineers should take into consideration position, electrode diameter, and thickness of material when specifying the included angle. Remember, it is the *included angle, not* the bevel angle, that is given on the welding symbol.

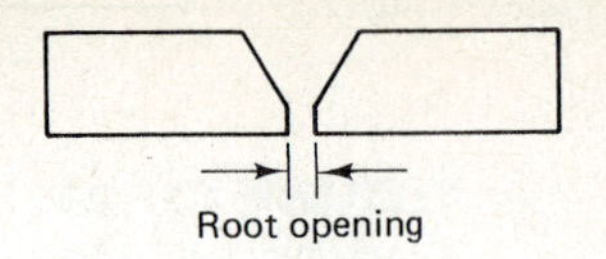

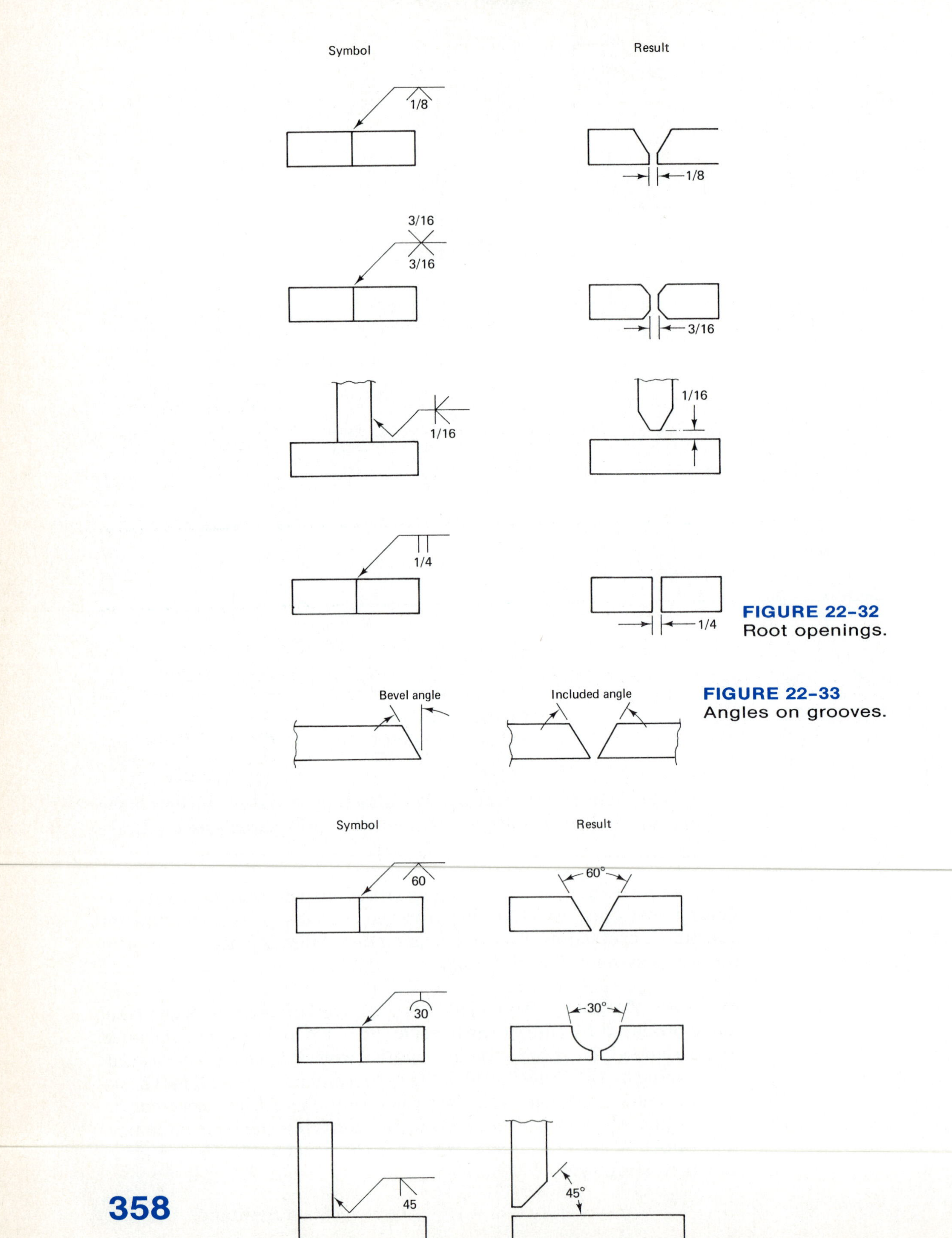

FIGURE 22-32
Root openings.

FIGURE 22-33
Angles on grooves.

Some typical groove weld symbol applications are shown in Figure 22–34.

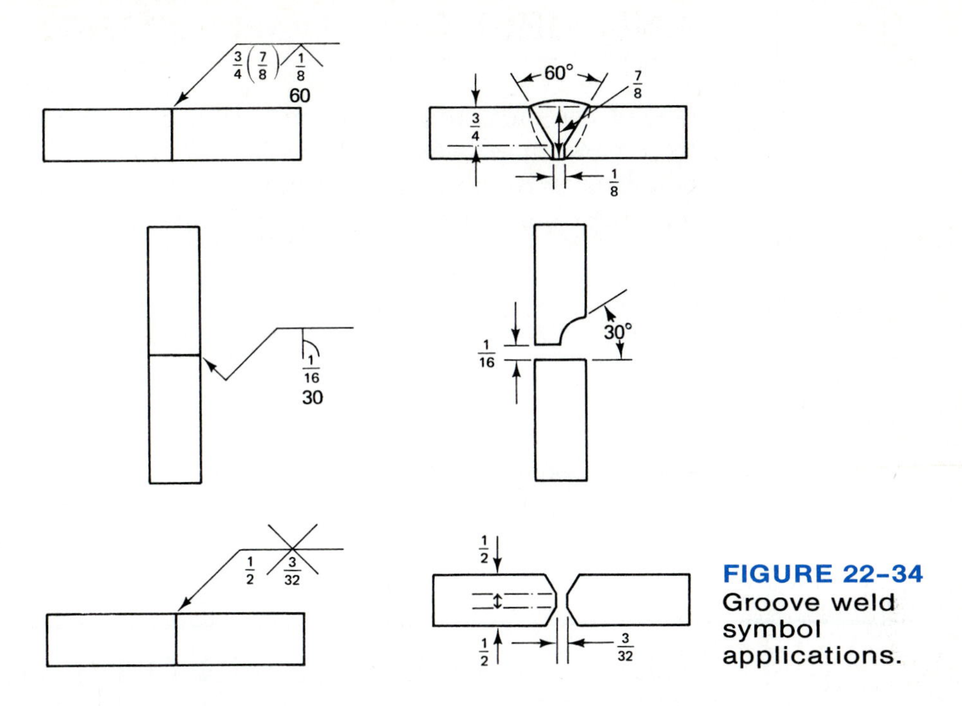

FIGURE 22–34 Groove weld symbol applications.

COMBINATION GROOVE AND FILLET WELD SYMBOLS

A bevel groove on both sides, capped off with fillets, is shown in Figure 22–35a; a J-groove on both sides, capped off with fillets, is shown in Figure 22–35b.

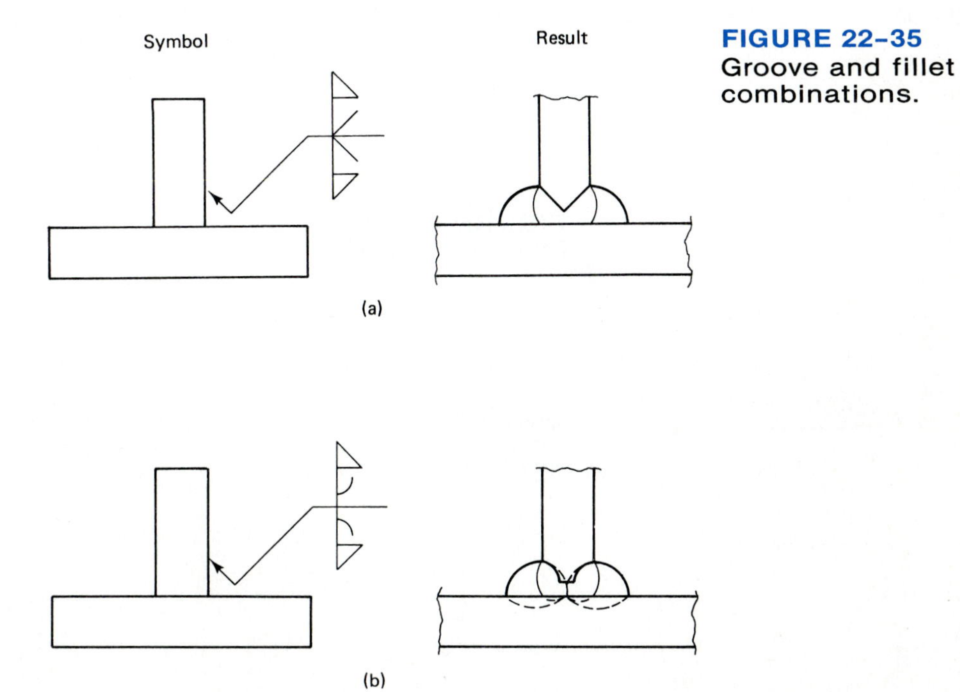

FIGURE 22–35 Groove and fillet combinations.

COMPLETE PENETRATION AND MELT-THROUGH SYMBOL

One of the most important factors in developing full weld strength is fully penetrated welds. On welding symbols we can indicate complete penetration by placing the letters "CP" in the tail of the symbol, or by using the melt-through symbol (Figure 22-36).

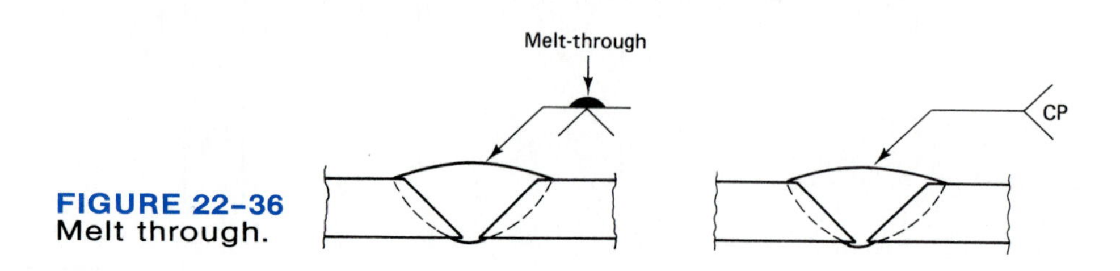

FIGURE 22-36
Melt through.

WELDS WITH BACKING OR SPACERS

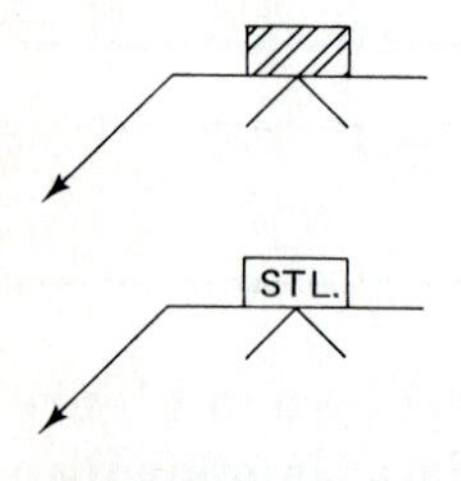

FIGURE 22-37
Symbols with backing.

Welds can be shown with weld backing or plate backing strips (Figure 22-37). The type of steel of which the backing is made can be indicated in the box.

A weld placed on the back of the groove joint is known as a back or backing bead. If it is a backing bead, it is put on *first,* then the bead is put on the back. The word "back" or "backing" can be placed in the tail (Figure 22-38).

Spacers are used on very thick sections to space apart groove welds (Figure 22-39). They are usually limited to groove welds under compressive stress. This is to prevent stress risers that could result from other types of stresses.

FIGURE 22-38
Back and backing.

FIGURE 22-39
Symbols with spacers.

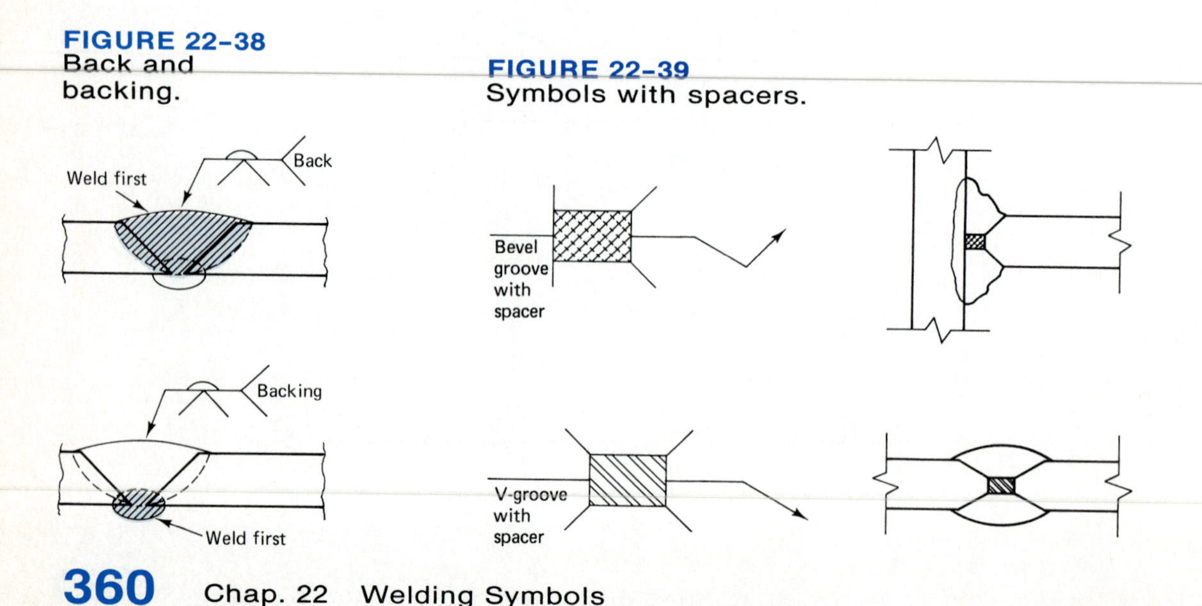

ADDITIONAL REFERENCE LINES

When it is necessary to show the sequence of weld operations, additional reference lines can be used. Always perform the welding operation on the reference line closest to the arrow. The next operation will be the second reference line back from the arrow, and so on (Figure 22-40). Welding symbols can also use one reference line, but stack weld symbols to show a combination of welds (Figure 22-41). When this system is used, always start from the reference line and work out when performing welding operations.

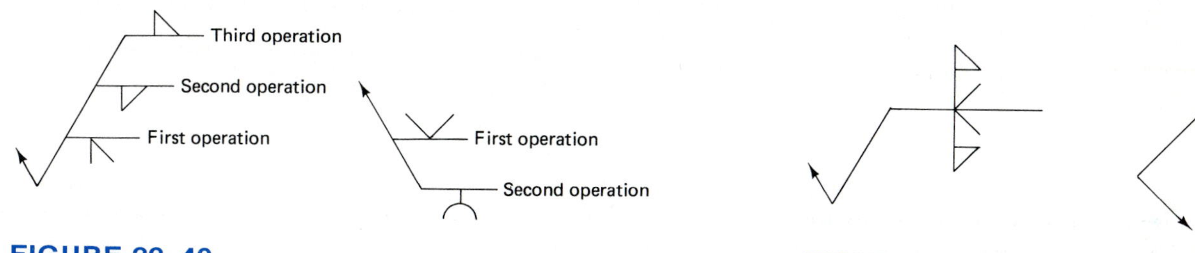

FIGURE 22-40
Sequence of application.

FIGURE 22-41
Stacked symbols.

PLUG AND SLOT WELDS

Plug and slot welds are used almost exclusively on lap joints or plates that are in lap joint configurations. A hole or slot is cut or drilled into one of the plates, then filled with weld (Figure 22-42). The welding sym-

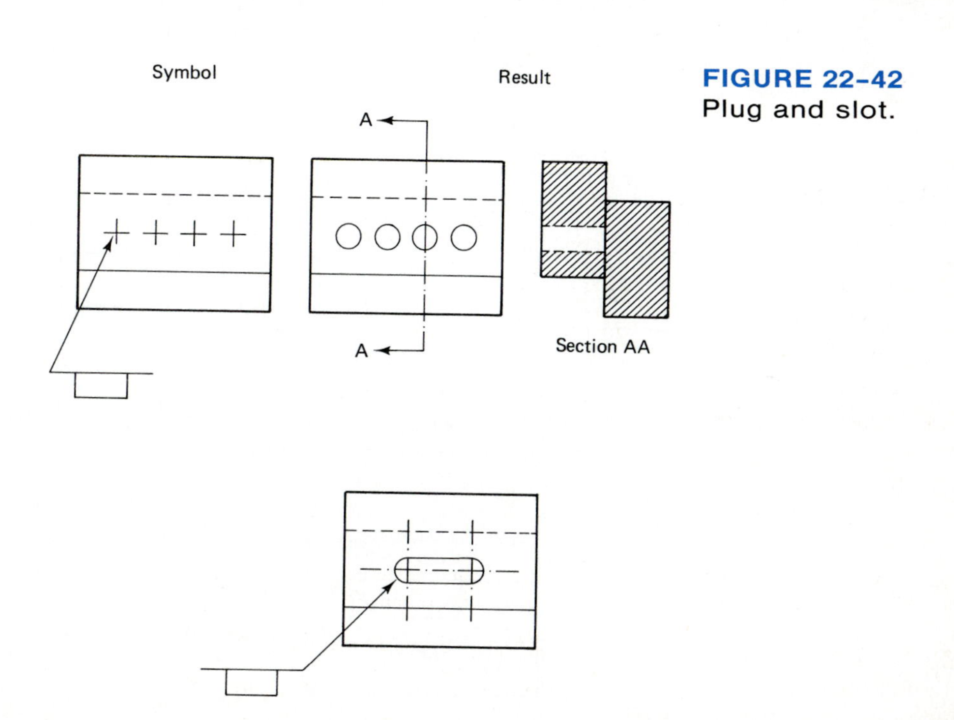

FIGURE 22-42
Plug and slot.

bol is the same for both a plug and a slot weld. Drawings using slot welds should include the details of the slot on the print, or have the word "slot" in the tail of the symbol (Figure 22-43). Plug and slot welds can include the following information on the symbols:

Plug: size (diameter at base), angle of countersink, depth of fill, pitch (center to center), contour, and finishing.

Slot: depth of fill, contour, and finishing

Note: if there is no depth of fill given, it is assumed that you fill the plug or slot completely.

It is also important to weld a plug or slot with the correct sequence. Starting from the outside edges, use a circular motion and work weld metal toward the center (Figure 22-44). Remove the slag from each layer before starting the next.

FIGURE 22-43
Plug and slot.

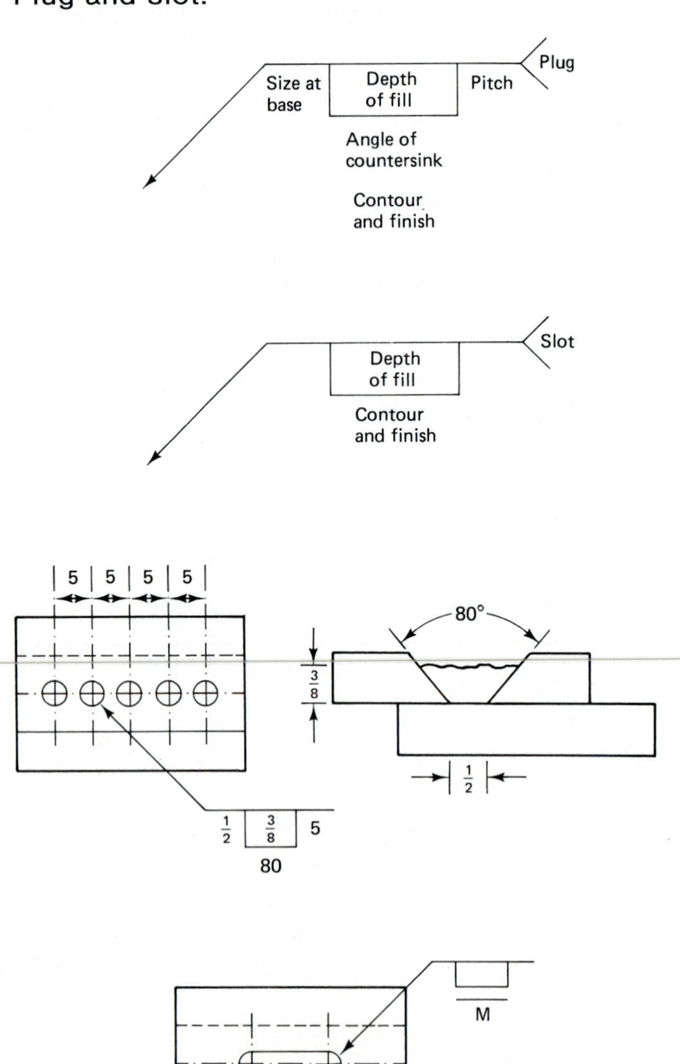

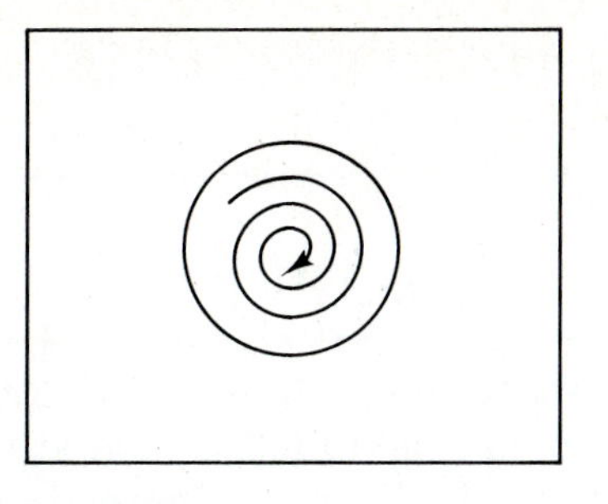

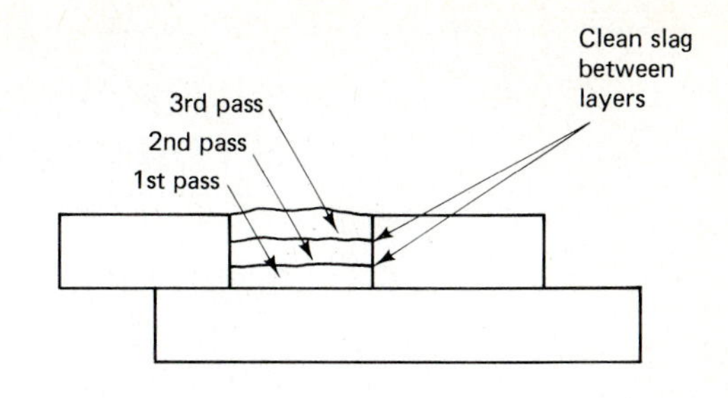

FIGURE 22-44
Making plug welds.

FLANGE WELDS

Welding sheet metal ($\frac{3}{16}$ in. and less) presents a problem when using conventional joints. Flange joints make welding much easier and more uniform in appearance. There are two types of flange joints: (1) the edge flange, and (2) the corner flange (Figure 22-45). Flange joints can be flanged just enough so that the extra is consumed, leaving the joint looking like a standard edge or corner joint (Figure 22-46).

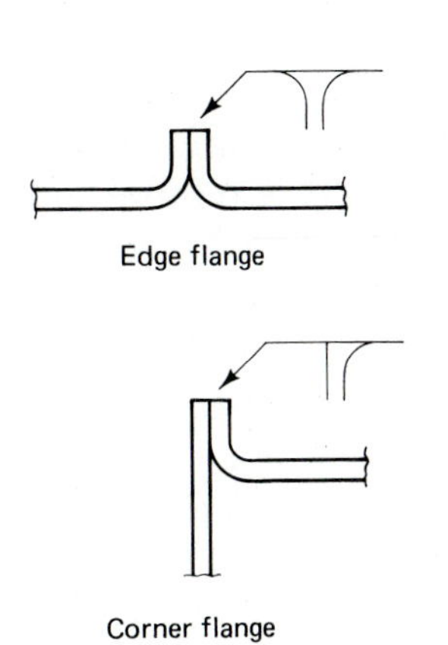

FIGURE 22-45
Edge flange and corner flange.

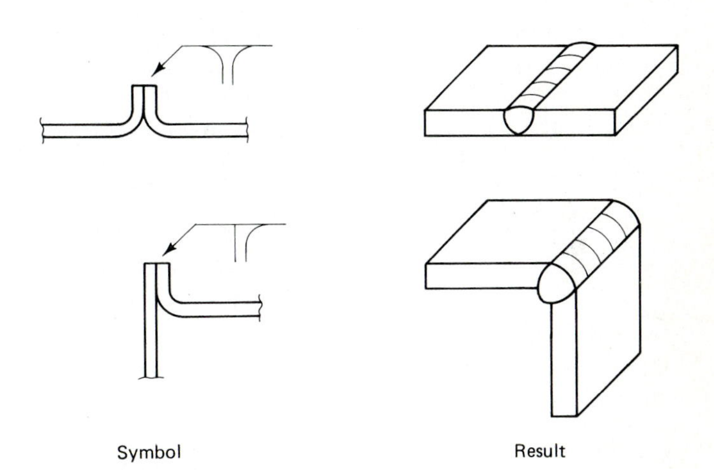

FIGURE 22-46
Edge flange and corner flange and results.

Sizing Flange Joints

Getting just the right amount of flange is critical in getting the joint to come out correctly. Radius, height above point of tangency, size of weld, and root opening are dimensions that can be added to flange symbols (Figure 22-47). The tangent is defined as touching a curve or radius. Notice that the height above tangency touches the radius curve.

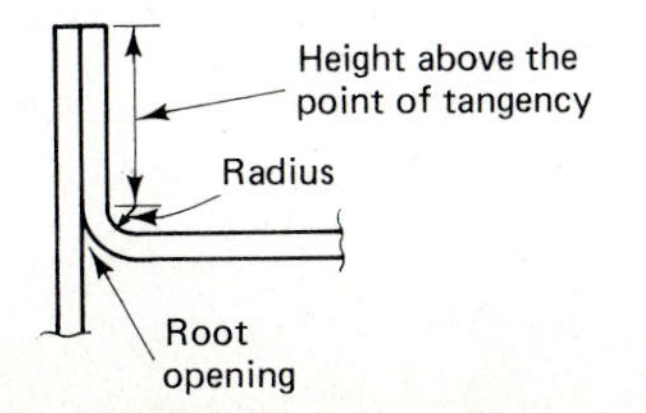

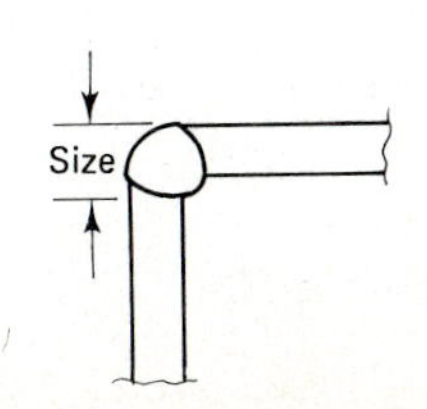

FIGURE 22-47
Sizes of edge and corner flanges.

SURFACING OR HARDFACING SYMBOLS

Welds that are applied to build up surfaces that are subjected to abrasion, impact, or high wear are known as surfacing or hardfacing welds. The extent of surfacing welds must be detailed on the drawing; however, the height of buildup can be added to the left of the symbol (Figure 22-48). More information about surfacing and hardfacing is included in Chapter 7.

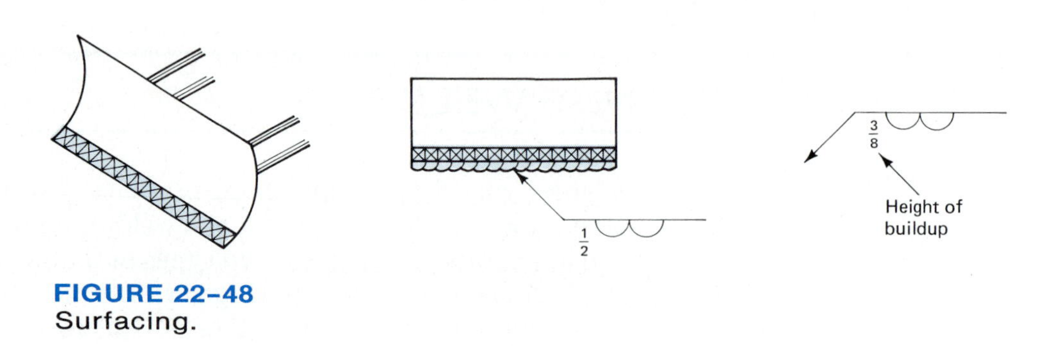

FIGURE 22-48
Surfacing.

SPOT WELDING SYMBOLS

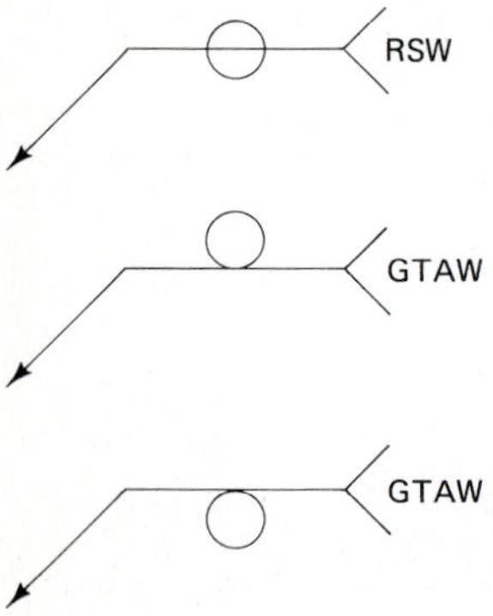

RSW

GTAW

GTAW

RSW: resistance seam weld
GTAW: gas tungsten arc weld
GMAW: gas metal arc weld

FIGURE 22-49
Spot.

The same symbol is used for both resistance and arc spot welding. However, if the symbol is referring to a resistance spot weld, it will have the neither side significance; if it is referring to an arc spot weld, it will have an arrow or "other side" significance (Figure 22-49).

A spot weld can contain the type of information shown in Figure 22-50, where $\frac{1}{4}$ is the diameter of the spot weld, 5 the number of spot welds, and 4 the pitch or center-to-center spacing between the spot welds. Shear strength or diameter of the spot may be placed on the left side of spot welding symbols (Figure 22-51; 200 = 200 lb of shear strength). Shear strength is used because spot welds under load are subjected to shear forces, not tension forces.

A variation of resistance spot welds are projection welds. They use the same symbol, but "RPW" (resistance projection weld) is put in the tail of the symbol (Figure 22-52). Resistance projection welds will have only an arrow or "other side" significance. The side the symbol is on is also the side the projection is placed on. A projection weld uses a

FIGURE 22-50
Size of spot.

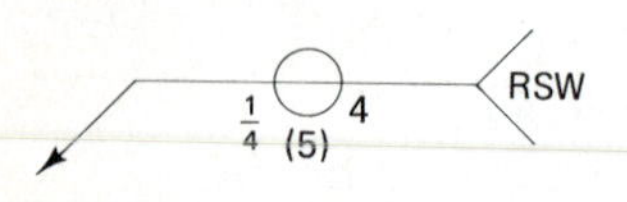

FIGURE 22-51
Spot symbols with shear strength.

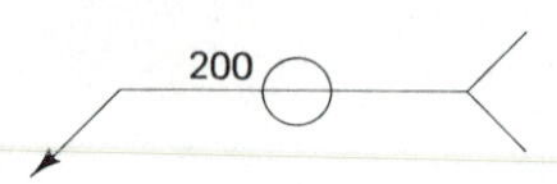

FIGURE 22-52
Resistance projection welds.

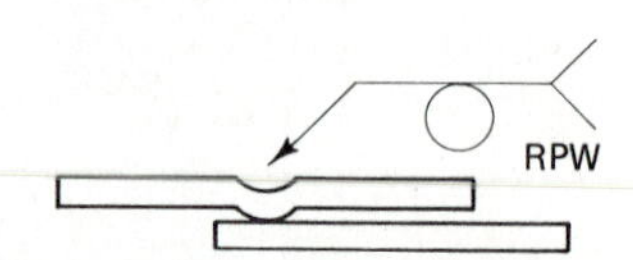

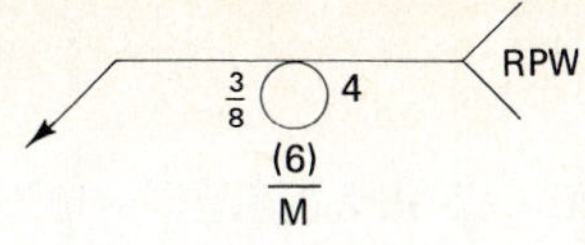

FIGURE 22-53 Resistance projection welds with sizes.

projection tool to place a small impression or projection on one of the two plates. The result is a weld similar to a spot weld, but one side is clear of weld marks. The same sizing information that is placed on spot welds can be placed on projection welds (Figure 22-53).

SEAM WELDING SYMBOLS

Seam welds are like spot welds in that they can be resistance or arc seam welds. Resistance seam welds (RSEW) will have the neither side significance. Arc seams will have "arrow" or "other side" significance to indicate on which side the weld seam is to take place (Figure 22-54).

Seam welds can contain the welding information on the symbol. See Figure 22-55, where $\frac{1}{2}$ is the width of seam, 2 the length of seam, and 5 the pitch of intermittent seams. The "width of seam" dimension can be substituted for by the "shear strength (in psi) of the seam." If there is only one dimension to the right of the symbol, it is the length of the seam (Figure 22-56; 300 = shear strength, 4 = length of seam).

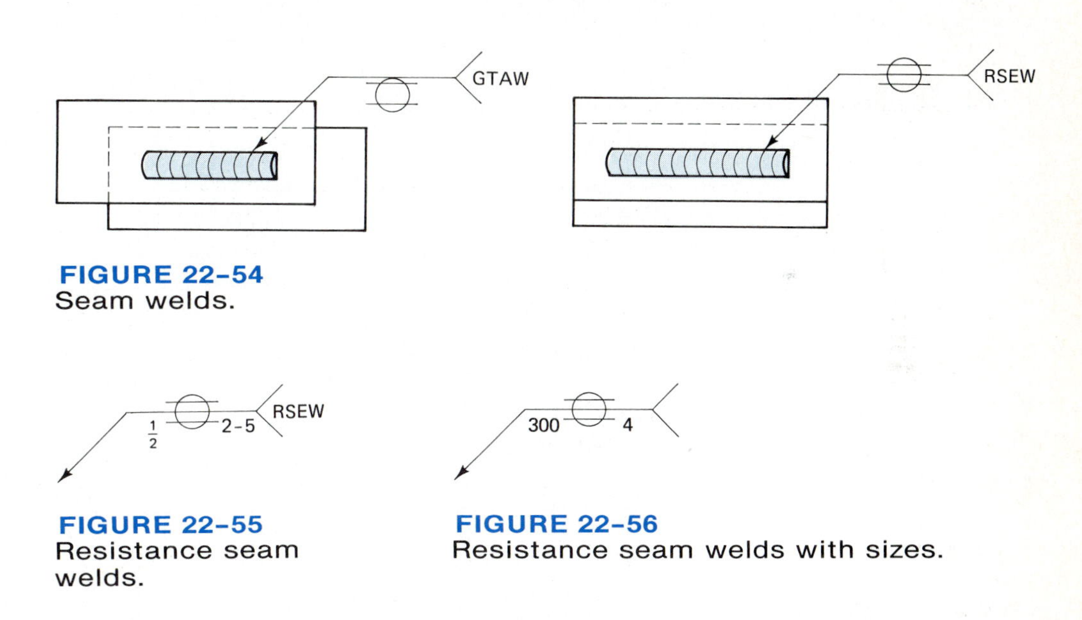

FIGURE 22-54 Seam welds.

FIGURE 22-55 Resistance seam welds.

FIGURE 22-56 Resistance seam welds with sizes.

CONTOUR AND FINISHING SYMBOLS

Convex Concave Flush

FIGURE 22-57 Finish contours.

Weld symbols may have three possible finish contours: (1) convex, (2) concave, or (3) flush (Figure 22-57). The most common are the flush and the convex; the concave symbol is somewhat rare, but does have some applications.

The method of making the desired contour is known as the finishing process. Commonly used finishing processes and their symbols are listed in Table 22-1. These symbols will be placed on top of the contour symbol when necessary to apply them. These contour and finishing symbols may be applied to almost any type of welding, from fillets and grooves to plug and slots, and even some spot welding applications. Some applications of contour and finishing symbols are shown in Figure 22-58.

TABLE 22-1
Symbols for Finishing Processes

SYMBOL	FINISHING PROCESS
C	Chipping (usually pneumatic chipping)
M	Machining (usually a milling process)
G	Grinding (usually a high-speed grinder or surface grinder)
R	Rolling (usually a hot-rolling process)

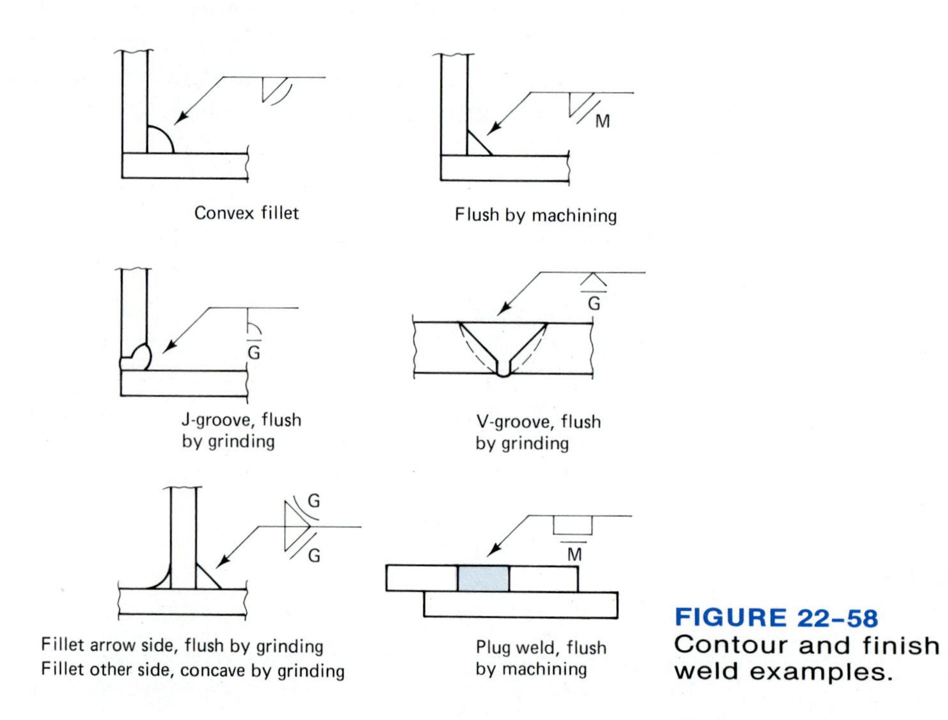

FIGURE 22-58 Contour and finish weld examples.

NONDESTRUCTIVE TESTING SYMBOLS

Nondestructive testing includes a group of testing methods that examine the material without destroying it. The commonly used NDT processes are listed in Table 22-2.

Many of the rules for standard welding symbols also apply to NDT symbols. Note also that standard welding symbols can be used in conjunction with NDT symbols, and on the same symbol (Figure 22-59). We can add in the tail some additional information that is used on conventional weld symbols, such as field weld, weld all around, and the number of tests (Figure 22-60).

With radiographic testing, we can also indicate the angle at

TABLE 22-2
NDT Processes

PROCESS	SYMBOL
Radiographic testing (x-ray and gamma ray)	RT
Ultrasonic testing	UT
Magnetic particle testing	MT
Penetrant testing	PT
Eddy current testing	ET
Visual testing	VT
Acoustic testing	AET
Leak testing	LT

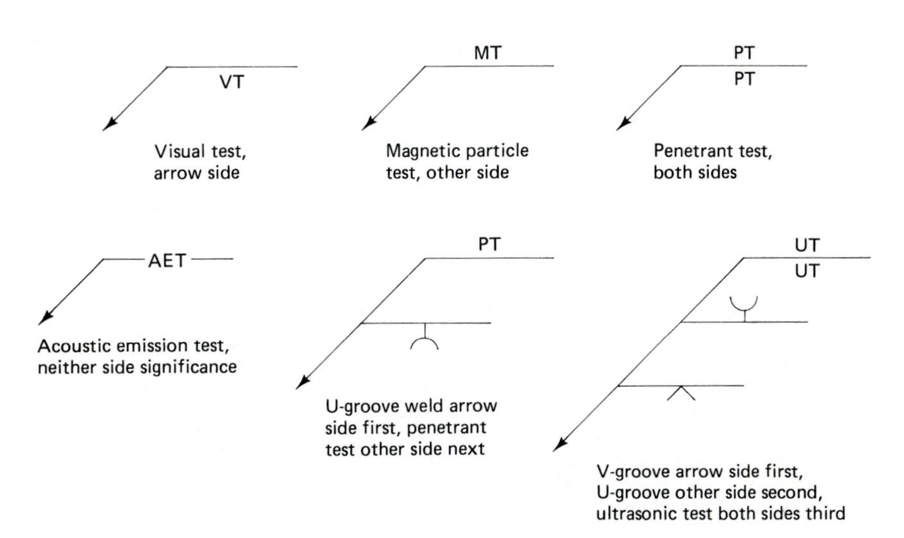

FIGURE 22-59
VT, MT, PT, AET, and UT.

FIGURE 22-60
All around NDT.

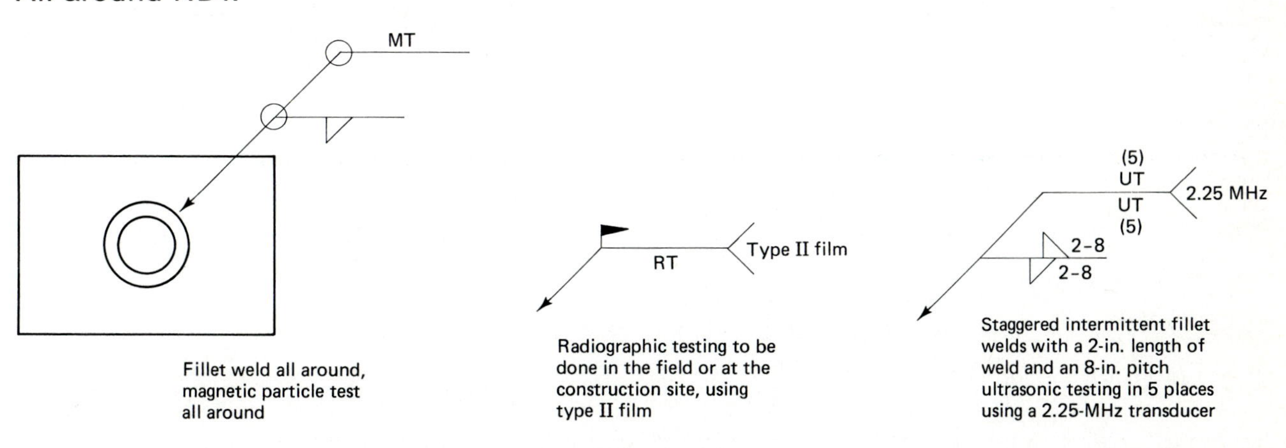

which the radiation source is to be aimed (Figure 22-61). In some cases this angle is critical for good radiographic quality and coverage.

The amount of testing required can be indicated on the welding symbol by giving the length or the percentage to be tested. The length

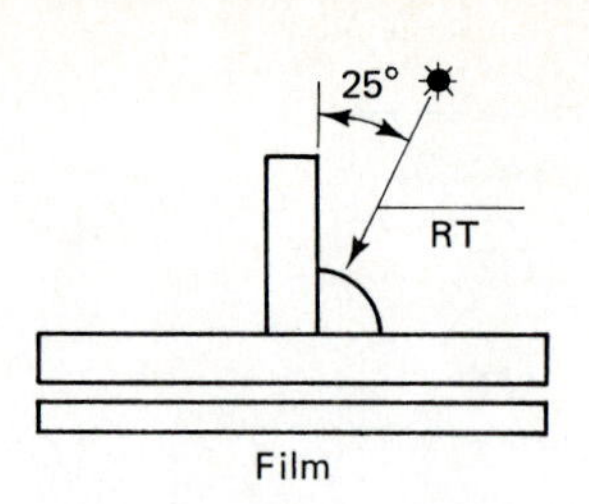

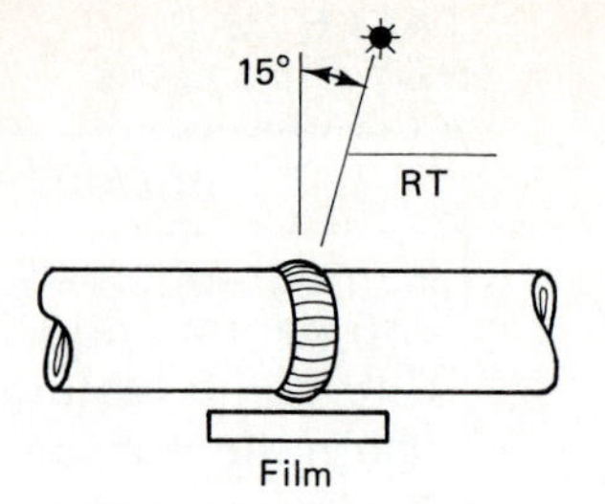

FIGURE 22-61
RT.

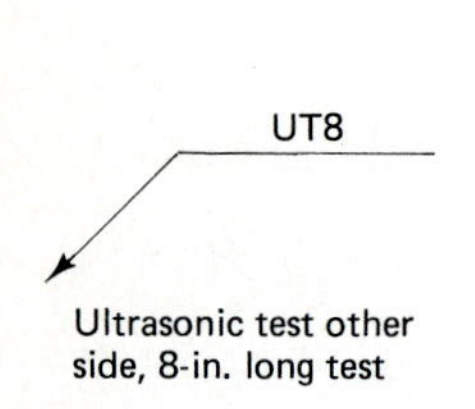

FIGURE 22-62
UT.

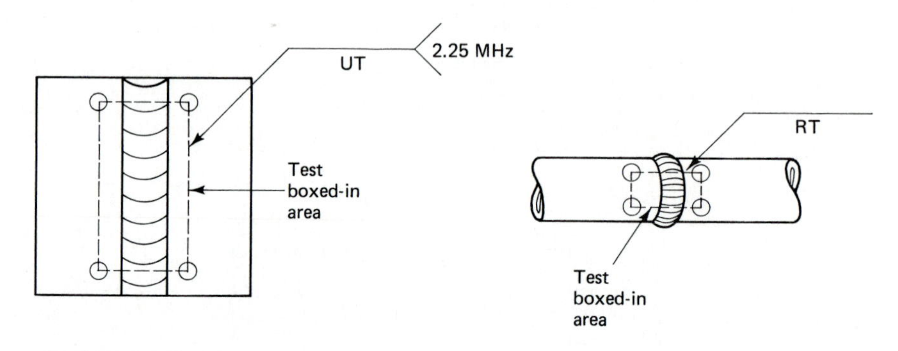

FIGURE 22-63
UT and RT in test areas.

to be tested is given to the right of the symbol in inches (Figure 22-62; metric prints may give the length in millimeters).

If it is necessary to show a more *specific area* to be tested, this is done by showing a dashed line around the area, with circles at the corners (Figure 22-63). This system is used to prevent the reader from mistaking the area or line as hidden lines or surfaces.

REVIEW QUESTIONS

1. Name the basic parts of the symbol shown in the accompanying figure.

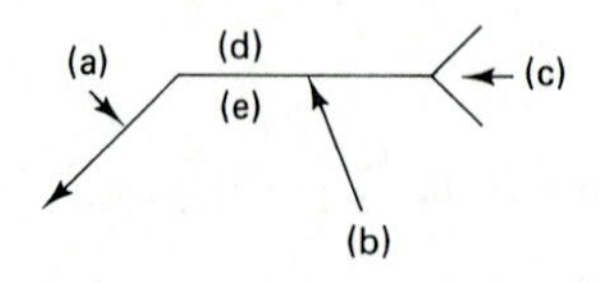

2. Identify the weld symbols shown in the accompanying figure.

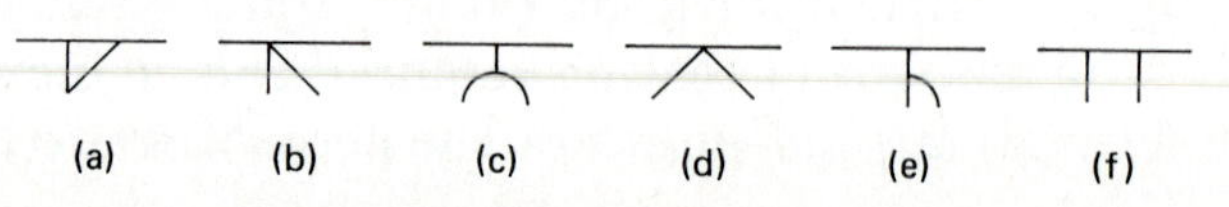

3. Draw in the welds where required.

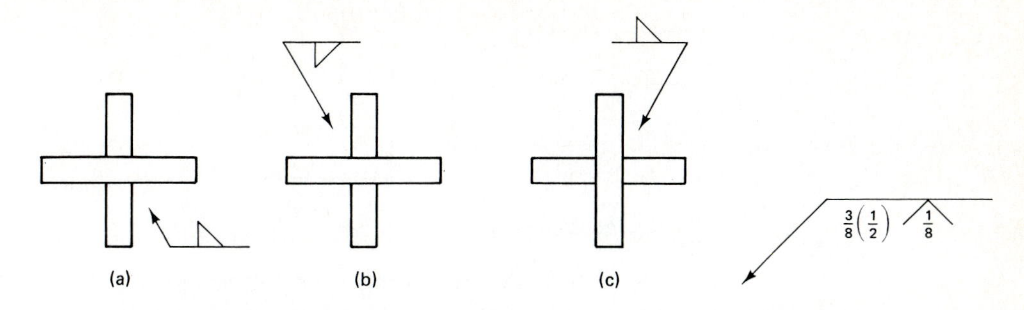

4. a. Identify the type of weld shown in the drawing.
b. What is the root opening?
c. What is the effective throat?
d. What is the depth of groove?

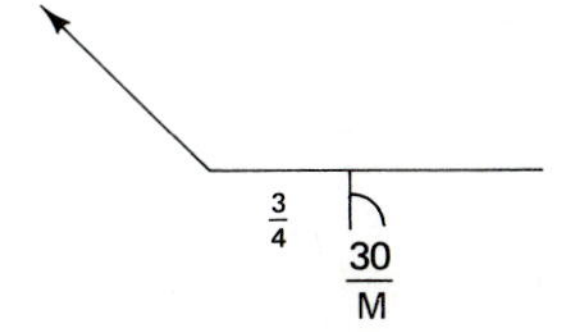

5. a. Identify the type of weld shown in the drawing.
b. What is the depth of groove?
c. What is the included angle?
d. What contour and finishing are required?
Extra credit: What is incorrect about this welding symbol?

6. Add correct sizes to complete the fillet weld symbol.

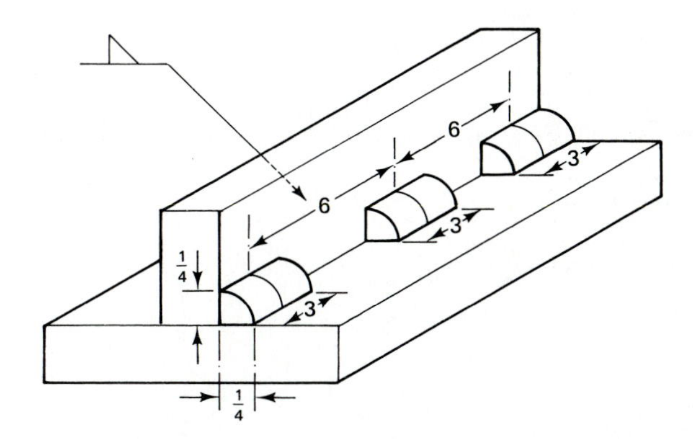

7. Complete the welding symbol; add the correct weld symbol and required dimensions.

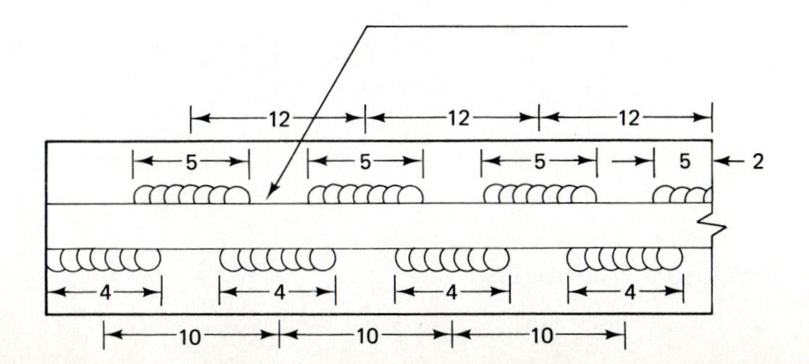

8. Draw in the required weld and the groove preparation into the joint.

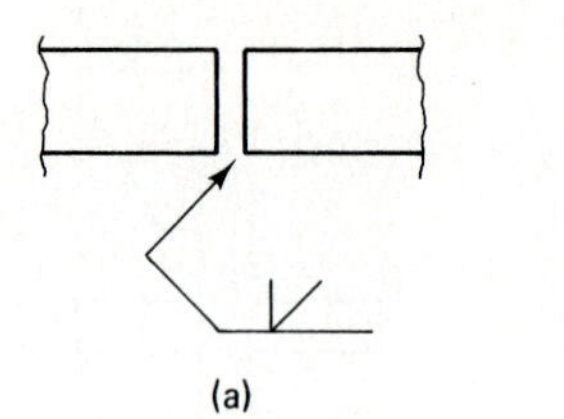

(a)

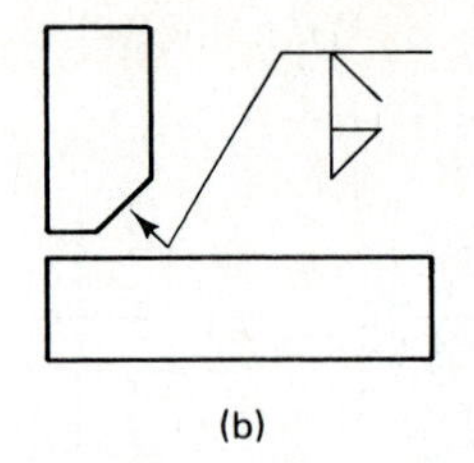

(b)

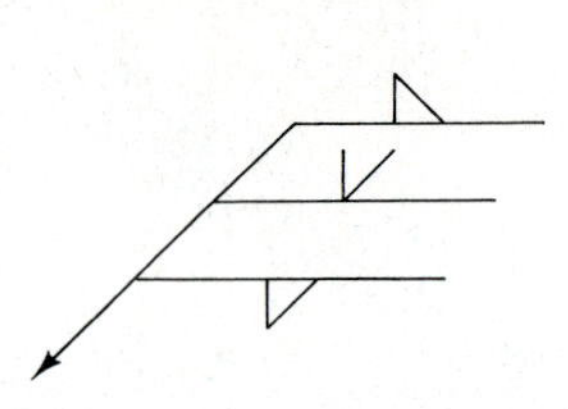

9. a. Identify the symbol sequence and side.
   b. First, ________ weld the ________ side.
   c. Second, ________ weld the ________ side.
   d. Third, ________ weld the ________ side.

10. Identify each symbol.

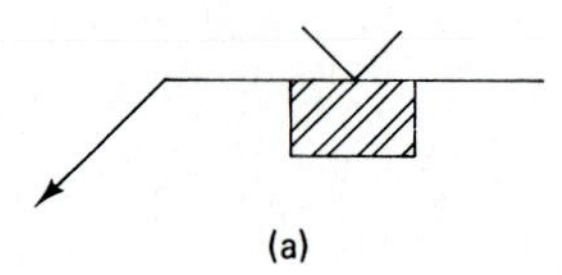

(a)

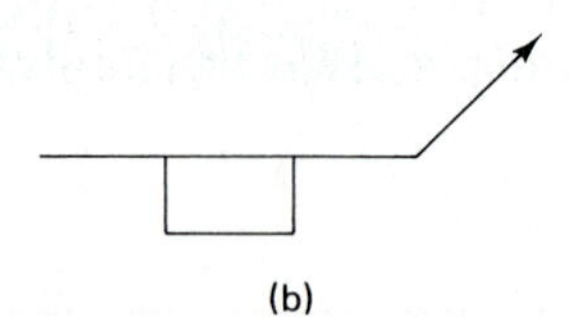

(b)

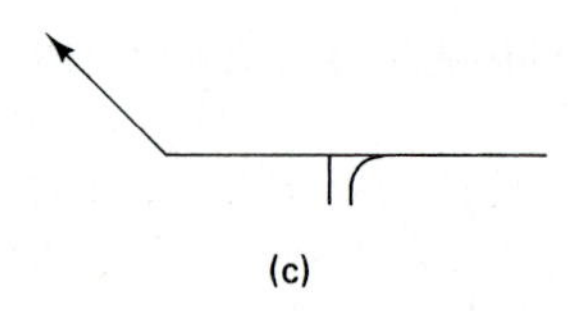

(c)

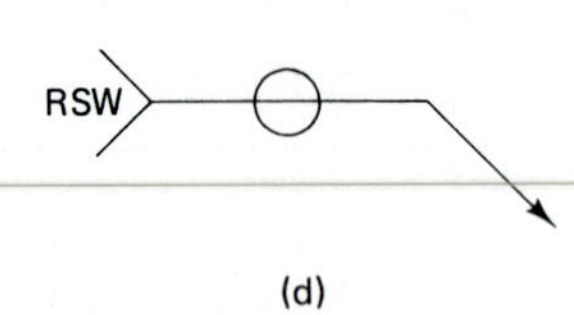

(d)

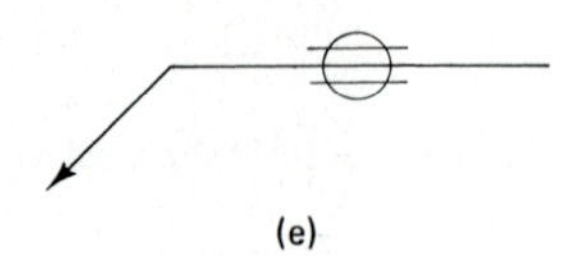

(e)

WELDING CODES AND STANDARDS

Welding codes are established when weld failure could result in loss of life, injury, or serious property damage. There are many welding codes in industry today, but you will find that in many areas they are similar in content. However, since buildings, bridges, boilers, pipelines, and other weldments each undergo different types of stresses, each code must differ slightly in these critical areas to satisfy these varying design differences.

PHOTO 23–1

The major welding-code-writing agencies in North America include:

- American Welding Society (AWS)
- American Society of Mechanical Engineers (ASME)
- American Petroleum Institute (API)
- American National Standards Institute (ANSI)
- Department of Defense (military codes, known as "mil specs")

Other welding codes exist, but the ones issued by these agencies are the most commonly used.

AWS CODES

The American Welding Society has written and published many welding codes and standards but is probably best known for its publication D1.1, Structural Welding Code. This standard, although it has been adopted as a standard for many types of weldments, is intended to apply to buildings, bridges, and related structures. The AWS D1.1 includes the major areas listed in Table 23-1. The American Welding Society has also published standards for welding filler metals. These standards set ranges for chemical compositions and strength requirements of electrodes and filler wires as well as fluxes.

TABLE 23-1
AWS D1.1 Code Sections

SECTIONS	NAME	TOPICS
1	General Requirements	Acceptable base and filler metals for use
1		Welding processes and safety precautions
1 and 2	Prequalified Joints	Drawings, symbols, and units of measurement
3	Workmanship	Weld sizes and profiles
3		Joint preparation, fit-up, dimension tolerances, and distortion control
3		Weld corrections (repairs) and stress-relief heat treatment
4	Techniques	Techniques including preheat and interpass requirements as well as special procedural requirements for SMAW, GMAW, SAW, and ESW
5	Qualification	Qualification of welders, tackers, welding operators, and welding procedures
6	Inspection	Inspection requirements, including NDT techniques
7	Stud Welding	Strengthening and repairing existing structures
8	Static Loaded Structures	Design criteria and special requirements for buildings
9	Dynamic Loaded Structures	Design criteria and special requirements for bridges
10	Tubular Structures	Design criteria and special requirements for tubular structures

Other AWS code publications:

- Structural Welding Code: Aluminum, D1.2
- Structural Welding Code: Sheet Metal, D1.3
- Structural Welding Code: Reinforcing Steel, D1.4
- Steel Hull Welding, D3.5
- Aluminum Hull Welding, D3
- Welded Steel Elevated Tanks, Stand Pipes, and Reservoirs for
- Water Storage, D5.2
- Automotive Welding Design, D8.4

Other AWS publications categories include:

- Safety and health
- Welding processes
- Inspection and qualification
- Metallurgy
- Educational and training

For a complete list of AWS publications, write

American Welding Society
P.O. Box 351040
Miami, FL 33135

ASME CODES

The American Society of Mechanical Engineers (ASME) series of codes are intended for use with boilers, pressure vessels, and nuclear reactors. As with AWS D1.1, the ASME code is used for similiar weldments in addition to boilers and pressure vessels; for example, tanks and containers that will contain dangerous substances may be welded to ASME standards.

The ASME boiler and pressure vessel code is divided into the sections listed in Table 23-2. The ASME code is one of the strictest codes; this is necessary because of the critical nature of most modern all-welded boilers and pressure vessels. Section IX is also used by other code-writing agencies for welder and procedure qualifications.

API CODES

The American Petroleum Institute (API) publishes standards for many areas of the petroleum industry. In the welding industry we are concerned with the API 1104, the standard for welding pipelines and re-

TABLE 23-2
ASME Code Sections

SECTION	TITLE
I	Power Boilers
II	Material Specifications Part A. Ferrous Materials Part B. Nonferrous Materials Part C. Welding Rods, Electrodes and Filler Metals
III	Subsection NCA. General Requirements for Division 1 and Division 2
III	Division 1 Subsection NB. Class 1 Components Subsection NC. Class 2 Components Subsection ND. Class 3 Components Subsection NE. Class MC Components Subsection NF. Component Supports Subsection NG. Core Support Structures Appendices
III	Division 2. Code for Concrete Reactor Vessels and Containments
IV	Heating Boilers
V	Nondestructive Examination
VI	Recommended Rules for Care and Operation of Heating Boilers
VII	Recommended Rules for Care of Power Boilers
VIII	Pressure Vessels Division 1 Division 2. Alternative Rules
IX	Welding and Brazing Qualifications
X	Fiberglass-Reinforced Plastic Pressure Vessels
XI	Rules for Inservice Inspection of Nuclear Power Plant Components. Division 1

lated facilities. This code is concerned with the welding of cross-country petroleum pipelines and natural gas pipelines. Major sections of API 1104 are listed in Table 23-3.

TABLE 23-3
API Code Sections

SECTION	TOPIC
1	General information
2	Qualification of welding procedures
3	Welder qualification
4	Design and preparation of a joint for production welding
5	Inspection and testing of production welds
6	Standard of acceptability—nondestructive testing
7	Repair or removal of defects
8	Radiograph procedure
9	Automatic welding

ANSI CODES

The American National Standards Institute (ANSI) approves many codes and standards. In the welding industry we are concerned with ANSI B31.1, the power piping code. This standard differs considerably from the API piping code in that "power piping" indicates used for high-pressure piping of steam or other substances for power generation. ANSI B31.1 is considered a fairly stringent standard, due to the dangerously high pressures used in power piping lines.

MILITARY CODES

Companies manufacturing for the military must follow the military specifications, or "mil specs." There are numerous military specifications covering almost every area of design, fabrication, and inspection. It is very important for the fabricator to communicate with the contract issuing agency to verify which mil spec(s) are to be followed.

USING THE CODES

Codes are often used in connection with each other. For example, ANSI B31.1 refers you to the ASME code, Section IX, for the Qualification of Welders and Procedures, and the API codes which refer the reader a number of times to the AWS code.

When studying codes, it is important to remember that it is not necessary, or even advisable, to memorize sections of a code. Rather, you need to know where to find the needed information quickly. For example, it is good to know that information on qualification of welding procedures is given in the first part of AWS D1.1, Section 5, or ASME Section IX, but memorizing specific sizes or variables of a code section may be a waste of time, because codes are amended fairly often. When any doubt exists regarding a specification, look it up in your code book.

An ability to use codes will greatly enhance your career. It is suggested that you obtain a copy of one of the welding codes and research various sections to become familiar with the weld quality levels required by codes. This will also give you important research skills that you will need if you plan to become a welding inspector.

REVIEW QUESTIONS

1. Why are welding codes established?
2. What welding code covers building and bridges?

3. What welding code covers welding of boilers, pressure vessels, and nuclear reactors?
4. What section of the AWS structural welding code covers qualifications of welding procedures, welders, and tackers?
5. What section of the ASME code covers welding and brazing qualifications?
6. What code covers the welding of petroleum pipelines?
7. What code covers the welding of power piping?
8. How should you study welding codes?

24 TESTING AND CERTIFICATION

WELDING PROCEDURES

Let's examine first the difference between welding procedures and welder qualification (certification). A welder qualification is a practical test given to a welder to verify his or her ability to produce sound welds. The type of test given the welder will depend on the welding process, type of metal, position, and other variables. The actual welder certification is a written document describing what the welder is certified to do and lists the test results.

A welding procedure is a test performed to verify that a weldment has the desired mechanical properties for a given job or loading conditions. The welding procedure will verify such things as the compatibility of base and filler metals, amperage settings, joint designs, preheat or interpass temperatures, and any variable that might affect the properties or integrity of the weld. Example procedures are shown in Figures 24-3 and 24-4 (pages 380 and 382).

Procedures should be performed before production starts. The time to find out that a base metal and filler metal are not compatible (or will not pass the desired mechanical properties) is not after you have produced 1500 parts, or whatever the case may be. Welding procedures will indicate these mechanical property difficulties prior to manufactur-

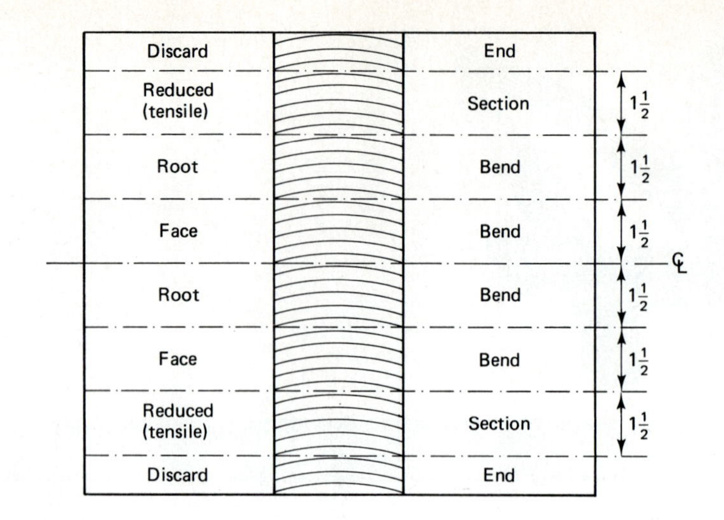

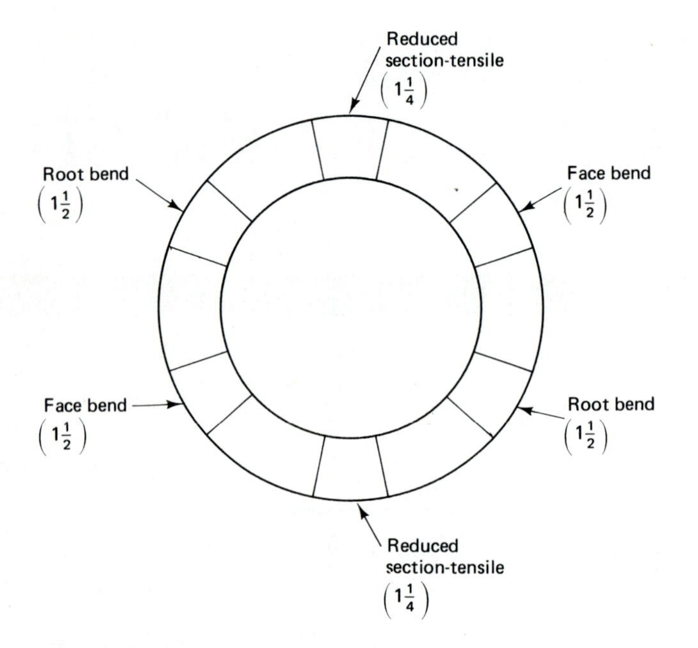

FIGURE 24–1 Procedure qualification requirements for plate and pipe $\frac{1}{16}$ to $\frac{3}{4}$ in. thick.

ing, while there is still time to change the procedure and correct the process.

There are usually a number of procedures that need to be qualified for a given job. The actual number of procedures that will be required depends on the essential variables. Any changes in these essential variables will require that a new procedure be qualified. The variables that require establishing a new procedure will vary from code to code.

Here is a list of typical essential variables.

1. Change in the welding process
2. Change in the type of metal (for example, mild steel to stainless steel)
3. Major classification change in an electrode or filler metal
4. Change in the joint design

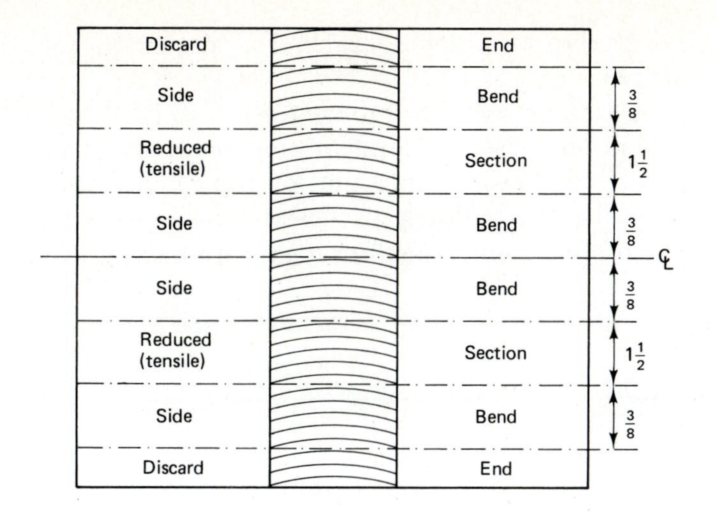

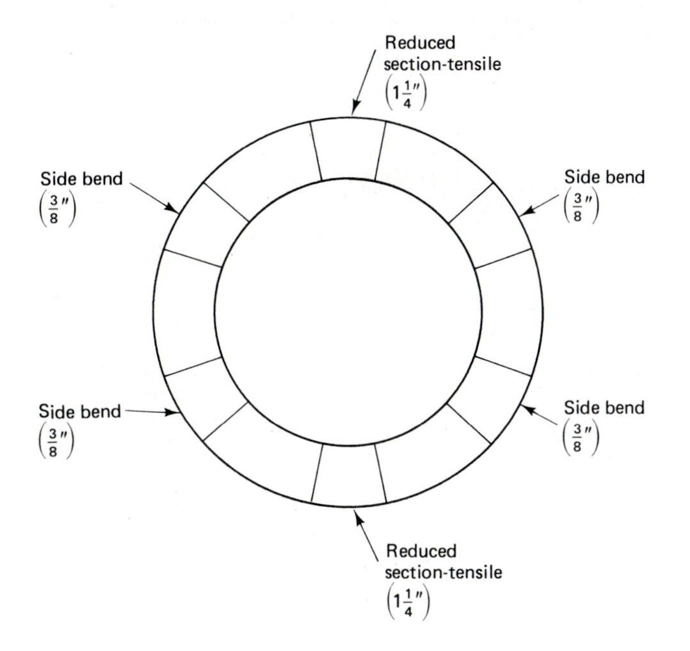

FIGURE 24–2 Procedure qualification requirements for plate and pipe over $\frac{3}{4}$ in. thick.

5. Change in the amperage preheat, or interpass, out of the range specified

The American Welding Society's structural welding code (AWS D1.1) allows fabricators and contractors to use "prequalified joints." These are procedures that if performed as specified in the code, will produce sound welds. If these procedures are used, only documentation of the welds used is required; mechanical testing may not be required.

Testing of welding procedures varies from code to code, but generally, destructive tests, including bend tests, tensile pulls, and nick break tests are used (Figures 24–1 and 24–2). Nondestructive testing of the procedure test plate or pipe is also required by some welding codes.

Most welding procedures require welding, testing, and recording of test data. However, they are intended to save the manufacturing fabricator money. Savings are realized in fewer product returns due to defective weldments, cracked welds, or unacceptable mechanical properties.

The *welding procedure forms* (Figures 24-3 and 24-4) contain all of the essential information required to make the weld and verify test results. This information may include the type of base and filler metals, the welding processes used, preheat, interpass, or postweld heat treatment shielding gases, and so on. (Continued on page 384.)

FIGURE 24-3 Welding Procedure Specifications (WPS)

Company Name ______________ By ______________
Welding Procedure Specification No. ______ Date ______ Supporting PQR No. (s) ______
Welding Process(es) ______________ Type(s) ______________

JOINTS (QW-402) DETAILS

Joint Design ______________
Backing (Yes) ______ (No) ______
Backing Material (Type) ______________

BASE METALS

Specification type and grade ______________

Mill Specifications
Thickness Range Covered

FILLER METALS

F-No: ______________

Code Spec ______

AWS No. (Class) ______________
Size of filler metals ______________

POSITIONS

Positions ______
Welding Progression: Up ___ Down ___

POSTWELD HEAT TREATMENT

Temperature Range ______________
Time Range ______________

PREHEAT

Preheat Temp. Min. ____________

Interpass Tem. Max. ____________

Preheat Maintenance ____________

GAS

Shielding Gas(es) ____________

Percent Composition (mixtures) ____________

____

Flow Rate ____________

____

____

ELECTRICAL CHARACTERISTICS

Current AC or DC ____________ Polarity ____________

Amps (Range) ____________ Volts (Range) ____________

Tungsten Electrode Size and Type ____________

Made of Metal Transfer for GMAW ____________

TECHNIQUE

String or Weave Bead ____________

Initial and Interpass Cleaning ____________

____

Contract Tube to Work Distance ____________

Multiple or Single Pass (per Side) ____________

____

Travel Speed (Range) ____________

Peening ____________

Other ____________

____

____

Weld Layers	Process	Filler Metal Class	Filler Metal Dia.	Current Type Polar.	Current Amp Range	Volt Range	Travel Speed Range	Other

FIGURE 24-3 (cont'd)
Department of Welding and Testing welding procedures qualifications.

Company Name ________________

Procedure Qualification Record No. ________________ Date ________________

WPS No. ________________

Welding Process(es) ________________

Types (Manual, Automatic, Semiauto.) ________________

JOINTS

Groove Design Used

BASE METALS

Material Spec. ________________

Type or Grade ________________

P No. ________ to P No. ________

Thickness ________________

Diameter ________________

Other ________________

FILLER METALS

Weld Metal Analysis A No. ________________

Size of Electrode ________________

Filler Metal F No.

SFA Specification ________________

AWS Classification ________________

Other ________________

POSITION

Position of Groove ________________

Weld Progression (Uphill, Downhill) ________________

Other ________________

PREHEAT

Preheat Temp. ________________

Interpass Temp. ________________

Other ________________

POSTWELD HEAT TREATMENT

Temperature ________________

Time ________________

Other ________________

GAS

Type of Gas or Gases ________________

Composition of Gas Mixture ________________

Other ________________

ELECTRICAL CHARACTERISTICS

Current ________________

Polarity ________________

Amps ________ Volts ________

Other ________________

TECHNIQUE

Travel Speed ________________

String or Weave Bead ________________

Oscillation ________________

Multipass or Single Pass (per Side) ________________

Other ________________

Tensile Test

Specimen No.	Width	Thickness	Area	Ultimate Total Load (lb.)	Ultimate Unit Stress (psi)	Character of Failure and Location

Guided Bend Tests

Type and Figure No.	Result

Toughness Tests

Specimen No.	Notch Location	Notch Type	Test Temp.	Impact Values	Lateral Exp.		Drop Weight	
					% Shear	Mils	Break	No Break

Fillet Weld Test

Result -- Satisfactory: Yes ____ No ____ Penetration into Parent Metal: Yes ____ No ____
Macro -- Results ____________________

Other Tests

Type of Test ____________________
Deposit Analysis ____________________
Other ____________________

Welder's Name ____________ Clock No. ________ Stamp No. ________
Tests conducted by ____________________ Laboratory Test No. ________

We certify that the statements in this record are correct and that the test welders were prepared, welded, and tested in accordance with the requirements of the governing code.

Manufacturer ____________________

Date ____________________ By ____________________

FIGURE 24-4 (cont'd)
Department of Welding and Testing procedure qualification record.

It also shows the results of testing, giving the tensile strength of the metal, and other mechanical tests, such as bend, fillet weld test, and macro etch tests. If nondestructive testing such as x-ray or ultrasonic testing is required, that will be included with its test results.

WELDER QUALIFICATION AND CERTIFICATION

Before welders can weld on code work (jobs being done under a welding code), they must first become certified. Usually, certification is done under one of the applicable welding codes. For example, building and bridge work may require that the welder be certified to the American Welding Society's structural welding code (D1.1). If welding boilers or pressure vessels, the welder may certify under the ASME boiler and pressure vessel code, Section IX. Cross-country petroleum or natural gas lines may be certified according to the American Petroleum Institute (API 1104). High-pressure power piping is usually done under the American National Standards Institute code (ANSI B31.1), which refers to ASME Section IX for qualification and certification of welders. Certification under one code may not necessarily qualify a welder under another code. It is usually up to the discretion of the engineer to accept or reject any previous qualifications that might be applicable on a current job.

As stated previously, a welder qualification, also known as a performance qualification, *is a practical welding test given to a welder to verify his or her ability to produce sound welds.* A welder certification is the written document describing the conditions and variables according to which the welder is certified. It also lists the test results.

Welders usually become certified to a specific code and the requirements of that code. Welder qualifications, like procedures qualifications, are subject to essential variables. For example, if a welder is qualified in the flat and horizontal positions but the current job requires welding in the vertical position, this position change would be considered an essential variable change and would require that the welder take the vertical qualification test. After testing, he would be certified in the vertical position.

Remember that a change in an essential variable requires that the welder take another qualification test to certify him or her to these changes. Once a welder is qualified and certified, he or she can weld within the limitations of that certification.

Typical essential variables for welder qualification include:

1. Changes in position
2. Changes in the welding processes
3. Changes in metals
4. Changes in joint designs
5. Change from plate to pipe

Welder certifications may be good for a limited period or may be good indefinitely. Most codes state that if for a long period (usually 3 or 6 months), a welder does not weld in the process in which he or she is certified, the welder must recertify. A certified welder's production welds are still subject to testing and inspection. As a matter of fact, a welder who consistently produces marginal welds may be asked to recertify.

The actual certification (qualification) test that will be taken depends on the code, joint design, position, and other variables required for the job. You can usually count on taking quite a few certification tests throughout your career, so keep practicing. The more you practice, the more comfortable and relaxed you will feel when taking an actual test. Some typical certification joints are shown in Figure 24-5.

The welding certification form (Figure 24-6) contains all of the essential information regarding how the test was performed, including the welding process, filler and base metals, the welding position, the test results, and other essential information.

FIGURE 24-5
Typical certification joints.

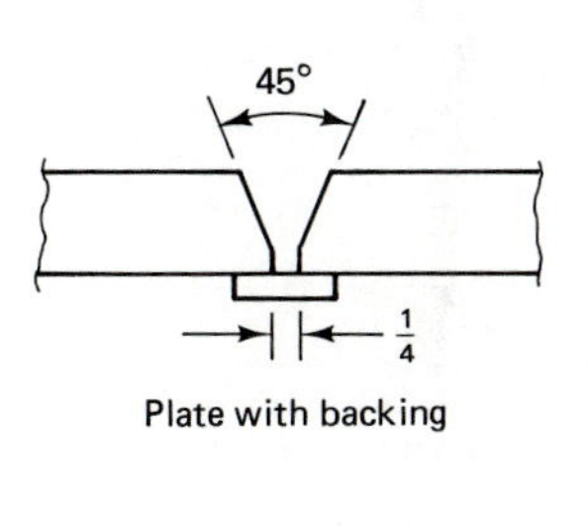

Plate with backing

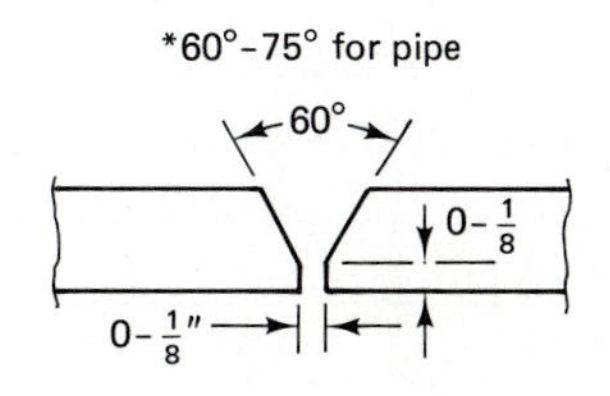

Plate or pipe without backing

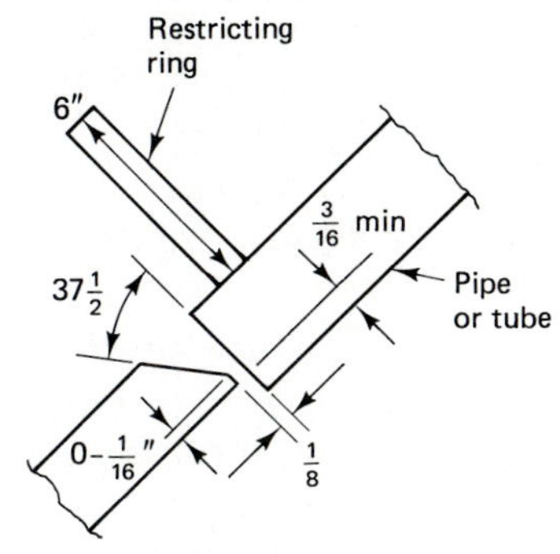

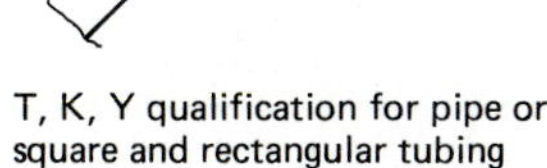
T, K, Y qualification for pipe or square and rectangular tubing

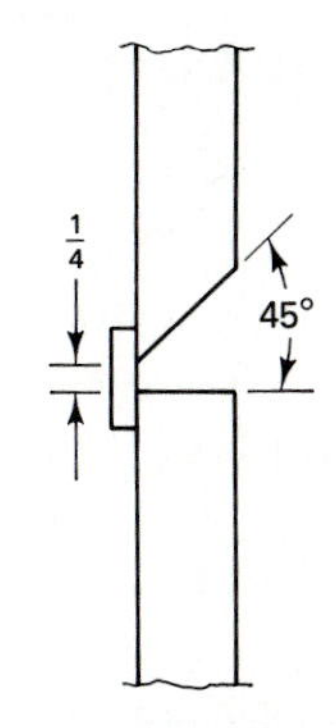

Horizontal plate or pipe with backing

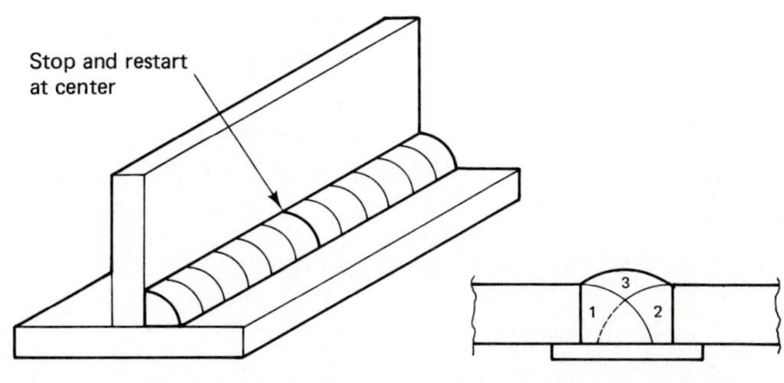

Fillet weld break test

Fillet weld root bend test (backing strip is air carbon arced off, and the specimen is bend tested)

Fillet weld tests

Welder Name ________ SS No. ________ Stamp No. ________
Welding Process ________ Type ________
In Accordance with Welding Procedure Specification (WPS) ________
Backing ________
Material Spec. ________ to ________ of P No. ________ to P No. ________
Thickness ________ Dia. ________
Filler Metal Spec. No. ________ Class No. ________ F No. ________
Other ________
Position (1G, 4F, 6G, etc.) ________
Gas Type ________% Composition ________
Electrical Characteristics (QW-409) Current ________ Polarity ________
Weld Progression ________
Other ________

Filler Metal Diameter and Trade Name ________
Submerged Arc Flux Trade Name ________
Gas Metal Arc Welding Shield Gas Trade Name ________

GUIDED BEND TEST RESULTS

Type and Fig. No.	Result

RADIOGRAPHIC TEST RESULTS
FOR ALTERNATIVE QUALIFICATION OF GROOVE WELDS BY RADIOGRAPHY

Radiographic Results: ________

FILLET WELD TEST RESULTS:

Fracture Test (Describe the location, nature, and size of any crack or tearing of the specimen)

Length and Percent of Defects ________ Inches ________%
Macro Test—Fusion ________
Appearance—Fillet Size (Leg) ________ in. × ________ in. Convexity ________ in.

Test Conducted By ________ Laboratory Test No. ________
We certify that the statements in this record are correct and that the test welds have been prepared, welded, and tested in accordance with the requirements of the governing code.
Organization ________
By ________
Supervised By ________
Date ________ Mechanical Test By ________

FIGURE 24–6
Department of Welding and Testing welder qualification test.

REVIEW QUESTIONS

1. What is the purpose of qualifying welding procedures?
2. When should companies qualify procedures for a given production run?
3. What is an essential variable?
4. How many procedures may need to be qualified for a given project?
5. What is the purpose of welder qualification (certification)?
6. What is the difference between welder and procedure qualification?
7. What does the welding position have to do with welder certification? Explain.
8. For how long are welder certifications usually in force?
9. What does the type of metal or the joint design have to do with welder certification? Explain.
10. What types of mechanical test are usually required for qualifying welders?

25 DESTRUCTIVE TESTING

As the name implies, destructive testing includes testing parts by means of a mechanical test that usually destroys the part. In production this may involve sample testing. Sampling is removing only some of the total batch for testing: for example, 1 unit out of 100, or 10 out of 1000. This system gives some idea of the reliability of a product, or at least helps to calculate the risk of failure. Examples of products that use destructive testing for sample testing are heat-treated bolts.

In weld testing we use destructive testing to qualify welders and welding procedures. The methods used are:

1. Tensile strength testing
2. Bend testing (side, face, and root bends)
3. Fillet weld break testing
4. Macro etch testing
5. Nick break testing
6. Notch toughness (impact) testing
7. Free bend testing

Some codes have additional tests. The test results described herein are only summaries of welding codes used in industry. We suggest that you consult the applicable code for exact requirements.

PHOTO 25–1
Tensile tester with microprocessor-assisted testing system. (Courtesy SATEC Systems, Inc.)

TENSILE STRENGTH TESTING

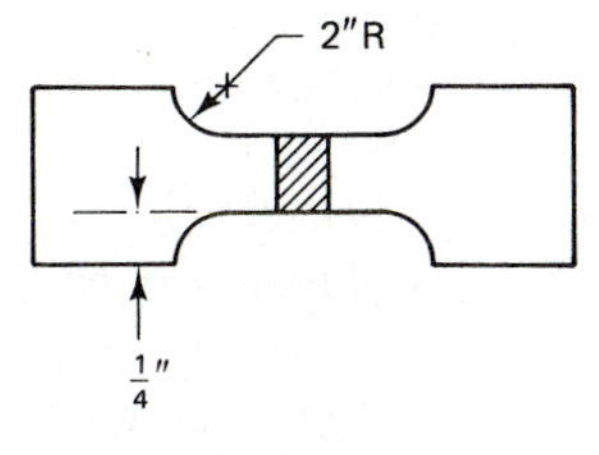

FIGURE 25–1
Tension specimen.

Tensile testing is also known as reduced section testing. Such tests involve machining a reduced cross-sectional area into a test strap at the weld area. Testing involves pulling the test straps with a tensile pull machine until they break. To pass, the tensile specimen (Figure 25–1) must pull to the minimum specified tensile strength of the base metal. Some codes allow a 5% reduction of tensile strength if failure occurs outside the weld area.

BEND TESTING

Bend tests may be free bends, side bends, face bends, or root bends, depending on how they are removed and bent. The face and root bends are usually removed from the test specimen as a $1\frac{1}{2}$-in.

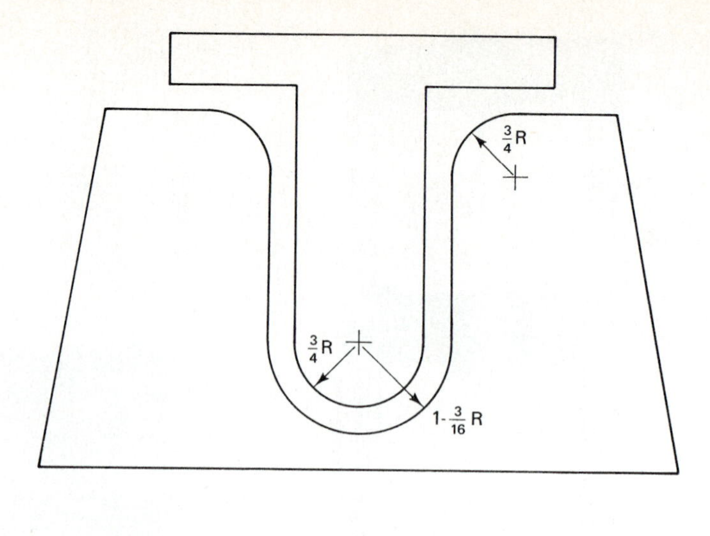

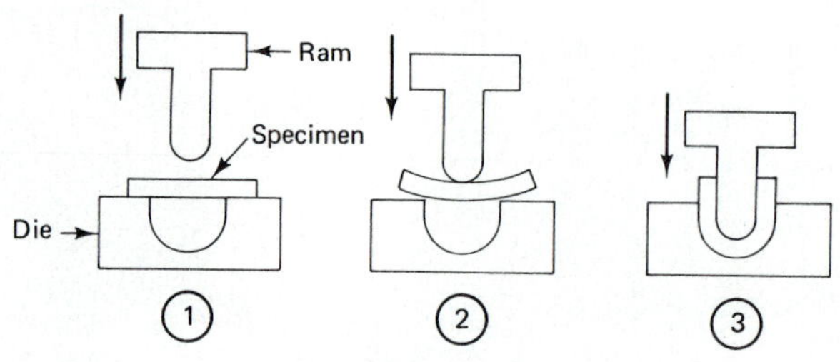

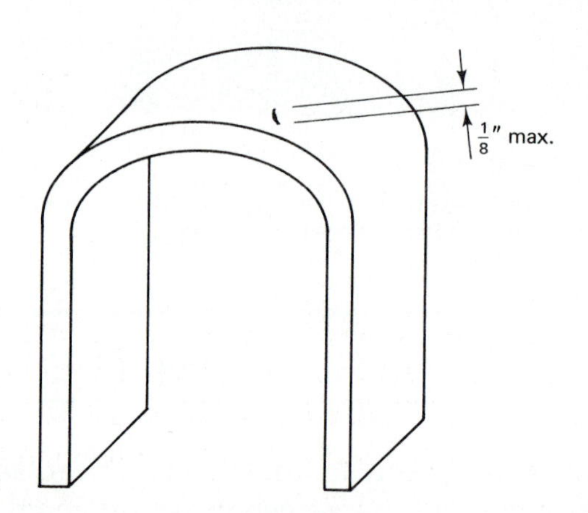

FIGURE 25-2
Bend test procedure and required results.

strap. The face bend is bent 180° exposing the face (or putting the face in tension). The root bend is done in the same way, but exposing the root of the weld. Side bends are removed as $\frac{3}{8}$-in. straps and usually bend 180°, exposing the side with the greater amount of discontinuity showing (Figure 25-3). Face, root, and side bends are usually bent using the guided bend technique. This technique forces the test specimen into a U-shaped dye (Figure 25-2).

FILLET WELD BREAK TESTING

The fillet weld break test requires the welder to deposit weld on one side of a tee joint. Stopping and restarting are usually required at the

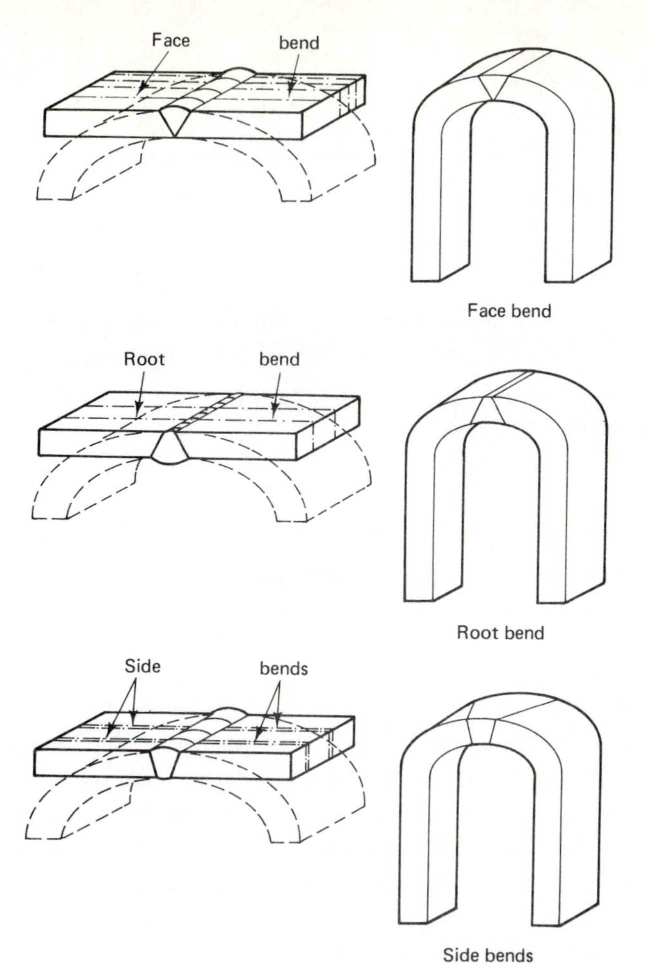

FIGURE 25-3
Bend test types: face, root, and side.

center of the joint. Testing is done by forcing the stem of the tee flat on the weld and plate (Figure 25-4).

To pass the fillet weld break test, the stem must bend flat on itself without breaking, or if it fractures, it must show complete fusion at the root. The broken edge should have no slag inclusions or porosity larger than $\frac{3}{32}$ in. If many small inclusions and porosity are present, they should not total more than $\frac{3}{8}$ in. in a 6-in.-long specimen.

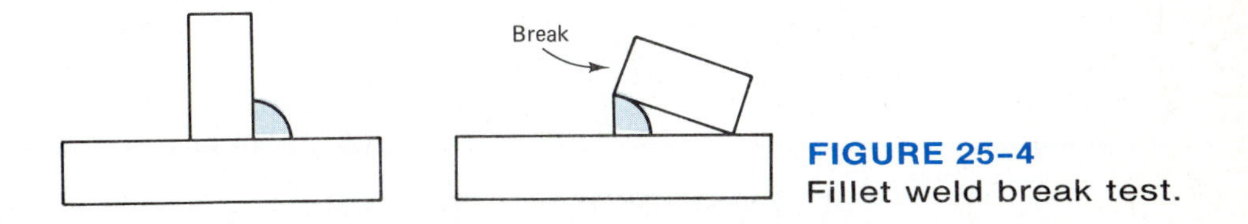

FIGURE 25-4
Fillet weld break test.

MACRO ETCH TESTING

Macro etch testing (Figure 25-5) is used to locate discontinuities and fusion zone areas. It is commonly used on fillet weld specimens. The procedure involves cutting and polishing a fillet weld section, then applying an acid etchent. The etching solution should give a clear def-

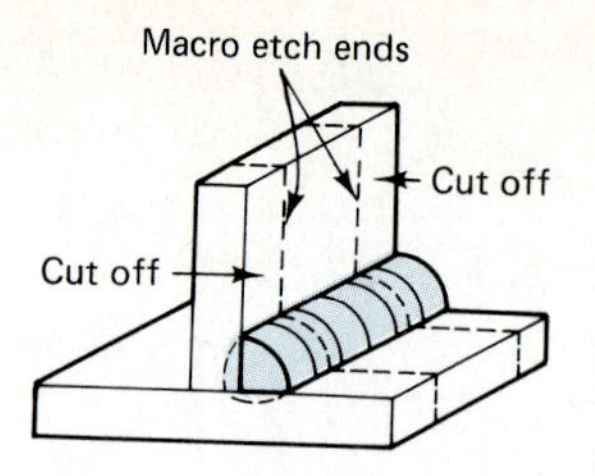

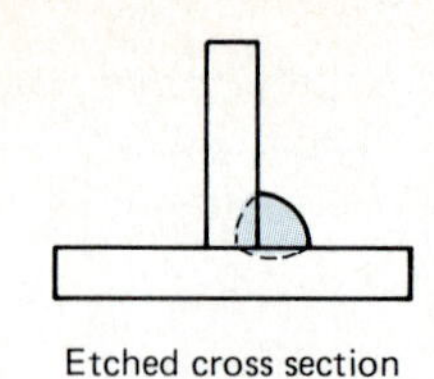

FIGURE 25-5
Fillet weld break test specimen.

inition of the weld and heat-affected zone. To pass the macro etch test the specimen should have equal leg length within $\frac{1}{8}$ in. of each other. It should show complete fusion to the root, and have no cracks or other visible discontinuities.

NICK BREAK TESTING

Nick break specimens are used to examine the interior of a weld (Figure 25-6). The procedure involves hacksawing notches into both sides of the test strap at the welds. The strap is then put in a vise and broken off with a hammer, which exposes the interior of the weld. The exposed interior of the weld should have no more than $\frac{1}{16}$ in. porosity or total more than 2% of the exposed surface area. Slag inclusions should not exceed $\frac{1}{32}$ in. in length, or $\frac{1}{8}$ in. in depth, or $\frac{1}{2}$ in. in wall thickness (whichever is least). There should be at least $\frac{1}{2}$ in. between any two slag inclusions that might be present.

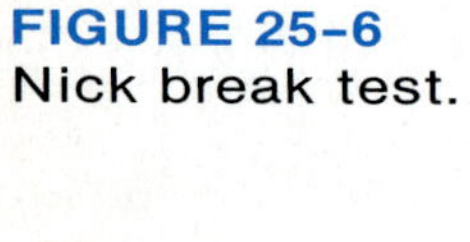

FIGURE 25-6
Nick break test.

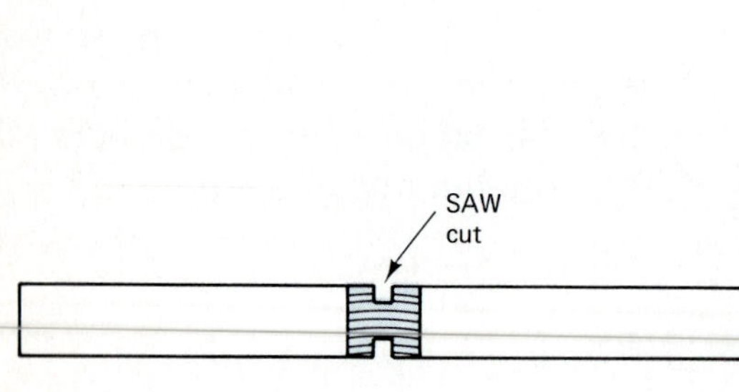

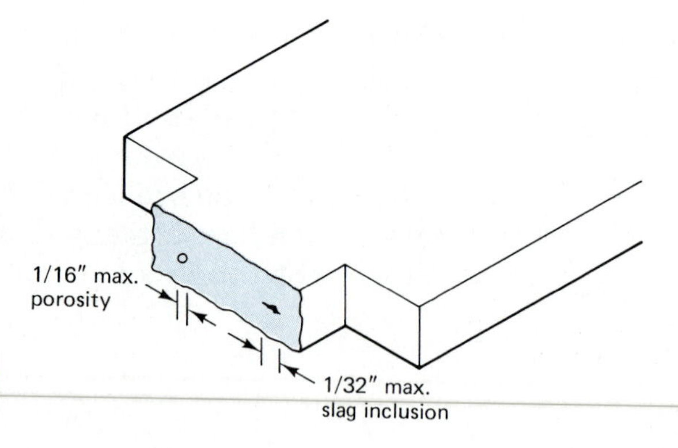

NOTCH TOUGHNESS OR IMPACT TESTING

Impact testing is required by some codes in certain metals to check for brittleness that may have developed during welding. There are three methods of impact testing; (1) the drop weight test, (2) the Izod, and (3) the Charpy method.

The drop weight type deposits a $2\frac{1}{2}$-in. weld on a plate, then cuts a small notch in the middle ($\frac{1}{16}$-in. saw cut). The specimen is supported at each end and a weight is dropped into the middle. Specimens are either bent and examined for cracks, or broken and the notched area is examined for weld, fusion zone, and base metal cracks. Two tests are usually performed to verify notch toughness.

The V-notch toughness test is done using the Izod method or the Charpy method. The procedure uses a 45° V notch cut in the test specimen. With the Izod method (Figure 25-7) the specimen is held tightly in a vise while a swing hammer strikes the specimen, putting the notch in tension and breaking the specimen. The impact strength is calculated by how far the hammer follows through after breaking the specimen. In the Charpy method (Figure 25-7) both ends of a V-notch specimen are supported and the hammer strikes and breaks the specimen at the center. Because impact strength varies with temperature, V-notch specimens must be brought to a specified temperature for testing.

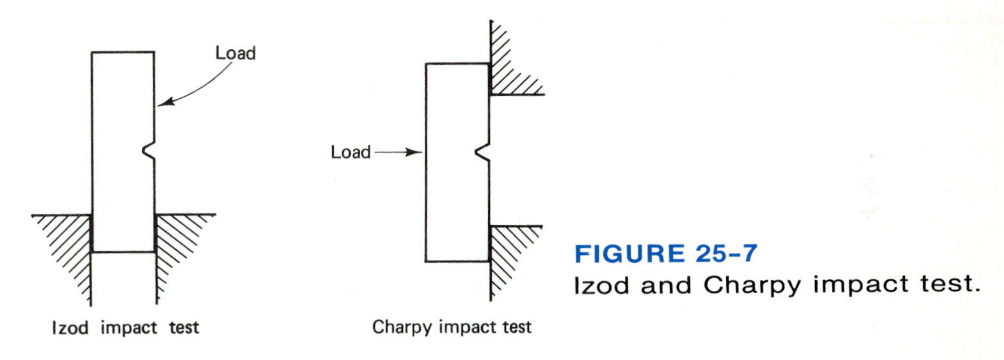

FIGURE 25-7
Izod and Charpy impact test.

PHOTO 25-2
Izod impact tester. (Courtesy SATEC Systems, Inc.)

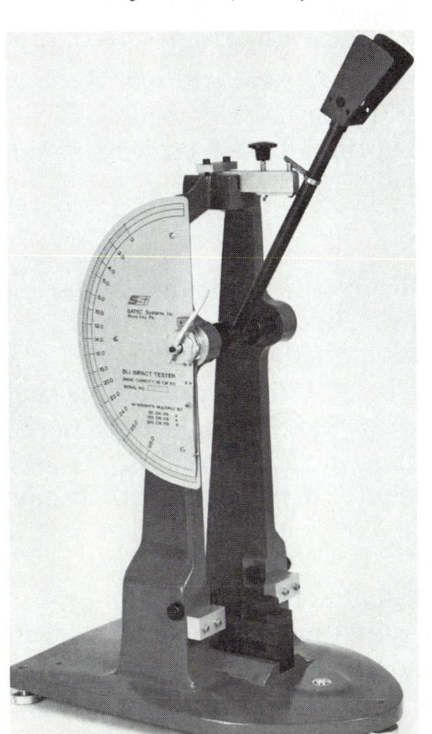

PHOTO 25-3
Impact tester. (Courtesy SATEC Systems, Inc.)

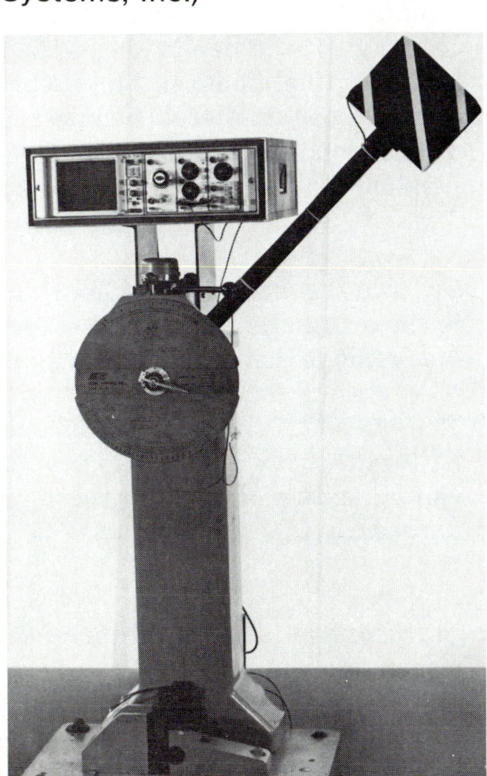

PHOTO 25-4
Drop weight tester. (Courtesy SATEC Systems, Inc.)

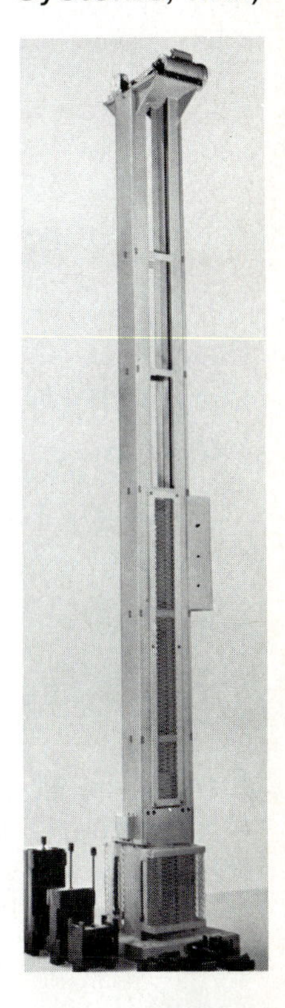

FREE BEND TESTING

Free bend tests are similar to guided bends, but no dye is used to shape specimens. Instead, a small bend called an initial bend or starter bend is put into the face side of the specimen. Then the specimen is bent further, to about 180°, by placing it in a vise or other compression device (Figure 25-8).

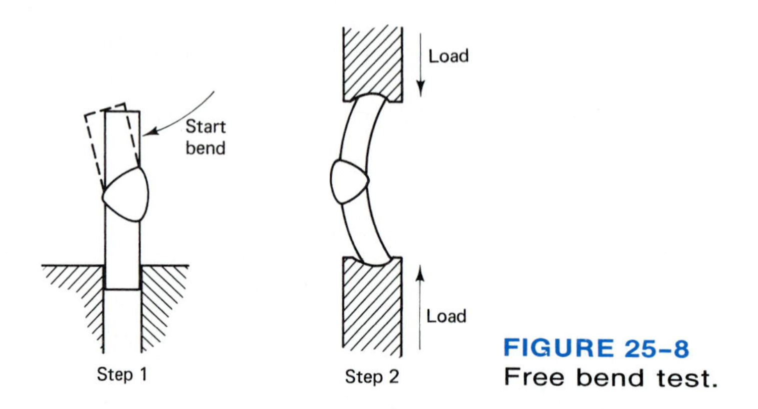

FIGURE 25-8
Free bend test.

GUIDED BEND TESTING

Practice joints, specifically butt joints, can be tested using the same test and accepting criteria that would be used if you were certifying. If your butt weld passes visual inspection, you are ready for the bend test. Either two side bends, or one face and one root bend, will be required for plate qualification. Metal thicknesses up to $\frac{3}{8}$ in. will use the face and root bends. Metal thicknesses $\frac{3}{4}$ in. and over use side bends. Thicknesses over $\frac{3}{8}$ in. and less than $\frac{3}{4}$ in. can optionally use either the side bend or the face and root bend system.

Bend Test Procedure for Face and Root

Cut two test sample sections $1\frac{1}{2}$ in. wide from either side of a centerline in the test plate. Bend each test section 180°, in the opposite directions, one against the face and one against the root (Figure 25-9).

Each test sample section should bend 180° without breaking. The surface of the bend should exhibit no cracks, slag inclusions, porosity, or other open discontinuity larger than $\frac{1}{8}$ in. (Figure 25-10). If each sample meets or exceeds these requirements, it is considered to have passed.

Bend Test Procedure for Side Bends

Starting from the centerline of the test plate, measure over 1 in. from either side and mark. Now measure an additional $\frac{3}{8}$ in. from the centerline and mark. Cut out the two $\frac{3}{8}$-in. sections that have been marked. Bend each sample section 180°, putting the side with the greater number of discontinuities face down into the die (Figure 25-11).

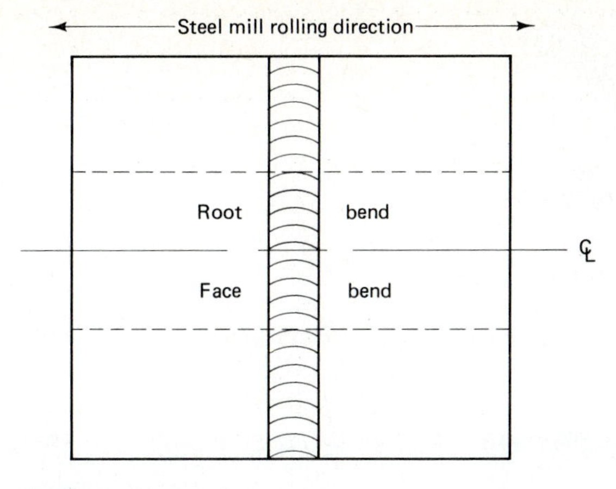

FIGURE 25-9
Root and face bend with rolling direction.

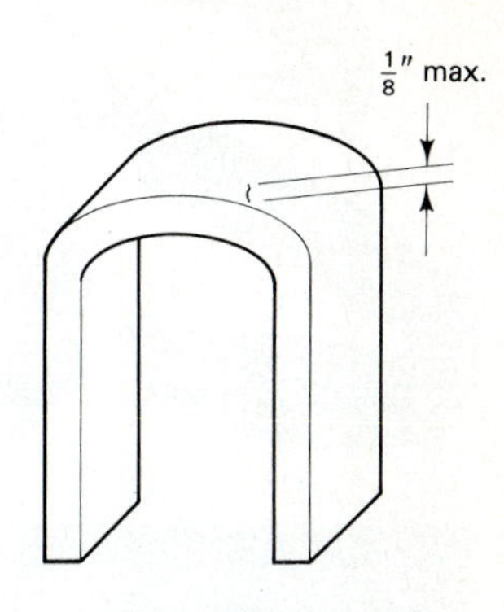

FIGURE 25-10
Bend test results required.

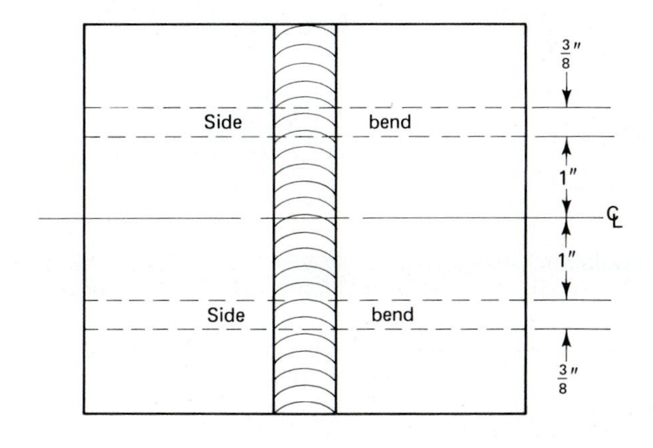

FIGURE 25-11
Side bend tests.

The surface of the bend section of the weld and heat-affected zone should exhibit no cracks, inclusions, porosity, or other open discontinuities larger than $\frac{1}{8}$ in. If each sample meets or exceeds these requirements, it is considered as having passed.

REVIEW QUESTIONS

1. What is destructive testing used for in production welding?
2. What types of forces are applied to tensile test?
3. List the three types of bend tests.
4. How far should guided bend samples be bent?
5. To pass the fillet weld break test, what is the maximum amount of slag inclusion allowed in a 6-in. specimen?
6. What size cracks are allowed on macro etch specimens?
7. How is the nick break specimen broken?
8. What is the maximum allowable porosity for nick break tests?
9. Why are notch toughness or impact tests performed?
10. Which side of a side bend test is placed down into the die?

26 NONDESTRUCTIVE TESTING (NDT)

Welds can be examined quite thoroughly and without destroying their end use, with today's modern nondestructive testing systems. These include:

1. Visual inspection (VT)
2. Dye penetrant testing (PT)
3. Magnetic particle testing (MT)
4. Ultrasonic testing (UT)
5. Radiographic testing (RT) (X-ray and gamma-ray testing)
6. Eddy current testing (ET)

There are other nondestructive testing (NDT) processes, such as leak testing, acoustic emission, hydrostatic (which can be classified as both destructive or nondestructive), and others. As a certified welding inspector (CWI) you will be qualified for visual inspection, but you may not be qualified to perform some of the other NDT processes. The CWI is, however, responsible for helping to interpret the results of these tests and for determining whether they meet the applicable code or specification requirements.

To perform the actual nondestructive test, such as radiography, most codes require that you meet the minimum requirements of the American Society for Nondestructive Testing (ASNT) recommended practice SNT-TC-1A. This document contains recommended practices for training, testing, and certification of nondestructive testing personnel. It is intended as a guide for employers who wish to qualify their employees in one of the NDT disciplines. We suggest that you consult the SNT-TC-1A if you would like detailed qualification information.

The ASNT breaks qualification of NDT personnel into three levels: Level I, Level II, and Level III. Level I personnel, working under the supervision of a Level II person, perform the specific calibration tests and evaluation in a specific NDT process from a minimum of written instructions (from the Level II).

Level II personnel are qualified to calibrate, test, interpret, and evaluate the applicable code or standard. They should be familiar with the limitations of the specific NDT process, and able to prepare written reports of their findings.

Level III personnel establish techniques and procedures for specific NDT process using the applicable codes and standards. They should be fully knowledgeable of the advantages and limitations of the NDT process, and be able to write these procedures and techniques clearly so that Level I and Level II personnel can work from them. Level III personnel should be able to train and test Level I and II personnel. The recommended training time and study categories are described in the SNT-TC-1A document.

VISUAL INSPECTION

Visual inspection (VT) is by far the simplest, fastest, and most economical method of nondestructive testing.

Most welds that are unacceptable will be rejected through visual

PHOTO 26–1
Visual inspection tools.

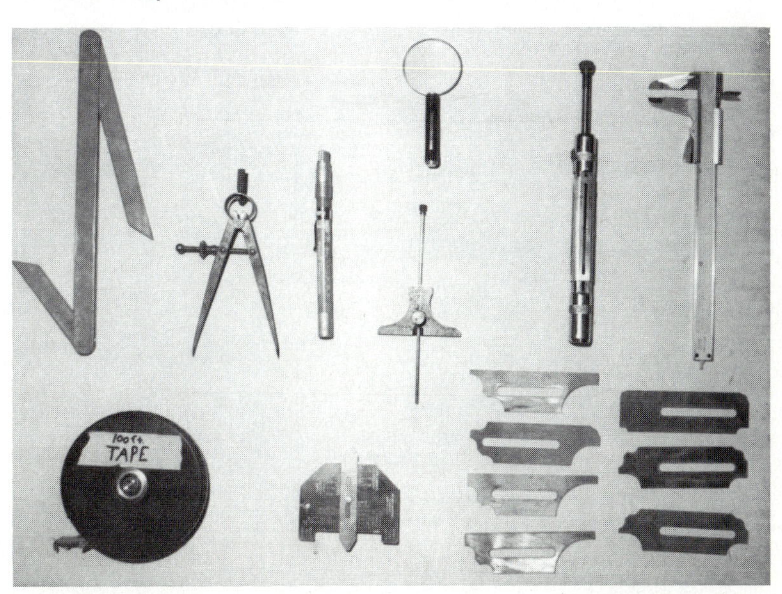

inspection. If the weld fails visually, there is no need for the expense of further nondestructive testing. As a CWI, you are qualified to perform most visual weld inspections.

Of course, the disadvantage of VT is its inability to examine the internal nature of the weld. Only discontinuities visible to the surface will be detected. Most welding codes require a minimum of visual testing on every weld.

The welding inspector can check joint fitup, alignment, dimensional accuracy, conformity to detail drawings, surface roughness, weld spatter, cracks, overlap, and many other defects. The inspector uses tools such as fillet weld gages, groove and contour gages, depth gages, calipers, and a magnifying glass.

PENETRANT TESTING

Another economical and effective method of NDT is liquid penetrant testing (PT). This process is especially effective on cracks or any discontinuities that are open to the surface of the part. The procedure is quite simple. A special penetrating dye is applied to the part and given time to penetrate into any flaws (discontinuities). Then the dye is removed. Due to capillary attraction, the dye is held in any discontinuities and bleeds out when the dye is removed. When developer is added, flaws become highly visible (Figure 26–1). A visible dye can be used or a more effective fluorescent dye, which is viewed under a black light (Figure 26–2). The only limitation to this process is that only surface cracks or discontinuities will show up that would not have been detected by simple visual inspection.

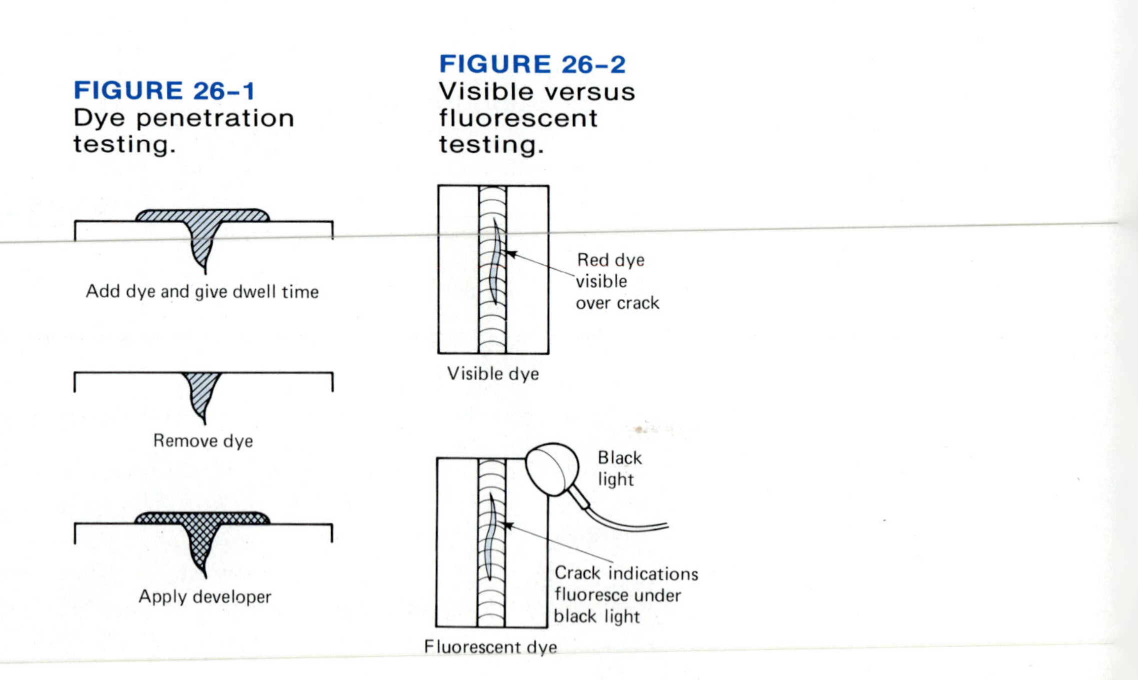

FIGURE 26–1 Dye penetration testing.

FIGURE 26–2 Visible versus fluorescent testing.

MAGNETIC PARTICLE TESTING

In magnetic particle inspection, a magnetic field is used which is induced into the part with a yoke, inspection prod, or other magnetizing device (Figure 26–3). Then iron powder is added to the part, which follows the magnetic lines of flux. If a discontinuity is present, a magnetic leakage field is set up, with opposite poles. Iron powder is attracted and outlines the defect. As with penetrant testing, magnetic particle testing can use a dry powder visible method or a fluorescent, oil-suspended powder system that is viewed under black light.

MT will locate any discontinuity on the surface or up to approximately $\frac{1}{4}$ in. into the subsurface (depending on the amperage or the magnetizing unit). The limitation of MT is that the material must be magnetic; thus aluminum, copper, and other nonferrous materials cannot be MT tested. MT has been used very effectively in construction to check beam-to-column connections or in many industries to inspect for cracks in engine blocks and castings.

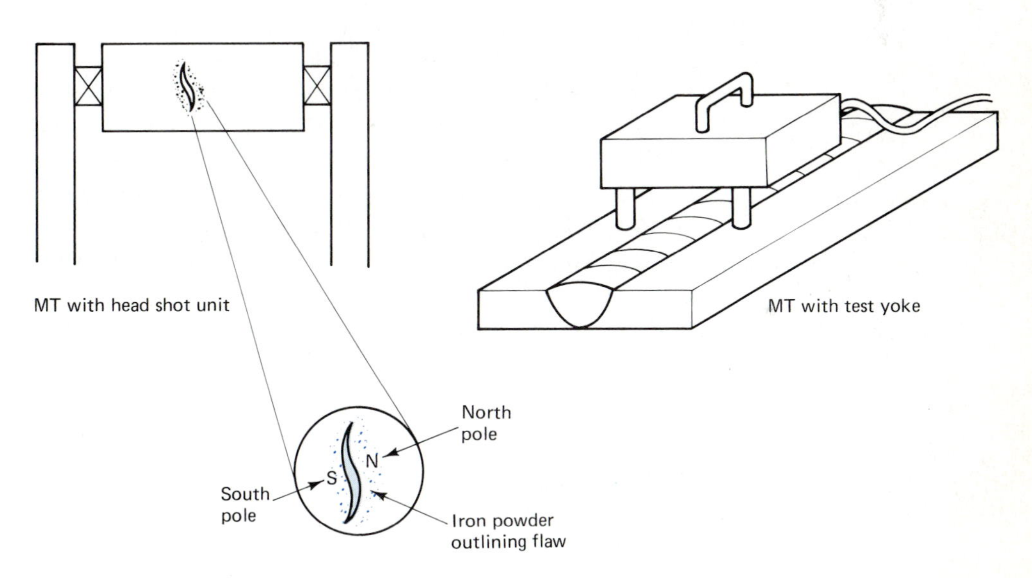

FIGURE 26–3
Magnetic particle testing.

ULTRASONIC TESTING

Ultrasonic testing (UT) uses ultrahigh-pitched sound waves above the frequency that can be heard by the human ear, to penetrate the part and reflect back the transmitted sound (Figure 26–4). A transducer containing a crystal is used to probe the surface to be inspected. A cathode-ray-tube (CRT) screen must be watched during scanning. If a defect is located, the sound will return faster than if it were to be reflected off

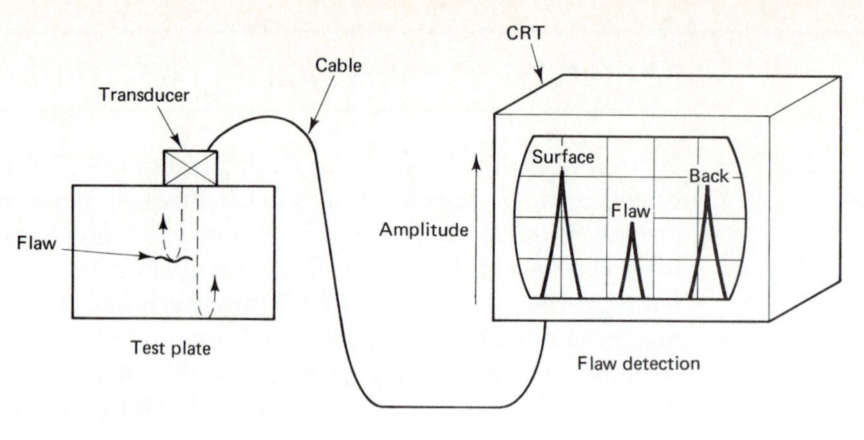

FIGURE 26–4
Ultrasonic pulsed echo testing.

the back side of the plate or pipe. Therefore, a blip appears on the screen and time baseline. Size, depth, and type of discontinuity can be detected by a skilled UT technician. This process can be used on almost any metal, and to inspect both fillet welds and groove welds with the use of shear and angle beam sound waves.

Most UT weld inspection uses the pulse echo or A-scan technique; however, with the B- and C-scan techniques, a permanent graphical readout can be made to record test results (Figure 26–5). B- and C-scan techniques are usually done by immersion; that is, the part to be inspected is immersed in a water tank. The water can then couples the sound between the part and the transducer. With A-scan sound, coupling is accomplished by spreading a thin layer of gel-type couplant over the surface to be tested.

There are UT calibration blocks used to measure the size and location of known defects. The navship block is used to calibrate for longitudinal testing. The DSC, or IIW block, is used in calibrating angle beam inspection techniques. A block with three closely drilled

PHOTO 26–2
Ultrasonic testing using angle beam. (Courtesy of Staveley NDT Technologies, Inc.)

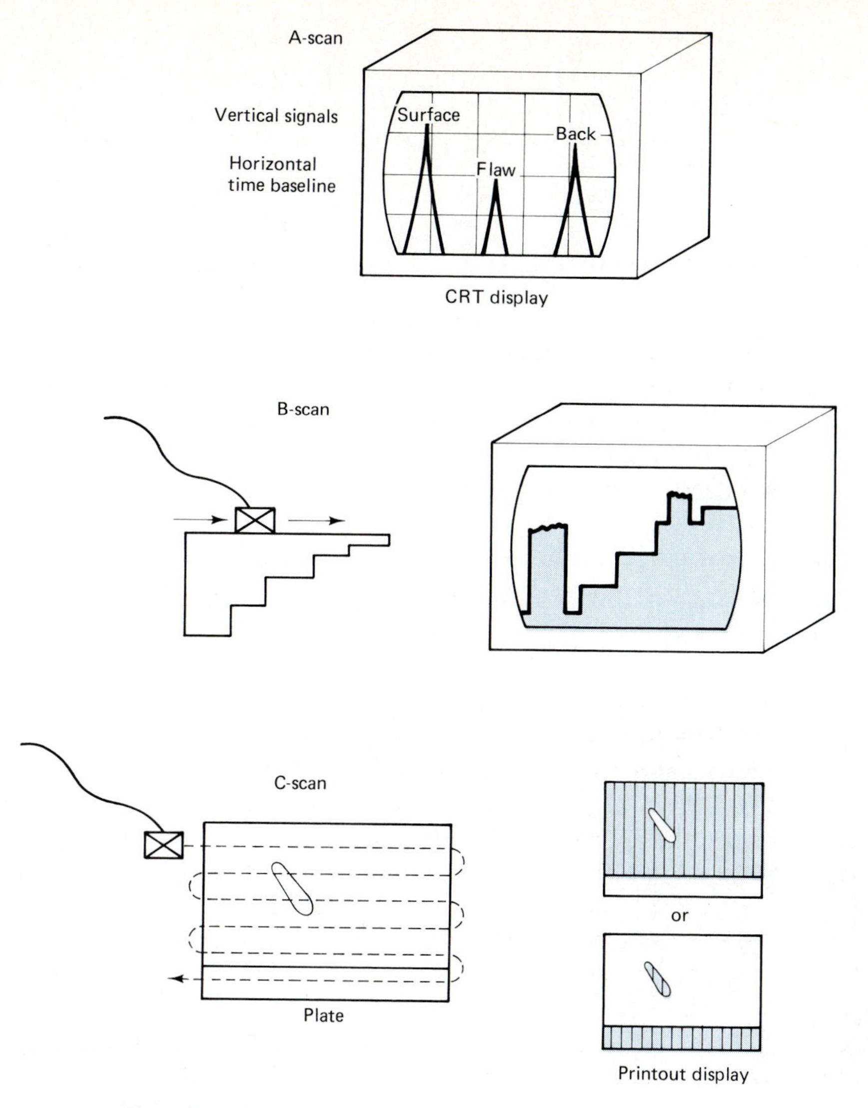

FIGURE 26–5
Ultrasonic testing display types.

holes is used to check resolution, the ability to distinguish between two defects in close proximity.

Another branch of UT is acoustic emission testing (AET). This method uses a graphic recorder to record any sound generated in the part during welding or hydrostatic testing (Figure 26–6). If a crack

PHOTO 26–3
Ultrasonic testing transducers. (Courtesy of Staveley NDT Technologies, Inc.)

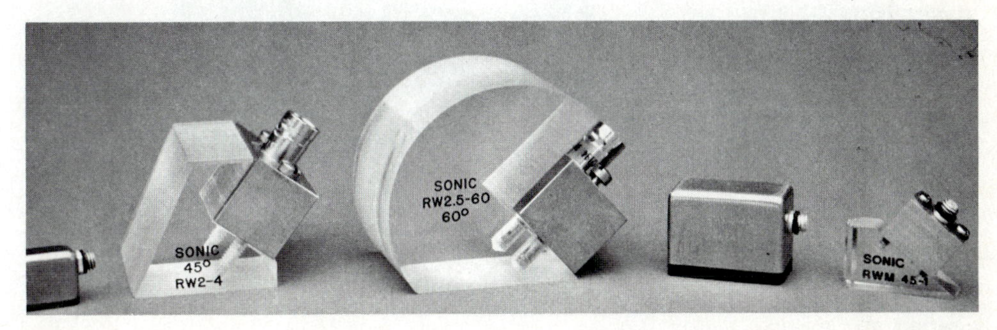

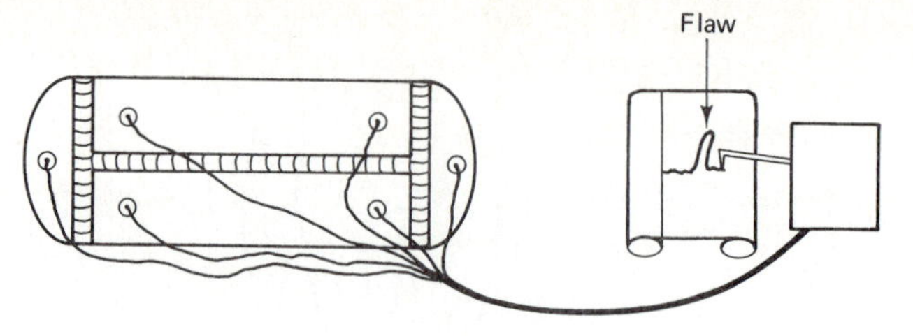

FIGURE 26-6
Acoustic emission testing.

were to result from heat or stress during welding or hydrostatic testing, it would produce a short burst of high-pitched sound. The transducer would then pick up and record the sound on the graphic recorder. Using multiple probes, some acoustic emission systems have the ability to triangulate and locate the vicinity of the flaw.

RADIOGRAPHIC TESTING

In radiography, also known as X-ray or gamma-ray testing, EMR or shortwave radiation is passed through the part to be tested, darkening the film behind it. The denser or thicker the part, the less EMR passes through, and the lighter the film. In contrast, the more radiation, the darker the film.

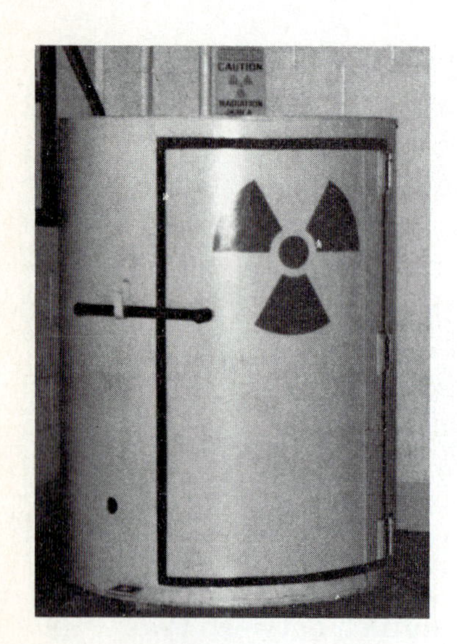

PHOTO 26-4
X-ray containment for radiograph testing of small parts. The containment vessel contains most of the dangerous X-rays. (Courtesy of Stan Roberts Photo Service.)

PHOTO 26-5
Interior of containment vessel. Film and test specimen are placed into the lead box in the center of the vessel. Note the X-ray tube is wrapped in lead to keep radiation levels low around the vessel. (Courtesy of Stan Roberts Photo Service.)

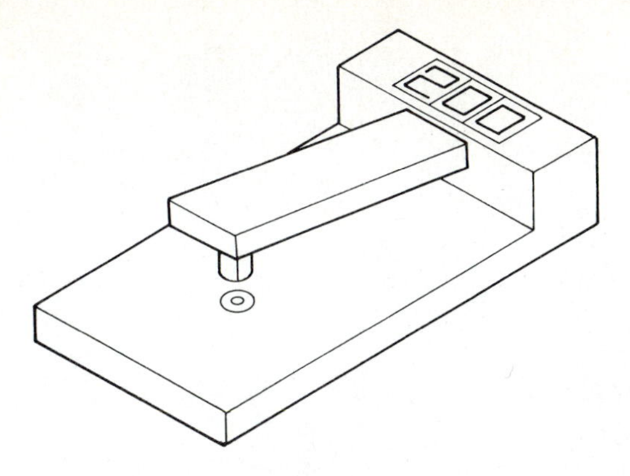

FIGURE 26–7
Digital densitometer.

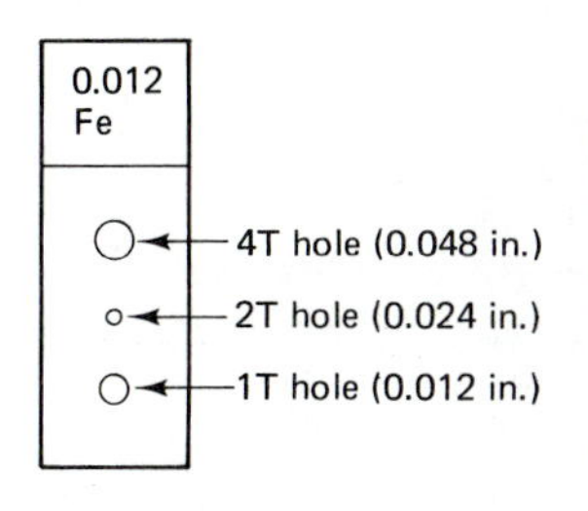

FIGURE 26–8
Penetrometer for radiographs of steel.

Radiography can pick up almost all discontinuities except plate laminations or similar defects that are at right angles to the X-ray beam. Radiography provides a permanent visual record, so the technician can shoot the part immediately and the CWI can interpret the film later to verify code requirements have been met.

One of the drawbacks of radiography is the radiation hazard. Technicians must observe strict regulations, and only qualified certified technicians may perform radiographic testing.

Safety devices when using RT include a survey meter to measure radiation in the air, and dosimeters which technicians wear when taking radiographs. Dosimeters must be recharged each day and their readings recorded in the technician's health file.

A major concern of industry has been the quality of radiographs. Special equipment has been developed to measure the quality of radiographs. If a radiograph is over- or underexposed, it may hide defects. Densitometers must be used to measure the film density (Figure 26–7). Applicable codes state the acceptable range of film densities.

Penetrometers or "pennies" are image-quality indicators. They measure how sensitively (how small a defect) a radiograph will pick up. Figure 26–8 shows a "12 thousands" (0.012) penetrometer. The three holes are the 1T (or 1 times the thickness), 2T (2 times the thickness), and 4T (4 times the thickness of the penetrometer). Most welding codes require that the 2T hole be visible on the radiograph. This means that the radiograph will detect discontinuities as small as 2% of the plate thickness. The figure shows a typical 1-in. plate radiograph using the AWS code requirements.

PHOTO 26–6
Radiography with gamma rays. (Courtesy of Tech/Ops, Inc.)

PHOTO 26-7
The restoration of the Statue of Liberty included many radiographs to check its structural integrity. Courtesy of Tech/ Ops, Inc.)

PHOTO 26-8
Radiography is used extensively on pipelines. (Courtesy of Tech/ Ops, Inc.)

EDDY CURRENT TESTING

Eddy current testing (ET) is popularly used in automated nondestructive testing. ET induces alternating current into the part. Flaws will set up "eddy currents" in the part, which is then detected by a search coil. ET is basically a go/no go testing procedure. As in ultrasonic testing, calibration is critically important with ET. Figure 26-9 shows a typical application of eddy current testing on weld seam pipe.

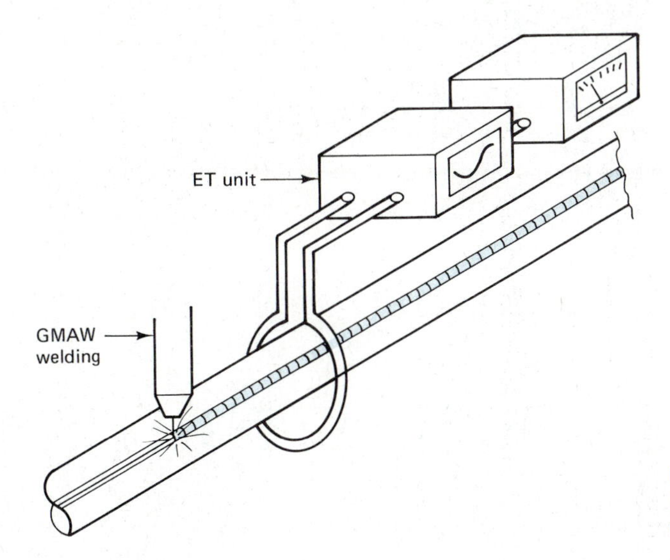

FIGURE 26-9
Eddy current testing on seam-welded pipe.

PROCESS SELECTION

It must be understood that all the NDT processes have unique advantages (Table 26-1). The selection of an NDT process should be based on the types of flaws that are common to the type of weldment or the types of flaws suspected. Some processes should be used in conjunction with each other. For example, ultrasonics and radiography complement each other. Ultrasonics easily picks up defects parallel to the surface of the plate, but has trouble picking up flaws perpendicular to the surface or in line with the beam. Radiography, on the other hand, easily catches flaws perpendicular to the surface, but seldom can show flaws parallel to the plate surface, such as plate laminations.

Welders, technicians, engineers, certified welding inspectors, and nondestructive testing personnel are all key people in an effective quality control program. Cooperation between all areas is essential. Cost savings are almost always realized through fewer rejects, low liability risk, and a better overall product and reputation.

TABLE 26-1
Advantages and Limitations of NDT Processes

	VT	PT	MT	UT	RT	ET
Accuracy of indications	Superior	Good	Good	Excellent	Excellent	Good
Cost of application	Very inexpensive	Very	inexpensive	Expensive	Expensive	Expensive
Speed of process	Rapid	Rapid	Fairly rapid	Fairly rapid	Slow	Rapid
Depth of thickness inspection on subsurface discontinuities	None	None	Fair	Good	Good	Fair
Cracks (general classes)	Fair	Excellent (surface only)	Excellent (surface only)	Excellent	Excellent	Good
Slag and tungsten inclusions	Good	Good (surface only)	Good	Excellent	Excellent	Good
Porosity (general classes)	Good	Good	Fair	Excellent	Excellent	Fair
Undercut	Superior	Good	Excellent	Poor	Fair/good	Fair
Laminations	N/A	N/A	Poor	Superior	Poor	Good
Overall effectiveness of process to detect all discontinuities	Fairly effective	Fairly effective	Fairly effecitve	Very effective	Very effective	Fairly effective

REVIEW QUESTIONS

1. What is the difference between destructive and nondestructive testing?
2. Who can perform nondestructive testing?
3. What are the advantages of visual inspection?
4. What is a disadvantage of dye penetrant testing?
5. What types of materials can be magnetic particle tested?
6. How is sound coupling to the test specimen accomplished with ultrasonic testing?
7. Would porosity show up on a radiograph as a light spot or a dark spot? Why?
8. What type of discontinuity might radiography miss?
9. What are penetrometers, and for what are they used?
10. On what should the selection of an NDT method be based?

WELDING INSPECTION AND QUALITY CONTROL

The best way to demonstrate what a quality control (QC) program can do is to show what can and has happened when one was lacking. Everything from costly rejects and repairs to major failures in bridges and buildings can result from a lack of quality control. Some codes require that each company develop a quality control program, including the development of a comprehensive QC manual, which describes in detail every step in quality monitoring, before, during, and after fabrication of the product.

Quality control is a cooperative effort of all departments (Figure 27-1) and sometimes requires a change of attitude by the work force, especially management. A "teamwork" attitude will work every time!

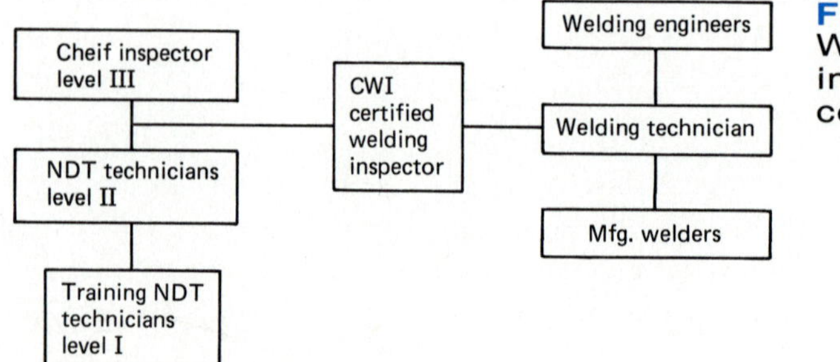

FIGURE 27-1 Where the CWI fits into the quality control program.

WELDING INSPECTOR

The welding inspector is the key person in any quality program. He or she is knowledgeable not only of the welding processes themselves, but also as to possible defects associated with each process. The inspector must know destructive and nondestructive testing processes and their limitations; he or she must be able to write clear, accurate reports and keep good records; must know how to use and interpret codes and standards; and most important, must be able to communicate effectively with welders, engineers, and technicians. Knowing the language and terminology of welding, as well as welding and nondestructive testing symbols, is absolutely essential. In some cases you will be working for the welding engineer or contractor, making sure that welds are produced as the designer has specified. In other cases you may work for a government agency or local municipality. As a welding inspector, the experience you gained as a welder will be invaluable when it comes to recognizing trouble areas.

The American Welding Society (AWS) sets standards for welding inspectors, conducts tests, and has published standards for the certification of welding inspectors. Welding inspectors are qualified under AWS on the following levels:

1. *Certified welding inspector (CWI).* The CWI performs, or oversees, all inspections required for a given project, under the applicable code, standard, or specification.
2. *Certified associate welding inspector (CAWI).* The CAWI performs inspections required for a given project, under the supervision of a CWI.
3. *Welding inspector specialist.* The welding inspector specialist performs the inspections required specifically for a company's or employer's needs. A person so classified is certified only by the company or employer, not by the AWS, and is valid only while employed by that company.

If you plan to become a welding inspector, we suggest that you consult AWS Code Section AWS QC1 in addition to studying this text. As a CWI you will be specifically responsible for the following areas:

- Verify that all work is being done in accordance with the applicable codes, standards, or specifications.
- Interpret drawings, specifications, and welding symbols.
- Verify that base metals and filler metals (electrodes, GMAW and FCAW wires, etc.) conform to specifications or applicable codes or standards, and are properly maintained.
- Verify that welding machines and equipment are in suitable condition to produce acceptable welds.
- Verify that all welders or welding operators are certified and current for the work they are performing.

- Verify that all applicable welding procedures have been qualified. (See Chapter 23 for a full explanation of welding procedures.)
- Verify that joint preparation and fit-up are as specified on the project drawings, and within tolerances.
- Inspect, evaluate, and mark all weld joints with a minimum of visual inspection.
- Review and evaluate other tests, such as destructive and nondestructive tests.
- Verify that welders are using the specified techniques for given applications, positions, or electrodes. (Technique examples would include weave or stringer, vertical up or down, etc.).
- Maintain required records and reports, as well as unstructured reports describing the overall quality and workmanship observed during construction or fabrication.

A welding inspector does not simply wander from weld to weld giving his or her approval or disapproval; it is not quite that easy. Let us trace the steps from start to finish of a typical certified welding inspector working a construction project.

Upon arrival at the construction site, the inspector should be furnished with the complete set of drawings and specifications for the project. If a complete set is not furnished, one should be obtained from the project engineer. The job cannot be done without knowing all the requirements and specifications.

Once the inspector knows what is required in the specification, he or she should check purchase orders to see that the correct metal, electrodes, and filler metals have been ordered. When these materials arrive at the construction site, the inspector should verify that the proper material has been shipped. The material should always be marked as it arrives and again immediately after it is cut or sheared. This will prevent loss of identity of unused pieces (Figure 27–2).

As fabrication begins, the inspector should monitor the overall assembly of the project, making sure that the correct parts or structural shapes are being cut as the specifications indicate. Prior to the start of welding, the inspector should check the joint edge preparation for both type and angle tolerances (Figure 27–3). Also, prior to welding, the inspector checks the tightness of the joint fit-up. Most codes spell out a minimum and a maximum root opening (Figure 27–4).

Once welding starts, the inspector must spotcheck the welders to make sure that specified techniques are being followed. The techniques could include specifying stringer beads instead of weaves for heat input control, or uphill progression on all vertical joints.

Once joints begin to be completed and "cool," they can usually be inspected immediately. However, some high-strength low-alloy, high-strength, or quenched and tempered metals require at least a 48-hour delay between welding and inspection. This is to allow any delayed cracks that may have developed to become visible to the inspector (Figure 27–5).

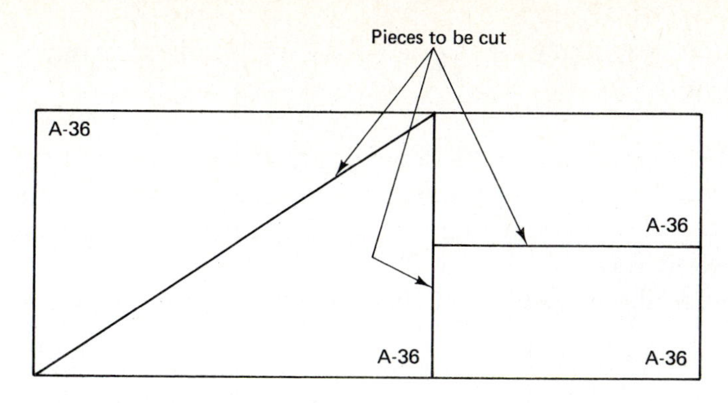

FIGURE 27-2
Identifying parts. Identify all cut or sheared pieces.

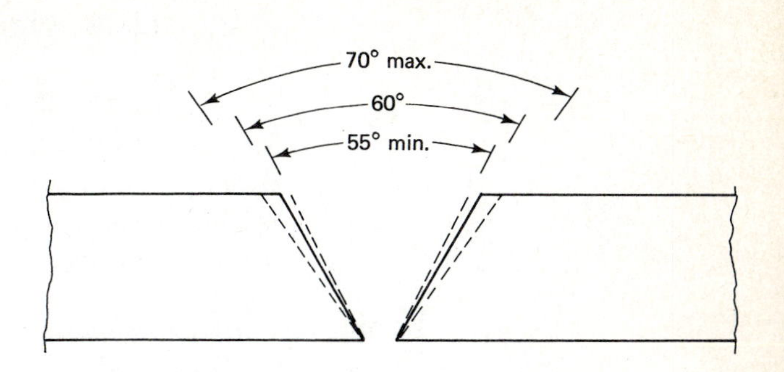

FIGURE 27-3
Angle tolerance. V-groove with +10°, −5° tolerance.

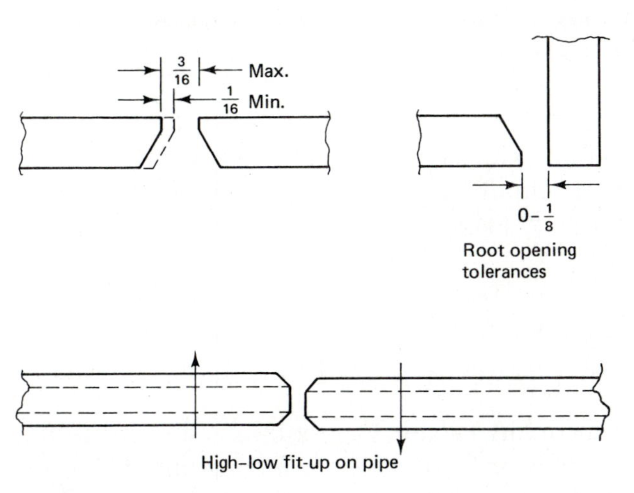

FIGURE 27-4
Root opening tolerances and high-low fit.

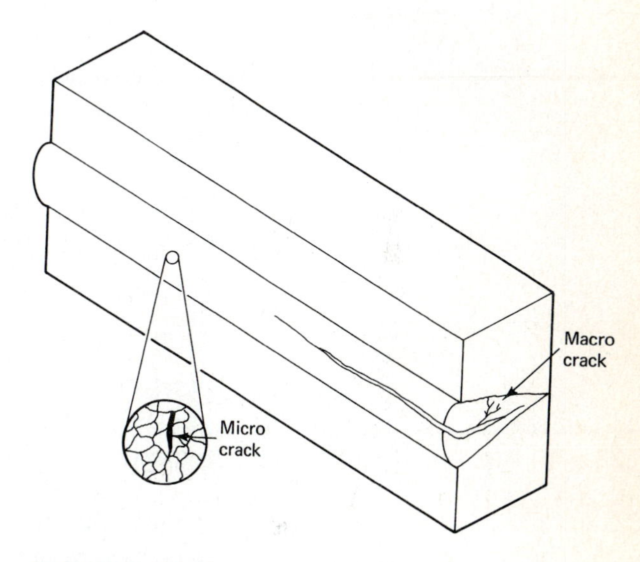

FIGURE 27-5
Micro and macro cracks.

INSPECTING THE FINAL WELDS

The inspector's final visual weld inspection for fillet welds should include:

1. A check of the leg length. Codes normally allow no more than a $\frac{1}{8}$-in. difference in leg length on equal-leg fillets.
2. A check on the convexity of the weld crown. Codes normally allow a flush to 0.100 in. convexity.
3. A check of the weld length.

For groove welds:

1. A check on the penetration at the root for complete fusion. No factory edge or unfused edge should be visible at the root.

2. A check on the convexity of the weld crown. Codes normally allow a maximum of $\frac{1}{8}$ in.

In addition, both groove and fillet welds must be examined for cracks, undercut, excessive spatter, porosity, and underfill. The applicable codes, standards, and specifications will spell out the exact allowable limits of these weld flaws.

Welding codes will evaluate weld flaws with three determining factors.

1. Type
2. Size
3. Location

Remember from the terminology chapter that a discontinuity is a flaw in the material; it may or may not be acceptable. A defect is a rejectable discontinuity because its type, size, or location exceeds the limits set by the applicable codes.

The *type* of flaw is a determining factor because some flaws are more serious than others. For example, slag inclusions and cracks are both serious flaws; however, small slag inclusions are usually acceptable, but usually, cracks are not acceptable, no matter how small they are.

Size is also a determining factor. On a weld X-ray small slag inclusions of less than $\frac{3}{32}$ in. (depending on the code or specification) are generally acceptable. Slag inclusions greater than $\frac{3}{32}$ in. are usually rejected.

Location is the third consideration. For example, discontinuities located on the corners or at the ends of welds are considered more severe than discontinuities located in the middle areas of the weld. Flaws found in or near the edges of a cut plate are considered more serious than flaws found in the center of plates.

DISCONTINUITIES

Examples of type, size, and location of discontinuity flaws are shown in Figure 27–6.

Spatter Codes normally state that excessive spatter (Figure 27–7) should be removed. However, continuous excessive spatter may warrant investigation into how the welding process is being applied. Voltages set too high or arc lengths that are too long can result in excessive spatter.

Undercut Codes may allow some limited undercut (Figure 27–8), depending on the depth and the direction of the undercut in relation to the primary stress applied to the joint. For example, as much as $\frac{1}{32}$ in. of undercut may be acceptable if it is longitudinal to the direction of

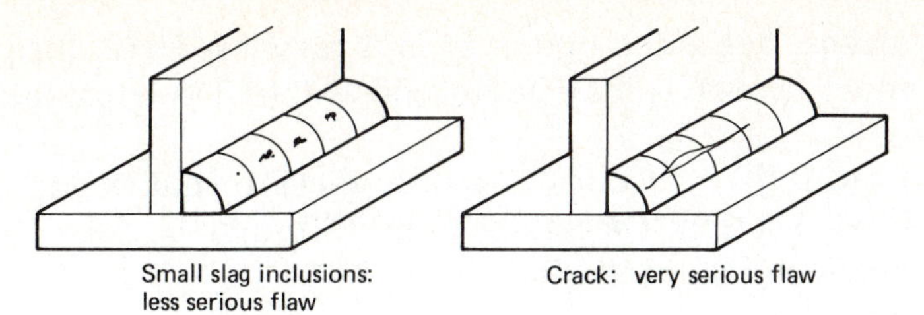

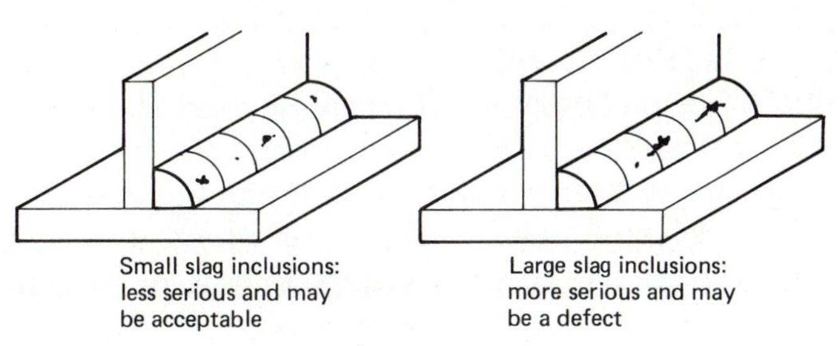

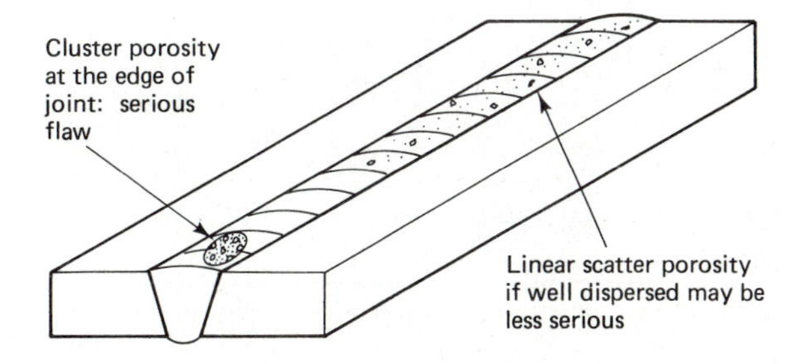

FIGURE 27-6
Type, size, and location of flaws.

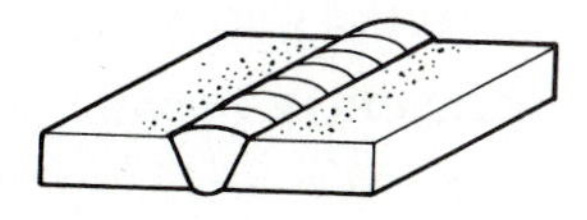

FIGURE 27-7
Excessive spatter.

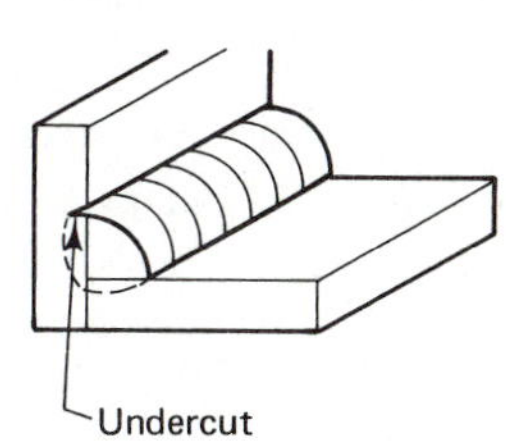

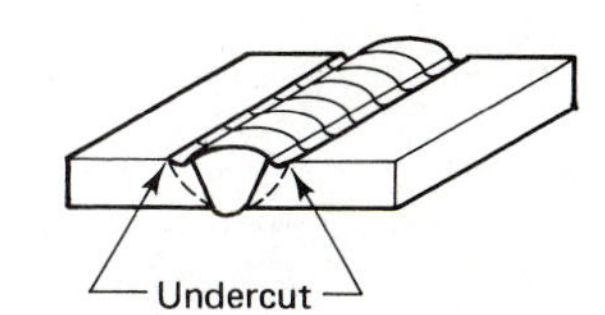

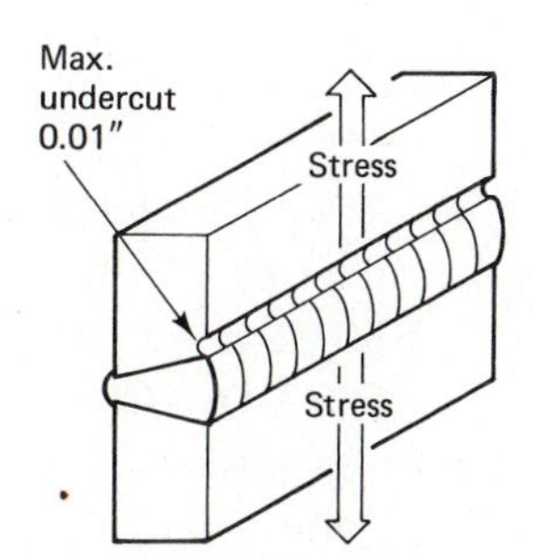

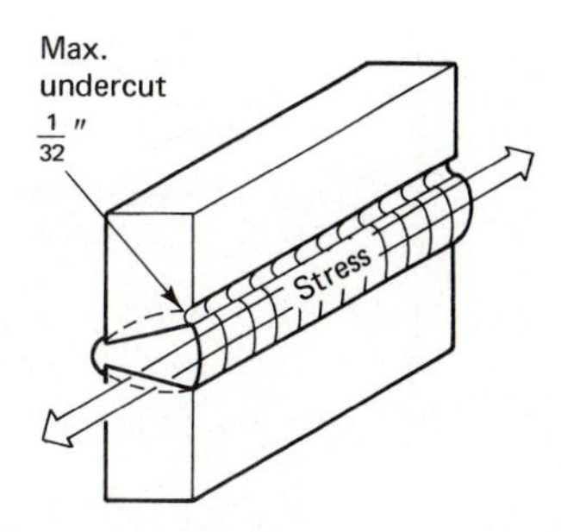

FIGURE 27-8
Undercut and maximum undercut.

stress, but if the undercut is transverse to (at a right angle to) the primary stress, it may be limited to 0.010 in. maximum, to be acceptable.

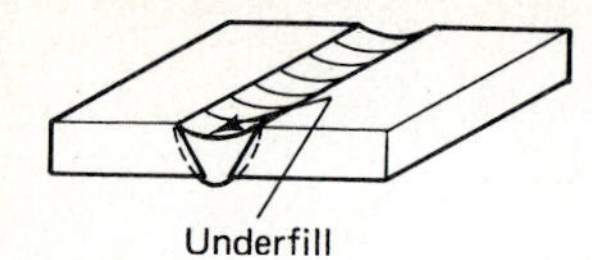

FIGURE 27-9
Underfill.

Underfill

Codes normally dictate that welds be filled to their full cross section; therefore, underfill must be filled to the required cross section (Figure 27-9).

Cracks

Welding codes do not allow any visible cracks in the final weldment. When cracks are found, they should be reported to the welding engineer. If the engineer approves, cracks can be repaired. The repair procedure usually requires that the crack be completely removed, plus a few inches of good metal beyond the end of the crack, and completely rewelded. It should be noted that there are many different types of cracks (crater cracks, throat cracks, toe cracks, hot cracks, cold cracks, micro cracks, etc.), all of which are unacceptable in the final weld. Figure 27-10 shows some examples of types of cracks.

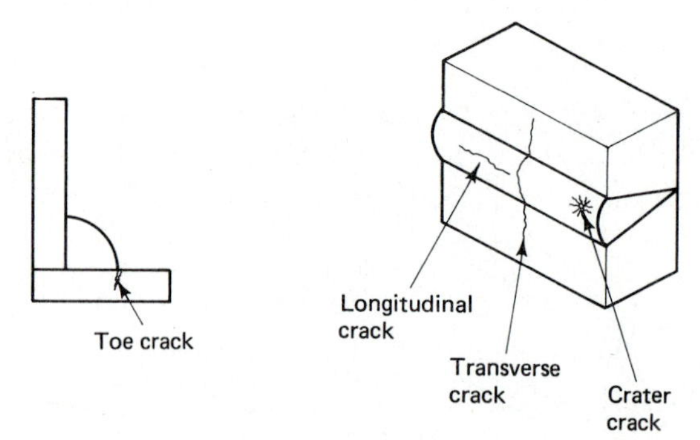

FIGURE 27-10
Toe crack, cracks, and craters.

Porosity

Gas can get trapped in the weld from oil or water in the plate or pipe, or oxygen attack from the atmosphere due to improper shielding. Codes do allow some porosity if it is well dispersed. Worm hole (piping porosity) and cluster porosity are not well dispersed (Figure 27-11), and must be removed and repaired if sufficiently large.

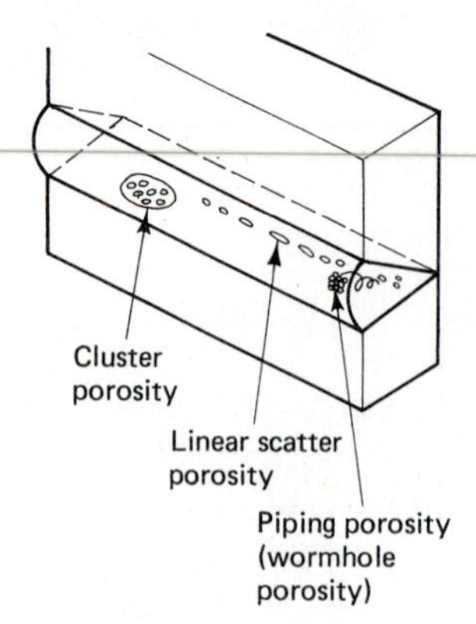

FIGURE 27-11
Porosity types.

Slag Inclusions

Most slag inclusions (Figure 27-12) must be removed if sufficiently large. Slag inclusions are dangerous to the weld because under stress, they can start a crack.

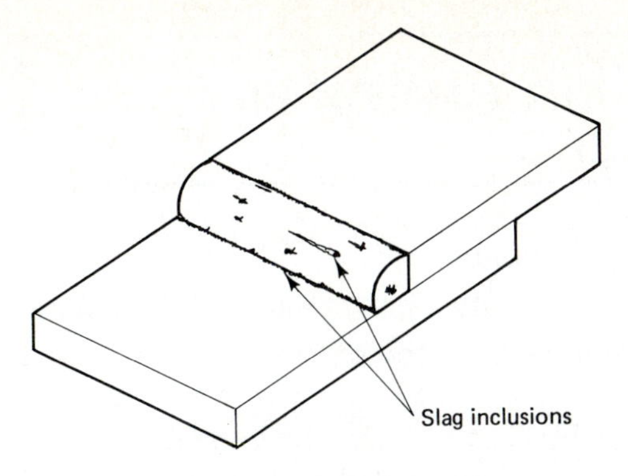

FIGURE 27-12
Slag inclusions.

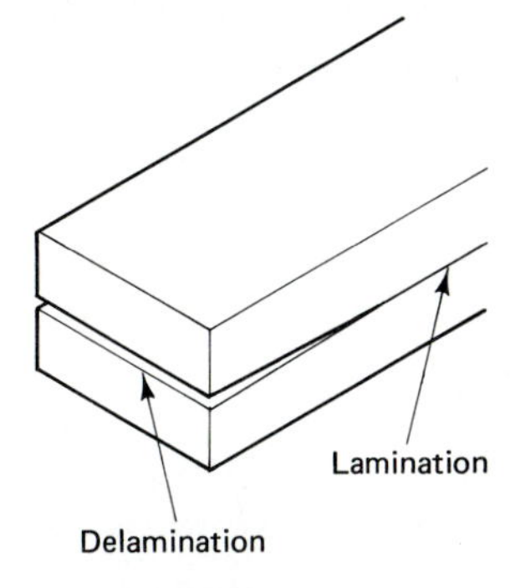

FIGURE 27-13
Delamination and lamination.

Laminations These discontinuities originate at the steel mill when two thin plates are hot rolled together to make one thick plate. Due to either insufficient roller pressure or insufficient heating of the steel, the plates may not completely fuse together in some areas. Delaminations usually occur when flame cutting or welding laminated plate. Delamination is the actual separation of laminated plate (Figure 27-13).

Lamellar Tears These result from overwelding and high stress or fatigue loading. They appear as "stair-step" cracks and extend well into the base metal, below the heat-affected zone (Figure 27-14).

Examples of acceptable welds are shown in Figure 27-15.

FIGURE 27-14
Lamillar tears.

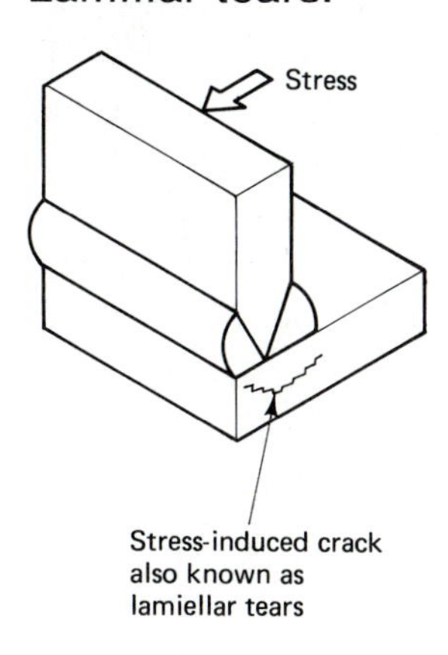

FIGURE 27-15
What an acceptable weld looks like.

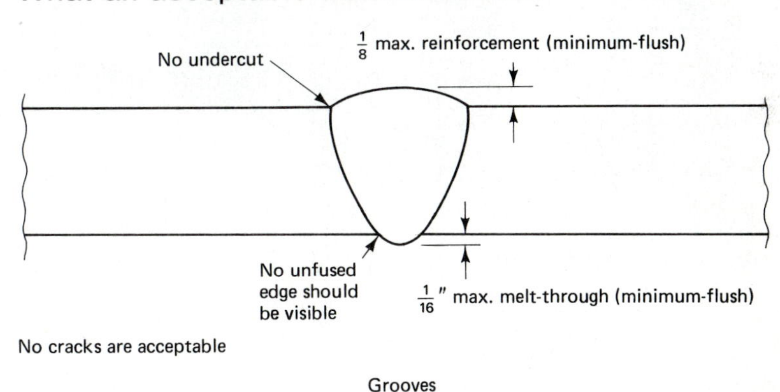

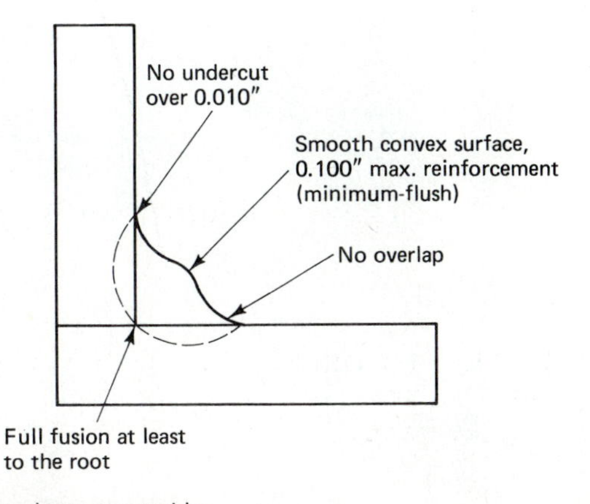

TOOLS FOR CHECKING WELD SIZES

1. Fillet weld gages
2. Combination fillet and groove weld gages
3. Visual inspection tools

Welding inspectors use these tools for visual inspection for leg sizes, convexity, and other weld dimensions.

Welds that are required to be visually inspected are recorded on a weld traveler. Note the weld traveler form contains space to record all the critical information about the weld. There should be a weld traveler form for each weld (see Figure 27-16).

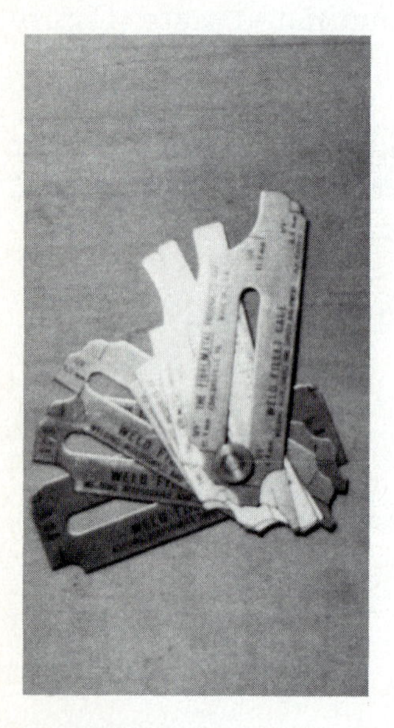

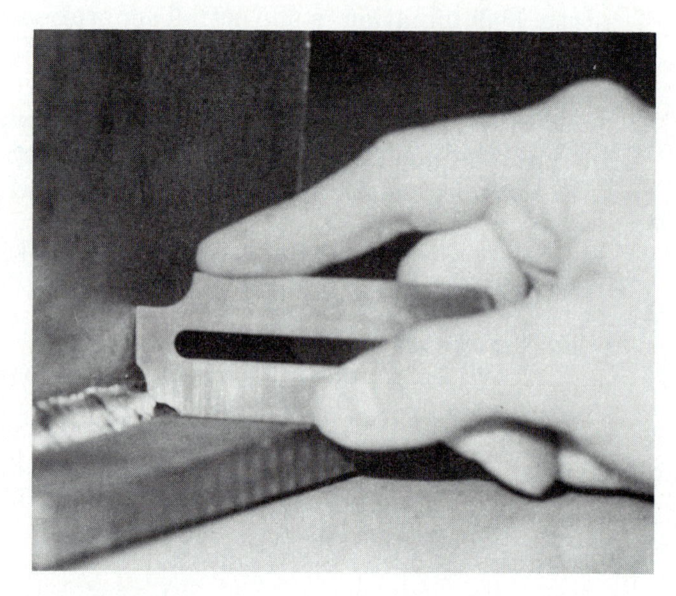

PHOTO 27-1
Fillet weld gages and their correct use.

PHOTO 27-2
Combination fillet-groove weld gage and its correct use.

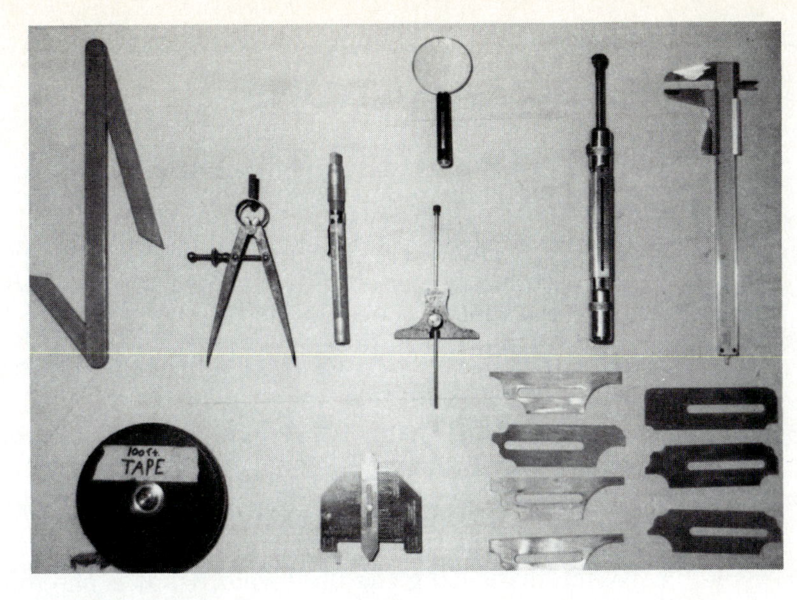

PHOTO 27-3
Welding inspector's tools for visual inspection.

FIGURE 27-16

Weld Number ______________ Material ______________ Acceptable +
Accepted: ☐ Filler Metal ______________ Unacceptable –
Rejected: ☐
As Per ______________

WELD TRAVELER

Comments

1. **Fit Up**
   Root Opening ______________
   Root Face ______________
   Alignment ______________
   Cleaning ______________
2. **Root Pass** (Groove Weld)
   Penetration, Convexity or Concavity ______________
   Penetration, Fused Edges ______________
   Oxidation ______________
3. **Filler and Cover Passes** (Groove Weld)
   Convexity ______________
   Fusion ______________
   Undercut ______________
   Underfill ______________
   Overlap ______________
4. **Fillet Weld**
   Leg Length ______________
   Convexity or Concavity ______________
   Undercut ______________
   Overlap ______________
   Underfill ______________
   Length ______________

CWI Name ______________________________ CWI Number ______________

*Feel free to use this for grading class welding projects.

REVIEW QUESTIONS

1. What must a welding inspector know other than the welding processes?
2. What are the three levels of AWS welding inspectors?
3. As a welding inspector, what are your responsibilities regarding filler metals?
4. As a welding inspector, what are your responsibilities regarding welding procedures?
5. As a welding inspector, what are your responsibilities regarding records and reports?
6. What can the welding inspector be doing before welding begins on a project?
7. When inspecting the final weld, what is the maximum difference in leg length size?
8. When inspecting the final weld, what is the maximum convexity of a groove weld?
9. Welding codes will evaluate weld flaws with what three determining factors?
10. What is the difference between a defect and a discontinuity?

STRUCTURAL STEEL FABRICATION

When working with structural steel or in the construction industry, it is important that you know not only the welding processes that you are applying, but the names of the components with which you will be working, and why they are used. Let's look at the basic components of buildings and bridges.

BEAMS AND COLUMNS

Columns are vertical members, so the concern is primarily with compression loads. These compressive loads can be very high. Imagine the compressive forces on the ground-floor columns of a 40-story building.

Columns are usually wide-flange or HP-beams. These beams are used because they absorb and evenly distribute the compressive loads.

The flanges on wide-flange beams are of the same thickness throughout, but the web is thinner. The flanges and the web are of the same thickness on HP-beams. The more uniform a beam's cross section, the better its resistance to buckling.

When sections such as S-beams or M-beams are used for compressive loads, they buckle faster under lower loads because of the tapered flanges, which gives these beams the characteristics of light weight but nonuniform cross section.

Beams are horizontal members, so they are concerned primarily with bending loads that result from gravity, but wind load is also considered. The strength of beams or girders is based on the following:

- Length of beam
- Depth of beam
- Area of the flanges

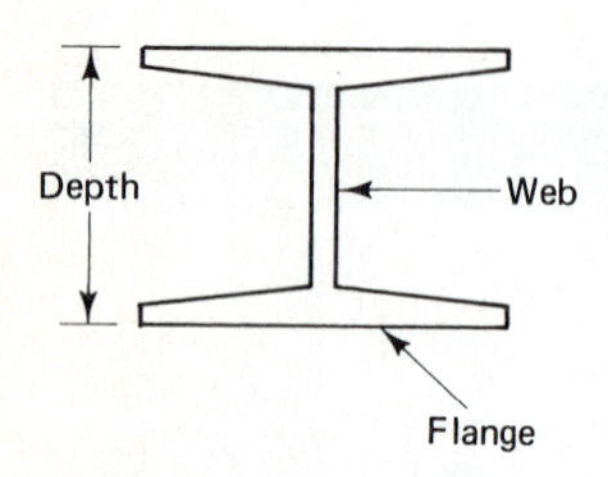

FIGURE 28-1
Parts of beams and columns.

The longer a beam or span, the more the weight of the beam itself (the dead load) becomes a factor. Small beams with long spans can support only light loads because a beam's strength is used to support itself—its dead load. Therefore, small beams are used for short spans and light loads.

A beam's depth is an extremely important factor in determining its strength. The deeper the beam, the stronger the beam. The area of the flanges is also important in determining its strength. The more square area in the flanges, the stronger the beam. Beam and column parts are shown in Figure 28-1.

FIGURE 28-2
Compression and tension.

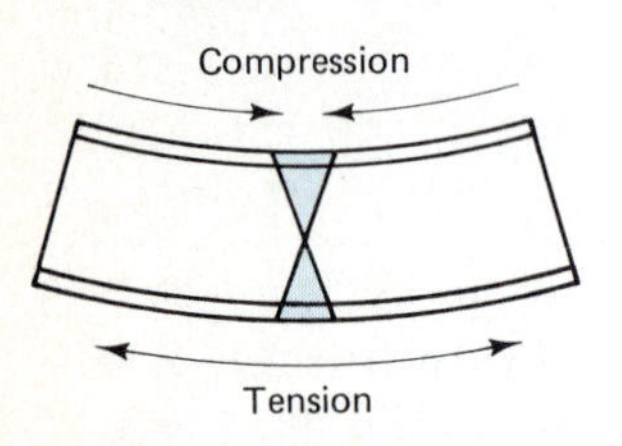

Why are beams shaped as they are? To answer that, let's analyze what happens when stress is applied to a beam (Figure 28-2). In the beam shown here, the top flange is being compressed and the bottom flange is under tension. In about the center of the beam in the web, there is no tension or compression. This area is known as the neutral axis. The farther you are from the neutral axis, the higher the tension or compression. If the flanges are placed far from the neutral axis, as with deep beams, the stronger the beam. Similarly, the more material or surface area we put out there to resist the tension and compression forces, the stronger the beam (Figure 28-3).

It is the placement and size of the beam flanges that give a beam its strength. The web is an important member of the beam design because it holds the flanges in place and resists shearing forces set up during deflection.

The I-shape used in beams offers the most strength, at the lowest weight, and is therefore the most economical structural shape for most structures.

The beams may be wide flange, S-beams (previously called I-beams), or M-beams (previously called H-beams). The "S" stands for

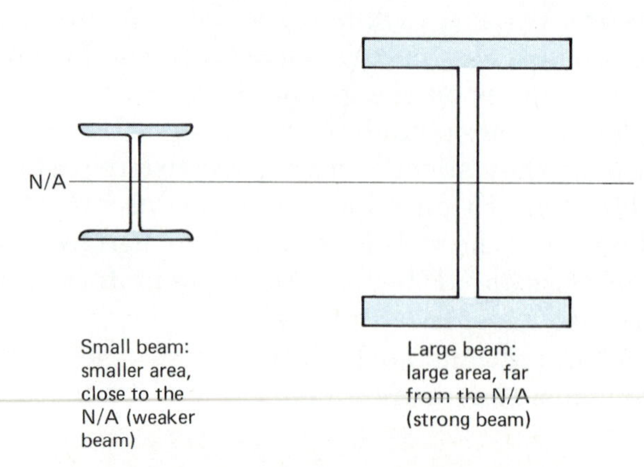

FIGURE 28-3
Small and large beams and neutral axis.

American standard beam, and the "M" stands for mill beam.

The S-beams (Figure 28–4) offer the most strength on long but fairly light, straight-down gravity loads. This is because they have good depth but short tapered flanges. The M-beam (Figure 28–5) is similar, but its flanges are longer. This gives it a little better resistance against wind loads (side-to-side loading). Wide-flange (WF) beams (Figure 28–6) are very popular for heavy loads. They have more surface

American standard beam S or I
D X L or D X wt/ft X L
Depth larger than flanges

S-beams are best designed for gravity loads

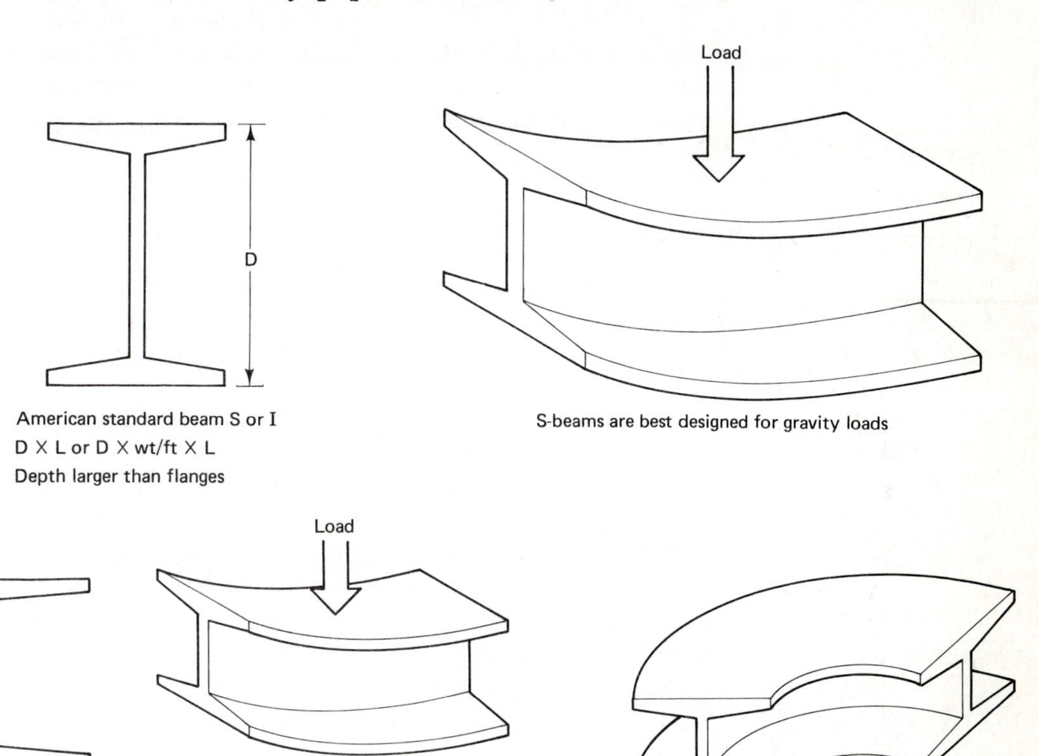

FIGURE 28–4
S and I beams.

Wind load

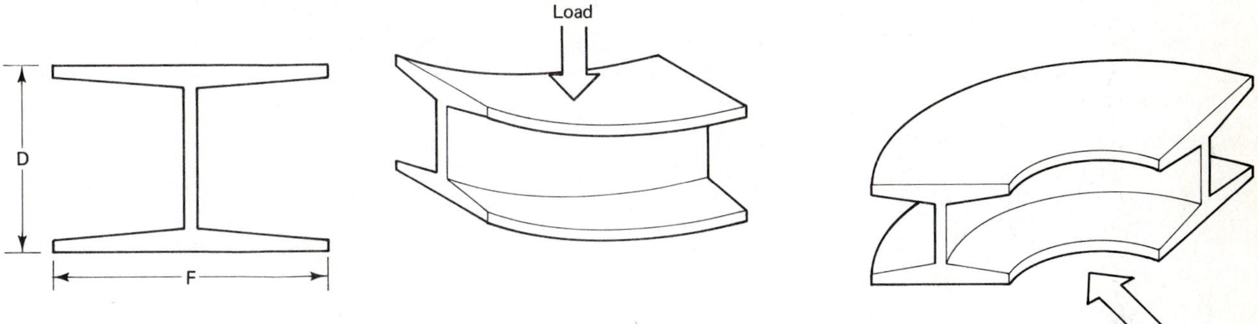

Mill beam M or H
D X F X L or D X wt/ft X L
Depth and flanges usually equal

M-beams are best designed for wind and gravity loads

FIGURE 28–5
M and H beams.

FIGURE 28–6
WF beams.

Load

WF (wide flange) sections

Good compressive and bending strength, used for columns and beams

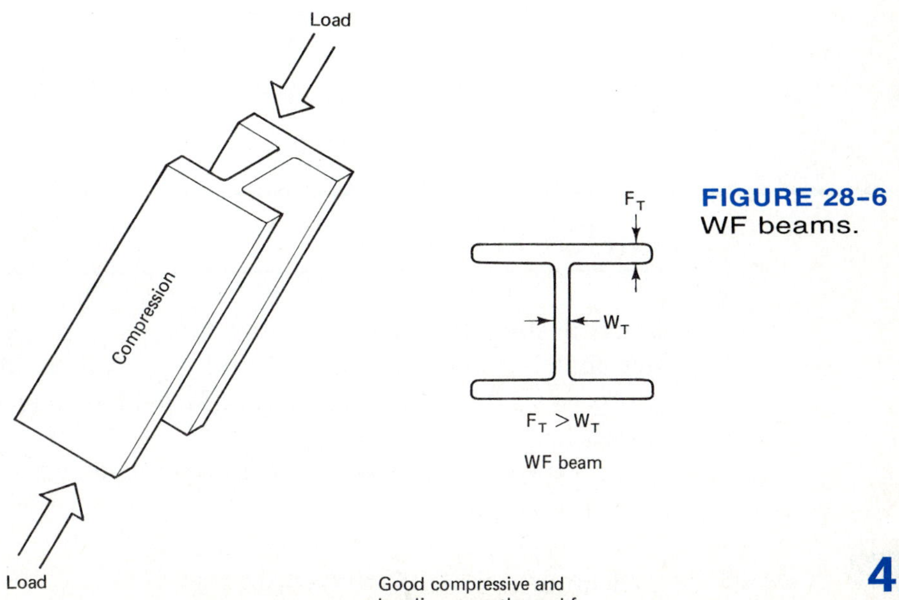

area in the flanges because they are not tapered. The wide-flange beam is the most commonly used section in building and bridge construction.

How Do Beams Fail?

Most beams and columns fail due to a phenomenon known as "buckling." As a beam begins to become overloaded, eccentric forces set up in the beam due to irregularities in its cross section. This induces a bending force. Because of the restraints at the ends of a beam or a column, simple bending will not occur, so a twisting and rolling action known as buckling occurs. Symmetrical sections will support much higher loads before buckling than will unsymmetrical sections (Figure 28–7).

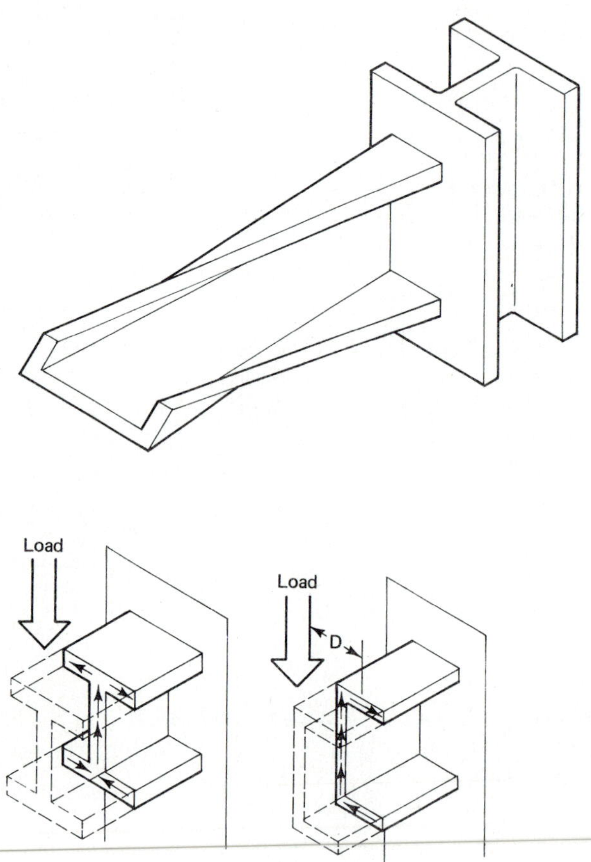

FIGURE 28–7
Buckling.

GIRDERS

What is the difference between a beam and a girder? A beam is a product that is rolled at the steel mill into WF, S, M, and other shapes; a girder is a fabricated section that is welded together from plate (Figure 28–8).

Girders are used when load requirements are higher than rolled

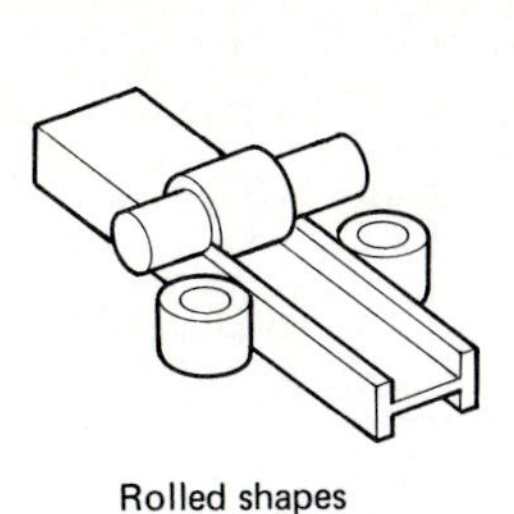

FIGURE 28–8
Rolled and fabricated shapes.

sections could support. Standard rolled section sizes go up to 36-in. depths, and some special sizes up to about 72 in. in depth are available.

Girders are usually found in high-stress load areas and relatively long spans on buildings and bridges especially. Girders come in the form of standard I-sections or "box girder" sections. These girders are common on highway bridges and highway entrance ramps. You will note that box girders have two flanges and two webs (Figure 28–9).

Girders are made up of flanges and webs just as rolled sections are, but it is common to add additional stiffening components to girders, such as transverse or longitudinal bearing stiffeners.

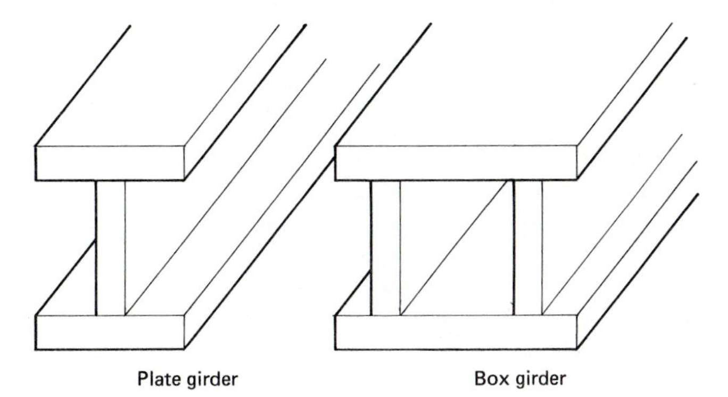

FIGURE 28–9
Plate and box girders.

How Do Stiffeners Work?

Transverse bearing stiffeners are used to help support the beam under critically high loads (Figure 28–10). They actually prevent the web from buckling under high shear forces. As the loads become critical and the web starts to buckle, the compression flange and the tension flange (top and bottom flanges) are forced together against the ends of the stiffener, thus supporting the web. The transverse stiffener is now under a compression load, thus absorbing some of the load. The girder is now acting more like a truss because the web is transferring tension forces diagonally to the top of each stiffener. The girder is now supporting a much higher load without failing, due to the transverse stiffeners.

Longitudinal stiffeners run longitudinal to the girder and are usually welded into place about one-fifth of the way down from the compression flange (top flange). They are common on bridges or very highly loaded spans on buildings. They are almost always used in combination with transverse stiffeners (Figure 28–11).

Cover plates (Figure 28–12) are long but slightly narrower plates which are welded to the tension and compression flanges. They

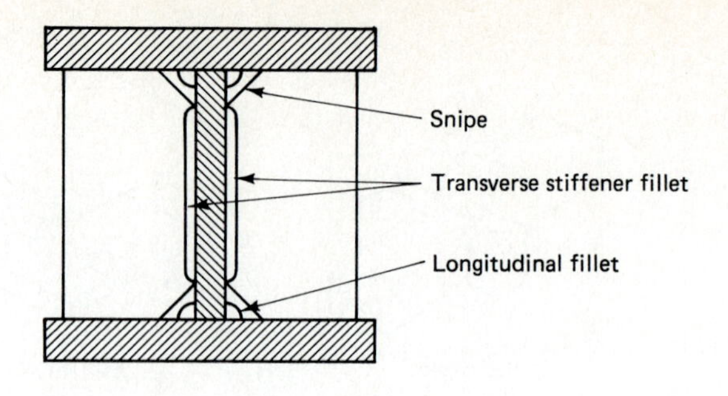

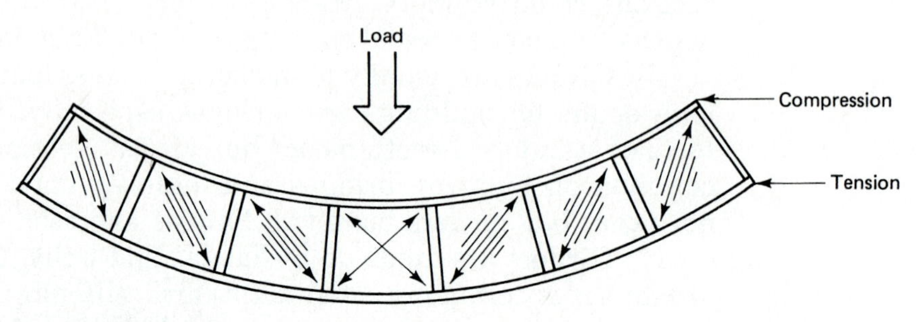

FIGURE 28-10
Girder parts.

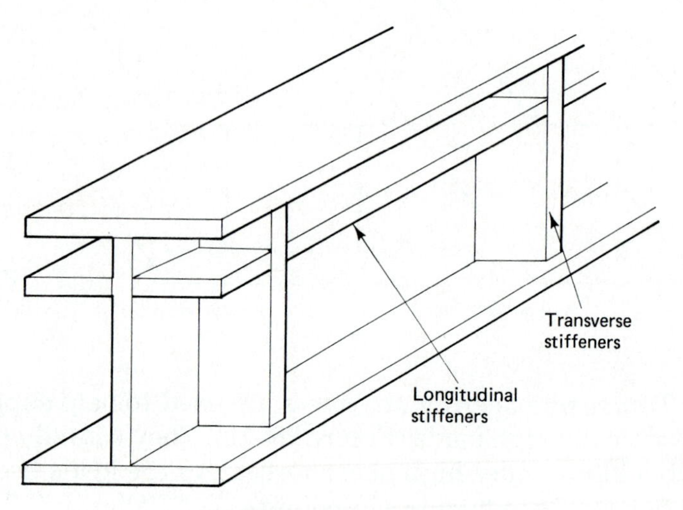

FIGURE 28-11
Girder stiffeners.

FIGURE 28-12
Cover plate.

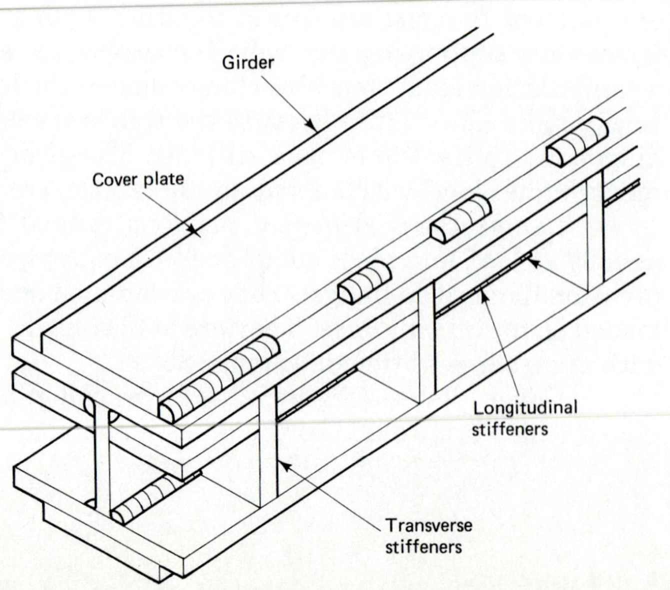

reinforce the beam or girder by adding more square area to the flanges. They may run full length or may be added only to the beam center or the region of high stress.

Welding Girders and Components

The decision as to where and how much weld to put on the girders is up to the welding engineer. Engineers use special formulas in calculating stress loads and required weld sizes and spacing between welds.

As a welder, technician, or inspector, you should always consult the job drawings and specifications to see exactly what weld sizes the engineer has specified.

Generally, you will find that flange-to-web joints have long continuous welds on the ends and shorter intermittent welds in the center (Figure 28–13). Highly stressed girders may have continuous welds throughout the joint.

Transverse stiffeners are fillet welded to the web. In some designs the stiffener is welded to the flange, and in others it is left unwelded, free to adjust to varying loads on the girder.

Longitudinal stiffeners are fillet welded to the web. They are interrupted where they meet transverse stiffeners. Cover plates are fillet welded to flanges.

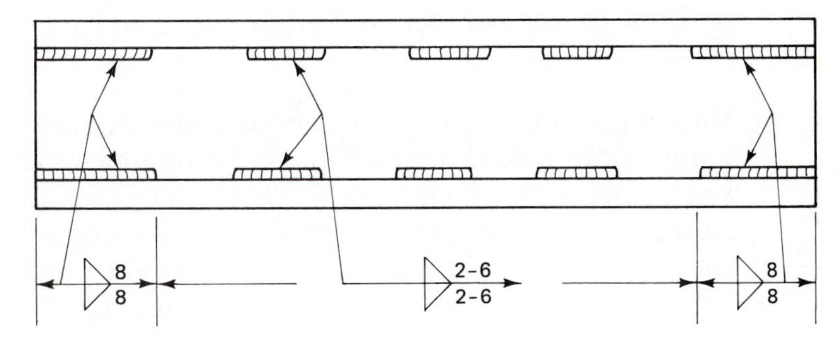

FIGURE 28–13
Girder welding.

TRUSSES

Trusses are used on long spans (usually over 50 ft) where moderate loads are encountered. A truss is made up of triangles, which gives the framework high rigidity (Figure 28–14).

The truss is considerably lighter than girders; therefore, over long spans the dead load (the weight of the structure itself) is much lower, allowing the truss to support much more live load. The live load is the additional load of the ceilings, floors, snow loads, and other material the structure was designed to support.

The fabrication of trusses is quite labor intensive and special attention must be given to distortion control; however, material costs are much lower because the trusses are lighter.

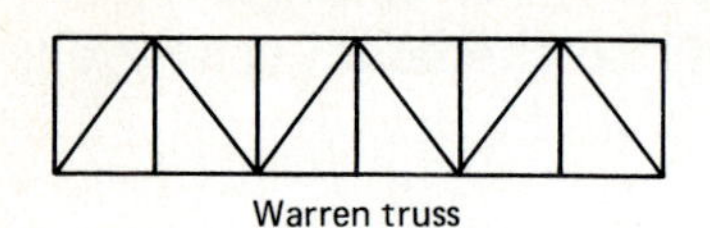

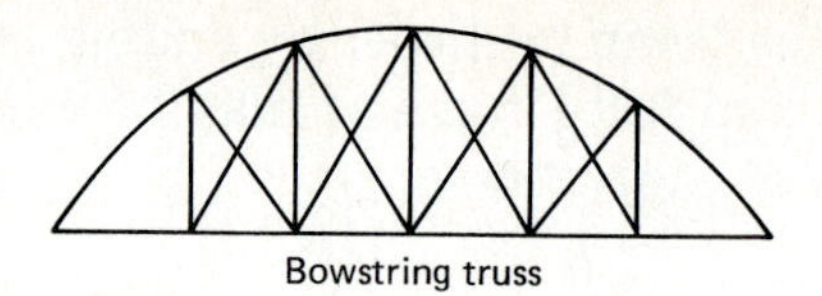

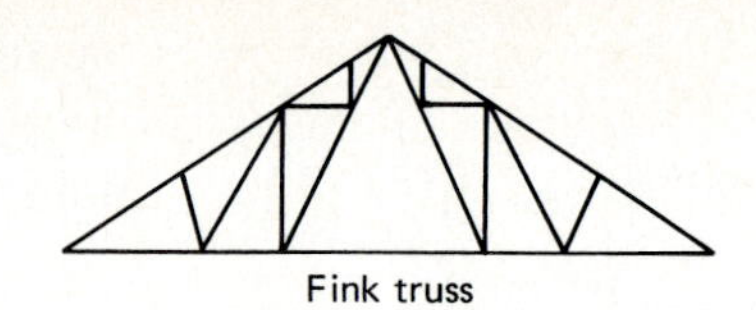

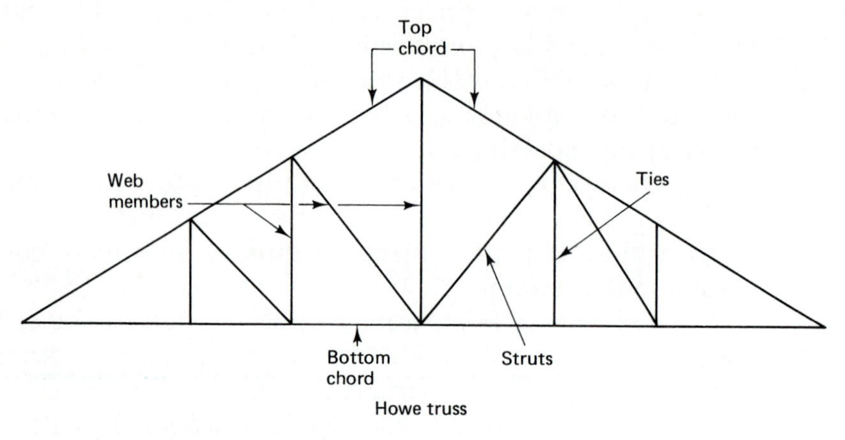

FIGURE 28-14
Trusses.

BASES FOR BUILDINGS AND BRIDGES

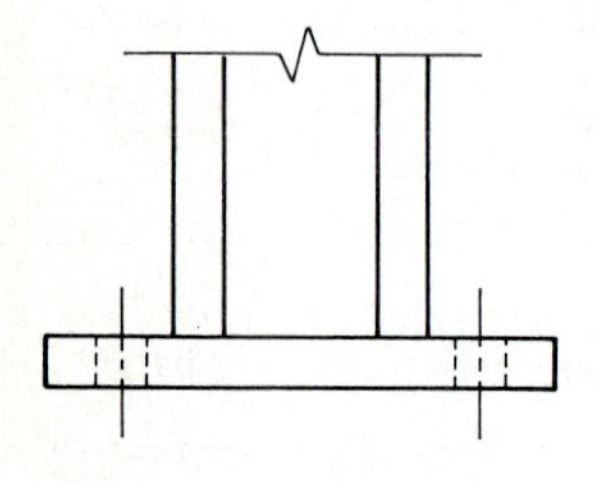

FIGURE 28-15
Column base.

Buildings require column bases that are designed for static (relatively nonmoving) loads (Figure 28-15). Bridge bases and connections must be flexible to absorb moving loads (dynamic loads) from traffic or other types of movement (Figure 28-16). If the bridge bases and connections were not flexible and allowed to move slightly under load, the connecting welds would eventually fatigue under this high unrelieved stress.

Column base plates are required to distribute the superhigh compression loads over a wider surface area. The high compression is reduced per square inch by distributing it into the base plate, then into the foundation (Figure 28-17).

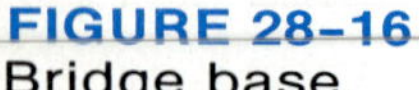

FIGURE 28-16
Bridge base.

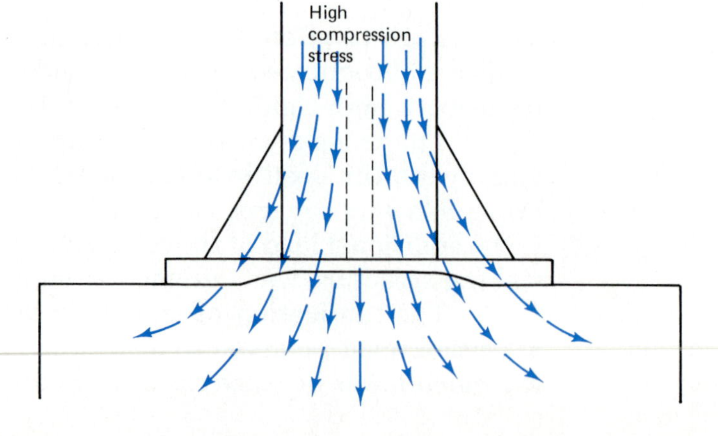

FIGURE 28-17
Column base diagram of stress.

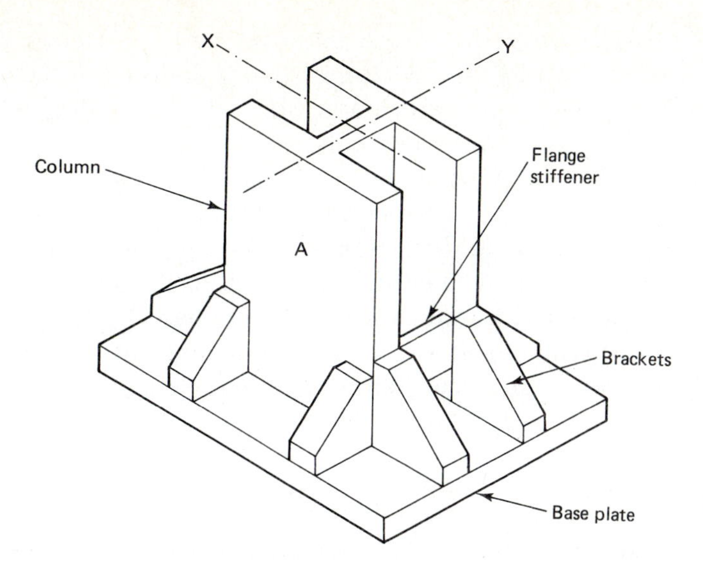

FIGURE 28-18
Column base parts.

BEAM-TO-COLUMN CONNECTIONS

Most of the structural steelwork on buildings and bridges is involved with connecting beams and columns together to shape the structure. Ends are not just butted together and beads deposited until it looks strong enough—it is not that easy. Instead, connections are made up of carefully sized connecting members and components that are critical to the structure's strength and flexibility. There are actually a number of systems that can be used to make beam-to-column connections. Let's look at a few typical designs and components used in these connections.

Simple Beam System

The simple beam, or fully flexible system, commonly uses an angle-iron seat and an angle-iron top connecting plate. This system leaves the beam quite free to deflect under high loads, thus putting little bending stress on the column. This allows column sizes to be moderate (Figure 28–19).

FIGURE 28-19
Simple beam connection.

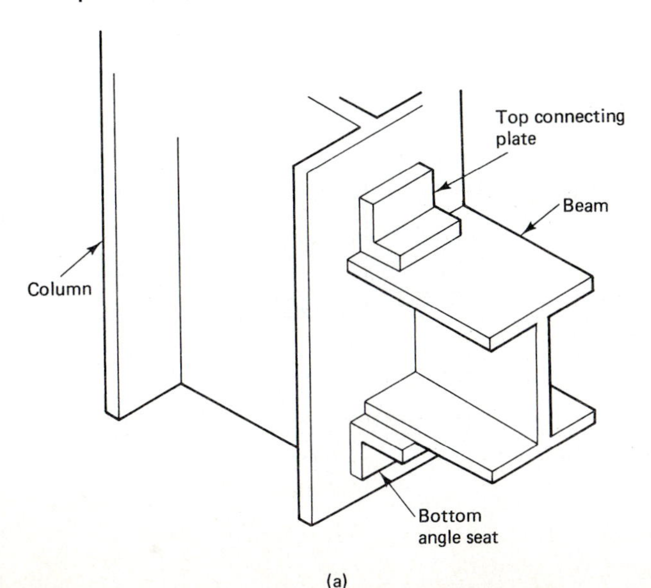

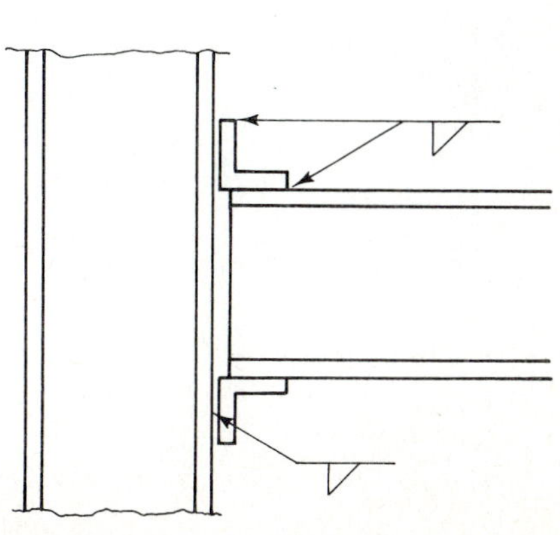

Fully Ridged Connection

The fully ridged system usually requires beams and columns to be welded directly to each other using little or no fastening components. When subject to high loads, this system transfers much of the beam bending stress into the column. Columns may have to be larger and stiffened with flange stiffeners for this design (Figure 28–20).

The semiridged connection combines some flexibility with a certain degree of rigidity. Semiridged connections usually use "tee seats" and semiridged top connecting plates. These top plates are designed to stretch as the beam deflects under load. Flange stiffeners are usually used to transfer stress through the column (Figure 28–21). Figure 28–22 shows some other connection components.

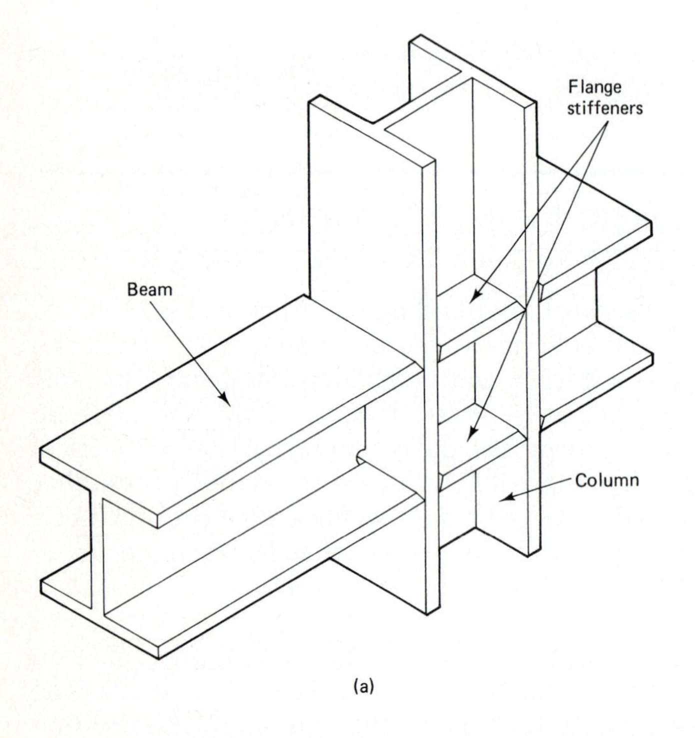

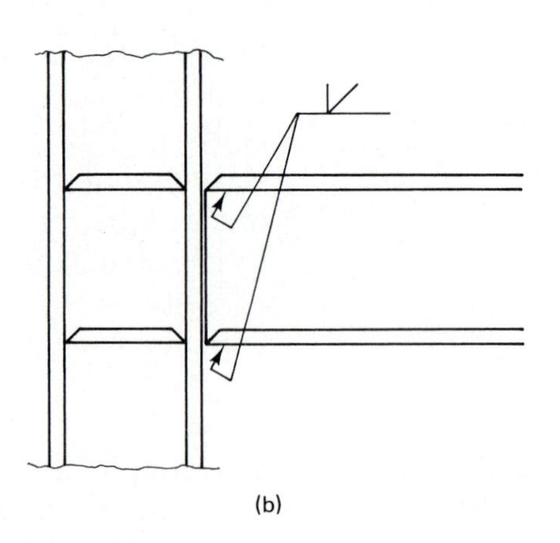

FIGURE 28–20
Fully ridged connection.

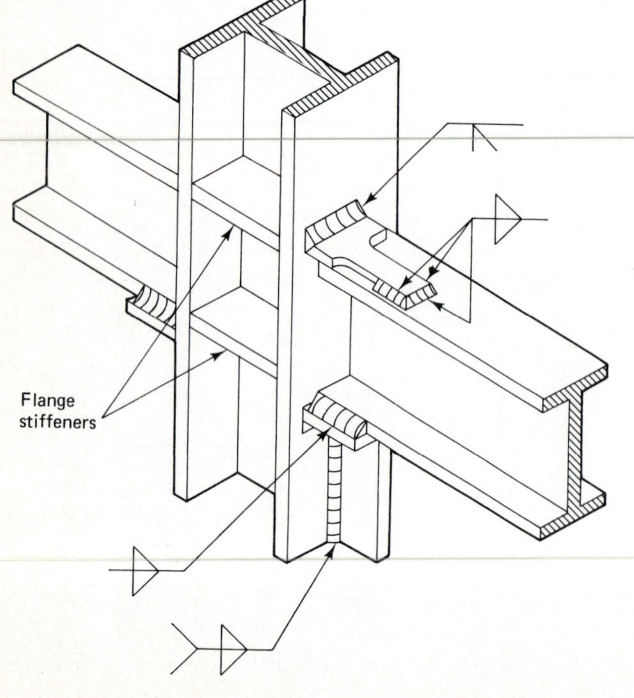

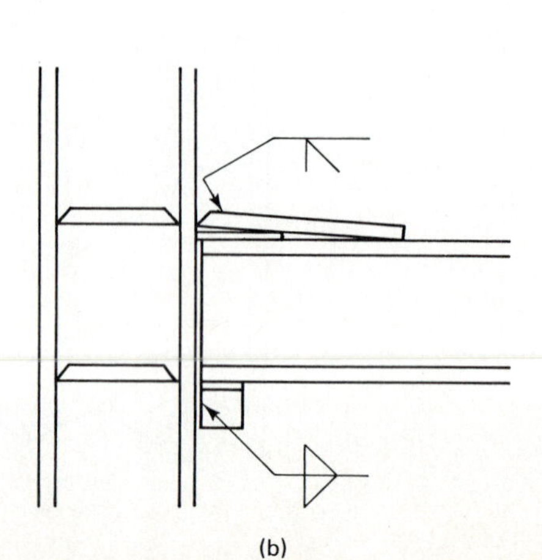

FIGURE 28–21
Semiridged connection.

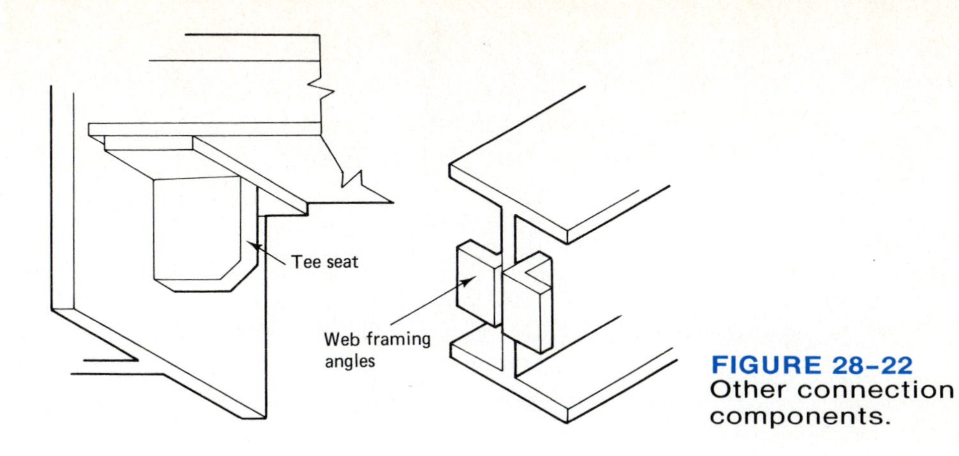

FIGURE 28–22
Other connection components.

STATIC AND DYNAMIC STRUCTURES

The two principal categories of structures are static structures and dynamic structures. Statically loaded structures are relatively nonmoving structures such as buildings (Figures 28–23 and 28–24). Dynamic loaded structures include bridges or parking garages.

The design for dynamically loaded structures is somewhat different than that of static structures. Engineers who design dynamically loaded structures must be concerned with fatigue, due to the high degree of movement. Components such as bridge base plates are designed to be flexible. Care must be taken to prevent notch effect or stress ris-

FIGURE 28–23
Bridge.

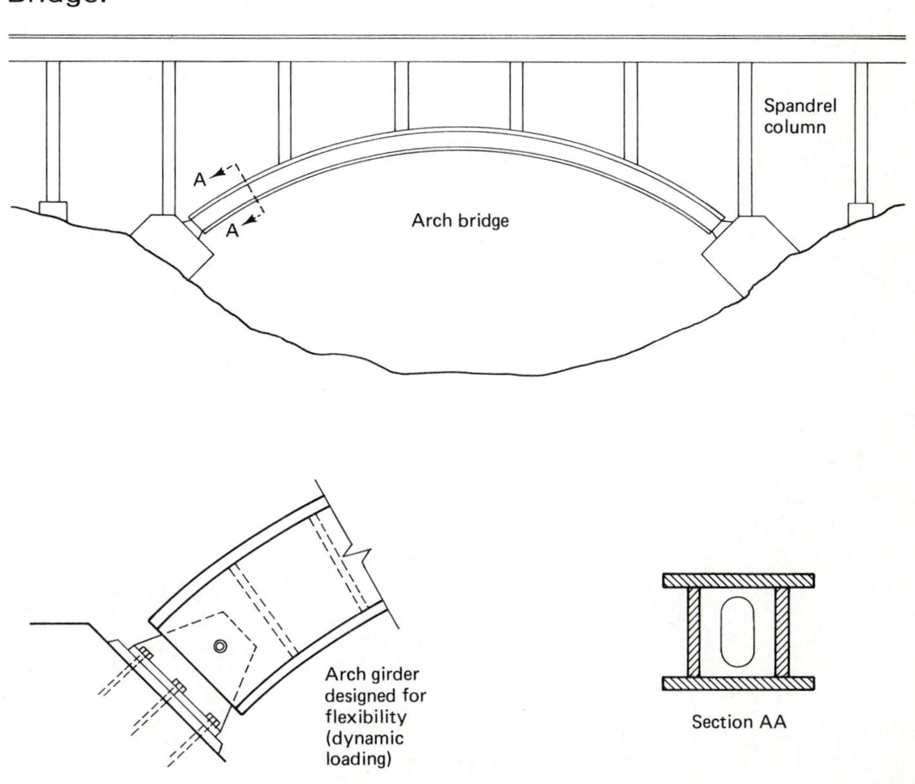

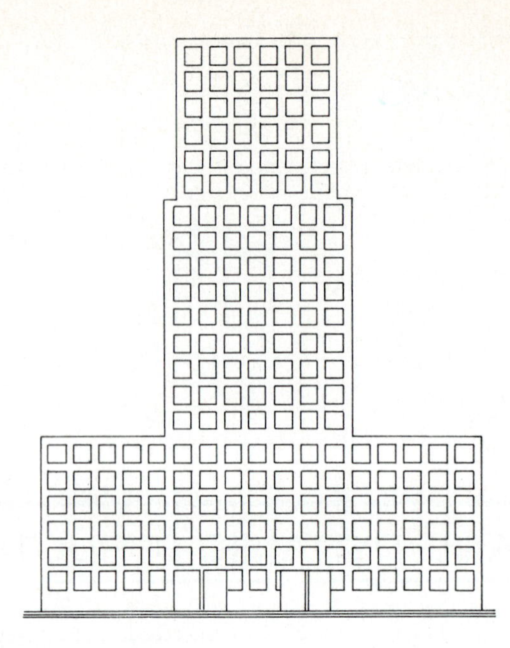

FIGURE 28–24
Building.

ers. Stress risers occur from poor workmanship or welding techniques, and under fatigue loading can cause cracks to form.

Overcut corners, sharp corners, incomplete weld fusion, and penetration all contribute to the possible development of stress risers (Figure 28–25).

We must use care not to accidentally produce a stress riser when cutting, and be careful not to leave any deep notches or gouges. Grind out and feather any rough-cut areas. When welding, make sure that there is no unfused roots, excessive overlap, or undercut.

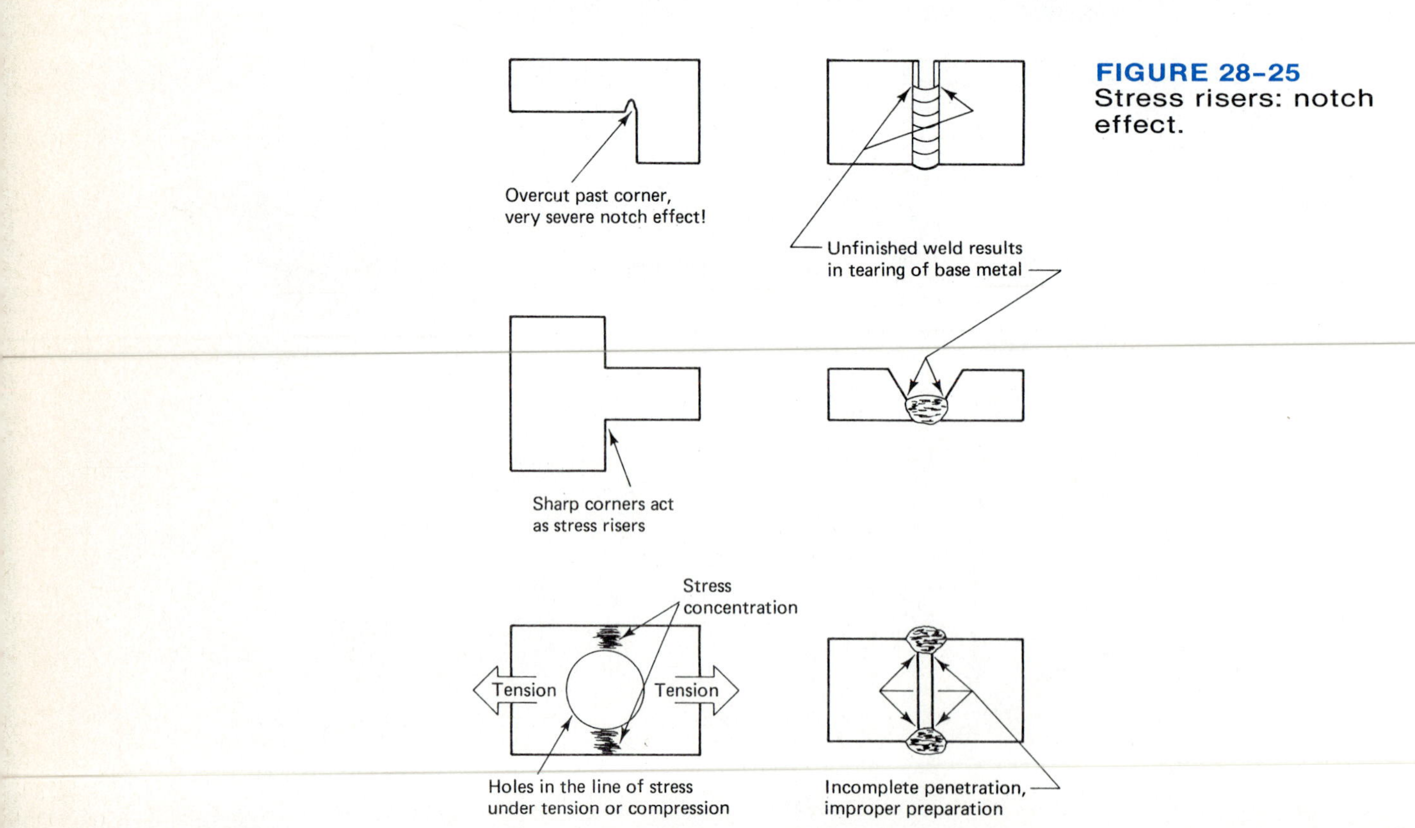

FIGURE 28–25
Stress risers: notch effect.

REVIEW QUESTIONS

1. What type of loads are columns under, and what type of structural members are used for these loads?
2. What factors determine the strength of a beam or girder?
3. What usually causes beams to fail?
4. What is the difference between a beam and a girder?
5. What may be added to a girder to strengthen it?
6. Where will trusses be used?
7. Why are column bases used on buildings?
8. List the three types of beam-to-column connections.

DISTORTION CONTROL

Always remember this basic fact: When materials are heated, they will expand, and when they are cooled, they will contract. The degree of expansion and contraction depends on the material and temperature to which it is elevated or cooled.

Distortion is a result of expansion and contraction. When we deposit weld, it is in the molten, expanded condition; when the weld metal cools, it contracts, pulling the surrounding base material in with it (Figure 29-1).

FIGURE 29-1
Plate bending before and after welding.

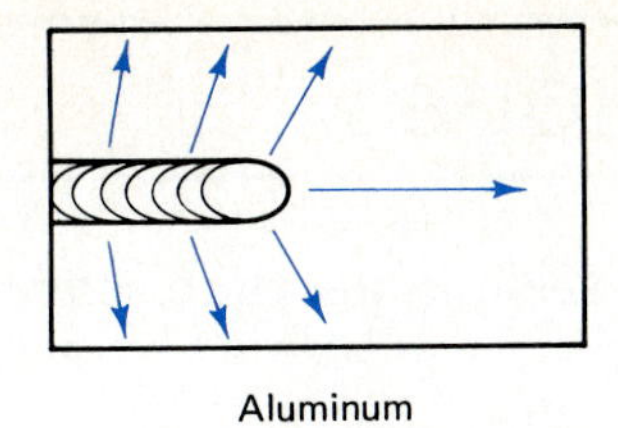

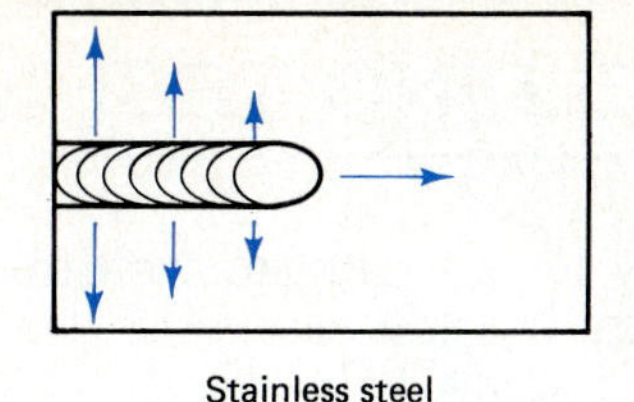

FIGURE 29-2
Heat transfer in aluminum and stainless steel.

Some metals will distort more than others. Metals with high heat (thermal) conductivity, such as aluminum and copper, expand fairly uniformally. Therefore, distortion of these parts is less; metals with low heat (thermal) conductivity, such as stainless steels and nickel steels, expand nonuniformally in various areas due to their uneven temperature distribution (Figure 29-2). Some areas are hot and expanding, and other areas are still cool and not expanding; the result is distortion. We may never be able to eliminate distortion totally, but by using distortion control techniques, we can at least reduce the effects.

JOINT DESIGN RULES

Let's first examine joint design rules that will result in lower distortion (Figure 29-3).

1. Keep the included angle as low as practical (Figure 29-3a).
2. Use a double V or double U, if possible (Figure 29-3b).
3. Preset parts in the direction opposite to the predicted distortion (Figure 29-3c).
4. Keep splices as close to the center of the beams and the weldments as practical (Figure 29-3d).

FIGURE 29-3
Joint design.

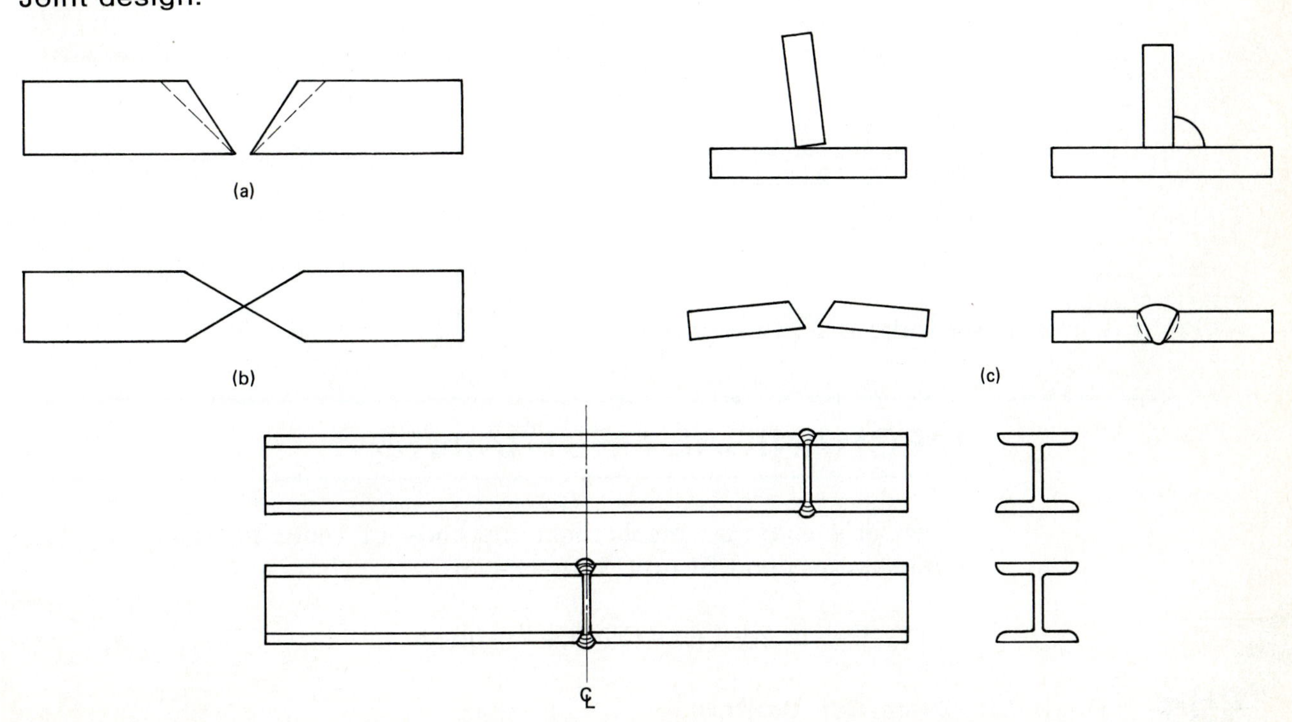

WELDING TECHNIQUES

Now let's examine welding techniques that reduce distortion (Figure 29–4):

1. Do not overweld (Figure 29–4a).
2. Deposit the minimum number of passes (Figure 29–4b).
3. Use a backstepping technique (Figure 29–4c).
4. Sequence the steps (Figure 29–4d).
5. Use intermittent steps (Figure 29–4e).
6. Sequence the sides (Figure 29–4f).
7. Keep the welding travel speeds high.

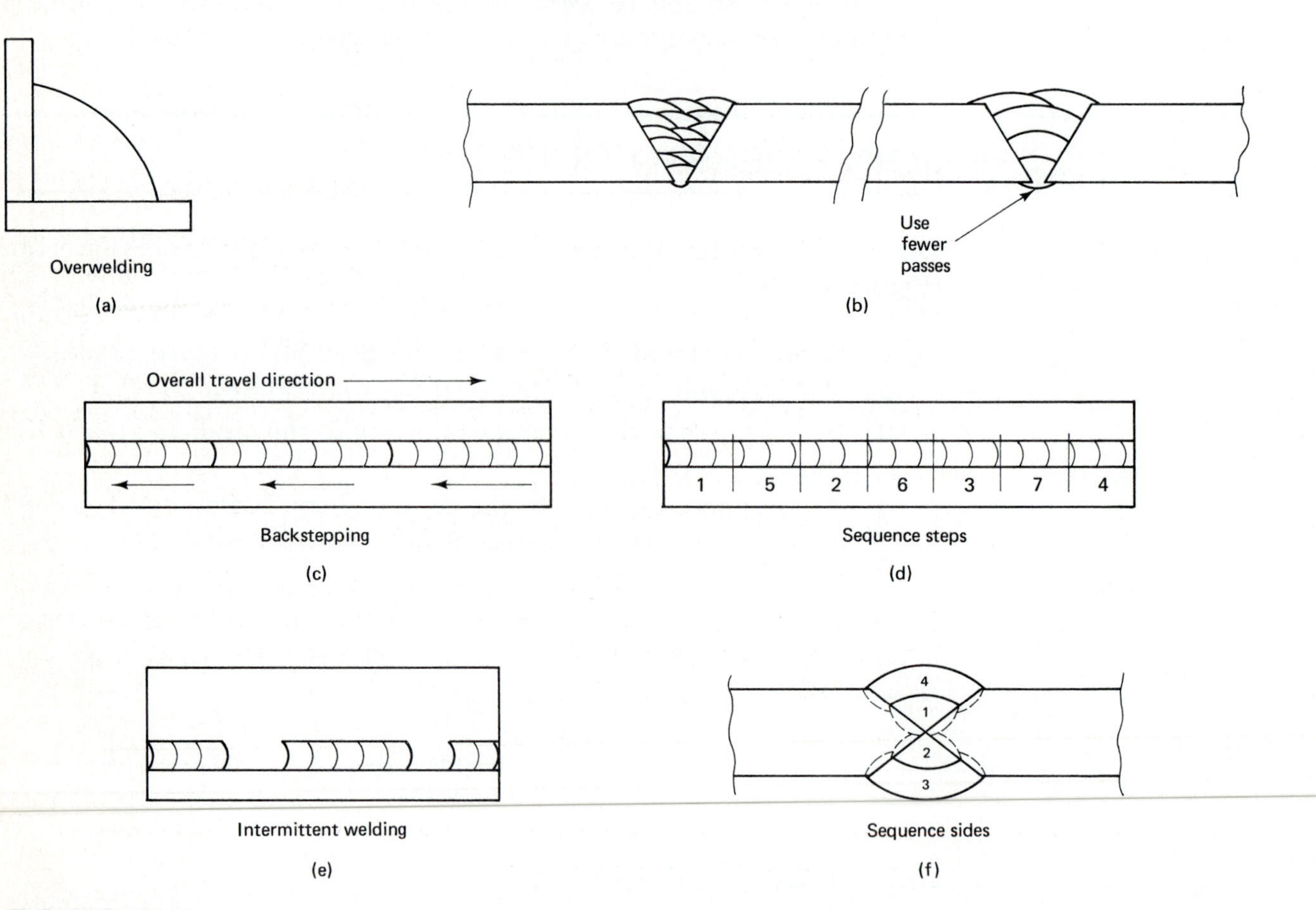

FIGURE 29–4
Welding techniques for distortion reduction.

MECHANICAL RESTRAINTS

Now let's examine mechanical methods of reducing distortion and maintaining good fit-up (Figure 29–5):

1. Use many tacks (Figure 29–5a).

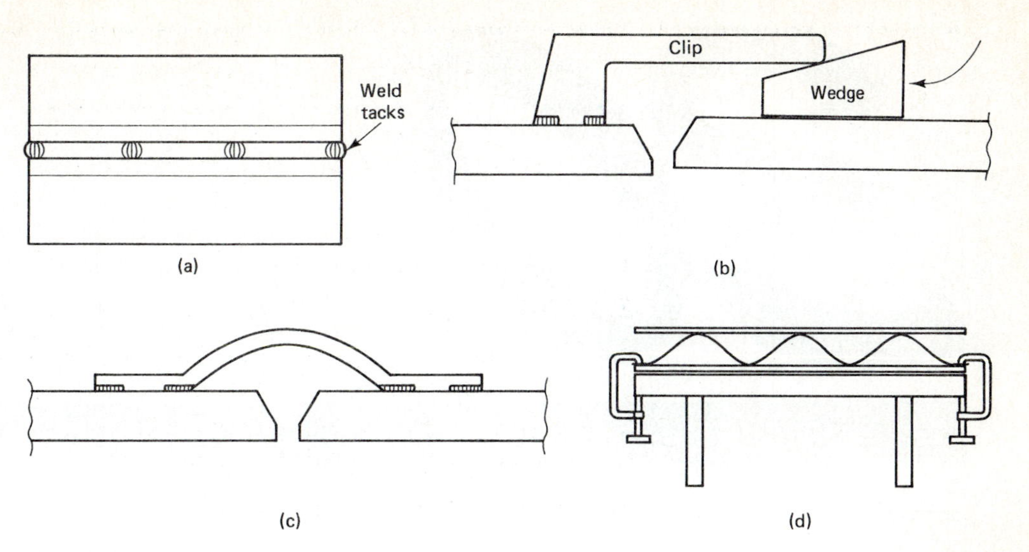

FIGURE 29-5
Mechanical methods for distortion reduction.

2. Use clips and wedges (Figure 29-5b).
3. Use strongbacks (Figure 29-5c).
4. Clamp the parts to a rigid fixture or table (Figure 29-5d).

Keep in mind that whenever parts are restrained during welding, a certain degree of residual stress develops. Distortion is actually the result of the relief of this stress. Therefore, some extremely hard or crack-sensitive parts should not be overrestrained.

It should also be noted that overall heat input plays a major part in distortion; therefore, limiting heat input greatly reduces distortion. You can calculate your heat input using this basic formula:

$$\text{heat input in (joules per inch)} = \frac{\text{volts} \times \text{amperes} \times 60}{\text{travel speed (ipm)}}$$

There are too many variables to state simply that a certain joules per inch rate will cause excessive distortion, but it does serve as a tool for comparison, especially if you weld similar joints and thicknesses often.

REVIEW QUESTIONS

1. What happens to metals when they are heated?
2. What happens to metals when they are welded?
3. How does aluminum react under heating, as compared to stainless steel?
4. How might one reduce distortion on stainless steels?
5. Explain back stepping technique.
6. What is a disadvantage of restraining welds during welding?
7. What is the heat input (in joules per inch) for a SMAW weld made at 95 amps, 24 volts, and 15 inches per minute (IPM)?

30 SHOP FABRICATION EQUIPMENT

Having the right tools and equipment will make the job easier and faster and will produce much higher-quality welds. *Weld positioners* allow us to move large weldments into the flat position for smooth, fast, high-quality welds. Positioners also speed up production of these weldments because the positioner can quickly rotate to the next weld seam. Positioners rotate the parts into various angles, which will put the weld seam as close to the flat position as possible.

Welding in the flat position is about three times faster than in other positions, because the high amperages and fluid puddles can be maintained. This results in very high deposition rates and high travel speeds.

Positioners can be electric, hydraulic, or mechanically controlled. They may be operated by hand controls or may be fully automatic microprocesser controlled. (Photo 30-1.)

Roll positioners are specifically designed for cylindrical weldments such as tubes, cylindrical tanks, and boilers. Positioners and roll positioners are often used in conjunction with automatic welding heads or robotic welding arms. This automates both the welding and positioning operations. (Photo 30-2 and 30-3.)

Shears are just about indispensable in welding and fabricating shops. Shears cut sheet and plate stock, and some are capable of cutting angle iron and punching. (Photo 30-4.)

PHOTO 30–1
Welding positioner. (Courtesy K. N. Aronson, Inc.)

PHOTO 30–2
Roll positioner working in conjunction with welding process. (Courtesy K. N. Aronson, Inc.)

PHOTO 30-3
Roll positioner working with welding head manipulator. (Courtesy K. N. Aronson, Inc.)

Metal benders or metal breaks bend corners or angles into plate and sheet stock. Most are manually operated using hydraulic, pneumatic, or manual mechanical power.

Metal rollers are used for rolling cylinder-shaped components or putting radiuses on components. They are usually hydraulically, manually, or mechanically gear driven. (Photo 30-5.)

A drill press is indispensable in the welding shop for producing holes that need to be centered precisely and to be within a few thousandths round.

Hand tools such as grinders, sanders, pneumatic chippers, drills, and hand metal shears are just a few of the tools that are needed to complete the fabricated work.

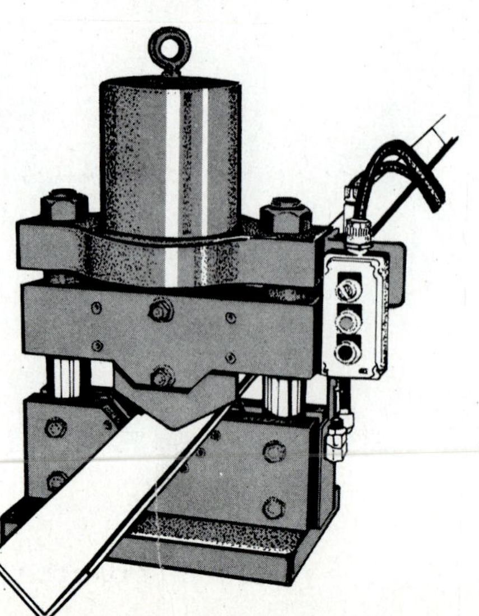

PHOTO 30-4
Angle iron shear. (Courtesy Franklin Manufacturing Inc.)

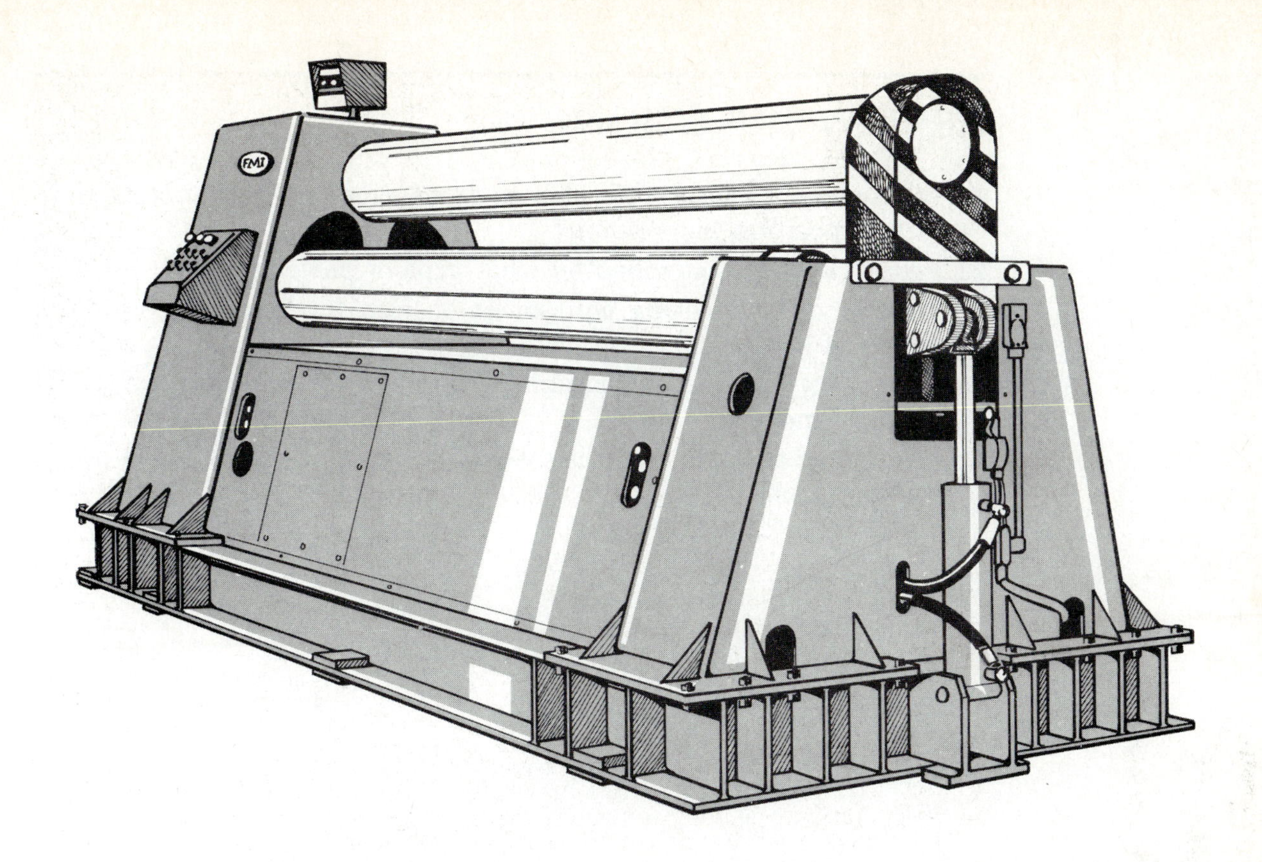

The way double initial pinch bending works:

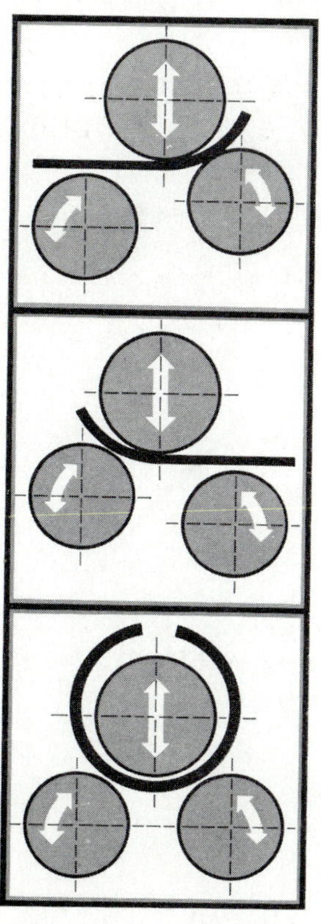

1. *The sheet is inserted and the power driven rear roll is raised to pre-bend the leading edge.*

2. *The rear roll is lowered and the sheet is rolled through the machine. The power driven front roll is then raised to pre-bend the trailing edge.*

3. *The material is centered and both lower rolls are brought into position for the desired radius and the cylinder is formed.*

PHOTO 30-5
Metal roller. (Courtesy Franklin Manufacturing Inc.)

REVIEW QUESTIONS

1. What type of shop fabrication equipment is used to put welds into correct position for welding?
2. Positioners are controlled by what type of power source?
3. What shop fabricating equipment is used for bending corners or angles of metal?
4. What shop fabricating equipment is used for creating a cylinder shape in metal?
5. What hand tools do you think would be necessary in a well-equipped welding shop?

ROBOTICS IN WELDING

Most welders fear the introduction of welding robots into industry because they feel they may lose their jobs to the robots. The facts show that robots will actually open many more opportunities and jobs than they will replace. Consider the following:

1. Most robots only replace workers in continuous and repetitious welding applications (monotonous jobs that most welders do not really want anyway).
2. Robots must be programmed and monitored continuously while in production.
3. A welding robot must be monitored by a welding technician who can troubleshoot welding problems and make prompt corrections.

A welding technician who plans to work in robotics must:

1. Learn to adjust travel speeds, voltages, amperages and wire feed rates, gas and flux flow rates, and so on (all welding variables).
2. Learn to identify welding and metallurgical problems and corrective actions.

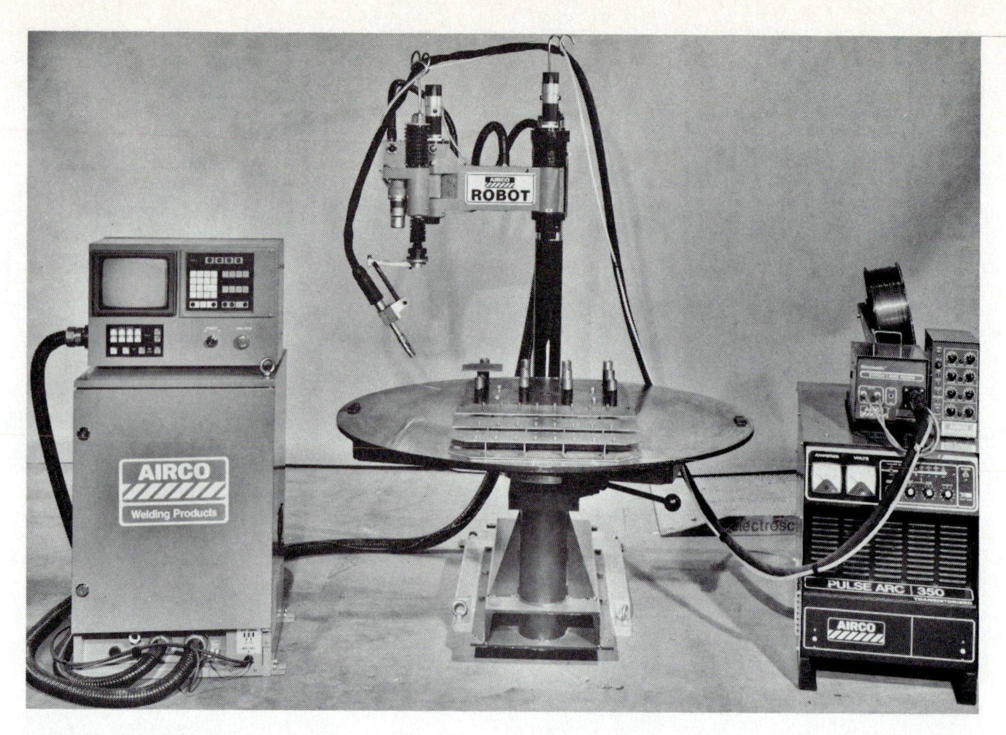

PHOTO 31-1
This system is equipped with a pulsed GMAW welding power source. It can be programmed to "slope out," giving very smooth uniform welds. (Courtesy of ESAB Welding Products, Inc.)

3. Learn efficient methods of production and robot cell designs.
4. Foresee weldability problems before welding starts.
5. Learn to adjust controls unique to automatic systems, such as slope in, crater out, and many new controls being added to modern welding power sources.

Most robotic technicians agree that what is essential in robotic training is the practical efficient application of the process they are applying (such as GMAW welding), not the machine assembly or assembly language of the processor.

Robots are much "friendlier" than they were even a few years ago. The necessity of learning the machine's assembly language is no longer necessary to get the most out of your robot.

This is not to say that a good practical background knowledge of electrical, electromechanical, or computer theory is not helpful or even necessary in some areas of robotics. It is similar to computers; years ago, the only one capable of programming a computer was the engineer familiar with its own machine language. But today, if you are familiar with BASIC (a simple computer language used in many home computers), you can program any number of manufacturers' computers. So you can learn your basic welding process variables and basic robot commands, but you don't have to be an electronics engineer to be a good welding robotics technician.

TODAY'S WELDING ROBOTS

The main purpose of an industrial robot is to speed production, lower the human error factor, increase productivity, and allow production in hazardous environments.

When a production device is considered a robot, it usually indicates that it has a high degree of flexibility. Today's robots are quite flexible; they can weld, paint, pick and place, and perform a variety of functions in a human-like fashion. They are controlled by hydraulics or electric servomotors or a combination of both. A typical robot system is shown in Figure 31-1.

Robots can be quite simple or extremely complex. The trend today is toward "smart systems." These systems can correct voltage, current, travel speed, and other variables as it travels along the joint. These "intelligent" systems are often controlled by laser seam tracking guidance systems and voltage sensor circuits that feed readings to the microprocessor, which then decides to correct, adjust, or continue welding the joint. Keep in mind that even these intelligent systems require adjustment and programming by skilled welding technicians.

Let's trace the steps of an advanced welding robotic system (Figure 31-2). This particular one is a laser tracking vision system interfaced to an "intelligent" robot.

PHOTO 31-2
Cyrovision real-time adaptive control sensor welding stainless steel. The line preceding the torch is a laser beam which is used to scan the weld joint. (Courtesy of Advanced Robotics Corp.)

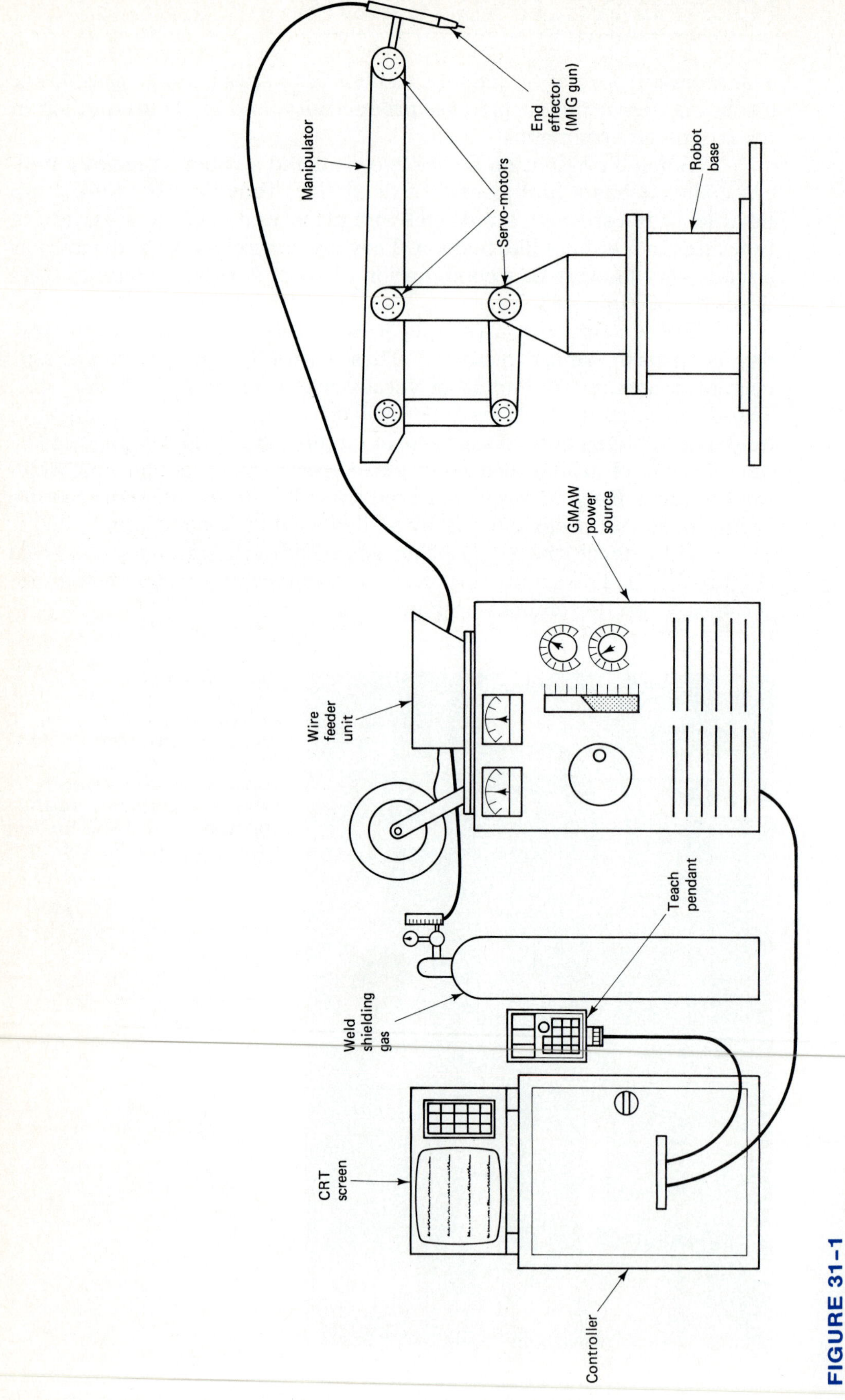

FIGURE 31-1
Typical robot system.

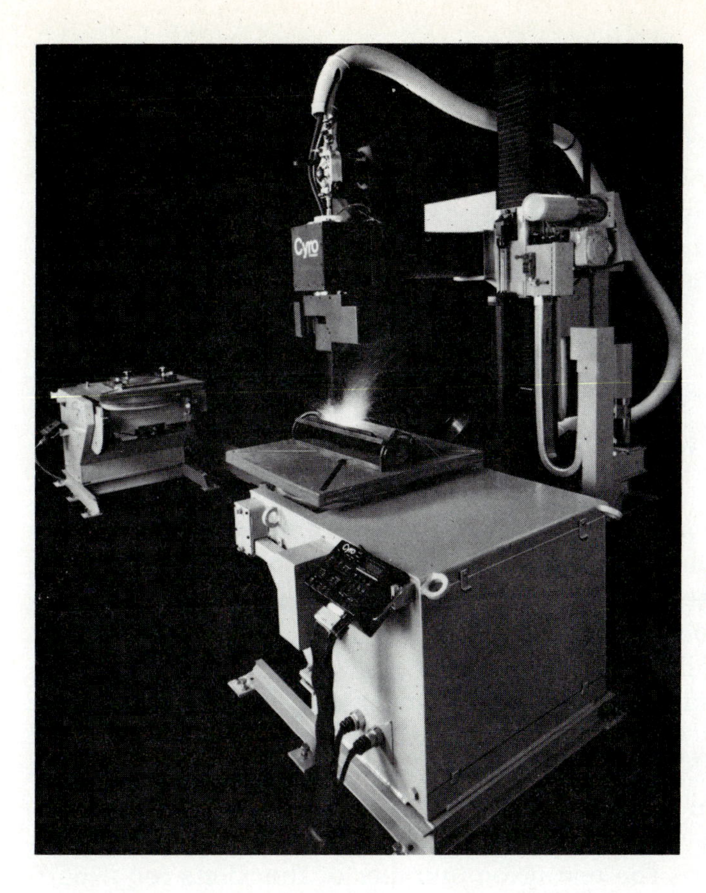

PHOTO 31-3
Five-axis Cyro 750 rectilinear robot with two Cyro RP positioners. Each positioner can tilt and rotate. Positioners manipulate the part to allow welding in the optional position. Touch pendant is attached to the front of the first positioner. Operator can unload and load one positioner while the robot is welding parts on the other. (Courtesy of Advanced Robotics Corp.)

FIGURE 31-2
Close-up of laser seam tracking system.

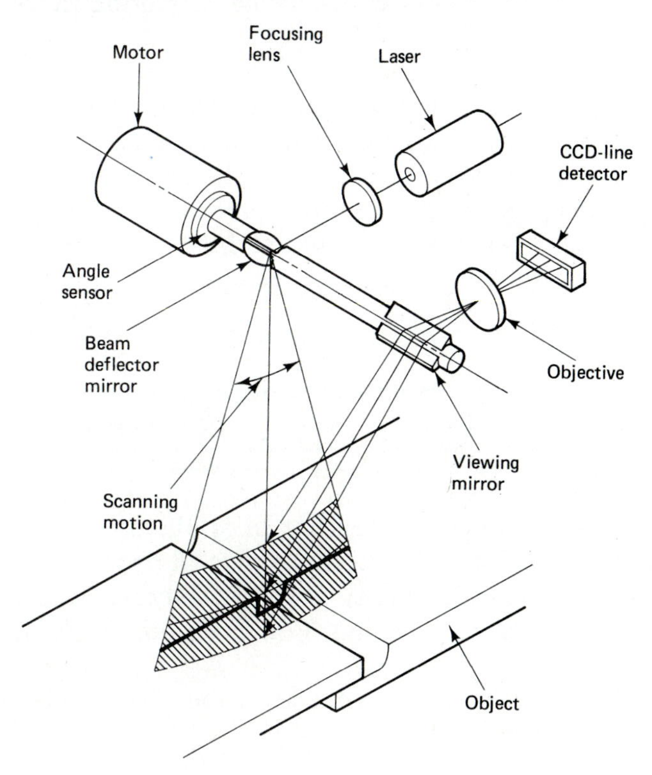

1. The laser scans for irregularities ahead of the seam and the data are sent to the buffer (data storage device).
2. The signals are stored until the precise moment when the welding arc reaches that point, and corrections are made.
3. Feed rate, amperage, voltage, and travel speed are computer controlled and make puddle adjustments. The results are a uniform precise weld.

ROBOT ECONOMICS

It is certainly not economical to place a sophisticated robot with many capabilities in a simple operation, or one that never changes. However, an operation that requires a high degree of repeatability may require a very sophisticated robot.

Not every application requires a five-axis articulated arm robot. Some simple pick-and-place operations that use a series of hydraulic cylinders and motors with a limit switch and a simple controller achieve fairly good results. The following factors should be analyzed when considering an industrial robot:

1. How complex is the operation (number of functions required)?
2. What is the degree of repeatability (preciseness of the operation)?
3. Does the operation change constantly?
4. Does the job require decision making in the middle of the operation?

In many cases a side beam or travel carriage with controllable volts, amperage, slope-in, and crater-out capabilities will be sufficient for the application. However, if the application requires decision making, very precise repeatability, and requires a complex series of movements, this is a job for an industrial robot. Payoff and return on a properly applied robotic manufacturing operation are reasonably quick.

ROBOT WORK CELL DESIGN

Cell design is very important in application of the robot. A good, efficient cell design will speed the overall production process and thus put less strain on the robot, which will increase the life of the robot and produce a safer operation. Points to remember in cell design:

1. Set up parts so that the robot has to do a minimum amount of moving.

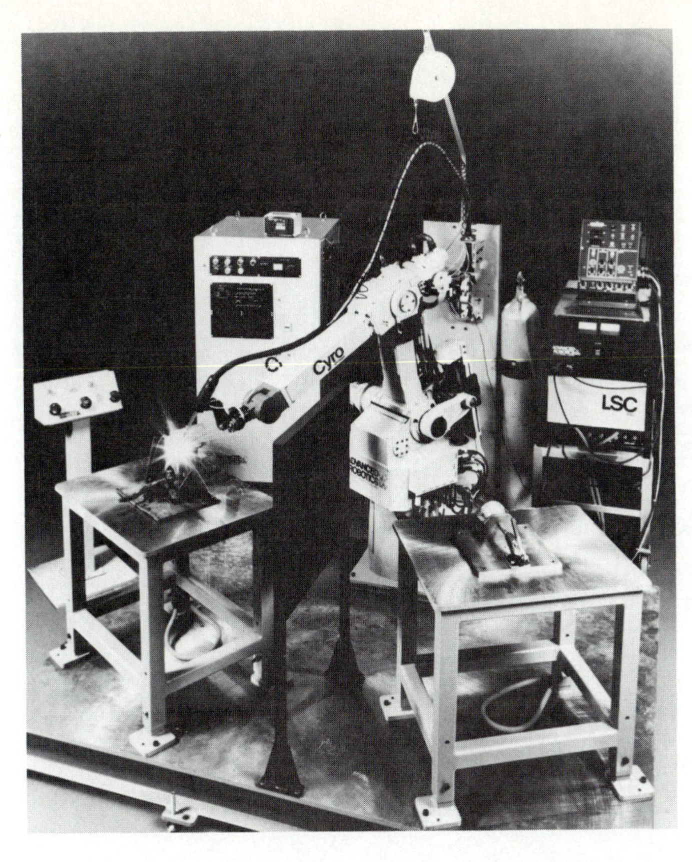

PHOTO 31–4
Cyro 820 robotic arc welding work cell, consisting of a five-axis articulated arm robot, a Cyro control, a GMAW process package, and two platen tables. Robot welds parts on one table while the operator loads and unloads parts on the other. (Courtesy of Advanced Robotics Corp.)

PHOTO 31–5
Cyro 1000 robotic arc welding work cell consisting of a five-axis Cyro 1000 robot, a five-axis Cyro RP positioner, a C-30 control, and a GTAW process package. Both tables on the positioner tilt and rotate, and the entire base rotates 180°. While operator is loading and unloading parts on one side, the robot welds on the other. (Courtesy of Advanced Robotics Corp.)

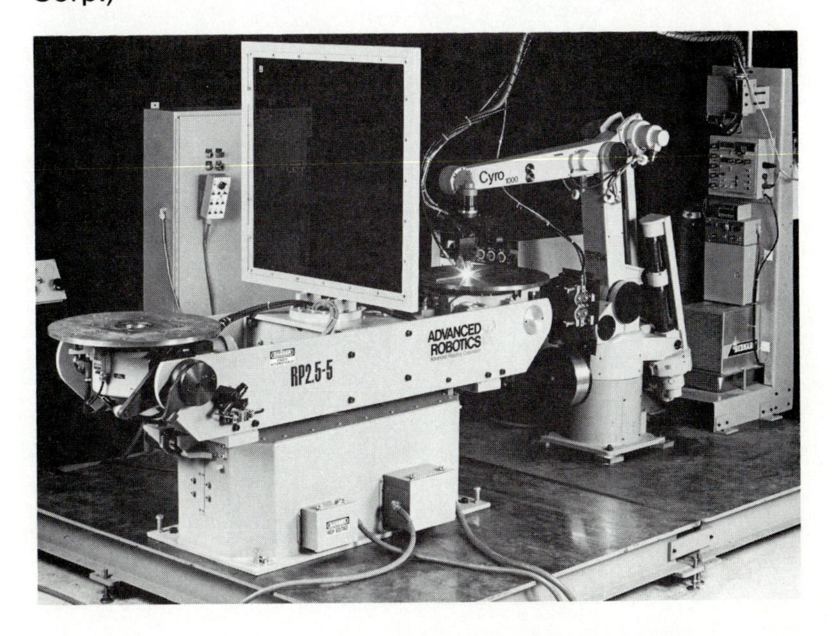

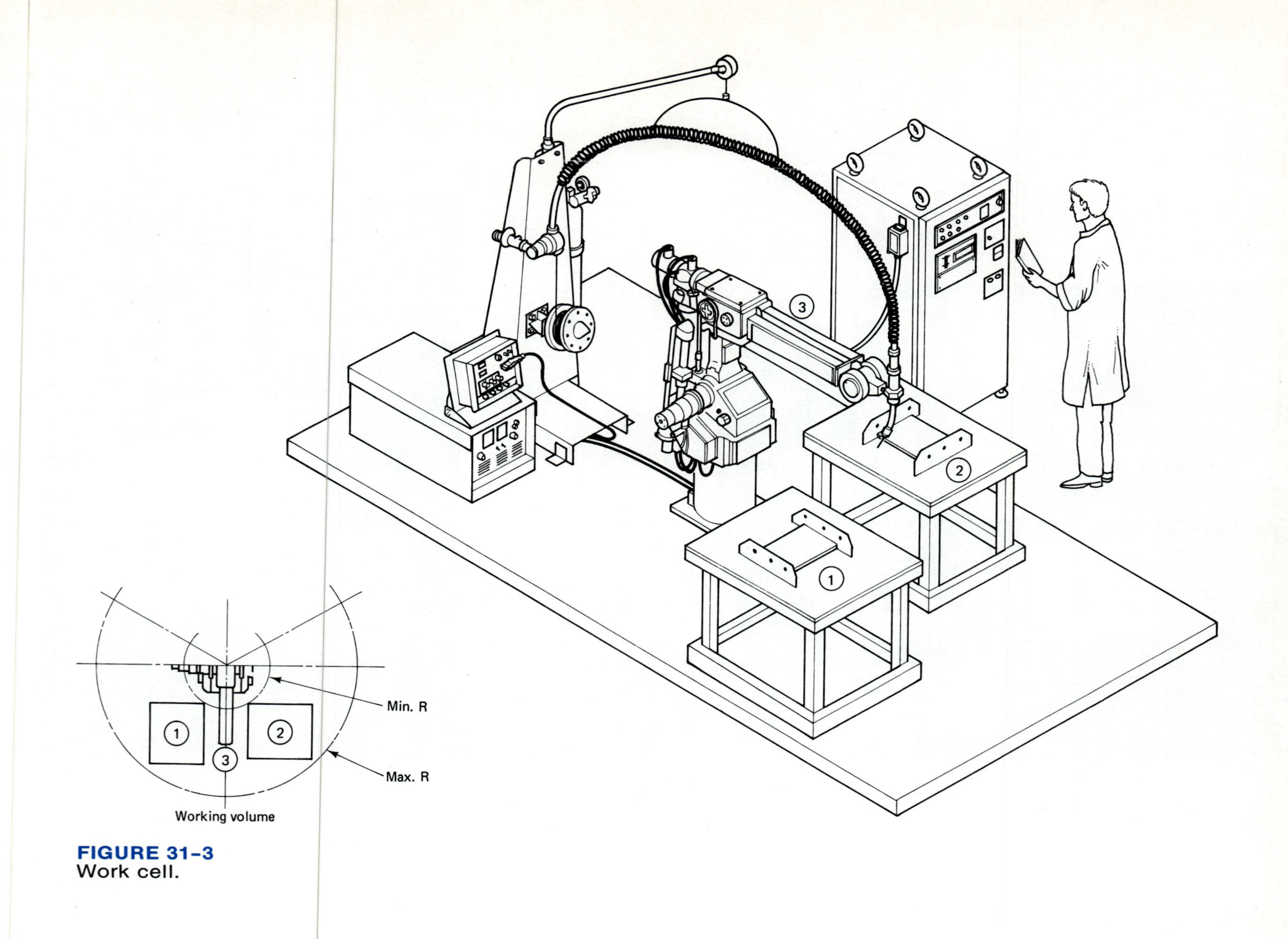

FIGURE 31–3
Work cell.

2. Locate the tip or torch cleaning devices as close as practical to the welding operation.
3. Utilize positioners that complement the positions and movement required for welding.

A typical work cell is shown in Figure 31-3.

ROBOTICS TERMINOLOGY

Axis The axis is the plane about which a robot manipulator moves. Most industrial robots have between two and five axes. Simple robots have only an *X* and a *Y* axis. Modern articulated robots have five to six axes, and some applications have required up to 10 axes. A five-axis robot is shown in Figure 31-4.

1. Waist axis (rotate)
2. Shoulder axis (up-down)
3. Elbow axis (up-down)
4. Wrist (rotate)
5. Wrist (up-down)

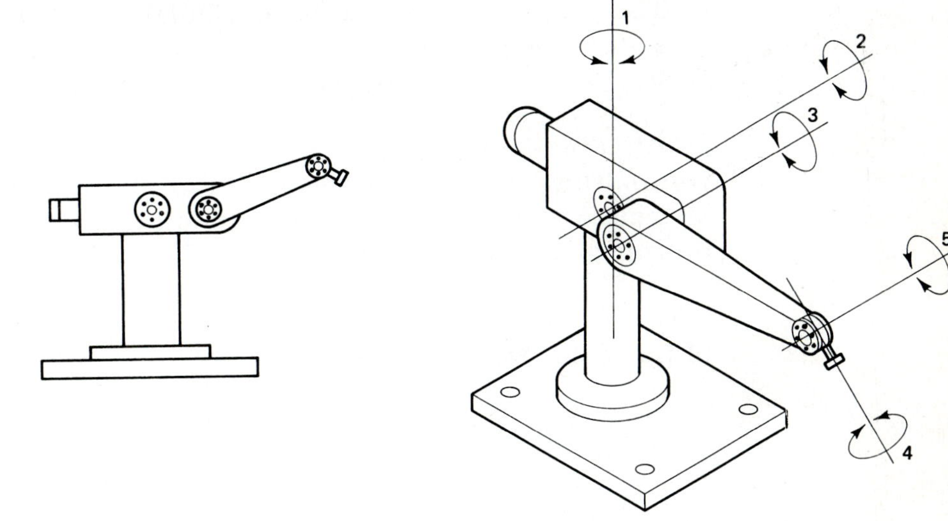

FIGURE 31-4
Five-axis robots.

Pick-and-Place Robots A pick-and-place robot is usually a simple robot used exclusively for the movement of parts from one point to another.

Manipulator The manipulator or robot arm is the exterior hardware that physically performs the function for which the robot was designed (Figure 31-5).

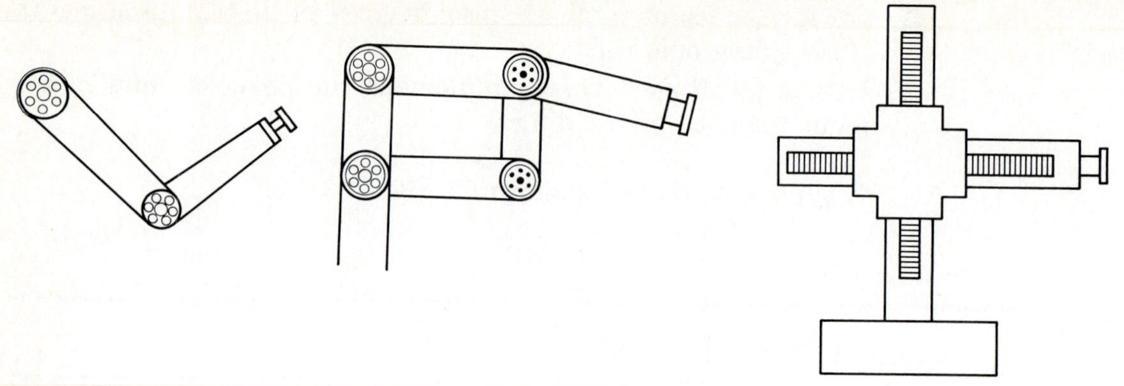

FIGURE 31-5
Manipulators.

End Effector An end effector is the tooling at the end of the arm that usually contacts the work. The end effector may be connected to a gripper, welding torch or gun, or spray paint gun.

Actuators Actutators actually move the manipulator to the position specified. Actuators for industrial robots are either electric servomotors, hydraulic cylinders, or hydraulic motors.

Servomotors Servomotors with *encoders* produce feedback signals that tell the robot the position of the manipulator (arm; Figure 31-6). These motors also drive the manipulator. Robots without servomotors or encoders must use mechanical limit switches to signal the robot to stop and start a movement.

Controller Also known as a processor controller or computer, the controller is the brains of the operation. All signals either originate or terminate at the controller. It is the main device for all robot operation, decisions, and control functions.

Work Envelope The work envelope is the total operational area of the manipulator (Figure 31-7).

FIGURE 31-6
Servo-controlled operation.

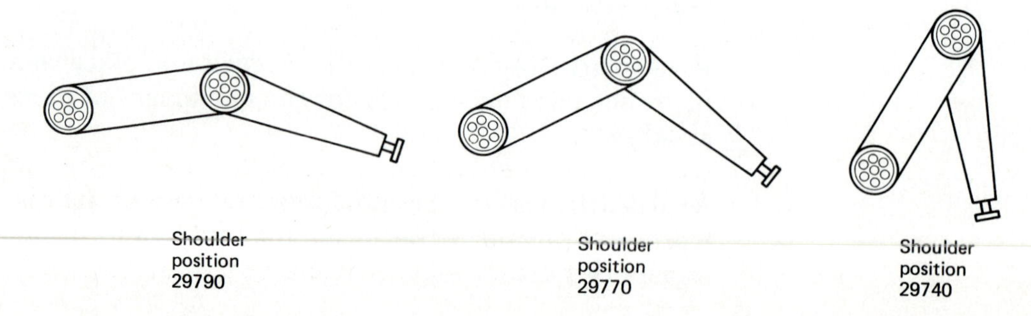

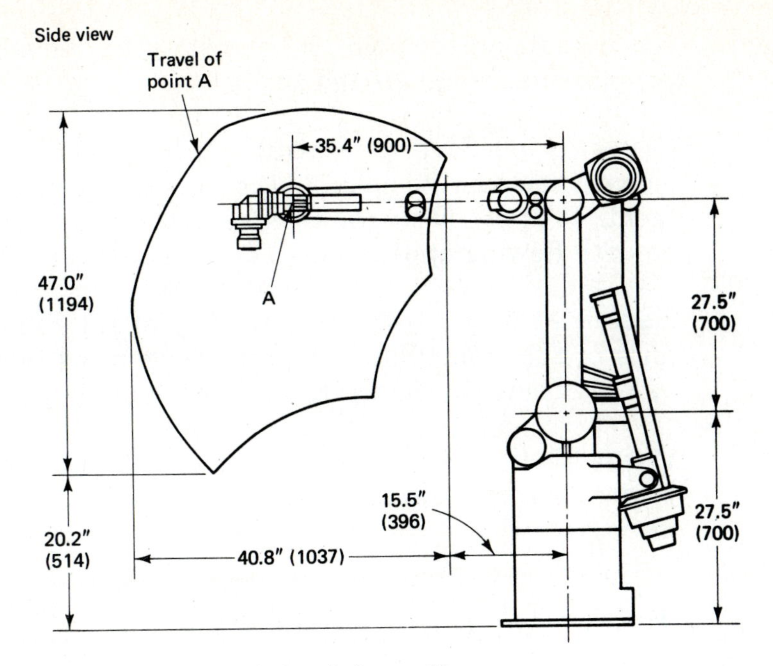

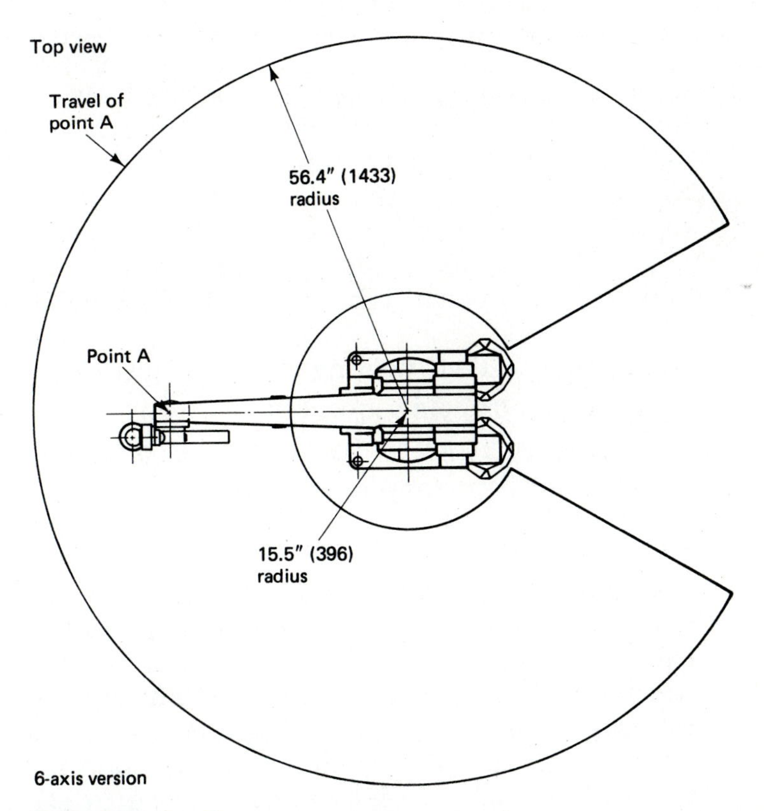

FIGURE 31–7
Work envelopes. (Courtesy of Advanced Robotics Corp.)

Point-to-Point Robot Mode **This method of getting the end effector from one point to another takes the shortest path from point A to point B. It is very quick, but abrupt in its motion. Care must be taken when in this mode not to interfere with other items, tooling, and so on, in the work envelope. Because the robot will take the most direct** route from point to point, it will not avoid items in its path unless it is programmed to go around them.

Continuous-Sweep Robot Mode This method is somewhat slower than point to point, but is a more graceful and human-like movement. It is excellent for jobs such as spray painting that require a smooth flowing application.

Repeatability The repeatability of the robot indicates the tolerance of the manipulator to perform required functions and return to various points in the work cell over a period of time.

Work Cell The area encompassing the robot, control, welding power source, positioner, and other production and supportive equipment.

INTERFACING AND SEQUENCING WELDING ROBOTS

Many industrial robots can be interfaced with a variety of additional job functions, such as grinding, machining, cleaning, or possibly with another feed robot that is supplying the welding robot with parts. Most robots have I/O (input/output) ports that can allow the main controller to sequence and synchronize all of these operations in unison.

The key to robot interfacing are the sequencing capabilities and number of I/O ports. Typical sequencing capabilities are:

1. Independent welding condition settings
2. Delay or hold sequences
3. Input/output control to another device (robot, grinder, etc.)
4. Branch or jump to another step

With *independent welding condition settings,* the robot can adjust to different voltage and amperage (or wire feed) requirements. For example, welding a thick section, then a thin section, or welding vertical up, then coming vertical down, all in the same program. With this control, when the robot comes to the thinner section it can be sequenced to lower the amperage and voltage, or vice versa.

With the *delay or hold sequence control,* the robot can be programmed to delay or hold until signaled to continue, or sent into a timed delay which will delay until it times out, at which point the robot will continue. This is most helpful in welds at the starts and stops to fill craters and flatten-out starts. Another area where delays are required is when another device is feeding parts to the welding robot. A delay would serve to hold the robot until the parts are fully in position for welding.

PHOTO 31-6
This system on the AR-1 robot allows the technician to set up and change the GMAW volt and ampere settings as the robot is welding or between steps. (Courtesy of ESAB Welding Products, Inc.)

With *input/output control* sequences we can activate other devices, such as signaling a conveyor to feed parts to the robot, or even signaling a computer to count parts. The possibilities are unlimited.

Branch or jump sequences are actual I/O commands that tell the robot controller to follow one of two or more predetermined paths based on whether or not the branch (or jump) is activated. Example: If the robot is in production producing a part with an optional component, the robot can be signaled whether or not the optional part is to be added. If the optional component is to be added, the extra steps to weld the component will be made. If the optional component is not to be added, the robot will jump past the extra steps.

CONCLUSION

Remember that any computerized manufacturing device such as a robot requires personal skill in that operation. In our case, skilled welding technicians familiar with all welding variables and perameters must program and monitor robotic applications. The most important step that you can take in preparation for this area of welding is learning the basic welding principles. Work on understanding what corrective action must be taken when weld beads are other than desirable; learn manipulating movements (weaves and stringers); and understand the effects of changing arc lengths, travel speeds, and machine adjustments. These areas are essential if you plan to work the welding robotics field.

REVIEW QUESTIONS

1. What areas are important to learn for a person planning to work in welding robotics?
2. Do welding robots have any of the five physical senses? Explain.
3. Can robots work in hazardous environments?
4. What computer language must you know to program most industrial welding robots?
5. Why might lasers be interfaced to welding robots?
6. What is the term used for the ability of a robot arm to return to each preprogrammed point within a few thousandths of an inch?
7. What is the term used for the total operational area through which the manipulator arm can move?
8. Can welding robots change the welding voltage, amperage, or travel speed while executing a program?

32 ADVANCED WELDING SYSTEMS

Advanced welding systems are futuristic welding processes and systems. That does not necessarily mean that they will not be found around various areas of the industry, but they are not as common as the traditional welding processes such as SMAW, GMAW, and GTAW. Many advanced welding systems are still limited to the laboratory, where their full potential is being researched.

Some of these systems are quite expensive, not only because expensive and sometimes rare components are used in the systems, but the high price compensates researchers and manufacturers for the many hours of research spent to develop the system. However, we must also weigh this cost against productivity. For example, electron beam systems can weld faster than 200 inches per minute. Production is increased tenfold over that of conventional welding processes. At that speed the machining would pay for itself in a short period of time. When these advanced welding systems become common in the welding industry, welded products will be produced faster and will be of higher quality.

ELECTRON BEAM WELDING

Electron beam welding (EBW) is one of the newest, most effective advanced welding systems today. Electron beam welding is a process whereby the welding of two metals is produced by heat, obtained from a concentrated beam of high-velocity electrons. These negatively charged particles of matter bombard the metal surface, causing such intense local heating that it almost instantly vaporizes a hole through the entire piece of metal on which it is focused. The walls of this hole are molten, and as the electron beam moves along the joint, the metal flows into the joint, producing an extremely deep and narrow weld. This method of welding is known as "keyholing." The electron beam produces a very intense, localized heat zone which provides the following advantages (Figure 32–1):

1. Extremely small heat-affected zone
2. Extremely deep, narrow weld deposit
3. High-speed welding
4. No V-groove preparation on butt joints
5. Clean, pure welds due to the vacuum (with medium- and high-vacuum systems)

The small heat-affected zone is the result of the intense localized heat input and the high travel speeds possible with EBW. The deep narrow penetration is the result of the extremely high power density of the narrow stream of electrons. Welding steel up to 6 or 8 in. is not uncommon. V-groove joint preparation is not necessary on a thick section. Thick sections can be square groove butt welded as long as they have good fit-up. This capability saves considerable preparation time and the cost of filler and a lot of filler metal.

Since the free electrons are negatively charged, they can be controlled and focused by charged focusing coils. This allows the technician to shape the beam into a fine point or a broad beam. Highly finished edges on thick square grooves would use a sharp narrow beam, whereas coarser finished or thinner sections use a wider, broader-focused beam. Whatever the beam width, it is usually focused at the middle of the part thickness (Figure 32–2).

FIGURE 32–1
EBW versus GTAW penetration.

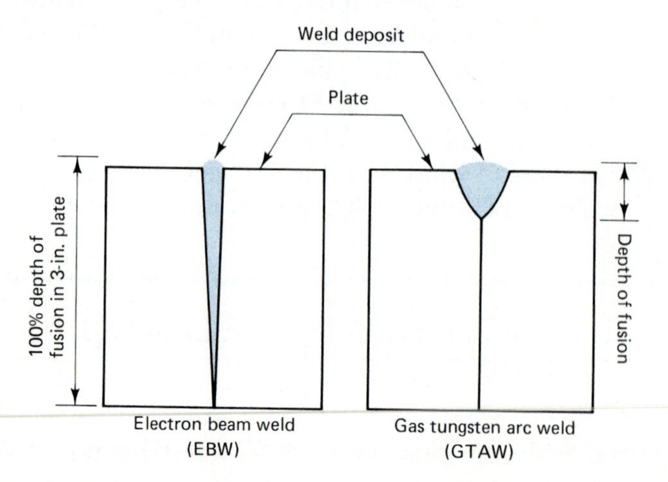

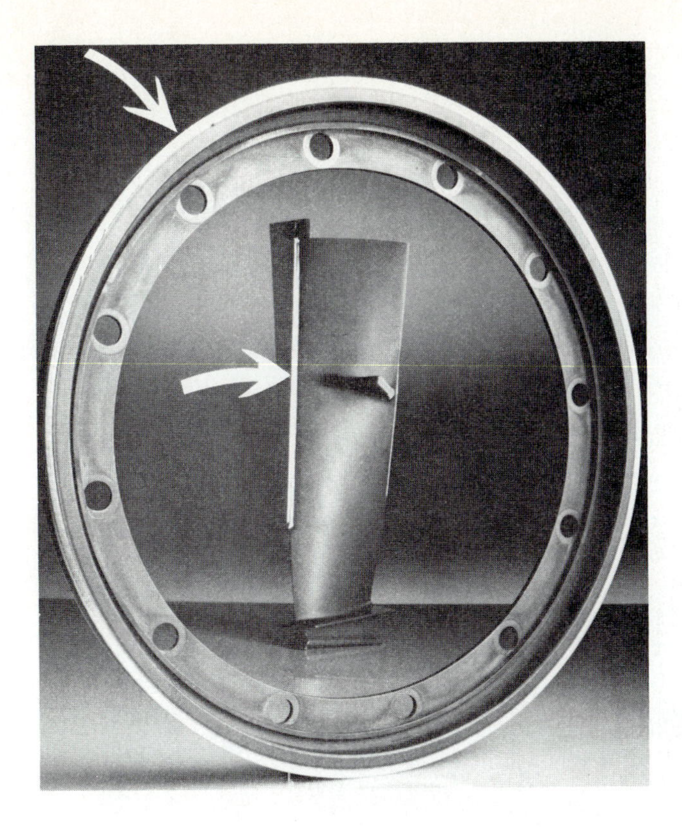

PHOTO 32–1
Electron beam welds on turbine blade leading edge and seal ring edge buildup. (Courtesy Ferranti Sciaky Inc.)

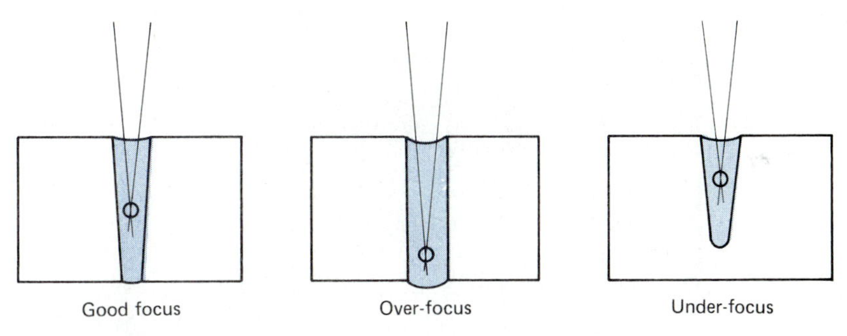

FIGURE 32–2
EBW focus.

Electron beam welding may be done in a high or medium vacuum or no vacuum. The vacuum is used because the electrons will interact with the gases in the atmosphere and defuse the beam. The beam actually defuses due to two principles:

1. Collision with the atmosphere
2. Mutual repulsion

The atmosphere is actually full of gases, such as oxygen, nitrogen, helium, argon, and others. At normal atmospheric pressure (as in nonvacuum EBW) these gases collide with the electron beam, causing the beam to disperse.

Mutual repulsion occurs because the electrons have like charges (electrons are negatively charged) and like charges repel. What is expe-

PHOTO 32–2
Electron beam gun with filler wire feed. (Courtesy Ferranti Sciaky Inc.)

PHOTO 32–3
Electron beam system with computer numeric controller and positioner. (Courtesy Ferranti Sciaky Inc.)

rienced is a slight dispersal of the narrow beam due to the electrons repelling each other.

High-vacuum EBW uses a vacuum chamber, where the air is pumped out of the chamber. This accomplishes two things. First, it provides a clear path for the electron beam. With most of the atmospheric substances evacuated, the electrons are free to travel without collision with air. Second, the vacuum provides a clean atmosphere for the molten metal. Harmful atmospheric contaminants such as oxygen, hydrogen, and nitrogen are removed in a vacuum.

A high-vacuum electron beam is used where deep penetration on a thick section of material is required, or reactive metals are to be welded. Medium-vacuum electron beam welding has the advantage of quickly creating the vacuum. This increases production speeds over that of high-vacuum EBW. The medium vacuum does, however, contain enough air to contaminate some reactive metals. It also produces a wider, more defused beam; therefore, penetration capabilities are less than with high-vacuum EBW.

Nonvacuum electron beam welding has the advantage of portability; welding is not restricted to a vacuum chamber. The beam is much broader and more diffused, so the electron beam gun must be brought into very close proximity to the workpiece or the beam will be too diffused to weld. Thicknesses that can be welded are limited, and shielding is required because of the lack of vacuum.

Technicians performing electron beam welding should be aware of the production of X-rays when welding some metals. Most electron beam welders and vacuum chambers have leaded glass windows, or closed-circuit TV sets for viewing the weld from a safe or shielded area. A cross section of an electron beam gun is shown in Figure 32–3.

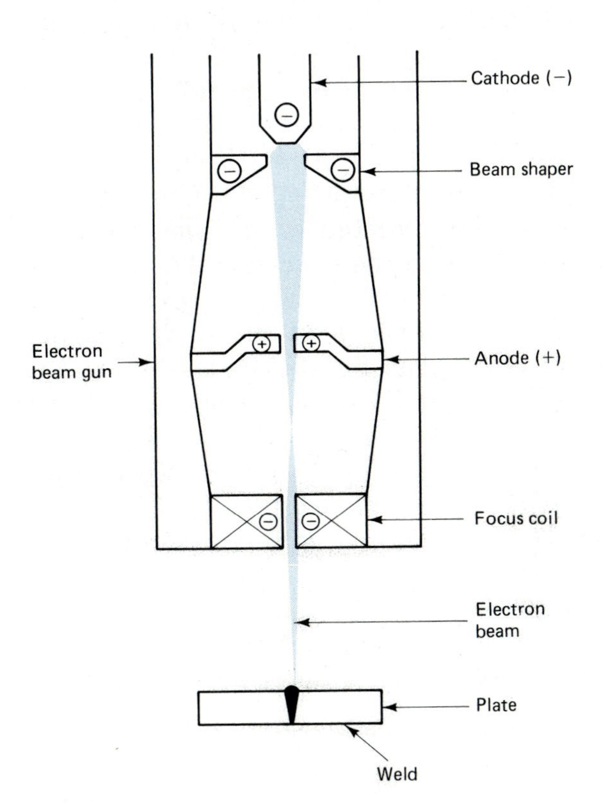

FIGURE 32–3
EBW gun.

LASER BEAM WELDING

The laser beam is another tool in today's welding industry which is gaining popularity. Lasers are used to weld "difficult to weld" materials, or hard-to-access areas, or extremely small components such as various electronic components. It is even used in the medical field to weld detached retinas back into place.

Laser light is actually a coherent (that is, of the same wavelength) light beam; an ordinary flashlight produces incoherent light (Figure 32–4). The production of laser light is a somewhat complex process, but read on and we will try to make sense of it. In a solid-state ruby laser, a neon flash tube emits light into a specially cut ruby crystal. As this ruby crystal absorbs light, its electrons become stimulated. As this stimulation increases, the electrons increase from their normal orbit (ground orbit) to an excited orbit. As yet more energy is pumped into the lasing medium (in this case, the ruby crystal), the energy that is absorbed exceeds its thermal equilibrium; in other words, it can no longer convert it to heat energy. At this point the electrons drop back to an intermediate orbit and a photon (light) is emitted. This is known as *spontaneous emission* (Figure 32–5).

As this sequence of events continues and more and more spontaneous emissions occur, these released photons tend to stimulate other excited photons into releasing their photons. This process is known as *stimulated emission* (Figure 32–6). Stimulated emission tends to cause the excited electrons to emit photons of the same wavelength; this is how the coherent (that is, of the same wavelength) light beam is produced.

The laser beam tube has a heavily silvered, totally reflective mirror at the back end, and a lightly silvered, partially reflective mirror at the opposite end. The coherent light reflects back and forth until it becomes dense enough to penetrate the partially reflective mirror. The laser beam is then directed to the workpiece by the use of mirrors, where it is focused on the exact point to be welded.

Lasers used in welding or cutting are solid state or gas lasers. These gas lasers are capable of higher-wattage outputs, so they are applied to thicker sections. Solid-state lasers may be used on thin sections or precision welding jobs. The primary difference between gas and solid-state beam generation is how the electrons are stimulated in the lasing medium. Solid-state lasers use light energy to stimulate their electrons into releasing photons. Gas lasers use an electrical charge to stimulate their electrons. Common types of lasers used for welding and cutting include:

Gas lasers: CO_2, N_2, and helium.
Solid-state lasers: ruby, neodymium—YAG

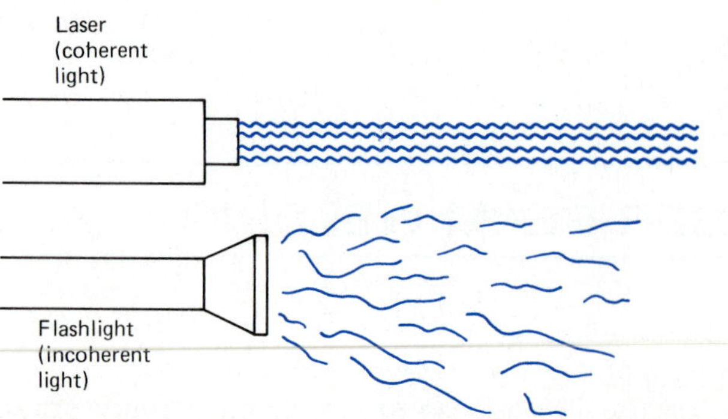

FIGURE 32–4
Laser versus incoherent light.

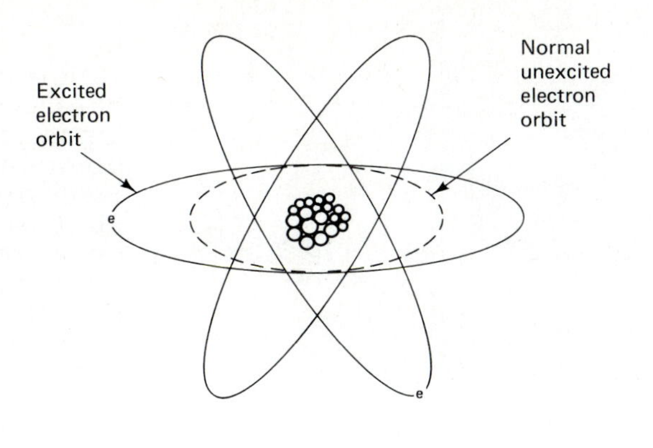

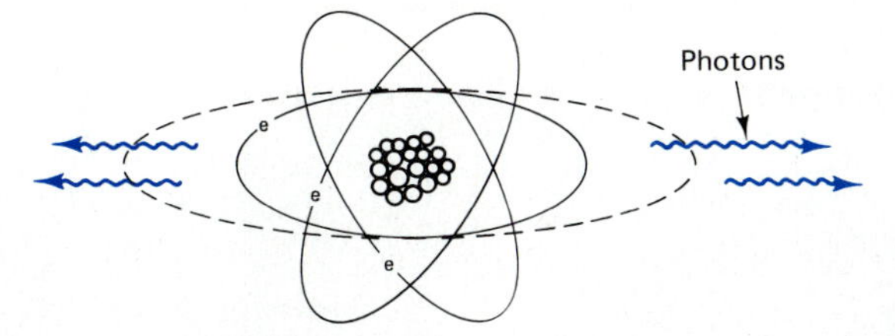

FIGURE 32-5
Spontaneous emission.

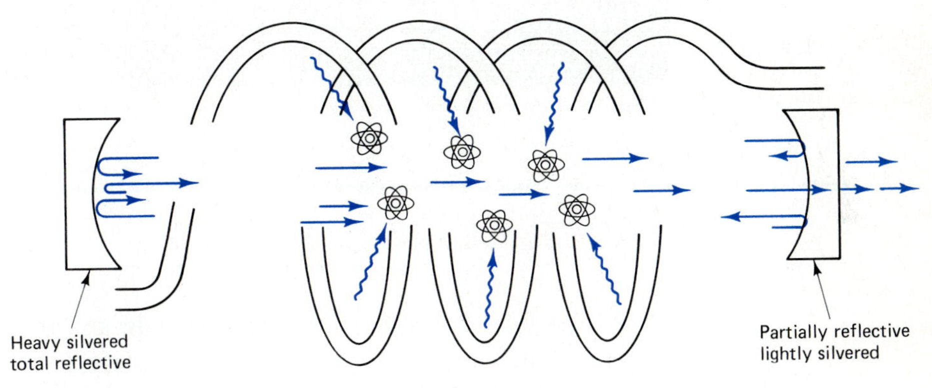

FIGURE 32-6
Stimulated emission and silvered mirrors.

The laser has many of the same advantages as those of the electron beam. It can make small, very narrow welds without overheating the surrounding base metal. This narrow beam of photon energy can penetrate up to about 1 in. of steel.

The laser beam can be directed over long distances by mirrors and lenses with very little loss in beam energy. The laser will not generate X-rays as the electron beam does; however, lasers do occasionally reflect off smooth or reflective surfaces and can cause burns. As with electron beam welding, laser beam welding uses the keyhole welding technique.

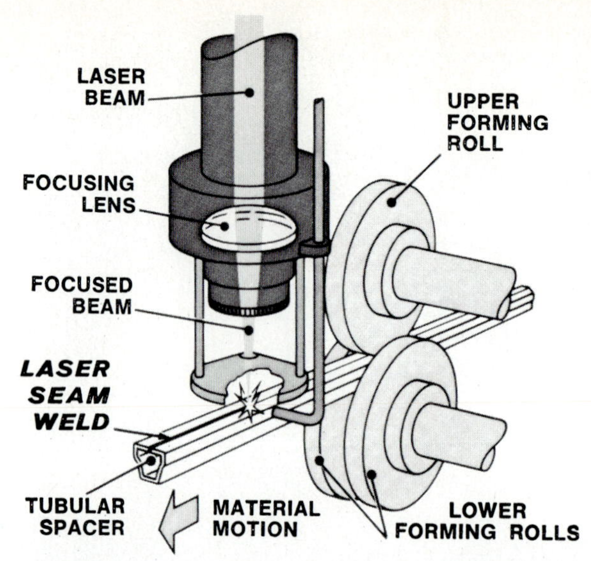

PHOTO 32-4 Automated laser welding. (b) Photon lasers are used to weld thousands of parts every day in a variety of manufacturing operations. Here, a time-lapse photo shows the completely automated laser process producing a circular, hermetic lap-weld on a precision temperature-sensing diaphragm. (c) The CO_2 laser beam is delivered to the focusing lens, condensed to high-density energy by the lens, and directed between the forming rolls to produce a fusion weld. (Courtesy of Lumonics, Inc.)

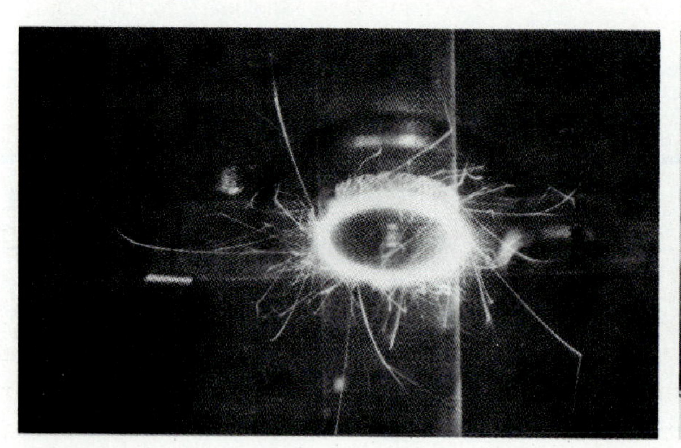

LASER BEAM CUTTING

The principles for generating the laser beam are the same in both laser cutting and welding. But with laser cutting we add an oxygen assist gas to help the cut. Other cutting assist gases, such as argon, helium, nitrogen, and CO_2, can be used. The inert gases give a clean cut on nonferrous metals such as aluminum because they do not oxidize the cut edge as oxygen does.

The beam leaves the generator tube at about $\frac{1}{2}$ to $\frac{3}{4}$ in. diameter. It is then directed to the work area using mirrors. At the work area the beam is angled down to the workpiece using a 45° mirror. The beam is then focused by one or a series of lenses. At the workpiece the beam is about 0.005 in. in diameter. Focusing takes the beam from about 1000 watts per square centimeter to about 10 million watts.

This intense energy initiates the cutting process. Now an assist gas, usually oxygen, is directed at the workpiece and through the sides of the beam to remove the vaporized metal in the kerf. Oxidation assists in continuing the progression of the cut.

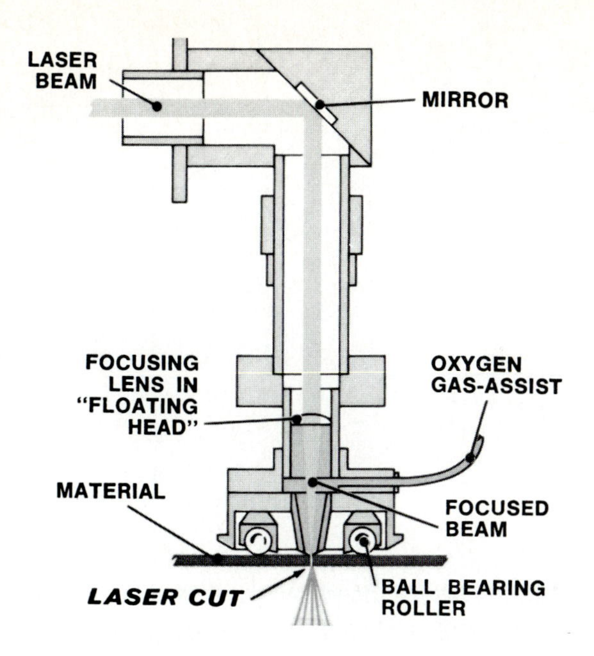

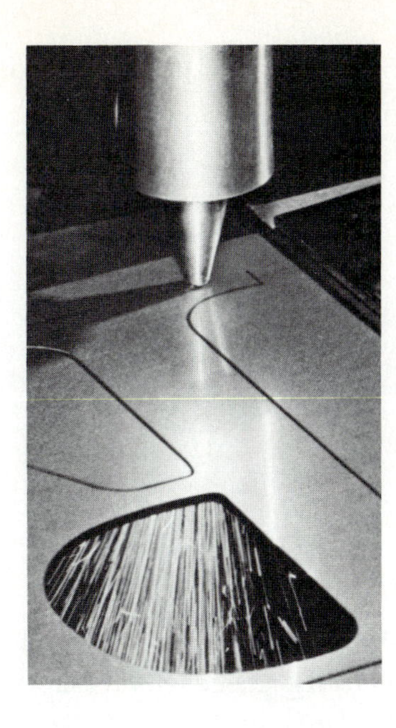

PHOTO 32–5
Laser cutting. (Courtesy of Lumonics, Inc.)

ULTRASONIC WELDING

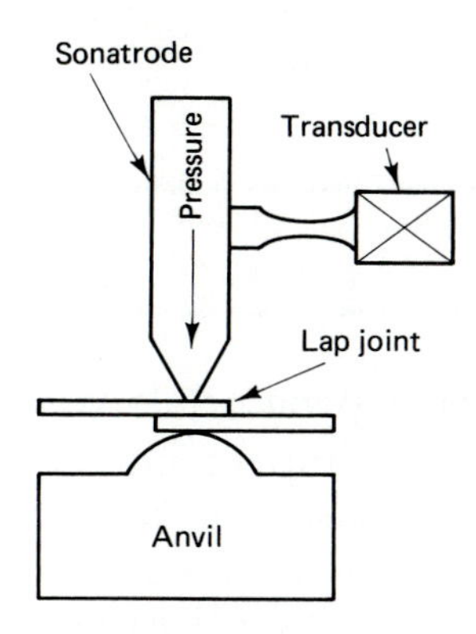

FIGURE 32–7
UT welding components.

Ultrasonic welding is a form of welding that uses high-pitched sound waves and pressure to join metal. Ultrasound consists of sound waves at frequencies higher than the human ear can hear. These sound waves are induced into the metal and create vibration and a friction effect that bonds the metal together. This process has some thickness limitations at this point. Only extremely light-gage metal can be joined and the pieces must be prepared into lap joint configurations. The metal must also be perfectly clean and must fit together smoothly.

Components of a typical ultrasonic machine include a transducer for producing the ultrasound, a sonotrode for carrying the sound vibrations and pressure to the tip, and an anvil and reed to support and add the pressure to the parts to be welded (Figure 32–7).

Ultrasonic welding is popular for welding small electrical components and wires together. It is especially useful for components that cannot be heated but need to be welded. Ultrasonic welding does have a promising future. It has been found to be very effective on aluminum alloys that are difficult to weld using other processes.

SOLID-STATE DEFUSION

Future welding processes will be done in the solid state, thus eliminating the undesirable effects of extreme heat. There will be no heat-affected zones, distortion, shrinkage, hot cracks, craters, slag inclu-

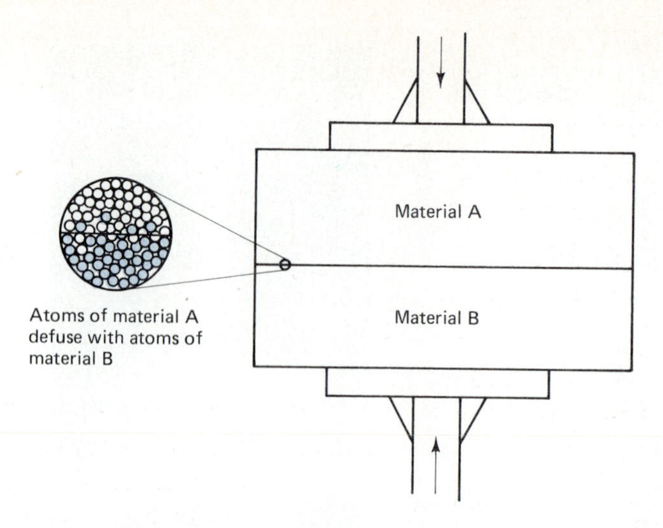

FIGURE 32-8
Solid-state defusion.

sions, porosity, or other problems associated with today's fusion welding processes. Solid-state defusion welding is to join parts with pressure and limited heating.

The surface of the parts must be very smooth and close fitting. The parts are then heated well below the melting point to increase molecular activity and promote diffusion. Pressure is then applied. The result is that some of the atoms of each part migrate into each other's molecules and produce a weld (Figure 32-8).

THERMAL SPRAYING

Thermal spraying is a method of surfacing or coating materials that require surface protection from wear, corrosion, or impact energy. Thermal spraying can be performed by flame or plasma spraying. Flame spraying (Figure 32-9) uses an oxyfuel flame to melt a filler material and propel it onto a surface (called the *substrate*). The filler mate-

FIGURE 32-9
Flame thermal spraying.

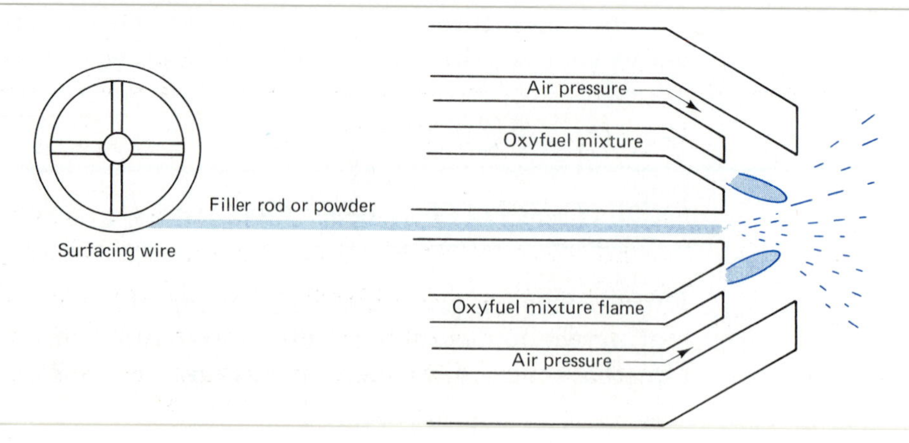

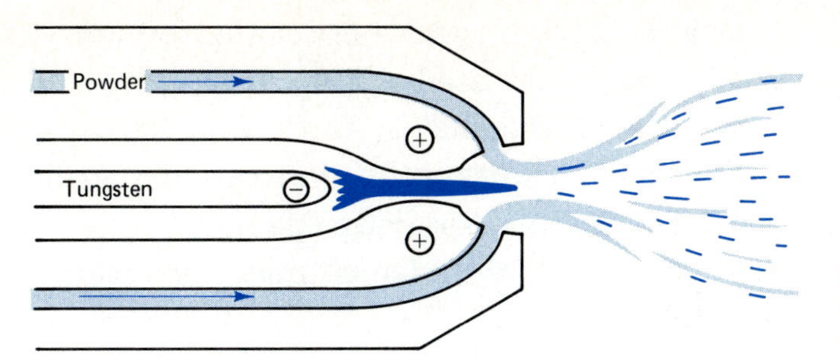

FIGURE 32–10 Plasma thermal spraying.

rial may be in the form of a powder or a solid filler wire. The device used to melt the filler and propel it to the surface is the spray gun. As the filler material (rod or powder) reaches the end of the nozzle, it is melted into small molten droplets by the oxyfuel flame. The droplets are then propelled to the substrate by air pressure. When the droplets reach the substrate they flatten out and join each other and form a bond to the substrate surface.

In plasma spraying (Figure 32–10) the heat for melting the surfacing material is obtained from a plasma arc (instead of an oxyfuel flame as in flame spraying). The intense heat of the plasma arc atomizes the powder or filler wire into extremely small molten particles. The particles are then projected to the substrate by the force of the plasma jet.

The surfacing material for plasma spray surfacing is usually in powder form. When the molten droplets reach the substrate they flatten out and bond to each other and the substrate surface. The plasma spray torch itself is usually of the nontransferred type. This means that the work or substrate is not part of the circuit. The torch actually has both negative (tungsten electrode) and positive (nozzle) poles contained in the torch head. This allows for plasma surfacing on nonconductor substrates.

Thermal spray surfacing is especially useful on surfaces or substrates that require even application or even heat distribution. Shafts are a typical example; arc welding shafts for buildup can leave them distorted and require extensive machining before they can be returned to service. But thermal spray surfacing gives even coverage with little or no distortion, and only a small amount of finish machining is required.

REVIEW QUESTIONS

1. Electron beam welding is done in what type of atmosphere?
2. How is heat obtained in electron beam welding?
3. What are the advantages of electron beam welding?
4. What type of penetration can electron bead achieve?
5. What causes the electron beam to diffuse?
6. Where are lasers used in welding?

7. What is the difference between laser light and visible light?
8. What gases are used to help lasers cut?
9. What is ultrasound?
10. What metals work well for ultrasound welding?
11. What is thermal spraying used for?
12. What is used for surfacing material with thermal spraying?

INDEX